HUMAN GERMLINE GENOME MODIFICATION
AND THE RIGHT TO SCIENCE

The advent of gene therapies and genome editing tools (ZFN, TALEN, CRISPR) are transforming not just science and medicine but also law. When the genome of germline cells is modified, the modifications could be inherited with far-reaching effects in time and scale. Legal systems are struggling with keeping up with the CRISPR revolution, and both lawyers and scientists are often confused about existing regulations. This book contains an analysis of the national regulatory framework in eighteen selected countries. Written by national legal experts, it includes all major players in bioengineering, plus an analysis of the emerging international standards and a discussion of how international human rights standards should inform national and international regulatory frameworks. The authors propose a set of principles for the regulation of germline engineering, based on international human rights law, which can be the foundation for regulating heritable gene editing both at the level of countries as well as globally. A companion website contains the legal instruments regulating human germline genome modification internationally and in each country surveyed in this book: www.freedomofresearch.org/ggel

Andrea Boggio is Professor of Legal Studies at Bryant University. He holds a J.S.D. from Stanford Law School.

Cesare P. R. Romano is Professor of Law and W. Joseph Ford Fellow at Loyola Law School, Los Angeles. He holds an LL.M. from NYU School of Law and a Ph.D. from the Graduate Institute of International Studies at the University of Geneva.

Jessica Almqvist is Associate Professor in Public International Law at the Autonomous University of Madrid and Senior Research Fellow at the Elcano Royal Institute. She holds a Ph.D. in law from the European University Institute and a diploma in Graduate Study in Political Science from the University of California, Berkeley.

Human Germline Genome Modification and the Right to Science

A COMPARATIVE STUDY OF NATIONAL LAWS AND POLICIES

Edited by

ANDREA BOGGIO

Bryant University

CESARE P. R. ROMANO

Loyola University, Los Angeles

JESSICA ALMQVIST

Autonomous University of Madrid

CAMBRIDGE
UNIVERSITY PRESS

University Printing House, Cambridge CB2 8BS, United Kingdom

One Liberty Plaza, 20th Floor, New York, NY 10006, USA

477 Williamstown Road, Port Melbourne, VIC 3207, Australia

314-321, 3rd Floor, Plot 3, Splendor Forum, Jasola District Centre, New Delhi - 110025, India

103 Penang Road, #05-06/07, Visioncrest Commercial, Singapore 238467

Cambridge University Press is part of the University of Cambridge.

It furthers the University's mission by disseminating knowledge in the pursuit of
education, learning and research at the highest international levels of excellence.

www.cambridge.org
Information on this title: www.cambridge.org/9781108718448
DOI: 10.1017/9781108759083

© Andrea Boggio, Cesare P.R. Romano and Jessica Almqvist 2020

First published 2020
First paperback edition 2022

A catalogue record for this publication is available from the British Library

Library of Congress Cataloging in Publication data
NAMES: Boggio, Andrea, editor. | Romano, Cesare (Cesare P. R.), editor. | Almqvist,
Jessica, editor.
TITLE: Human germline genome modification and the right to science :
a comparative study of national laws and policies / edited by Andrea Boggio, Bryant
University; Cesare P.R. Romano, Loyola University, Los Angeles; Jessica Almqvist,
Universidat Autonoma, Madrid.
DESCRIPTION: Cambridge, United Kingdom ; New York, NY : Cambridege
University Press, [2019] | Includes bibliographical references and index.
IDENTIFIERS: LCCN 2019016367 | ISBN 9781108499873 (alk. paper)
SUBJECTS: LCSH: Genetic engineering – Law and legislation. | Human genome. |
Gene editing.
CLASSIFICATION: LCC K3611.G46 H86 2019 | DDC 344.04/196—dc23
LC record available at https://lccn.loc.gov/2019016367

ISBN 978-1-108-49987-3 Hardback
ISBN 978-1-108-71844-8 Paperback

To Maya, who came along on a lucky day (A.B.)
To Bourbon, who kept me company and waited for me patiently while I was writing yet another sentence (C.R.)
To Sebi, for always trying to see the best in everything around us, and to Rebe, for constantly striving for all of us to do better (J.A.)

Contents

Figures

Tables

Notes on Contributors

Editors

JESSICA ALMQVIST is Associate Professor in Public International Law at Madrid's Universidad Autónoma (Autonomous University of Madrid, UAM) and Senior Research Fellow at the Elcano Royal Institute. She holds a Ph.D. in law from the European University Institute (Florence), a Diploma in graduate studies of political science from the University of California, Berkeley, and Juris Candidatis (LL.B. and LL.M.) from Lund University.

She specializes in international human rights law, international criminal law, and international humanitarian law. She has authored *Human Rights, Culture and the Rule of Law* (Hart, 2005), coedited *The Role of Courts in Transitional Justice: Voices from Latin America and Spain* (Routledge, 2012), and published extensively on other topics, including international judicial independence, universal jurisdiction, the protection of civilian populations, the rights of victims of terrorism and counterterrorism measures, and the recognition of states.

Prior to joining the UAM, she has held research positions in the Centre on International Cooperation at the New York University (NYU), *Fundación para las relaciones internacionales y el diálogo exterior* (FRIDE), and the Centre for Political and Constitutional Studies in Madrid, and has been a visiting fellow at Columbia Law School in New York City (NYC). She has taught courses at the New School for Social Research in NYC and the Ortega y Gasset Institute in Madrid.

ANDREA BOGGIO is Professor of Legal Studies at Bryant University. He specializes in science and health policy and has authored numerous publications on legal and ethical issues in reproductive medicine, genomics, research with human subjects, and the management of communicable diseases, including *Compensating Asbestos Victims. Law and the Dark Side of Industrialization* (Ashgate, 2013) and *Health and Development: Toward a Matrix Approach* (Palgrave Macmillan, 2009).

Dr. Boggio has been a visiting scholar at universities in Austria, Canada, Italy, and Vietnam, and has advised international bodies (the UN Committee on Economic, Social and Cultural Rights, WHO, UNDP, and OECD) and national governments (Canada and Italy) on issues of science and health policy.

He received his J.S.D. from Stanford Law School and completed his postdoc at the Institute for Ethics, History, and the Humanities at the University of Geneva (Switzerland). Before joining Bryant University in 2006, he was a lecturer at the Centre for Professional Ethics at Keele University (United Kingdom). He is a member of the New York State Bar.

CESARE P. R. ROMANO is Professor of Law and W. Joseph Ford Fellow at Loyola Law School, Los Angeles. He holds an M.A. (Laurea) in Political Science, University of Milano (1992); a D.E.S. (Diplôme des Études Superieures), Graduate Institute of International Studies, University of Geneva (1995); an LL.M., New York University Law School (1997); and a Ph.D., Graduate Institute of International Studies, University of Geneva (1999).

His expertise is in public international law, and in particular international human rights, international courts and tribunals, international environmental law, and international criminal and humanitarian law. Between 1996 and 2006, he created, developed, and managed the Project on International Courts and Tribunals, a joint undertaking of the Center on International Cooperation, NYU, and the Centre for International Courts and Tribunals at University College London, becoming a world-renowned authority in the field.

In 2011, Professor Romano founded the International Human Rights Clinic at Loyola Law School. Since then, he has led his students in the litigation of dozens of cases before the Inter-American Commission of Human Rights and specialized UN human rights bodies, including the first individual communication to the UN Committee on Economic, Social and Cultural Rights dealing with restrictions on research in human embryos.

Contributors

GALI BEN-OR is the National Legislation Database Director in the Israeli Parliament (the Knesset). In 2013, she initiated the establishment of the database and leads the continuous process of its design and construction. Previously, she held various positions in the Legal Advice and Legislation Department of the Ministry of Justice for almost twenty years. In her last position she was Senior Director of the Judicial Review and Integrity Team. She holds an LL.B. from the Law Faculty of the Hebrew University of Jerusalem.

Ben-Or was in charge of legal advice and legislation on a wide range of subjects including genetics, cloning, surrogacy, posthumous reproduction, forensic databases, genetic databases, administrative courts, nominations in the civil service, conflicts of interest, etc. She coordinated the team that drafted the government resolution that established the Israeli National Bioethics Council.

She took part in drafting many laws, including the Israeli surrogacy law and the police forensic database law. She led the drafting of the Israeli cloning law, the genetic information law, and more. She lectured about these subjects in numerous conferences held by government ministries and universities. Ben-Or has published several articles and coauthored several chapters in books in the area of bioethics.

ALESSANDRO BLASIMME holds a degree in philosophy and a master's in bioethics from La Sapienza University of Rome (Italy), as well as a PhD in bioethics from the European School of Molecular Medicine at University of Milan (Italy). He has held research appointments at the French National Institute of Health and Medical Research (INSERM) and the University of Zurich (Switzerland) before joining the Swiss Federal Institute of Technology (ETH Zurich) as a senior scientist in bioethics. In 2013, he received a Fulbright-Schuman scholarship to undertake research at Harvard University (United States). His activities revolve around ethical and regulatory issues in biomedicine and biotechnology. In particular, his work covers areas such as genomics, stem cell research, cell therapy, gene therapy, precision medicine, digital health, and the use of artificial intelligence in medicine.

DOROTHÉE CAMINITI received a *Juris Doctorate* (J.D.) from the Université Catholique de Louvain (Belgium), a joint *Legum Magister* (LL.M.) in intellectual property law and technologies from the University of Leicester (United Kingdom) and the Université de Liège (Belgium), as well as a master of bioethics (M.B.E.) from the Harvard Medical School (United States). She was an attorney in the Litigation and Arbitration Department at Cleary Gottlieb Steen & Hamilton in Brussels, before joining the Swiss Federal Institute of Technology (ETH Zurich) as a research fellow in bioethics. Dorothée's work focuses on questions at the intersection of bioethics and law. She is particularly interested in bioethical issues raised by genomics, big data, and personalized medicine.

CHARLOTTE DE KLUIVER is Lecturer at the Department of Legal Theory of VU University Amsterdam, the Netherlands. After studying Law at VU University, she studied health law at the University of Amsterdam and legal research at Utrecht University. In both her research and teaching, she focuses on the legal, ethical, and governance implications of biotechnological developments. Currently, she is involved in research projects on biohacking and genomic negligence.

After almost three years as an editor of Dutch law journal *Ars Aequi*, she is now a member of the board of the journal's publishing house. Her publications include edited volumes on judicial activism by the Dutch Supreme Court and on flaws in the Dutch Civil Code, as well as publications on doctor–patient confidentiality and the legal meaning of autonomy. Recently, she has coauthored a report on the rule of law in alternative dispute resolution, commissioned by the Dutch Research and Documentation Centre.

IÑIGO DE MIGUEL BERIAIN is Distinguished Researcher at the University of the Basque Country and Research Professor at IKERBASQUE . He has written five books and over one hundred articles and book chapters on bioethics and biolaw, for which he has won several international awards. He has participated in ten European research projects on these topics.

TIMO FALTUS studied biology and law in Frankfurt, Germany. Since 2016, he is a research fellow and coordinator of the interdisciplinary joint research project GenomELECTION on ethical, legal, and social questions of genome editing, located at Martin-Luther-University of Halle-Wittenberg and at the Natural History Museum of Berlin, funded by the German Federal Ministry of Education and Research.

CALVIN W. L. HO is Associate Professor at the Faculty of Law of The University of Hong Kong, and was Assistant Professor at the NUS Centre for Biomedical Ethics, Co-Head of the WHO Collaborating Centre for Bioethics in Singapore, and Co-Head of the Accountability Policy Task Team of the Global Alliance for Genomics and Health. He is also the Editor-in-Chief of the journal *Asian Bioethics Review* (published by Springer Nature) and an ethics board member of Médecins Sans Frontières (Doctors without Borders). Additionally, he served as an assistant director with the Legal Aid Bureau of the Ministry of Law (Singapore), as well as on the Singapore Nursing Board and advisory committees for transplantation and for genetic testing of the Ministry of Health (Singapore). He has published on biomedical law and ethics, regulatory governance and health policy and is the coeditor of *Bioethics in Singapore: An Ethical Microcosm* (World Scientific, 2010) and *Genetic Privacy* (Imperial College Press, 2013); the author of *Juridification in Bioethics: Governance of Human Pluripotent Cell Research* (Imperial College Press, 2016); and an author of the *WHO Guidelines on Ethical Issues in Public Health Surveillance* (2017).

HEIDI CARMEN HOWARD is a senior researcher at the Centre for Research Ethics and Bioethics, Uppsala University. She has expertise in genetics (Ph.D., McGill) as well as Bioethics (M.Sc., KULeuven), and her multidisciplinary team studies the policy, ethical, legal, and social issues (ELSI) of new technologies and approaches, particularly in genomics. Currently, her main areas of research are the ELSI of new genomic technologies, including genome modification and high throughput sequencing. She also studies policy aspects of genomics, uncertainty in genetics and genomics, direct-to-consumer genetic testing, public health genomics, participant-centric research initiatives, and the role of public engagement in genomics.

Dr. Howard is Work Package leader of a H2020 SWAFs Project called SIENNA (2017–2021, www.sienna-project.eu) studying the ELSI of different technology areas including Genomics, Robotics/AI, and enhancement. She is the principal investigator on a VR-funded-grant (2018–2021) on the ELSI of gene editing in humans. She is also an active and productive member of the European Society of Human Genetics ELSI group (the Public and professional policy committee) as well as of the Common Service ELSI team of BBMRI-ERIC. She is on the ethics committee of EIT Health (www.eithealth.eu/), is chair of the ethics group of the Genome Medicine Sweden initiative, and is Associated Researcher at Ethics and Society Group, Wellcome Trust Genome Campus, UK.

ROSARIO ISASI, J.D., M.P.H., is a Research Assistant Professor at the Dr. John T. Macdonald Foundation Department of Human Genetics at the University of Miami Miller School of Medicine. She holds multiple appointments, including at the Institute for Bioethics and Health Policy, as well as the John P. Hussman Institute for Human Genomics and the Interdisciplinary Stem Cell Institute.

Professor Isasi's research is devoted to identifying and analyzing the social, ethical, and policy dimensions of novel and disruptive genetic technologies. She has built an international reputation as a scholar with particular expertise in the area of international comparative law and ethics regarding genomics and regenerative medicine.

She holds many leadership roles in major international initiatives. Since 2017, Professor Isasi has been appointed as the President's International Fellow of the Chinese Academy of Sciences (CAS) and Adjunct Professor at the CAS' Institute of Zoology. She is Co-Investigator and Ethics/Regulatory/Communications lead for the SouthEast Enrollment Center (SEEC) consortium member of the NIH's All of Us Research Program. In addition, she serves as the Ethics/Policy Advisor of the European Commission's European Human Pluripotent Stem Cell Registry (hPSC^{REG}) and is a member of the American Society for Human Genetics (ASHG) Task Force on "Gene Editing." Finally, she is a member of the Ethics and Policy Committee of the International Society for Stem Cell Research (ISSCR) and is the Chair of the International Stem Cell Forum (ISCF) Ethics Working Party, a consortium of funding agencies for regenerative medicine.

TETSUYA ISHII is Professor at the Office of Health and Safety, Hokkaido University. He holds a Ph.D. in bioscience from Hokkaido University. He worked at Japan Science and Technology Agency, while completing a program officer program in US NIH. Subsequently, he worked at the Center for iPS Cell Research and Application at Kyoto University. Currently, he is studying bioethics in the biomedical field and was an invited speaker of the International Summit on Human Gene Editing at the U.S. NASEM in 2015 and the Second International Summit on Human Genome Editing at the University of Hong Kong in 2018.

YANN JOLY, Ph.D. (DCL), FCAHS, Ad.E., is the Research Director of the Centre of Genomics and Policies (CGP). He is Associate Professor at the Faculty of Medicine, Department of Human Genetics cross-appointed at the Bioethics Unit, at McGill University. He was named "advocatus emeritus" by the Quebec Bar in 2012 and Fellow of the Canadian Academy of Health Sciences in 2017. Professor Joly is a member of the Canadian Commission for UNESCO (CCU) Sectoral Commission for Natural, Social and Human Sciences. He is the current Chair of the Bioethics Workgroup of the International Human Epigenome Consortium (IHEC) and of the Ethics and Governance Committee of the International Cancer Genome Consortium (ICGC). He is also a member of the Human Genome Organization (HUGO) Committee on Ethics, Law and Society (CELS). Professor Joly's research interests lie at the interface of the fields of scientific knowledge, health law (biotechnology and other emerging health technologies), and bioethics. He has published his findings in over 120 peer-reviewed articles featured in top legal, ethical, and scientific journals. He served as a legal advisor on multiple research ethics committees in the public and private sectors. Professor Joly also sits on editorial committees and acts as a reviewer for a wide range of publications in his field. In 2012, he received the Quebec Bar Award of Merit (Innovation) for his work on the right to privacy in the biomedical field.

HANNAH KIM (M.D., Ph.D.) is a research assistant professor at the Department of Medical Law and Ethics, Division of Medical Humanities and Society, College of Medicine at Yonsei University, South Korea. She is also a visiting scholar at the Centre of Genomics and Policy, McGill University. She studied medicine and holds a Ph.D. in healthcare law from Yonsei University. She worked as a fellow at the Department of Medical Law at Yonsei University in 2016–2017. She was a member of Consultation Committee for DTC-GT under the Ministry of Health and Welfare in South Korea and a member of Institutional Review Board in Severance

Hospital, Yonsei University. Since 2011, she has focused on ethical, legal, and social issues surrounding genomics and advanced biotechnologies, from four ethical, legal, and social implications (ELSI) projects.

ERIKA KLEIDERMAN, B.Sc., LL.B., is a lawyer and an academic associate at the Centre of Genomics and Policy (McGill University). She holds a civil law degree (LL.B.) from the Université de Montréal, as well as a B.Sc. in Psychology from McGill University. She was called to the Quebec Bar in 2014. Her research focuses on the ethical, legal, and social implications surrounding access to data and genetic information, as well as the regulation of stem cells, regenerative medicine, and new reproductive technologies. She is also working in the context of anti-doping with a focus on the potential applications of gene therapy and gene modification for performance enhancement (gene doping), particularly in minors. Erika is a member of the McGill University Health Centre Research Ethics Board.

Erika is engaged in the stem cell and regenerative medicine community through her involvement with the Stem Cell Network's Trainee Communications and Training and Education Committees, as well as through her coordination role in the pan-Canadian initiative aimed at assessing the adequacy of existing regulatory frameworks and considerations for reframing the Assisted Human Reproduction Act, in light of evolving reproductive technologies. She is also the Coordinator of the Canadian International Data Sharing Initiative (Can-SHARE) and the Access Officer of the Canadian Partnership for Tomorrow Project (CPTP).

JAMES LAWFORD DAVIES is a lawyer and partner at Hill Dickinson LLP in London, specializing in regulation and litigation in the life science sector. He studied at Durham, Oxford, and Cardiff before qualifying as a lawyer in 2000 after training with a specialist healthcare firm in London. Following qualification, he became increasingly specialized in the regulation of reproductive technologies and embryo research, and subsequently in broader human tissue and cell-based therapies and research. He later joined the litigation team at Clifford Chance before setting up his own niche life sciences firm in 2010 and joining Hill Dickinson in 2019.

Mr. Lawford Davies advises a large number of clinics, hospitals, universities, and research centers regulated and licensed by the Human Fertilisation and Embryology Authority (HFEA) and the Human Tissue Authority (HTA). He also advises in relation to the regulation of medical devices and pharmaceuticals by the Medicines and Healthcare Products Regulatory Agency (MHRA), with a particular interest in cell and gene therapies.

He has been involved in most of the leading cases relating to reproductive technologies, embryo, and stem cell research. He is also an experienced litigator and has been involved in cases at every level in the UK courts.

Much of his work concerns the judicial review of regulatory bodies and cases arising as a result of regulatory action.

In addition to his private practice, Mr. Lawford Davies lectures widely on his specialist areas and is an honorary lecturer in the Department of Biochemical Engineering at the University College London.

LINGQIAO SONG (LL.M., B.Sc.) is a Chinese lawyer and an academic associate at the Centre of Genetics and Policy at the McGill University Faculty of Medicine. She is a member of the Data Access Office of the International Cancer Genome Consortium. She serves as an IRB member at the Faculty of Medicine and a fellow at the Research Group on Health and Law at the Faculty of Law. Her research lies in the interdisciplinary study of genetics, ethics, and law and, in particular, genetic policy related to the Chinese regulatory framework. She has participated in a variety of projects ranging from genetic discrimination, innovative genetic test for salmonella, genomic data sharing, to human gene editing.

RICK MAAS is Lecturer at the Department of Legal Theory of VU University Amsterdam, the Netherlands. After studying Law at VU University, he studied Private Law in Leiden and Philosophy of Law at the VU. In his research he explores the legal, ethical, and governance implications of biotechnological developments. Currently, his research focuses on liability issues arising from artificial reproductive techniques, more specifically genomic negligence. His most recent research concerns legal and ethical implications of wrongful life and wrongful birth lawsuits.

KERRY LYNN MACINTOSH is the Inez Mabie Distinguished Professor of Law and a High Tech Law faculty member at Santa Clara University School of Law. She received her B.A. from Pomona College, graduating summa cum laude. She obtained her J.D. from Stanford Law School and was elected to the Order of the Coif. Professor Macintosh has published extensively in the field of law and biotechnology. Her works include three academic books: *Enhanced Beings: Human Germline Modification and the Law* (2018), *Human Cloning: Four Fallacies and Their Legal Consequences* (2013), and *Illegal Beings: Human Clones and the Law* (2005). In addition, her many book chapters and articles explore legal and policy issues related to infertility, assisted reproduction, human cloning, embryonic stem cell research, and human–animal hybrids. Professor Macintosh is a member of the American Law Institute, a law reform organization.

MARÍA DE JESÚS MEDINA-ARELLANO is a Ph.D. in Bioethics and Medical Jurisprudence. Full-time research professor at the Legal Research Institute in the National Autonomous University of Mexico (UNAM), she is also a member of the National Research System, SNI-level 1, in Mexico's CONACyT.

She is qualified lawyer who graduated from the Autonomous University of Nayarit (UAN), Mexico in 2004. In January 2008, she graduated with the equivalent of an M.Phil. (with honors) from the Postgraduate Law Division at the National Autonomous University of Mexico (UNAM). In 2010, she was awarded the UNAM-Alfonso Caso silver medal for outstanding achievement, as the best M.Phil. thesis of the 2008 class in the area of social sciences. From 2013 to 2014, she worked as Deputy of the Judicial School in the Local Court of Justice in Nayarit.

Currently, she is the Deputy of the College of Bioethics (civil association that gathers scientists and physicians working in the area of bioethics from practical and academic perspectives) and Member of the Mexican Society for Stem Cell Research, the Council Board of the National Commission of Bioethics (CONBIOÉTICA) in Mexico, and the Board of Directors of the International Association of Bioethics (IAB).

DIANNE NICOL is Professor of Law at the University of Tasmania in Australia and Director of the Centre for Law and Genetics (CLG), which is housed in the Law Faculty. The broad theme of the CLG's research is the regulation of biotechnology, human genetics and genomics, and stem cell technology. Dianne's research particularly focuses on the legal and social issues associated with the commercialization of genetic knowledge and patenting of genetic inventions. She has held a number of Australian Research Council (ARC) discovery grants and currently leads two ARC-funded projects, one on the legal, research ethics, and social issues associated with genomic data sharing and the other on the regulation of innovative health technologies. The research presented in her chapter was funded by Australian Research Council grant DP180101262. In 2012, Dianne was appointed to a three-member expert panel to review pharmaceutical patenting in Australia. She was a member of two principal committees of the Australian National Health and Medical Research Council, the Australian Health Ethics Committee, and the Embryo Research Licensing Committee in the triennium from 2015 to 2018 and the Gene Technology Ethics and Community Consultative Committee of the Office of the Gene Technology Regulator from 2017 to 2018. In 2018 she was appointed as chair of the Embryo Research Licensing Committee. It should be noted that the views and interpretations expressed in her chapter are entirely her own and should in no way be connected with any of these agencies. She is a fellow of the Australian Academy of Law.

GUIDO PENNINGS is Full Professor of Ethics and Bioethics at Ghent University (Belgium) where he is also the director of the Bioethics Institute Ghent (BIG). He has published extensively on ethical problems associated with medically assisted reproduction and genetics.

He was Affiliate Lecturer in the Faculty of Politics, Psychology, Sociology and International Studies at Cambridge University from 2009 till 2013 and is Guest Professor on "Ethics in Reproductive Medicine" at the Faculty of Medicine and Pharmaceutical Sciences of the Free University Brussels since 2009. He was also former coordinator of the Special Interest Group on Ethics and Law of the European Society of Human Reproduction and Embryology (ESHRE), former member of the Ethics Committee of the European Society of Human Reproduction and Embryology (ESHRE), member of the National Advisory Committee on Bioethics in Belgium, member of the Federal Commission on Scientific Research on Embryos in vitro, and chair of the Ethics Committee of the Faculty of Arts and Philosophy at Ghent University.

LUDOVICA POLI is Assistant Professor of International Law at the University of Turin, Department of Law, where she teaches Public International Law and European Court of Human Rights case law. She holds a Ph.D. in Public International Law (2008), a master's degree in Peacekeeping Management (2004) and a master's degree in Bioethics and Clinical Ethics (2016). Her main field of research is human rights law; in particular, she is interested in studying the impact of developments in medical science on fundamental rights and, more in general, the intersection between human rights protection and bioethics. She has also explored other fields of international law, including humanitarian law, international criminal law, and the law of international organizations. Her publications cover a number of issues, such as the role of regional organizations in peace and security maintenance; the responsibility to protect and humanitarian intervention; women's rights, gender, and sex crimes in international law; artificial reproductive technologies and human rights. Ludovica is a member of the Italian Society of International Law (SIDI) where she coordinates the SIDI Interest Group on Bioethics and International and European Biolaw. She is also a member of the Italian Society for International Organization (SIOI); the International Institute of Humanitarian Law (IIHL); and the Interdisciplinary Center for Research and Studies of Women and Gender of the University of Turin (CIRSDE).

VARDIT RAVITSKY is Associate Professor at the Bioethics Program, School of Public Health, University of Montreal, and Director of Ethics and Health at the Center for Research on Ethics. Ravitsky is Vice-President of the International Association of Bioethics and member of the Standing Committee on Ethics of the Canadian Institutes of Health Research (CIHR) and of the Institute Advisory Board of CIHR's Institute of Genetics. She is also a member of the National Human Genome Research Institute's (NHGRI) Genomics and Society Working Group. Previously, she was faculty at the Department of Medical Ethics at the

University of Pennsylvania and a fellow at the Department of Bioethics of the NIH and at the National Human Genome Research Institute (NHGRI).

Ravitsky's research focuses on the ethics of genomics and reproduction and is funded by CIHR, FRQSC, SSHRC, and Genome Canada. She has published over 120 articles and commentaries on bioethical issues. Her research interests in bioethics include genetics, reproductive technologies, health policy, and cultural perspectives. She is particularly interested in the various ways in which cultural frameworks shape public debate and public policy in the area of bioethics. Born and raised in Jerusalem, Ravitsky brings international perspectives to her research. She holds a B.A. from the Sorbonne University in Paris, an M.A. from the University of New Mexico, the United States, and a Ph.D. from Bar-Ilan University, Israel.

CARLOS MARÍA ROMEO CASABONA is a professor of criminal law at the University of the Basque Country (UPV/EHU), where he leads a research group on bioethics and biolaw. He has written over 10 books and 200 articles and book chapters on these topics. He has been part of the main ethics committees in Europe.

SANTA SLOKENBERGA (LL.D.) is a postdoctoral researcher in public law at Lund University (Sweden) and a researcher at Uppsala University (Sweden). Her postdoctoral research focuses on regulating the standards of medical care and scientific uncertainty. Her other research interests include legal issues in human genetics and genomics. Since 2011, Santa has been teaching in the fields of EU law and medical law at Uppsala University and at Riga Stradins University (Latvia) for both undergraduate and graduate students. Since 2014, Santa has been teaching the summer school course "International Human Rights in Healthcare" at Yale University Sherwin B. Nuland Summer Institute in Bioethics (United States).

BRITTA VAN BEERS is Associate Professor at the Department of Legal Theory of VU University, Amsterdam, the Netherlands. In her research, she explores the legal-philosophical aspects of the regulation and governance of biomedical technologies. She is particularly intrigued by the legal and philosophical questions raised by the regulation of assisted reproductive technologies, such as wrongful birth and wrongful life actions, selective reproduction and reproductive tourism. In more recent work, she focuses on the governance of personalized medicine and human gene editing.

After studying law and philosophy at the University of Amsterdam and New York University School of Law, she obtained a Ph.D. degree at VU University Amsterdam for her dissertation "Person and Body in the Law: Human Dignity and Self-Determination in the Era of Medical Biotechnology" (The Hague, 2009). For this book she received prizes from the Dutch Health Law Association and the Praemium Erasmianum Foundation.

Recent publications include the volumes *Personalised Medicine, Individual Choice and the Common Good* (coedited with Sigrid Sterckx and Donna Dickenson; Cambridge University Press, 2018), *Symbolic Legislation and Developments in Biolaw* (coedited with Bart van Klink and Lonneke Poort; Springer, 2016), and *Humanity across International Law and Biolaw* (coedited with Wouter Werner and Luigi Corrias; Cambridge University Press, 2014).

EFFY VAYENA is a full professor of bioethics at the Swiss Federal Institute of Technology (ETH Zurich). She has completed her education as a social historian with a Ph.D. in Medical History from University of Minnesota. She then worked at the World Health Organization on ethical issues in reproductive medicine and research. Upon her return to academia, she helped establish and coordinated the Ph.D. program in Biomedical Ethics and Law at University of Zurich and was subsequently awarded a professorship by the Swiss National Science Foundation. She founded the Health Ethics and Policy Lab at the University of Zurich in 2015 before moving to the Swiss Federal Institute of Technology (ETH Zurich) in 2017. She is also a visiting professor at the Center for Bioethics at Harvard Medical School and a Faculty Associate at the Berkman Klein Center for Internet and Society at Harvard University, where she was previously a fellow.

Foreword

Human Germline Genome Modification and the Right to Science: A Comparative Study of National Laws and Policies by Andrea Boggio, Cesare P. R. Romano, and Jessica Almqvist constitutes a tour de force resource for those seeking more explicit normative guidance on this controversial subject. Indeed, what distinguishes this book is not only the quality and rigor of its comparative, legal methodology, but the human rights lens it employs as its conceptual framework. The "right to science" is both the leitmotif and the lens through which the current laws and policies of eighteen countries, spanning North America, Asia, Europe and other Western countries, and the European regulatory framework, are filtered. The common template of analysis applied to each chapter addresses the questions and issues raised by human germline modification throughout the translational pipeline from basic, animal, and clinical research to possible clinical applications thereby allowing for cross-country comparisons. Moreover, the international discussion of this hitherto largely dormant human right to science has the power to refocus the current largely "prohibitive" stance found across the world. Indeed, even today pro-scriptive and technique-specific language has effectively shut down debate on human germline modification. The issue was seen as settled, since largely banned and accompanied by criminal sanctions justified by reference to human genetic identity, integrity, and immutability. And then came Clustered Regularly Interspaced Short Palindromic Repeats (CRISPR).

Hopefully, this awakening of the right to science – its freedoms and its benefits – will serve to place the debate on germline modification outside the binary and polarized clichés of "slippery slopes" and "playing God." It can catalyze, activate, and shape the contours for radical innovation. It can move the debate away from imagining hypothetical, harmful applications to considering evidence-based benefits. Perhaps somatic gene therapies will lead the way and help develop the quality and safety foundations for the acceptability of clinical research into germline modifications. Most importantly, the legal actionability of the right to science could begin immediately with the creation of national and international governance structures that guide public debate and receive and monitor applications. Reflexive governance could also provide the common ethical guideposts for a responsible translation of the rights of its citizens, both individually and collectively, to say nothing of the rights of future generations.

Professor Bartha Maria Knoppers

Canada Research Chair in Law and Medicine
Director, Centre of Genomics and Policy
Faculty of Medicine, Department of Human Genetics
McGill University

Preface

As this book's manuscript was finalized, the media announced that twin girls, whose DNA had been genetically modified with CRISPR tools, were born. The news was shocking, not only because this was the first case of human germline genome modification resulting in a birth, but also because it happened outside the boundaries of the law.

The scientist who made the shocking announcement on the eve of an international conference in Hong Kong on CRISPR, Dr. He Jiankui, of the Southern University of Science and Technology, in Shenzhen, People's Republic of China, said he and his colleagues altered embryos for seven couples during fertility treatments, with two female volunteers becoming pregnant as a result. One gave birth to female twins, named Lu Lu and Na Na, and the other is still pregnant. He said his goal was not to cure or prevent an inherited disease but to try to bestow a trait that few people naturally have – an ability to resist possible future infection with HIV, the AIDS virus. Dr. He Jiankui declared to Associated Press: "I feel a strong responsibility that it's not just to make a first, but also make it an example. Society will decide what to do next."[1] What we now know is that Dr. He's experiments involved serious violations of national regulations and international standards.

When we launched this book's project, we knew the application of CRISPR to human germline cells was going to take place, somewhere, somehow. We certainly did not foresee that it would happen so soon and within an overall legal framework that almost invariably prohibits clinical applications of germline genome modification. Dr. He's stunt will certainly jolt lawmakers across the globe, and that is good and overdue. However, the risk is that the scientific and human rights implications of this extremely complex field of scientific

---

[1] M. Marchione, "Chinese Researcher Claims First Gene-Edited Babies," Associated Press, November 26, 2018.

research will be glossed over in the vain search for quick and easy solutions. We hope our project can help in shaping the debate that will certainly take place in the coming years.

Like many projects, this one is largely the result of personal connections and intersecting research agendas. For a few years, Cesare had been litigating a case (communication) regarding research on human embryos before the Committee on Economic, Social and Cultural Rights together with the Associazione Luca Coscioni for the freedom of scientific research. Andrea had been a member of the board of the Associazione for years, supporting its work with a project mapping indicators of freedom of research. The two had become interested in the topic of the right to science and the rights of science, starting work on a book together. Cesare had worked with Jessica on international justice in New York, at New York University, at the Project on International Courts and Tribunals in the early 2000s. When Jessica contacted Cesare saying she intended to apply for a research grant and was shopping for ideas, Cesare had just finished discussing human genome modification with Andrea. Andrea reached out to Bartha Knoppers, the Director of the Centre of Genomics and Policy, Faculty of Medicine, Human Genetics, of McGill University. Bartha kindly gave us access to her vast professional network. The rest, as they say, is history.

At the outset of the project, it became apparent to us that, as it often happens with disruptive technological breakthroughs, states are struggling to keep up with developments and to regulate research and applications, creating a patchwork of national legislations. Some have equipped themselves with fairly sophisticated legislation and regulatory bodies to ensure research can advance but within acceptable limits. Others have instead opted to restrict research and applications as much as possible to ward off any dangers. Many have not yet adopted any national legislation, and keep on relying on outdated legislation that is unsuitable to regulate gene editing in general, and surely germline editing in particular. They are waiting to see in which direction most other states are going. Often, the boundaries of what is legally permitted are unclear. Gaps and unresolved legal issues abound. The only clear pattern is that clinical research is prohibited or under a moratorium in all countries. In fact, so far, no law seems to permit the implant of a genetically modified embryo (or an embryo created by using genetically modified germ cells) in uterus, whether for research purposes or to start a pregnancy.

At the international level, bar the Oviedo Convention and the European Clinical Trials Regulation, there are no clear, global legal standards on germline genome editing. However, international human rights standards, as codified in the Universal Declaration of Human Rights and the twin Covenants (civil and political rights, and economic, social, and cultural rights), do provide the four corners within which regulatory frameworks of heritable gene editing must be placed and developed.

Within this fragmented and incomplete regulatory environment, this book explores both levels of regulation (domestic and international) and makes a case for using the human right to "benefit from advances in science and technology" (the right to science) and to "freedom indispensable for scientific research" (the rights of science) as a guiding framework to regulate germline engineering. The book accomplishes these goals in three steps. First, it maps national legislation of germline engineering in eighteen states and one region (Europe). The mapping exercise is not only descriptive but also normative, to the extent that it allows for the identification of best practices and the guidance of states that are still in the process to adapt their national legislation.

Second, the book provides a comparative analysis of the laws of the chosen jurisdictions. This analysis identifies patterns and trends as well as areas where policy work needs to be done. Our findings show a pattern of prohibition of any form of clinical research and clinical applications of germline genome modifications. With regard to basic research, the picture is much more fragmented with regulations taking a number of different paths. However, for the most part, these regulations are obsolete, incomplete, and unclear about what research can and cannot accomplish lawfully.

Third, the book connects national legislation to the core international obligations all states have as a matter of international human rights law. In particular, it puts at the center of the discussion the so-called right to science as a source of normative guidance in this complex area.

Although the "right to science" and the "rights of science" are some of the oldest human rights, dating back to the late 1940s, they are probably the least known, discussed, and enforced of all international human rights. However, over the past few years, they have been the object of much renewed attention. The Committee on Economic, Social and Cultural Rights is in the process of drafting a General Comment on them, and a number of nongovernmental organizations and scholars are busy organizing symposia and writing books on the matter.

No book, however, has applied the right to science and the rights of science to germline engineering. Our argument as to how human rights law applies to germline genome modifications is twofold. First, the current patchwork of national legislation infringes upon scientists' freedom to engage in basic research that has the potential to revolutionize how disease is construed and treated. Second, the right to science mandates that the material benefits of science are shared. Our reading is that a blank prohibition against clinical research infringes upon the right of patients to have access to the benefits of science. These arguments are novel and readers will certainly benefit from a deeper understanding of the extent to which international human rights law can inform debate on germline engineering.

Andrea Boggio, Cesare P. R. Romano, and Jessica Almqvist

Acknowledgments

The book would not have been possible without the invaluable input from a number of people. Our gratitude goes, first of all, to the two dozen contributors who gave their time and energy to write the national chapters of this book and graciously put up with some inquisitive and pushy editors. However, this book would have been so much more difficult to accomplish without Bartha Knoppers, Director of the Centre of Genomics and Policy, Faculty of Medicine, Human Genetics at McGill University, who helped us identify them and graciously wrote the Foreword. We also want to thank Rosario Isasi and Erika Kleiderman for contributing with ideas and suggestions at a kick-off meeting for the project, which has shaped the format of the book. We also want to thank Robin Lovell-Badge and George Church for their feedback, particularly on the sections discussing the science of gene editing.

At Loyola Law School, Los Angeles, we need to thank Alda Merino-Caan (JD 2019) for copy-editing the manuscript; Caitlin Hunter, Reference Librarian/Foreign and International Law Librarian, for handling fast and efficiently random requests about national legislation of disparate countries; and the Dean and Associate Deans for giving Cesare a sabbatical to complete this and other projects.

At Autonomous University of Madrid, we are indebted to the Institute of Human Rights, Democracy and Culture of Peace and Non-Violence (DEMOSPAZ) for helping us organize and host an international expert seminar – The Right to Benefit from Science: Revisiting Human Germline Genome Modification – which took place in the UAM Law Faculty on June 7 and 8, 2018, to prepare the book. We want to thank those who attended the seminar for their valuable feedback.

At Bryant University, we want to thank the Office of the Provost and the Department of History and Social Sciences for their support.

We are grateful to Banco Santander, for its financial support of this project, and whose funding was received through the UAM (project for interuniversity cooperation UAM-Banco Santander with the United States 2017/EEUU/02) and the Office of the Provost at Bryant University.

We want to thank Podhumai Anban and Judieth Sheeja at Integra Software Services, who copyedited the book, and Birgitte Necessary, who compiled the book index.

Last but not least, we would like to thank John Berger, commissioning editor at Cambridge University Press, for believing in this project and offering us the assistance of a fine publisher to circulate our ideas and Chloe Quinn and Jackie Grant for assisting diligently throughout the book production process.

Abbreviations

ABM	Biomedicine Agency (Agence de la Biomédecine) (France)
ACC	Animal Care Committees (Canada)
ACHR	American Convention on Human Rights
AHRA	Assisted Human Reproduction Act (Canada)
ANM	National Academy of Medicine (Académie Nationale de Médecine) (France)
ANR	National Research Agency (Agence Nationale de la Recherche) (France)
ARCs	Assisted Reproduction Centres
ART	Artificial/Assisted Reproductive Technologies
ASRM	American Society for Reproductive Medicine
AUGMENT	AUtologous Germline Mitochondrial ENergy Transfer
BAC	Bioethics Advisory Committee (Singapore)
BMBF	Federal Ministry of Education and Research (Bundesministeriums für Bildung und Forschung) (Germany)
BOE	Official Gazette (Boletín Oficial del Estado) (Spain)
BSA	Bioethics and Safety Act (Republic of Korea)
BVerfG	Federal Constitutional Court (Bundesverfassungsgericht) (Germany)
C. civ.	Civil Code (Code civil) (France)
C. pén.	Criminal Code (Code pénal) (France)
C. rur.	Rural Code (Code rural) (France)
CAHBI	Ad hoc Committee of Experts on Bioethics (Council of Europe)
Cas9	CRISPR associated protein 9

CBER	Center for Biologics Evaluation and Research (United States)
CCAC	Canadian Council on Animal Care
CCMO	Central Committee on Research involving Human Subjects (Centrale commissie voor medisch-wetenschappelijk onderzoek op mensen) (Netherlands)
CCNE	National Ethical Consultative Committee for Life and Health Sciences (Comité Consultatif National d'Éthique pour les sciences de la vie et de la santé) (France)
CDA	Christian Democratic Appeal (Het Christen-Democratisch Appèl) (Netherlands)
CDBI	Steering Committee on Bioethics (Council of Europe)
CEC	Clinical Ethics Committees (Mexico)
CEDAW	Convention on the Elimination of All Forms of Discrimination against Women
CENATRA	National Centre for Transplants (Centro Nacional de Trasplantes) (Mexico)
CEPAFIC	Spanish Committee for the Protection of Animals Used for Scientific Research (Comité español para la protección de los animales utilizados con fines científicos) (Spain)
CFDA	Chinese Food and Drug Administration (PRC)
CIBIOGEM	Inter-Secretariat Commission on Biosecurity of Genetically Modified Organisms (Comisión Intersecretarial de Bioseguridad de los Organismos Genéticamente Modificados) (Mexico)
CIHR	Canadian Institutes of Health Research
CINVESTAV	Centre for Research and Advanced Studies (Centro de Investigacion y Estudios Avanzados) (Mexico)
CJEU	Court of Justice of the European Union
CNDH	National Commission of Human Rights (Comisión Nacional de los Derechos Humanos) (Mexico)
CNRS	National Center for Scientific Research (Centre National de la Recherche Scientifique) (France)
CNTS	National Centre for Blood Transfusion (Centro Nacional de Transfusión Sanguínea) (Mexico)
CoE	Council of Europe
COFEPRIS	Federal Commission for the Prevention against Sanitary Risks (Comisión Federal de Prevención Contra Riesgos Sanitarios) (Mexico)

COGEM	Commission on Genetic Modification (Netherlands)
CONACYT	National Council for Science and Technology (Consejo Nacional de Ciencia y Tecnología) (Mexico)
CONBIOÉTICA	Comisión Nacional de Bioética (National Commission of Bioethics) (Mexico)
CPEUM	Political Constitution of the Mexican United States (Constitución Política de los Estados Unidos Mexicanos) (Mexico)
CRISPR	Clustered Regularly Interspaced Short Palindromic Repeats
CSP	Public Health Code (Code de la santé publique) (France)
CSTI	Council for Science, Technology and Innovation (Japan)
D66	Democrats 66 (Democraten 66) (Netherlands)
DCC	Civil Code (Burgerlijk Wetboek) (Netherlands)
DH-BIO	Council of Europe's Committee on Bioethics
DHHS	Department of Health and Human Services (United States)
DIA	Drug Inspection Administration (PRC)
DNA	Recombinant Deoxyribonucleic Acid
EC	European Council
ECHR	European Convention on Human Rights and Fundamental Freedoms
ECJ	European Court of Justice
ECtHR	European Court of Human Rights
EGE	European Group on Ethics in Science and New Technologies
ELSI	Ethical, Legal and Social Aspects
EMA	European Medicines Agency (EU)
EPC	European Patent Convention
EPO	European Patent Office
ERC	European Research Council
ERLC	Embryo Research Licensing Committee (Australia)
ERRIH	Ethical Review of Research Involving Humans (Lag om etikprövning av forskning som avser människor) (Sweden)
ESchG	Embryo Protection Act (Embryonenschutzgesetz) (Germany)
ESHRE	European Society for Human Reproduction and Embryology
EU	European Union

FCCyT	Consultative Body on Science and Technology (Foro Consultivo de Ciencias y Tecnología) (Mexico)
FDA	Food and Drug Administration (United States)
G20	Group of 20
GAP	Good Animal Practice (Canada)
GDP	Gross Domestic Product
GenTG	Genetic Engineering Act (Gentechnikgesetz) (Germany)
GfH	German Society of Human Genetics (Deutsche Gesellschaft für Humangenetik)
GG	Grundgesetz (Basic Law) (Germany)
GHC	General Health Council (Mexico)
GHL	General Health Law (Mexico)
GIA	Genetic Integrity Act (Lag om genetisk integritet) (Sweden)
GMO	Genetically Modified Organism
GMP	Good Manufacturing Practices
GT Act	Gene Technology Act 2000 (Australia)
GTR	Gene Technology Regulator (Australia)
HBRA	Human Biomedical Research Act (Singapore)
hESCs	Human Embryonic Stem Cells
HFC	Health and Family Planning Commission (PRC)
HGRMO	Human Genetic Resources Management Office (PRC)
HIV	Human Immunodeficiency Virus
HLA	Human Leucocyte Antigen
HSA	Health Sciences Authority (Singapore)
IACHR	Inter-American Commission of Human Rights
IACtHR	Inter-American Court of Human Rights
IBC	Institutional Biosafety Committee
ICB	Italian Committee for Bioethics (Comitato Italiano di Bioetica) (Italy)
ICCPR	International Covenant on Civil and Political Rights
ICESCR	International Covenant on Economic, Social and Cultural Rights
ICH	International Conference on Harmonisation
ICSI	Intracytoplasmic Sperm Injection
IIJ-UNAM	The Institute for Legal Research at the National Autonomous University of Mexico (Instituto de Investigaciones Jurídicas de la Universidad Nacional Autónoma de México)

IMSS	Mexican Social Security Institute (Instituto Mexicano del Seguro Social) (Mexico)
INAI	National Institute for Transparency and Access to Information (Instituto Nacional de Acceso a la Información) (Mexico)
IND	Investigational New Drug (United States)
INEGI	National Institute of Statistics and Geography (Instituto Nacional de Estadística y Geografía) (Mexico)
INMEGEN	National Institute for Genomic Medicine (Instituto Nacional de Medicina Genómica) (Mexico)
INSERM	Ethics Committee of the French National Institute for Health and Medical Research (Institut National de Santé et Recherche Médicale) (France)
IPN	National Polytechnic Institute (Instituto Politécnico Nacional) (Mexico)
iPSCs	Induced Pluripotent Stem Cells
IRB	Institutional Review Board
ISSCR	International Society for Stem Cell Research (United States)
ISSFAM	Institute for Social Security and the Mexican Army Forces (Instituto de Seguridad Social para las Fuerzas Armadas de México) (Mexico)
ISSSTE	Institute of Social Security and Services for Civil Servants (Instituto de Seguridad y Servicios Sociales para los Trabajadores del Estado) (Mexico)
IVF	In Vitro Fertilization
IVG	In Vitro Gametogenesis
IVM	In Vitro Maturation
JSOG	Japan Society of Obstetrics and Gynecology
LAA	Laboratory Animal Act (Republic of Korea)
LAGH	Federal Act on Human Genetic Testing (Loi fédérale sur l'analyse génétique humaine) (Switzerland)
LANGEBIO	National Laboratory of Genomics for Biodiversity (Laboratorio Nacional de Genómica para la Diversidad) (Mexico)
LBOGM	Biosafety Law on Genetically Modified Organism (Ley de Bioseguridad de los Organismos Genéticamente Modificados) (Mexico)

LGG   Federal Act on Non-Human Gene Technology (Loi féd-
                    érale sur l'application du génie génétique au domaine
                    non humain) (Switzerland)
LPA   Animal Welfare Act (Loi fédérale sur la protection des
                    animaux) (Switzerland)
LPMA   Federal Act on Medically Assisted Reproduction (Loi
                    fédérale sur la procréation médicalement assistée)
                    (Switzerland)
LRCS   Federal Act on Research Involving Embryonic Stem
                    Cells (Loi fédérale relative à la recherche sur les cellules
                    souches embryonnaires) (Switzerland)
LRH   Federal Act on Research involving Human Beings (Loi
                    fédérale relative à la recherche sur l'être humain)
                    (Switzerland)
MAR   Medically Assisted Reproduction
MDFS   Ministry of Drug and Food Safety (Republic of Korea)
ME   Ministry of Economy (Mexico)
MEXT   Ministry of Education, Culture, Sports, Science, and
                    Technology (Japan)
MHLW   Ministry of Health, Labour, and Welfare (Japan)
MMT   Mitochondrial Manipulation Techniques
MoH   Ministry of Health (PRC) (Mexico) (Singapore)
MoHW   Ministry of Health and Welfare (Republic of Korea)
MoST   Ministry of Science and Technology (PRC)
MRT   Mitochondrial Replacement Therapy
mtDNA   Mitochondrial DNA
NAFTA   North American Free Trade Agreement
NEK/CNE   Swiss National Advisory Commission on Bioethics
                    (Commission nationale d'éthique dans le domaine de la
                    médecine humaine) (Switzerland)
NHC   National Health Commission (PRC)
NHI   National Health Institutes (Institutos Nacionales de
                    Salud) (Mexico)
NHMRC   National Health and Medical Research Council
                    (Australia)
NHS   National Health Services (Mexico)
NICE   National Institute for Health and Care Excellence
                    (Belgium)
NIH   National Institutes of Health (United States)

NOMs	Official Mexican Norms or Standards (Normas Oficiales Mexicanas) (Mexico)
NPC	National People's Congress (PRC)
NRT	New Reproductive Technologies
NSERC	Natural Sciences and Engineering Research Council of Canada
NWO	The Netherlands Organization for Scientific Research (Nederlandse organisatie voor Wetenschappelijk Onderzoek)
OAGH	Ordinance on Human Genetic Testing (Ordonnance sur l'Analyse Génétique Humaine) (Switzerland)
OAS	Organization of American States
OClin	Ordinance on Clinical Trials in Human Research (Ordonnance sur les essais cliniques dans le cadre de la recherche sur l'être humain) (Switzerland)
OECD	Organisation for Economic Co-operation and Development
OFSP	Federal Office of Public Health (Office fédéral de la santé publique) (Switzerland)
OPAn	Animal Welfare Ordinance (Ordonnance sur la Protection des Animaux) (Switzerland)
OPECST	Parliamentary Office of Evaluation of Scientific and Technological Options (Office Parlementaire d'Évaluation des Choix Scientifiques et Technologiques) (France)
OPMA	Reproductive Medicine Ordinance (Ordonnance sur la procréation médicalement assistée) (Switzerland)
ORCS	Ordinance on Research involving Embryonic Stem Cells (Ordonnance relative à la recherche sur les cellules souches embryonnaires) (Switzerland)
OSAV	Federal Food Safety and Veterinary Office (Office Fédéral de la sécurité alimentaire et des affaires vétérinaires) (Switzerland)
OUC	Ordinance on Handling Organisms in Contained Systems (Ordonnance sur l'utilisation des organismes en milieu confiné) (Switzerland)
PAA	Pharmaceutical Affairs Act (Republic of Korea)
PAHO	Pan American Health Organization
PAN	National Action Party (Partido Acción Nacional) (Mexico)
PatG	Patent Act (Patentgesetz) (Germany)
PEMEX	Mexican Petrol (Petróleos Mexicanos) (Mexico)

PGD	Pre-implantation Genetic Diagnosis
PHCR Act	Prohibition of Human Cloning for Reproduction Act 2002 (Australia)
PMI	Precision Medicine Initiative (PRC)
PNT	National Platform for Transparency (Plataforma Nacional de Transparencia) (Mexico)
PRC	People's Republic of China
PRI	Institutional Revolutionary Party (Partido Revolucionario Institucional) (Mexico)
PvdA	Labour Party (Partij van de Arbeid) (Netherlands)
R&D	Research & Development
RCOG	Royal College of Obstetricians and Gynaecologists (Belgium)
REC	Research Ethics Committee
RF	The Instrument of Government (Regeringsformen) (Sweden)
RIHE Act	Research Involving Human Embryos Act 2002 (Australia)
RNEC	National Registry of Clinical Trials (Registro Nacional de Estudios Clínicos) (Mexico)
ROK	Republic of Korea
S&T	Science and Technology
SAMR	State Administration for Market Regulation (PRC)
SC	State Council (PRC)
SCJ	Science Council of Japan
SCJN	Mexican Supreme Court (Suprema Corte de Justicia de la Nación) (Mexico)
SCNT	Somatic Cell Nucelear Transfer
SCOC	Stem Cell Oversight Committee (Canada)
SEP	Ministry of Education (Secretaria de Educación Pública) (Mexico)
SNI	National System of Researchers (Sistema Nacional de Investigadores) (Mexico)
SSHRC	Social Sciences and Humanities Research Council of Canada
STA	Science and Technology Act (Mexico)
StGB	Criminal Code (Strafgesetzbuch) (Germany)
Swissmedic	Swiss Agency for Therapeutic Products (Swissmedic, Institut suisse des produits thérapeutiques) (Switzerland)

TAB	Office of Technology Assessment at the German Parliament (Büro für Technikfolgen-Abschätzung beim Deutschen Bundestag)
TALENS	Transcription Activator-like Effector Nucleases
TCART	Toronto Center for Assisted Reproductive Technologies (Canada)
TCPS 2	Tri-Council Policy Statement: Ethical Conduct for Research Involving Humans (Canada)
TEU	Treaty on the European Union
TFEU	Treaty on the Functioning of the European Union
TIARAS	Tissue and Research Application System
TierSchG	Animal Protection Act (Tierschutzgesetz) (Germany)
TierSchVersV	Animal Protection – Experimental Animal Ordinance (Tierschutz-Versuchstierverordnung) (Germany)
UDHR	Universal Declaration on Human Rights
UN	United Nations
UNAM	National Autonomous University of Mexico (Universidad Nacional Autónoma de México)
UNESCO	United Nations Educational, Scientific and Cultural Organization
UNRISD	United Nations Research Institute for Social Development
US	United States of America
USD	United States Dollars
VDI	Association of German Engineers (Verein Deutscher Ingenieure) (Germany)
VVD	People's Party for Freedom and Democracy (Volkspartij voor Vrijheid en Democratie) (Netherlands)
WB	World Bank
WHO	World Health Organization
WMA	World Medical Association
WMO	Medical Research on Human Subjects Act (Wet wetenschappelijk onderzoek met mensen) (Netherlands)
ZFN	Zinc-Finger Nuclease
ZON	Dutch Healthcare Research (ZorgOnderzoek Nederland) (Netherlands)

1

Introduction

Andrea Boggio, Cesare P. R. Romano, and Jessica Almqvist

As early as 2000, Gregory Stock and John Campbell noted that the achievement of the capacity to make changes to human germline cells was "inevitable."[1] "The real question," they postulated, was not "whether the technology will become feasible, but when and how it will."[2] The advent of the CRISPR/Cas9 family of genome-editing tools answered the question of how. As to when, it is now: humanity has already entered a new age. How it will unfold depends on the regulatory frameworks that each state, and all of them collectively, will be able to provide in response to this pathbreaking scientific and technological development.

The primary goal of this book is to analyze how human germline genome modification is currently regulated in a selected, but fairly broad and representative, set of developed countries, and to assess their national governance practices in the light of the existing and emerging international legal obligations states have. These include the obligation to respect international human rights, and specifically the so-called right to science and the rights of science. It is obvious, but not always recalled, that the mosaic of national regulatory frameworks does not exist in a vacuum. There is a broader international framework within which the various national frameworks need to be placed. The human rights framework constraint governs and guides states in their law-making activities.

Until relatively recently, international human rights standards had not featured prominently in bioethical analyses, even though most international documents relating to bioethics issued during the past two decades are framed on a rights-based approach and attach utmost importance to the notion of

[1] G Stock and J Campbell, *Engineering the Human Germline: An Exploration of the Science and Ethics of Altering the Genes We Pass to Our Children*, Oxford University Press, 2000, p. 6.
[2] Ibid., p. 5.

human dignity, which is a concept used also in human rights law.[3] However, bioethical analyses do not necessarily take all relevant human rights into account or use the concept of dignity in the same way as human rights law does. In this book, we link human rights law to the regulation of human germline genome modifications at the national level.[4] For all its limitations and weaknesses, the existing human rights system, with its extensive body of international standards and wide range of mechanisms, cannot be ignored. As it has been said, the human rights framework provides a "more useful approach for analysing and responding to modern public health challenges than any framework thus far available within the biomedical tradition."[5] Indeed, there are few mechanisms available other than human rights to function as "a global ethical foundation, a *Weltethik.*"[6] In fact, we conclude this book by showing how international human rights standards are applicable and can be used to respond to the challenge of regulating human germline genome modifications.

Because this book speaks to a very wide and diverse audience, one made of legal scholars and biomedical scientists, nongovernmental and governmental actors, we believe it is necessary to begin by introducing some basic concepts and terms related to human germline modification that might very well be familiar to some readers but not necessarily to others. In this Introduction (Chapter 1), we discuss the science of human genome modification in general, how it relates to human reproduction, and the specific advances that CRISPR/Cas9 represents and the family of tools it has generated to date. We then explain the methodology we followed preparing this book, including how countries were selected and what we chose to focus on and why.

This is necessary background for a discussion of the national and international governance frameworks for human germline genome engineering. Chapter 2 sets out the emerging international governance framework. The aim is to introduce the reader to the key actors (international organizations,

---

[3] On the intersection between international bioethics law and international human rights law, see, in general, R Andorno, *Principles of International Biolaw: Seeking Common Ground at the Intersection of Bioethics and Human Rights*, Bruylant, 2013; S Holm, *The Law and Ethics of Medical Research: International Bioethics and Human Rights*, Cavendish Publ., 2005.

[4] For a condensed version of the data and arguments presented in this book, see, A Boggio, B M Knoppers, J Almqvist, and C P R Romano, "The Human Right to Science and the Regulation of Human Germline Engineering," *The CRISPR Journal*, Vol. 2 (2019), pp. 134–142.

[5] J Mann, "Health and Human Rights: Protecting Human Rights is Essential for Promoting Health," *British Medical Journal*, Vol. 312 (1996), p. 924.

[6] D Thomasma, *Proposing a New Agenda: Bioethics and International Human Rights*, Cambridge Quarterly of Healthcare Ethics, 2001, Vol.10, pp. 299–310, p. 300.

international agencies, and learned societies), both at the global and regional level, as well as the most important legal standards they have articulated so far, including in the area of human rights law. This discussion is not limited to treaties, of which there are few, but also takes so-called soft law, such as international declarations and guidelines, into account. Then Part III, the bulk of this volume, presents an analysis of the national regulatory frameworks of eighteen selected countries and one region: Europe (Chapters 3 to 21). Each "national chapter" follows a template that we explain in the "methodology" section below.[7] Finally, we conclude the book by looking at the existing national and international regulatory frameworks in light of the five foundational principles that a reading of international bioethics law combined with international human rights standards suggests. They are: (i) freedom of research, (ii) benefit sharing, (iii) solidarity, (iv) respect for dignity, and (v) the obligation to respect and to protect the rights and individual freedoms of others (Chapter 22). We conclude by offering our vision of an international governance framework that promotes science and technological development while being mindful and respectful of international human rights standards, as well as the different sensitivities with which citizens from different parts of the world approach this complex problem.[8]

I THE SCIENCE

At least since the discovery of DNA, biomedical researchers have been on a quest to develop tools to make targeted changes to the genome of living organisms for scientific, therapeutic, and economic reasons. Efficient, reliable, and low-cost genome-editing tools can be used, and are already used, for different purposes. They can help advance science, for instance by understanding what functions each gene controls, and be used to prevent or cure genetic disorders, for instance by correcting errors in the genome. Or, they can be used to improve and develop certain traits in a given cell or organism. Scientists modify the DNA of animals not only to observe how changes affect the animal itself but also to predict how the same changes in the human genome might affect human health. Or, they might seek to propagate a particular suite of genes throughout a species (the so-called gene drive) to drive to extinction insects that carry pathogens (such as mosquitoes that transmit malaria), control invasive species, or eliminate herbicide or pesticide

---

[7] See, in this chapter, Section II.
[8] The views expressed in this and the final chapter do not necessarily reflect the views of the authors who contributed the national chapters to this book.

resistance.[9] In plant bioengineering, gene editing is used as a tool to improve resistance against fungi (corn), crop yields during drought stress (maize), and accelerate growth to boost yield (tomato and rice).[10]

A series of genome-reading (sequencing) and genome-editing tools, including ZFN, TALEN and CRISPR (Cas9 and Cas12a) gave scientists the instruments to modify the genome precisely, efficiently, and at much lower cost. This scientific breakthrough has opened unprecedented and hitherto unthinkable research and therapeutic scenarios. It is also raising major ethical and regulatory dilemmas, especially when the genome that is being edited is the one of human germline cells, as this means changing the genes passed on to future generations.

Germline cells are one of the two kinds of cells that mammals have. Gametes, that is, oocytes (eggs) in females and sperm in males, are our germline cells. All other cells that make up the human body are "somatic cells." Modifications made to somatic cells stay within the body of the person whose cells have been modified. However, modifications made to germline cells, under certain circumstances, are heritable. They might be passed on to the next generation and into the future, potentially altering the genome of entire populations.

There are valid reasons for wanting to modify the genome of entire populations, if not the whole of humanity. The modification of the genome of human germline cells allows attaining two major therapeutic goals. First, it could be used to greatly reduce the burden of hundreds of hereditary genetic diseases that condemn millions to considerable pain or death every year, much as vaccines have eliminated some viruses that have plagued humanity for millennia. Second, it allows curing genetic diseases before they manifest their often-devastating effects on the human body. That is a more efficient and ethically superior goal than allowing a patient to get sick first and then cure (if there is a cure).

To achieve these therapeutic goals, two paths are being considered: editing gamete precursor cells, or gametes before fertilization, or editing embryos after fertilization. Either can be done outside the human body (in vitro or ex vivo) or directly inside the body (in vivo). The difference is whether the mutation is corrected before fertilization (editing of gamete and precursor cells) or after

---

[9]   N Kofler and others, "Editing Nature: Local Roots of Global Governance," *Science*, Vol. 362 (2018), p. 527.

[10] C Gao, "The Future of CRISPR Technologies in Agriculture," *Nature Reviews Molecular Cell Biology*, Vol. 19 (2018), p. 275.

(editing of embryos). In the end, the goal of either method is similar: allowing parents to procreate a genetically related offspring who does not carry an undesirable mutation.[11] Since both methods necessarily involve human reproduction, we introduce some basic concepts of human reproduction and embryology before turning to the science of gene editing.

1 Germline Cells, Reproduction, and Embryo Development

Sexual reproduction involves a process that begins with the fertilization of an egg (Figure 1.1). This occurs when the male gamete (sperm) and the female gamete (oocyte) fuse to form the "zygote," a cell that carries two complete sets of chromosomes: one set matching the sperm DNA and the other the oocyte DNA. Subsequently, the zygote (which scientists sometimes call a "one-cell embryo") develops into a ball of cells and undergoes a series of divisions, called "cleavages." As this ball of cells develops, it assumes different names. At the first stage of development, it is called a "morula," consisting of sixteen cells. The morula then develops into a "blastocyst" when the internal cells are pushed outward forming a cavity. The outer cells become epithelial tissue, which will eventually form the placenta that supports the embryo in its growth in the uterus.

It is important to note that the cells of a blastocyst, known as "inner cell mass," are undifferentiated and pluripotent. This means that they have not yet committed to becoming any particular type of cell and that they can still become any tissue of the human body, including gametes. It follows that, if the cells of blastocysts are edited, any modification will have effects on the germline of the human being who will develop from a modified blastocyst and their progeny.

Upon entering the uterus (or being artificially transferred into the uterus, if fertilization occurred outside, in vitro), the blastocyst implants in the endometrium, the mucous membrane lining the uterus, and develops into a "gastrula." From a single-layer structure, the cells of the ball are reorganized into a multilayered one. The transition from blastocyst to gastrula, known as "gastrulation," is characterized by the formation of the primitive streak, which is a structure that forms on the back of the balls of cells. After the appearance of the primitive streak, the internal cells are reorganized in three germ layers (endoderm, ectoderm, and mesoderm), each of which will produce a distinct cell lineage that form the various types of tissue that form the human body.

Gastrulation, which takes place about fourteen days after fertilization, is a turning point in cell development because, after the formation of germ

---

[11] N Kofler and K L Kraschel, *Treatment of Heritable Diseases Using CRISPR: Hopes, Fears, and Reality*, Seminars in Perinatology, Vol. 48 (2018), pp. 515–521.

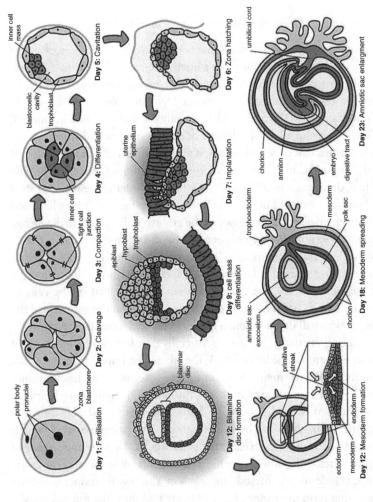

FIGURE 1.1 The first few weeks of embryogenesis in humans: beginning at the fertilized egg, ending with the closing of the neural tube Zephyris – SVG version, CC BY-SA 3.0, https://commons.wikimedia.org/w/index.php?curid=10811330

layers, cells are no longer pluripotent. They are differentiated, meaning that they are able to develop only into certain types of tissue. This happens during the next stage of development, when the outermost layer of cells of the embryo (ectoderm) thickens, activating the formation of the neural plate, which will give rise to both the spinal cord and the brain. The outcome of this process, called "neurulation," is the transformation of the gastrula into a "neurula." The other two germ layers, interacting with each other and the emerging neural system, contribute to the development of all individual organs by way of cell movements and cell differentiation (organogenesis). These organs include the gonads, the reproductive glands that produce the gametes.

Theoretically, gametes or their precursor cells can be edited in vivo, cells in the fetus in utero, within the testis, or early embryos within the oviduct before implantation. This scenario is hypothetical because of technical and legal limitations. In fact, until recently, scientists have not been able to culture embryos in vitro during gastrulation, and in vitro experimentation on embryos beyond fourteen days has been prohibited for the past forty years across the globe. However, in May 2016, labs in the United States and the United Kingdom reported being able to sustain human embryos in vitro for up to thirteen days.[12] This means that what is now just a mere hypothetical scenario may become reality.

Fertilization and some of the early stages of reproduction can also take place in vitro, outside the female body. Since 1978, when the first child resulting from an in vitro fertilization (IVF) process was born, scientists have been refining techniques. During IVF, the gametes are retrieved from the parents-to-be and, after proper handling, combined using various techniques. The resulting zygote is then cultured in vitro and eventually is implanted in the uterus. If not immediately implanted, the morula or the blastocyst is cryopreserved for varying lengths of time for various medical reasons. Sometimes, frozen morulae or blastocysts end up not being used by the woman or couple. When that happens, some are donated to other infertile couples. Some are used in research (supernumerary or spare embryos). Those that are not donated for reproduction or research are destroyed or, in some countries, kept frozen in perpetuity.

2 The Emergence of the CRISPR Tools

We now turn to the technology that is seen as the most promising for gene editing: the CRISPR tools. CRISPR (the acronym for clustered regularly interspaced short palindromic repeats) is a family of DNA sequences found within

---

[12] J Rossant and P P L Tam, "Exploring Early Human Embryo Development," *Science*, Vol. 360 (2018), pp. 1075–1076.

the genomes of prokaryotic organisms such as bacteria and archaea. It is the mechanism of self-defense that cells possess to fend off viral attacks. When a virus attacks, the bacteria capture snippets of DNA from invading viruses and store it to remember the identity of the invader so that, in the event of a second attack, the bacteria can activate its defense mechanism. This mechanism entails releasing an enzyme called "Cas9" (short for "CRISPR-associated 9"), which acts as a pair of "molecular scissors" and cuts the DNA chain at a specific location. As a defense mechanism, Cas9 is used to disable the attacking virus. Researchers have figured out that Cas9 can be used to cut DNA of human cells to edit it. To do so, Cas9 is injected in a cell along with a piece of RNA, known as "guide RNA" or "gRNA," that is engineered to have an end that is complementary to the DNA sequence that Cas9 must cut. Once the gRNA is injected, Cas9 is guided to the right spot by the guide, which is looking for the target DNA, and cuts the DNA sequence where gRNA. Once the DNA chain is cut, scientists can add or remove DNA sequences (Figure 1.2).

To everyone but those who practice it, it might sound like magic, but it is not. It is the result of a long process of scientific research and discovery that started toward the end of the twentieth century. It is a rather long and intricate story, involving scientists in Asia, Europe, and North America, in universities and the private sector, producing a complex network of knowledge from bioinformatics, genetics, and molecular biology. It begins with some rather obscure research on bacteria and microbes. In 1987, Yoshizumi Ishino and colleagues at Osaka University discovered an unusual series of repeated sequences in DNA sequences outside of coding regions, later called clustered regularly interspaced short palindromic repeats (CRISPR), that contribute to the process of programmed cell death – the deliberate suicide of an unwanted cell – in *E. coli* bacteria.[13] Then, in the early 1990s, in Spain, Francisco Mojica made the first attempt to describe the function of CRISPR. The Spanish scientist published first a paper in 1993 describing a repeated sequence of thirty bases in the genome of the microbe *Haloferax mediterranei*.[14] Intrigued by the discovery, Mojica kept studying these curious sequences in his lab at the University of Alicante. This research led to the publication, in 2005, of the first paper presenting CRISPR's likely function: to encode the instructions for an adaptive immune system that

---

[13]   Y Ishino and others, "Nucleotide Sequence of the Iap Gene, Responsible for Alkaline Phosphatase Isozyme Conversion in Escherichia Coli, and Identification of the Gene Product," *Journal of Bacteriology*, Vol. 169 (1987), p. 5429.

[14]   F Mojica, G Juez and F Rodriguez-Valera, "Transcription at Different Salinities of Haloferax Mediterranei Sequences Adjacent to Partially Modified PstI Sites," *Molecular Microbiology*, Vol. 9 (1993), p. 613.

CRISPR-Cas9

How the genome editor works

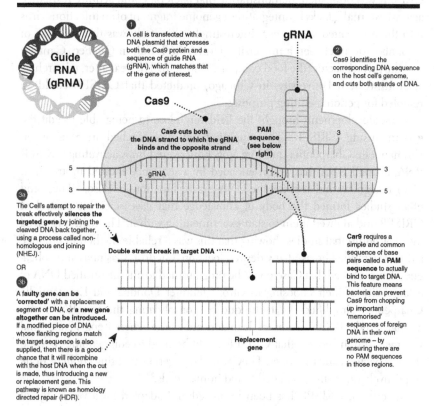

① A cell is transfected with a DNA plasmid that expresses both the Cas9 protein and a sequence of guide RNA (gRNA), which matches that of the gene of interest.

gRNA

② Cas9 identifies the corresponding DNA sequence on the host cell's genome, and cuts both strands of DNA.

Guide RNA (gRNA)

Cas9

Cas9 cuts both the DNA strand to which the gRNA binds and the opposite strand

PAM sequence (see below right)

③a The Cell's attempt to repair the break effectively **silences the targeted gene** by joining the cleaved DNA back together, using a process called non-homologous end joining (NHEJ).

OR

③b A faulty gene can be 'corrected' with a replacement segment of DNA, or **a new gene altogether can be introduced.** If a modified piece of DNA whose flanking regions match the target sequence is also supplied, then there is a good chance that it will recombine with the host DNA when the cut is made, thus introducing a new or replacement gene. This pathway is known as homology directed repair (HDR).

Double strand break in target DNA

Replacement gene

Car9 requires a simple and common sequence of base pairs called a **PAM sequence** to actually bind to target DNA. This feature means bacteria can prevent Cas9 from chopping up important 'memorised' sequences of foreign DNA in their own genome – by ensuring there are no PAM sequences in those regions.

FIGURE 1.2 CRISPR-Cas 9: how the genome editor works
Adapted from J Lewin W, CRISPR-Cas 9: How the Genome Editor Works, https://commons.wikimedia.org/w/index.php?title=File:CRISPR-Cas9-biologist.jpg&oldid=269431713.

protected microbes against specific infections.[15] As Eric Lander reports, the review of the paper was painstaking, with journal editors failing to appreciate the significance of Mojica's discovery.[16]

[15] F Mojica, J García-Martínez and E Soria, "Intervening Sequences of Regularly Spaced Prokaryotic Repeats Derive from Foreign Genetic Elements," *Journal of Molecular Evolution*, Vol. 60 (2005), p. 174.
[16] E S Lander, "The Heroes of CRISPR," *Cell*, Vol. 164 (2016), p. 18.

Since then, knowledge development accelerated. Several labs started working on CRISPR and, in 2007, the first description of the biological functions of CRISPR was published, demonstrating that bacteria could acquire resistance against external attacks by integrating a genome fragment of an infectious virus into these repeated sequences. Interestingly, this study was the outcome of a collaboration between a university (Université Laval, in Quebec, Canada) and the private sector (the food company Danisco). The year after, researchers at Northwestern University, in Chicago, predicted that CRISPR could be retooled for genome-editing purposes.[17]

These developments brought the field very close to being able to edit the genome using CRISPR. The tipping point was reached in 2012, when Emmanuelle Charpentier and Jörg Vogel discovered *trans*-activating CRISPR RNA, a small RNA that, when encoded by a sequence located upstream in Cas9, recognizes and activates the DNA-cutting functions of Cas9.[18] These and other studies formed the body of knowledge that gave scientists control of CRISPR and allowed them to start experimenting with it. The road was paved by a few additional studies showing that this was a reliable method to produce a double-stranded break at any desired location. Giedrius Gasiunas and colleagues demonstrated that Cas9 could be programmed to cut the purified DNA of a target site of their choosing in vitro.[19] Jennifer Doudna and Emmanuelle Charpentier simplified the technique by consolidating the two RNAs into a single-guide RNA.[20] Another leap forward came when Feng Zhang and George Church showed that CRISPR could be used to edit genomes beyond those of bacteria and microbes. They proved that genome editing could be done respectively on mammalian cells[21] and human cells.[22]

Since then, CRISPR has been improved and adapted for a vast range of applications and is being used in countless labs worldwide, in all kinds of life sciences. Between 2012 and 2018, more than 10,000 scientific papers have

[17] L Marraffini and E Sontheimer, "CRISPR Interference Limits Horizontal Gene Transfer in Staphylococci by Targeting DNA." *Science*, Vol. 322, (2008), pp. 1843–1845.

[18] E Deltcheva and others, "CRISPR RNA Maturation by *Trans*-Encoded Small RNA and Host Factor RNase III," *Nature*, Vol. 471 (2011), p. 602.

[19] G Gasiunas and others, "Cas9–crRNA Ribonucleoprotein Complex Mediates Specific DNA Cleavage for Adaptive Immunity in Bacteria," *Proceedings of the National Academy of Sciences*, Vol. 109 (2012), E2579-E2586.

[20] M Jinek and others, "A Programmable Dual-RNA–Guided DNA Endonuclease in Adaptive Bacterial Immunity," *Science*, Vol. 337 (2012), pp. 816–821.

[21] L Cong and others, "Multiplex Genome Engineering Using CRISPR/Cas Systems," *Science*, Vol. 339 (2013), p. 819.

[22] P Mali and others, "RNA-Guided Human Genome Engineering via Cas9," *Science*, Vol. 339 (2013), pp. 823–826.

mentioned CRISPR.[23] The result is an impressive body of scientific knowledge, which has brought us closer to bringing gene editing to the bedside. Listing all accomplishments is rather futile. By the time this book is published, the advancements listed here will have likely been shadowed by new progress. However, it is worth taking notice of the major achievements in the field and the likely trajectory of gene editing with an eye on its application to human germline genome engineering.

3 Current Advancements in Germline Gene Editing

When CRISPR is used on human germline cells, the ultimate goal is therapeutic: treating or preventing genetically inherited diseases.[24] We still lack the knowledge necessary to develop clinical tools even though the now-disgraced Chinese scientist He Jiankui stunned the world when he announced the first use of CRISPR-Cas9 genome editing in human embryos.[25] For legal and ethical reasons explained in depth in this book, germline gene therapies have not yet been legally tested clinically on humans anywhere.[26]

However, basic research has made progress toward achieving the therapeutic goals of CRISPR-based germline gene therapies. At this point in history, scientists have formulated five hypothetical paths to CRISPR-based clinical applications. First, immature gametes could be edited into gene-corrected mature ones that can be subsequently used in assisted reproduction.[27] So far, basic scientists have provided proof of concept studies demonstrating that pre-fertilization gamete editing is possible.[28] What is emerging from these studies is that editing mature gametes is not simple (sperm) or desirable (eggs) and that CRISPR tools should target gamete's early precursor (spermatogonial stem cells and germinal vesicle oocytes).[29]

---

[23] In October 2018, we ran PubMed searches of papers mentioning "CRISPR" and "CRISPR Cas9" after January 1, 2012. The searches retrieved 10,813 and 6,988 papers respectively.

[24] This book focuses on the regulation of human germline engineering for therapeutic purposes. We do not discuss genetic enhancement.

[25] Z Zhihao, "Initial Investigation Finds Scientist Evaded Oversight," China Daily January 22, 2019.

[26] A probe by the Guangdong health ministry concluded that He had violated government bans and regulations. XINHUA, Guangdong Releases Preliminary Investigation Result of Gene-Edited Babies, January 21, 2019, www.xinhuanet.com/english/2019-01/21/c_137762633.htm accessed January 25, 2019.

[27] R Vassena and others, "Genome Engineering Through CRISPR/Cas9 Technology in the Human Germline and Pluripotent Stem Cells," *Human Reproduction Update*, Vol. 22 (2016), pp. 411–419.

[28] Kofler and Kraschel, "Treatment of Heritable Diseases Using CRISPR" (n 10), p. 515.

[29] Vassena "Genome Engineering Through CRISPR/Cas9 Technology in the Human Germline and Pluripotent Stem Cells" (n 26), pp. 413–416.

Second, CRISPR can be used to modify induced pluripotent stem cells (iPSCs) to generate gametes (in vitro gametogenesis). iPSCs are derived by reprogramming skin or blood *cells* back into an embryonic-like *pluripotent* state that enables the development of any human cell. Scientists have successfully performed this procedure on mice and on somatic cells. iPSCs of mice were differentiated into the primordial germ cell of both sperm and eggs to further develop them into a fully functional gamete.[30] Also, offspring from two sets of maternal genomes and two sets of paternal genomes were produced.[31] In humans, iPSCs derived from patients with cystic fibrosis were edited using CRISPR tools to produce cells free of the mutations that cause the disease.[32] While in vitro gametogenesis shows promise as a path to the clinical application of CRISPR, an assessment of its real impact will only be possible once methods for in vitro culture and protocols for gamete differentiation improve.

Third, CRISPR can be used to modify embryos in vitro and then implant the edited embryos in utero. Progress on this path is more substantial than on in vitro gametogenesis. Human zygotes have already been successfully edited. The first successful case of human zygote editing was reported in 2015 when Junjiu Huang and his colleagues at the Sun Yat-sen University in the People's Republic of China corrected a mutation in nonviable zygotes.[33] This study, however, has been the subject of intense scrutiny, and its findings challenged. For instance, other researchers applied a similar method to mouse embryos and found a large number of genetic deletions, which led them to question the validity of Huang's study.[34] Two years later, Shoukhrat Mitalipov and colleagues at the Oregon Health and Science University edited viable embryos.[35]

[30] K Hayashi and others, "Offspring from Oocytes Derived from In Vitro Primordial Germ Cell–like Cells in Mice," *Science*, Vol. 338 (2012), pp. 971–975; K Hayashi and others, "Reconstitution of the Mouse Germ Cell Specification Pathway in Culture by Pluripotent Stem Cells," *Cell*, Vol. 146 (2011), pp. 519–532; C Yamashiro and others, "Generation of Human Oogonia from Induced Pluripotent Stem Cells In Vitro," *Science*, Vol. 362 (2018), pp. 356–360.

[31] Z Li and others, "Birth of Fertile Bimaternal Offspring Following Intracytoplasmic Injection of Parthenogenetic Haploid Embryonic Stem Cells," *Cell Research*, Vol. 26 (2015), pp.135–138; Z Li and others, "Generation of Bimaternal and Bipaternal Mice from Hypomethylated Haploid ESCs with Imprinting Region Deletions," *Cell Stem Cell*, Vol. 23 (2018), pp. 665–676.

[32] A L Firth and others, "Functional Gene Correction for Cystic Fibrosis in Lung Epithelial Cells Generated from Patient iPSCs," *Cell Reports*, Vol. 12 (2015), pp. 1385–1390.

[33] P Liang and others, "CRISPR/Cas9-mediated Gene Editing in Human Tripronuclear Zygotes," *Protein & Cell*. Vol. 6 (2015), pp. 363–372. The experiments resulted in changing only some genes and had off-target effects on other genes.

[34] F Adikusuma and others, "Large Deletions Induced by Cas9 Cleavage," *Nature*, Vol. 560 (2018), E8–E9.

[35] H Ma and others, "Correction of a Pathogenic Gene Mutation in Human Embryos," *Nature*, Vol. 548 (2017), pp. 413–419.

This study has also been scrutinized,[36] but in this case its data have been confirmed by Ma and colleagues[37] and its findings replicated in subsequent studies.[38] The issue that scientists are focusing on now is reducing or eliminating off-target mutations that may appear as the result of embryo editing. These are modifications of the genome in spots that the CRISPR-based intervention was not meant to hit. Germline engineering cannot advance unless this problem is addressed and resolved. Readers must also take notice of the substantial volume of animal studies (on rats, cattle, sheep, dogs, and pigs) showing how embryos can be created and modified.[39] It is worth noting, for instance, the editing of mice zygotes to prevent the onset of cataract in offspring[40] and the production of an artificial mice embryo, wholly created in a lab.[41]

Fourth, CRISPR can be used to modify embryos and fetuses in vivo (fetal gene therapy). For the time being, in vivo editing remains "purely theoretical" as no attempt to correct disease-causing mutations in vivo on humans has been reported.[42] However, in vivo editing of animal embryos has already been successfully performed. Proof-of-principle experiments to prevent or cure inherited diseases, including hemophilia and metabolic disorders, in mice were published in 2018.[43] A team of scientists led by Simon Waddington and Ahad Rahim from the University College London successfully transferred a corrective gene of a severe form of Gaucher disease, a condition in which the body does not produce a key enzyme that breaks down a certain fat, in mouse fetuses while in utero.[44] Success was measured by the fact that, once born, the animals were indistinguishable from mice that had undergone the same treatment at birth or the day after being born. To achieve this result,

---

[36] D Egli and others, "Inter-Homologue Repair in Fertilized Human Eggs?," *Nature*, Vol. 560 (2018), E5–E7.

[37] H Ma and others, "Ma et al. Reply," *Nature*, Vol. 560 (2018), E10.

[38] J Wilde and others, "Efficient Zygotic Genome Editing via RAD51-Enhanced Interhomolog Repair," bioRxiv (2018).

[39] Vassena "Genome Engineering Through CRISPR/Cas9 Technology in the Human Germline and Pluripotent Stem Cells" (n 26), p. 413.

[40] Y Wu and others, "Correction of a Genetic Disease in Mouse via Use of CRISPR-Cas9," *Cell Stem Cell*, Vol. 13 (2013), pp. 659–662.

[41] B Sozen and others, "Self-Assembly of Embryonic and Two Extra-Embryonic Stem Cell Types into Gastrulating Embryo-Like Structures," *Nature Cell Biology*, Vol. 20 (2018), pp. 979–989.

[42] Kofler and Kraschel, "Treatment of Heritable Diseases Using CRISPR" (n 10), p. 519.

[43] L Villiger and others, "Treatment of a Metabolic Liver Disease by In Vivo Genome Base Editing in Adult Mice," *Nature Medicine*, Vol. 24 (2018), p. 1519; A C Rossidis and others, "In Utero CRISPR-Mediated Therapeutic Editing of Metabolic Genes," *Nature Medicine*, Vol. 24 (2018), pp. 1519–1525.

[44] G Massaro and others, "Fetal Gene Therapy for Neurodegenerative Disease of Infants," *Nature Medicine*, Vol.24 (2018), pp. 1317–1323.

a virus expressing the correct gene had been injected in the brain of the fetus. CRISPR was not used in the process, but one can easily see how the correction could be done using gene editing, too. If implemented on humans, this strategy will likely be considered a form of somatic cell therapy because modifications will likely target exclusively somatic cells of the embryo. However, one could also envision the (remote) possibility of in vivo editing of gametes.

Finally, CRISPR could be used as postnatal gene therapy. Similarly to fetal gene therapies, these interventions are likely to focus on somatic cells. Therapies can be conducted ex vivo and in vivo. In ex vivo interventions, cells are taken from the patient, edited in vitro, and then transferred back into the patient. In vivo interventions entail the direct delivery of CRISPR components to a patient, often by injecting into tissues that can be easily targeted (e.g., eye or skeletal muscle). CRISPR-based somatic gene therapy has already reached the clinical stage. Ex vivo therapies based on genetically modified somatic cells are being experimented on humans in several countries. Trials are testing therapies based on knocking out certain receptors and adding different ones, usually carrying a cancer-associated surface protein.[45] A clinical trial testing an experimental CRISPR-Cas9 therapy for the blood disorder β-thalassemia was launched in the summer of 2018.[46] One treatment was approved in the United States in 2017. After the completion of a clinical trial with a rate of success of 83 percent of patients being cancer-free after three months, the FDA approved the use of genetically modified blood cells for the treatment of patients with a type of acute lymphoblastic leukemia.[47] In vivo interventions are yet to be tested in clinical trials. Based on animal study, the most promising candidates for in vivo clinical trials are therapies to correct hearing loss, inherited blindness, muscular dystrophy, and liver disorders.[48]

---

[45]   S Williams, "CRISPR Inches Toward the Clinic," *The Scientist*, August 1, 2018.

[46]   "A Safety and Efficacy Study Evaluating CTX001 in Subjects With Transfusion-Dependent β-Thalassemia" (ClinicalTrials.gov Identifier: NCT03655678), https://clinicaltrials.gov/ct2/show/NCT03655678 accessed October 21, 2018.

[47]   S Williams, "First CAR T-Cell Therapy Approved in U.S.," *The Scientist*, 30 August 2017.

[48]   Kofler and Kraschel, "Treatment of Heritable Diseases Using CRISPR" (n 10), p. 515. X Gao and others, "Treatment of Autosomal Dominant Hearing Loss by In Vivo Delivery of Genome Editing Agents," *Nature*, Vol. 533 (2017), pp. 217–221; M Tabebordbar and others, "In Vivo Gene Editing in Dystrophic Mouse Muscle and Muscle Stem Cells," *Science*, Vol. 351 (2016), pp. 407–411; C Long and others, "Correction of Diverse Muscular Dystrophy Mutations in Human Engineered Heart Muscle by Single-Site Genome Editing," *Sciences Advances*, Vol. 4 (2018), eeap9004.

In closing, therapy is not the only legitimate goal of the use of CRISPR on human germline cells. The advancement of science is also a worthwhile goal. CRISPR gives scientists the opportunity to gain fundamental knowledge on how the human embryo develops.[49] Gene-editing technology allows scientists to shut down specific genes so that the function of these genes in the early stages of embryo development can be assessed with precision. This has been done on human embryos by suppression of the gene encoding protein OCT4, which is involved in the self-renewal of undifferentiated embryonic stem cells.[50] The hope, according to George Church, is to "reconstruct a full lineage tree all the way back to its single-cell stage."[51] This will have "a direct impact in clinical applications for reproductive medicine, including the rational design of treatments for developmental diseases and improvement of fertility programs."[52]

II METHODOLOGY

The primary goal of this book is to analyze existing international and national standards on human germline genome modifications through the lens of international human rights standards and, in particular, of the so-called human right to science and the rights of science. In this section, we explain the multistage methodology that we designed and implemented to achieve this goal.

The first stage was ideation. We started by developing a vision of what germline engineering is and how to study it from a legal perspective that involves both national and international legal instruments. First, we construed "germline engineering" as all practices that, along the "bench to bedside" continuum, involve modifications of human germline cells.[53] Traditionally, this continuum is conceptually divided into four stages: basic research, animal studies, clinical research, and clinical applications. Basic research is characterized by a quest for fundamental understanding of biological process.[54]

---

[49] A Ruzo and A H Brivanlou, "At Last: Gene Editing in Human Embryos to Understand Human Development," *Cell Stem Cell*, Vol. 21 (2017), pp. 564–564.

[50] N Fogarty and others, "Genome Editing Reveals a Role for OCT4 in Human Embryogenesis," *Nature*, Vol. 550 (2017), pp. 67–73.

[51] L Brownell, "Genetic Barcodes Trace Cells Back to Single-Cell Origin," *The Harvard Gazette*, August 9, 2018.

[52] Ruzo and Brivanlou, "At Last" (n 48), p. 565.

[53] N Evitt, S Mascharak and R Altman, "Human Germline CRISPR-Cas Modification: Toward a Regulatory Framework," *The American Journal of Bioethics*, Vol. 15 (2015), pp. 25–29.

[54] D. Stokes, *Pasteur's Quadrant: Basic Science and Technological Innovation*, Brookings Institution Press, 1997, p. 73.

Animal studies (or preclinical research) focus on testing drugs or therapies on living animals. In clinical research, drugs or therapies are tested on living humans. Clinical applications consist of interventions on patients in a clinical setting. Based on the characterization of the four stages – our first organizational principle – we produced a definition of what each stage entails as it relates to human germline engineering based on the literature on gene editing reviewed earlier in this chapter. The result is the following four definitions:

(a) *Basic research* involves in vitro or ex vivo studies of germline tissue of humans, animals, or of the two in combination, to produce a fundamental understanding of the biological mechanisms of germline genome modification.

(b) *Preclinical studies* involve the modification of animal germline genome, either in vivo or ex vivo, with subsequent transfer in the animal (i.e., a modified embryo is implanted in a female animal). These studies are conducted for a variety of reasons, such as using animals as models to study how genome modifications affect the animals and testing the safety and effectiveness of drugs and therapies before they are tested on humans. The regulation of *preclinical research* involving modification of animal germline genomes is mostly permissive, so we will not discuss it extensively in this book in the interest of brevity.

(c) *Clinical studies* involve a living person whose germline tissue is genetically modified in vivo, or who receives germline tissue that was modified ex vivo (i.e., by transferring a modified embryo in the uterus of a research participant), to test the safety and efficacy of germline genome engineering.

(d) *Clinical applications* involve the delivery of drugs or therapies, developed using germline genome engineering, to patients in a clinical setting.

We asked authors to discuss the regulation of mitochondrial replacement treatment (although not-Cas9 based) but excluded unintentional genome modifications, which are common in the event of chemotherapy. Segmenting germline engineering in four stages helped with research and organizing the material.

The "bench to bedside" continuum was our first organizing principle to analyze the national governance frameworks consistently throughout the book. The second one is that both international and national sources of regulations must be taken into account when analyzing the rules governing different stages in germline engineering. National legal systems are not only sources of much regulation but also the vehicles for the implementation of international law, including human rights law. The fundamental insight is that, while the national and international levels are conceptually distinct, their separation is blurred when it comes to analyzing the rules that germline engineering practitioners must follow. The legal space in which scientists and clinicians operate is shaped by rules emanating from both the national and the international legal orders. In the practice of germline engineering, from the bench to the bedside, the two levels merge and form the body of rules that regulate these activities.

This insight has led us to shape the content of the book around three major buckets: one describing existing and emerging international legal standards regulating human germline genome modification (Chapter 2), a second, much longer, part discussing how the object of our study is regulated nationally, in a series of selected countries (Chapters 3 to 21), and a third outlining the interplay between national regulations and the human rights framework. Relevant international legal standards were easy to identify because the number of binding instruments connected with germline engineering is still small. However, since we also wanted to take stock of *emerging* international standards that have not yet become hard law, international declarations, guidelines, and statements by international agencies and learned societies are considered as well. Moreover, given the primary goal of the book, we wanted to explore in more detail the meaning and scope of the right to science and the rights of science, as well as the obligations states have as a result of them.

The study of how germline engineering is regulated nationally was necessarily more complex. The first step was to identify a set of countries whose experiences might be representative of the general state of affairs of germline engineering, from both the regulatory and the practical points of view. This required including in the mix states where germline engineering is prohibited and others where it is not, or only partially. Also, it meant including countries from different regions of the world to reflect differences in the interpretation of international law at the regional level, as well as cultural differences at the level of domestic laws and regulations. In the end, we settled for a set of eighteen countries and one region. The countries we chose are: Australia; Belgium; Canada; France; Germany; Israel; Italy; Japan; Mexico; the Netherlands; People's Republic of China; Singapore; South Korea; Spain; Sweden; Switzerland; the United Kingdom; and the United States. The region we decided to focus on is Europe. Eight out of eighteen are European countries, seven of which are members of the European Union and only one, Switzerland, of the Council of Europe. Because both the Council of Europe and the European Union have adopted legal instruments that affect research on human germinal genome modification, we decided to include a chapter discussing them separately from the chapters dedicated to some of their members.

Outside of Europe, the book profiles three North American countries (Canada, the United States and Mexico), four Asian countries (the People's Republic of China, Japan, the Republic of Korea (South Korea), and Singapore), and two other countries: one in the Middle East, Israel, and one in the Pacific, Australia. Of the eighteen surveyed states, all are members of the United Nations Economic and Social Council, and sixteen are members of the Organisation for Economic Co-operation and Development (the exceptions are the People's Republic of China and Singapore). Our sample includes states that are internally organized as federations or confederations (e.g., Australia, Belgium, Canada, Germany, Mexico, Switzerland, the United Kingdom, the United States) and unitary states, with varying degrees of centralized control. As far as the relationship between national law and international law is concerned, our sample includes both mostly "monist" states (e.g., Switzerland, the Netherlands, or Spain) and mostly "dualist" states (e.g., the United Kingdom or Sweden), and a score in between at various points in the spectrum.

We believe the sample captures the bulk of states that have the technological capacity, at present time, to engage in significant human genome modification research and that it is sufficiently diverse to include the whole range of variations of national practices and governance regimes. However, ours is just a sample. It is not a comprehensive study of the practices of all countries that are significantly engaged in human genome modification. A truly comprehensive study would have included Russia, India, Brazil, South Africa (the so-called BRICS countries), and perhaps even Argentina, Pakistan, Turkey, or Saudi Arabia. Although we could not identify national legal experts to write those chapters, there is little indication that the governance experience of those countries departs significantly from that of the countries we selected.

We need to stress that this is not the first attempt to compare national regulatory practices regarding human genome modification in general and germline in specific.[55] However, this is the first book that attempts to map

---

[55]   See, R Isasi, E Kleiderman and B M Knoppers, "Editing Policy to Fit the Genome?," *Science*, Vol. 351 (2016), pp. 337–339; R Isasi and B M Knoppers, "Mind the Gap: Policy Approaches to Embryonic Stem Cell and Cloning Research in 50 Countries," *European Journal of Health Law*, Vol. 13 (2006), pp. 9–25; T Ishii, "Germ Line Genome Editing in Clinics: the Approaches, Objectives and Global Society," *Briefings in Functional Genomics*, Vol. 16 (2017), pp. 46–56; I De Miguel Beriain, "Legal Issues Regarding Gene Editing at the Beginning of Life: an EU Perspective," *Regenerative Medicine*, Vol. 12 (2017), pp. 669–679; R Yotova, The Regulation of Genome Editing and Human

a significant large sample of countries in depth, at book length, using a fairly consistent template, and taking into consideration not only national legislation but also the international obligations the countries surveyed have.

The second step was designing a template that would give authors guidance on the issues to be analyzed in all studies, to ensure coherence and consistency. We identified key questions that each country report needed to address. The goal was to provide up-to-date accounts of the rules that germline engineering practitioners must follow at the national level and, at the same time, connecting them to international sources and institutions. The aim was to determine whether existing and emerging international standards, especially the right to science and the rights of science, have been translated into national practice or what is needed to bring this practice into line with them.

In its final form, our template was divided into three sections. The first asked authors to highlight the most important regulatory instruments and bodies affecting research involving human germline genome editing, and specifically to identify the legal instruments (law, regulations, and other legal instruments), as well as policies and guidelines, that regulate or guide human germline genome editing. We also asked them to identify governmental and nongovernmental bodies that oversee research and clinical applications, and funding opportunities. These are the specific questions that we asked:

(a) *Regulatory framework*: What are the applicable laws, policies, and guidelines (in that hierarchical order) at the international and national levels? Discuss legislation as well as key policies and guidelines (opinions issued by national ethics committees, professional organizations, or scientific bodies) focusing specifically on genome editing or directly relevant to germline engineering. In addition to in-text discussion, this information should also be presented in the form of a table listing key instruments.

(b) *Definitions*: Does the law and/or the other relevant regulatory instruments define the key terms (i.e., embryos, gametes, genome editing)?

(c) *Oversight bodies*: Are research and clinical applications involving human germline genome editing under the authority of oversight bodies? What kind and at what level (national, local, institution, specialized agencies)? What are the requirements set by these bodies (i.e., pre-approvals, disclosures, data sharing)?

Reproduction Under International Law, EU Law and Comparative Law, Nuffield Council on Bioethics, 2017.

(d) *Funding*: Can research involving human genome editing be funded using public money? Is research being funded? Which bodies fund research involving human genome editing?

The second section of the template focused on substantive legal provisions relating to basic research, preclinical studies, clinical research, and clinical applications. These are the specific questions that we asked:

(a) *Regulation of basic research*: Discuss whether basic research in humans or embryos/gametes using germline genome-editing technologies is permitted, prohibited, or restricted.
(b) *Regulation of preclinical research*: Discuss whether preclinical research involving germline genome editing in nonhumans is permitted, prohibited, or restricted.
(c) *Regulation of clinical research*: Discuss whether clinical research using germline genome-editing technologies is permitted, prohibited, or restricted.
(d) *Regulation of clinical applications*: Discuss whether findings of research using germline genome-editing technologies can be used in a clinical setting (i.e., to initiate a pregnancy with edited embryos or with edited gametes).

The third and final section required authors to discuss current perspectives and future possibilities. Specifically, we asked them to evaluate the current regulatory framework in their country and provide insights with regard to future developments at the regulatory level. These developments could be both likely and recommended future developments.

We then identified experts who could write the country reports and sent a letter of invitation to write a book chapter along with the template, expected length of the chapter, and deadlines. Authors were asked to submit a first draft for review. We reviewed first drafts on an ongoing basis. Each of the three book editors reviewed all drafts. An editor read them all as first reader, passing it to the second reader, followed by the third reader. For the first draft, we focused on adherence to the template and clarity. To ensure that authors covered all of the ground that they were expected to cover, we commented on the structure of the chapter, failure to discuss certain issues, sections that were underdeveloped, and sections that were either redundant or not essential. We also commented on clarity of the draft, stressing that the intended audience of the book is diverse, including legal scholars interested in science policy and practitioners (scientists and clinicians) interested in legal issues. In the subsequent weeks, authors submitted a second draft, which we reviewed in the same order. The reading of the second draft focused on ensuring that the queries and suggested edits based on the first draft had been addressed and on footnotes (sufficient support for factual statements and proper citation of sources). After receiving our edits, authors submitted the final version of their chapters, which was reviewed by an editorial assistant, who cleaned the submissions and ensured proper formatting of citations.

The conclusions (Chapter 22) discuss the existing national and international regulatory frameworks in light of the five foundational principles that a reading of international bioethics law combined with international human rights standards suggests. They are: (i) freedom of research; (ii) benefit sharing; (iii) solidarity; (iv) respect for dignity; and (v) the obligation to respect and to protect the rights and individual freedoms of others. We identified these principles by looking at key international bioethics instruments, and in particular the three UNESCO declarations (on the human genome and human rights (1997), human genetic data (2003), and bioethics and human rights (2005)),[56] while also taking into account the key provisions of the International Bill of Rights that concern science, including Article 15 of the International Covenant on Economic, Social and Cultural Rights (ICESCR) and Article 27 of the Universal Declaration of Human Rights.[57]

This chapter is divided into three sections: (I) evidence, where we summarize what emerges from a legal and comparative analysis of the national chapters included in this volume; (II) analysis, in which we discuss the extent to which the current national regulatory standards are consistent with the five foundational principles we identified; and, (III) recommendations, where we offer our vision of an international governance framework that promotes science and technological development while being mindful and respectful of international human rights standards as well as the different sensitivities with which citizens from different parts of the world approach this complex problem.

Of course, the views expressed in this final chapter do not necessarily reflect the views of the authors who contributed the national chapters to this book. We delivered the manuscript to the publisher on January 31, 2019.

---

[56] See, in this book, Chapter 2, Section 2.1.
[57] Ibid., Section 3.

2

The Governance of Human (Germline) Genome Modification at the International and Transnational Level

Cesare P. R. Romano, Andrea Boggio and Jessica Almqvist

The core of this book is a discussion of how human germline genome modification is currently regulated at the national level in a selected number of states. However, national regulations can be properly understood and assessed only by keeping in mind the larger international and transnational framework within which these national legal regimes exist. Although bioethics and international human rights law were born out of the same horrors of the Second World War, for most of the remaining part of the twentieth century they developed in parallel without significant crossovers. Human genome modification has traditionally been discussed under the heading 'bioethics', using its concepts, terminology and discourse. However, at the beginning of the twenty-first century bioethics and human genome modification started being discussed within the wider international human rights framework and the even wider international law framework.

Before discussing international norms and the legal instruments that contain them, for the benefit of those readers who are not familiar with international law and relations, its concepts and terminology, we will introduce in Section 1 some key terms and a quick primer to international law and international human rights. Those who are already familiar with them should skip directly to Section 2, where the relevant norms of international bioethics law are discussed. Section 3 discusses international human standards that are most relevant to the field of human germline genome modification, and in particular the so-called right to science and the so-called rights of science. Finally, in Section 4 we discuss how these rights can contribute to the emerging international regulatory framework. In particular, we will highlight five key principles associated with the 'right to science' and the 'rights of science' and their connection to the other international instruments we have reviewed in this chapter: freedom of research, benefit sharing, solidarity, respect for human dignity and respect of the human rights of others.

I KEY TERMS AND A QUICK PRIMER OF INTERNATIONAL LAW AND INTERNATIONAL HUMAN RIGHTS LAW

1 Key Terms

Throughout this volume terms such as 'international', 'transnational', 'supra-national', 'multinational', 'regional' and 'global' are used. They are not syno-nyms. Each has a specific meaning. 'International' refers to relations *between* sovereign states and the laws regulating their interactions (mostly but not exclusively). 'Transnational' refers to relations across national borders between non-governmental actors. 'Supranational' refers to a phenomenon whereby several states have transferred part of their sovereignty to common interna-tional institutions. The difference between 'international' and 'supranational' is a matter of degree of transfer of sovereignty, with 'supranational' defining legal phenomena in which the transfer of sovereignty is greater. 'Multinational' is used to indicate an organization, often a corporation, oper-ating in multiple sovereign states at once. Finally, the term 'global' indicates a phenomenon that occurs in multiple states, on multiple continents at once, while the term 'regional' is used to refer to a phenomenon that occurs only in a portion of the globe (e.g. the European Union or the African continent).

We also speak of 'governance' and 'regime'. The term 'regime' describes not only a set of norms focusing on a given subject-matter but also the decision-making machinery to create, update and enforce them.[1] Thus, the international legal regime regulating human germline genome modification comprises both norms and the institutions that articulate them. In a very broad sense, 'governance' is the 'act of governing' or administering a community or an issue. More specifically, it is used to indicate the processes of interaction and decision-making among the actors involved in a collective problem that lead to the creation, reinforcement or reproduction of both norms and institutions.[2]

2 Basic Notions of International Law

International law is the set of norms that regulates the life of the so-called international community. This community is composed mainly of sovereign

---

[1] S Krasner (ed.), *International Regimes* (Cornell University Press 1983).

[2] John Gerard Ruggie defined 'governance' as 'systems of authoritative norms, rules, institutions, and practices by means of which any collectivity, from the local to the global, manages its common affairs'. J Ruggie, 'Global Governance and "New Governance Theory": Lessons from Business and Human Rights' (2014) 20 *Global Governance* 5.

states, but it includes also international organizations (i.e. associations of sovereign states created to pursue together some shared goals), natural persons (i.e. individuals) and legal persons, that is to say organizations sharing a common goal (when they are not-for-profit, they are called 'non-governmental organizations' (NGOs); if their purpose is profit, they are 'corporations').

Although international law was initially born to regulate relations between sovereign states only, towards the second half of the nineteenth century, international organizations having a legal personality separate from that of their members started appearing. As a result, international law evolved to include the regulation of these organizations as well as their relationship to sovereign states. Later, in the wake of the Second World War, states realized that the protection of individuals could not depend on their citizenship of a certain country but instead had to be based on their status as humans. Indeed, the task of protecting human rights could not be left to the country of nationality only, especially since too often the source of the threat to individual rights and liberties was exactly the person's own government. This led to the adoption of a special set of international norms, called 'international human rights', whereby states recognize a set of basic rights to be enjoyed by everyone, without any discrimination.

International law has three main sources, of which two are arguably the most important:[3] The first one is 'treaties', which are written agreements concluded between two or more sovereign states with the intent to create legal obligations between the parties.[4] Some treaties are bilateral and function like 'contracts', creating mirror-like, tit-for-tat, obligations. By and large, the legal concepts applying to contracts domestically apply also to treaties. However, some treaties are multilateral and more akin to 'constitutions': They create a structure and a process to create new international law.

---

[3]   Article 38 of the Statute of the International Court of Justice is regarded as a codification of the sources of international law. It reads: '1. The Court, whose function is to decide in accordance with international law such disputes as are submitted to it, shall apply: a. international conventions, whether general or particular, establishing rules expressly recognized by the contesting states; b. international custom, as evidence of a general practice accepted as law; c. the general principles of law recognized by civilized nations; d. ... judicial decisions and the teachings of the most highly qualified publicists of the various nations, as subsidiary means for the determination of rules of law.'

[4]   Vienna Convention on the Law of Treaties, 1155 *UNTS* 331, art. 2.1.a: '"Treaty" means an international agreement concluded between States in written form and governed by international law, whether embodied in a single instrument or in two or more related instruments and whatever its particular designation.'

The Covenant on Economic, Social and Cultural Rights, where the right to science is codified, is an example.[5]

The second source is 'customary international law'. Customary international law is the tradition of the 'international community'.[6] It is what most states do, or do not do, most of the time. International legal scholars identify two components of this custom: 'practice' and *opinio juris*. Practice is the objective element. It is what states do and do not do, say or do not say. To find it, one looks at the way states behave every day, on any given topic. National laws, decisions, orders, actions, public statements and the like are all things international legal scholars peruse to determine what the prevailing practice of the international community is on any given subject. Then there is *opinio juris*, which is Latin for 'the opinion of what the law is'. This is the subjective element of custom. It is what states think their legal obligations are. To find the subjective element, one looks at the same documents one looks at to find practice but tries to go deeper and determine whether states follow a certain practice out of a sense of legal duty as opposed to courtesy, convenience or expediency. The subjective element of custom is present only when a practice is based on a sense of legal duty.

The key difference between treaties and customary international law is that treaties bind only those states or international organizations that ratified them. As a general rule, they do not bind third parties (*pacta tertiis nec nocent neque prosunt*).[7] Conversely, customary international law binds all members of the 'international community'. New states, upon creation, inherit the whole body of customary international law obligations created by the international community. There is no opt-out, but for the case of the so-called persistent objector. If a state persistently objects to the creation of a norm of customary international law, as it emerges, it might successfully argue the norm does not apply to it. However, the 'persistent objector' is a unicorn in international law. There are very few cases of states that have been able to successfully call themselves out of a new norm of customary international law. It is very costly politically to be the nay-sayer for years and decades.[8]

---

[5] See, in this chapter, Sections III.1 and 2.

[6] On customary international law, see, in general, T Treves, 'Customary International Law' (*Max Planck Encyclopedia of Public International Law*, last updated November 2006).

[7] Vienna Convention on the Law of Treaties (n 4), arts. 34–38.

[8] It should be noted that, occasionally, principles enshrined in a treaty acquire the status of customary law and thus become binding on all members of the 'international community', including states or international organizations that have not ratified the treaty. For instance, this is the case of the obligation not to send any person to a country where there is a real risk that he or she may be exposed to arbitrary deprivation of life. See, Human Rights Committee, 'General Comment No 24: Issues relating to reservations made upon ratification or accession to

The third main source of international law, which is less important in practice, is 'general principles of law' common to nations.[9] These are legal precepts that one can find in the national legal system of most states, such as the principle that 'contracts must be respected' (*pacta sunt servanda*) or 'you cannot try the same person twice for the same crime' (*ne bis in idem*). However, because these principles tend to be extremely vague, they are rarely invoked when trying to ascertain what international legal obligations exist on any given subject.

Finally, there is a fourth source, one that is not considered officially a source as such but that plays an important role in the life of the international community: 'soft law'.[10] The term 'soft law' is used to indicate a very broad and diverse set of standards that are included in documents that are not binding *per se*. This includes the many 'declarations', 'final acts', 'communiqués' and the like that states often issue, individually or in concert with international organizations. These documents are not binding. They are not meant to create immediate obligations for states. However, since customary international law is what most states do most of the time, soft law is a very good place to find out what the prevailing practice and *opinio juris* are. Soft law documents are, in a sense, customary international law in the making. This source of international law is important for the topic of this book because, as we will discuss later in this chapter, various soft law instruments dealing with bioethical issues have been adopted in past decades.

To see how 'soft law' can become 'hard law', let's consider the case of the Universal Declaration of Human Rights (UDHR).[11] Nowadays, it is considered the keystone of the whole international human rights regime, but it was not always like that.[12] It was adopted in December 1948 by the General Assembly of the United Nations. Out of fifty-eight members of the United Nations at that time, forty-eight voted yes, zero no, eight abstained, and two did not vote.[13] At adoption, it was a non-binding aspirational document, setting out not a list of obligations but rather a list of what states hoped could become

the Covenant or the Optional Protocol is thereto, or in relation to declarations under Article 41 of the Covenant' (1994) UN Doc CCPR/C/21/Rev.1/Add.6, para 8.

[9] See, in general, G Gaja, 'General Principles of Law' (*Max Planck Encyclopedia of Public International Law*, Oxford University Press, last updated May 2013).

[10] On soft law, see, in general: D Thürer, 'Soft Law' (*Max Planck Encyclopedia of Public International Law*, Oxford University Press, March 2009).

[11] Universal Declaration of Human Rights, adopted 10 December 1948, UNGA Res 217A (III) (UDHR).

[12] H Charlesworth, 'Universal Declaration of Human Rights (1948)' (*Max Planck Encyclopedia of Public International Law*, last updated February 2008).

[13] Ibid.

obligations. However, by virtue of its large endorsement within the General Assembly, and because over time states kept on referring back to it, either by incorporating some or the whole of it in their national constitutions and mentioning it during discussions at the United Nations, it gradually hardened into customary international law.[14] Practice, supported by *opinio juris*, created a new set of norms. What is more, parts of it are even considered 'jus cogens', which is customary international law that cannot be derogated, under any circumstance.[15] It is the core law of the 'international community'.

3 Basic Notions of International Human Rights Law

The Universal Declaration of Human Rights is foundational of international human rights law. Soon after it was adopted in 1948, it became clear that more was needed to provide more robust protection of those rights. Ideally, the answer would have been a treaty that created binding obligations for states to protect the human rights of those within their jurisdiction. The problem was that by then the Cold War had started and the world had split into two opposing camps, the East and the West, each with very different views about what human rights are and which ones should be given priority.[16] In a nutshell, the East preached that humans could not be free unless they were equal in wealth, opportunities and power. The West replied that every-one is equal in individual freedom. The East championed heavy governmental intervention in the life of citizens to provide them 'from cradle to grave' with what they needed. The West believed the government should stay out of the life of its citizens and just ensure them protection and security and leave it to the markets to satisfy their needs and wants.

Thus, when East and West sat down to think about a single human rights treaty to complement the non-binding Universal Declaration, their views started looking irreconcilable. The result was the parallel negotiation of two separate treaties. One, the International Covenant on Civil and Political Rights (ICCPR), put the accent on freedoms, on the duty of states to stay out of the life of their citizens, on the harm they should not do to them (e.g. deprive of life, torture, enslave, not give a fair trial, arrest arbitrarily, invade privacy, stifle expression, etc.).[17] The other, the International Covenant on

---

[14] Ibid.

[15] J Frowein, 'Jus Cogens' (*Max Planck Encyclopedia of Public International Law*, last updated March 2013).

[16] See, in general, J L Gaddis, *The Cold War: A New History* (Penguin 2005).

[17] International Covenant on Civil and Political Rights (adopted 16 December 1966, entered into force 23 March 1976), 999 UNTS 171 (ICCPR).

Economic, Social and Cultural Rights (ICESCR), stressed individual rights and the duty states have to provide for their citizens, the goods they have to give them (e.g. education, work, health, social security, culture, 'adequate conditions of living', etc.).[18] Both covenants were adopted in 1967 and, predictably, Western states flocked to ratify the former and shunned the latter, while Eastern states did the opposite.

Although this ideological split eventually lost much of its significance as many Western states adopted a 'welfare state' model and Eastern states gradually eased restrictions to the enjoyment of civil and political freedom in their societies, it is a cleave that endures to this day. For instance, the United States has ratified the Covenant on Civil and Political Rights but not the Covenant on Economic, Social and Cultural Rights, while the People's Republic of China has ratified the Covenant on Economic, Social and Cultural Rights but not the Covenant on Civil and Political Rights. Most states of the world, however, have ratified both, and now these two treaties, together with the Universal Declaration of Human Rights, are considered to form the so-called International Bill of Rights.[19] It is not a stretch to say that perhaps the triad has acquired customary international law value, although the devil is in the details.

Indeed, with the exception of jus cogens, there is no hierarchy between international human rights. For instance, the right to freedom of expression is not more important than the right to education. They are obviously linked, as one requires the other. However, there is a fundamental difference between civil and political rights, on the one hand, and economic, social and cultural rights, on the other. Civil and political rights mostly create immediate obligations that must be fulfilled by states here and now (obligations 'of result'). Many of them are binary. Either one is held in slavery or not. Either one is subjected to medical or scientific experimentation with consent or without. It is also easy to hold every state to the same standards, since many of the civil and political rights oblige states not to do something, such as kill or torture. However, because economic, social and cultural rights require states to do something, and the things they must do are often broad in scale and very expensive and states have very different amounts of resources, material and human, the obligation is only 'of means'. That is states must make progress towards fulfilling them, but progress is measured depending on their available

---

[18]   International Covenant on Economic, Social and Cultural Rights (adopted 16 December 1966, entered into force 3 January 1976) 993 UNTS 3 (ICESCR).

[19]   UN Office of the High Commissioner for Human Rights, 'Fact Sheet No.2 (Rev.1), The International Bill of Human Rights' www.ohchr.org/Documents/Publications/FactSheet2Rev .1en.pdf accessed 8 November 2018.

national resources. It is more difficult to have a single standard for all. They are rights to be realized 'progressively', and progress is a never-ending process.

Of course, these are broad generalizations. There are civil and political rights that require large expenditure and have the same 'work in progress' nature of economic, social and cultural rights. A fair and efficient judicial system is neither cheap nor can be constructed in a day.[20] And there are economic, social and cultural rights that can be implemented here and now.[21] Denying someone access to healthcare on discriminatory grounds cannot be justified by scarce national resources. Yet, these are important concepts to keep in mind as we discuss international human standards relating to human germline genome modification.

Globally, states have built the edifice of international rights under the aegis of the United Nations. In parallel fashion, groups of states have reiterated many of those rights or, in certain cases, provided for more or other rights at the regional level to accommodate for social, cultural, historical and political regional circumstances. Thus, five regional human rights systems emerged over time: in Europe (starting early 1950s); in the Americas (starting late 1960s); in Africa (starting in the 1980s); in the Arab world (starting late 1990s to early 2000s); and in South East Asia (2010s). The same structure we find at the global level (a declaration and two key treaties, one on civil and political rights and the other on economic, social and cultural rights) is also found in each of these regional organizations, with some variations to adjust for different timelines and historical, social and political preferences.[22] As a result, human rights are protected internationally through the United Nations at the global level, and by organizations such as the Council of Europe, the African

---

[20]   S Holmes and C R Sunstein, *The Cost of Rights: Why Liberty Depends on Taxes* (New York, W.W.W. Norton & Company Ltd 1999).

[21]   For an analysis of whether the International Covenant on Economic, Social and Cultural Rights creates 'obligations of conduct' and 'obligations of result', see, UN Committee on Economic, Social and Cultural Rights, 'General Comment No 3: The Nature of States Parties' Obligations under the Covenant' UN Doc E/1991/23 (art. 2, para 1 of the Covenant).

[22]   For instance, the human rights system of the Organization of American States features a general declaration (American Declaration of Human Rights), a treaty focusing on civil and political rights (the Inter-American Convention on Human Rights) and another on social, economic and cultural rights (the Protocol of San Salvador). The human rights system of the Council of Europe has no declaration but two core treaties, one focusing on civil and political rights (the European Convention on Human Rights) and another on social, economic and cultural rights (European Social Charter). The one on the African Union has a treaty focusing on civil and political rights (the Banjul Charter) but no declaration and no treaty focusing on social, economic and cultural rights. The human rights system of the Arab League has a treaty on civil and political rights but no declaration, nor a treaty focusing on social, economic and cultural rights. Finally, the Association of Southeast Asian Nations (ASEAN) features a declaration but no treaty yet.

Union, the Organization of American States, Arab League and the Association of Southeast Asian Nations (ASEAN) at the regional level. Global and regional regimes overlap significantly, with the various regional regimes reinforcing each other, and together form customary international law.

II INTERNATIONAL BIOETHICS LAW

The broader international framework within which the various national governance regimes exist includes both international norms specifically referring to human genome modification, which are part of the corpus of international bioethics law, and international human rights standards. As it was said earlier, human genome modification has traditionally been discussed under the heading 'bioethics', using its concepts, terminology and discourse. However, at the beginning of the twenty-first century bioethics and human genome modification started being discussed within the wider international human rights framework and the even wider international law framework. Thus, first, in this section (II) we will present the relevant international bioethics norms. In the next (III), we will discuss some specific international human rights norms that, we believe, must be taken into consideration when discussing the overall international legal framework for human germline genome modification.

However, before we turn to that we need to make it clear that any attempt to present the legal standards regulating human genome modification in general, and germline genome modification in particular, presents three main challenges that must be addressed. The first one relates to the scope of analysis. Governance of human *germline* genome modification is a crucial but narrow facet of the larger question of the governance of human genome modification *tout court*, which, in turn, is a subset of a broader field, international bioethics law, which is itself a specialized branch of international law. A discussion focusing exclusively on the current state of governance of human *germline* genome modification would be very short and much of it would also be speculative. More importantly, it would fail to take into account all relevant standards that come into play when addressing the question of the governance of germline modification from bench to bedside. Indeed, what the relevant laws and standards are can be properly understood only if they are connected to the broader field of international bioethics law, which is made of hundreds of instruments that have been adopted since the end of the Second World War.[23]

---

[23] For a list of hard law and soft law international bioethical instruments, see, Human Rights Library, 'Bioethics and Human Rights Links' (University of Minnesota) http://hrlibrary .umn.edu/links/bioethics.html accessed 7 November 2018. On international bioethics law,

The second challenge to our inquiry is the number and nature of the normative instruments that seek to regulate or govern the modification of human germline or somatic cells. So far, in international law there is no binding legal instrument dedicated to human genome modification. All there is are a few soft law instruments, of which the 1997 UNESCO Universal Declaration on Human Genome and Human Rights is the most important. These instruments will be analysed in this chapter. However, as one zooms out to the larger field of bioethics law, some binding legal instruments with regional scope become relevant. These include the Council of Europe's Convention for the Protection of Human Rights and Dignity of the Human Being with regard to the Application of Biology and Medicine (Oviedo Convention) and some EU directives and regulations. These instruments are discussed in detail in a separate chapter on the 'European' regulatory regime.[24]

The third challenge is the nature of the actors. The international governance of human genome modification, whether that be germline or somatic, is not just the province of states and intergovernmental organizations, but it is also one where dozens of non-governmental actors, generally called 'civil society', participate. Between 2015 and early 2018, at least sixty-one ethics reports and statements have been crafted by more than fifty countries and civil society organizations across the globe, many of them academic, including the American and European Societies of Human Genetics, the European Society of Human Reproduction and Embryology, and the International Society for Stem Cell Research, the Nuffield Council on Bioethics (UK), the Danish Council on Ethics, industry groups and organizations including the Biotechnology Innovation Organization and various genome-editing bio-tech companies, and political groups such as the '2015 White House'. However, for the legal scholar who aspires to describe this growing field, this richness and diversity of participants complicate the task of assessing the legal value of the declarations they issue.[25] It is beyond the scope of this book to discuss them all, but we will point out a few that, in our opinion, might

see, in general: R Andorno, 'Towards an International Bioethics Law' (2004) 2–3 *Journal International de Bioéthique* 131, 131–149.

[24] See, in this book, Part 2, Section II, Chapter 6. We recommend readers who want to explore more in-depth broader international bioethical standards to read further. For a bibliography, see, F Molnár-Gábor, 'Bioethics' (*Max Planck Encyclopedia of Public International Law*, last updated January 2015).

[25] For a survey of 'ethical statements' on the matter, see, in general, C Brokowski, 'Do CRISPR Germline Ethics Statements Cut It?' (2018) 1:2 *The CRISPR Journal* 115, 115–125. On Brokowski's survey, see, in this chapter, Section II.4.

become the basis for government-sanctioned international legal standards in the future.

Besides the Council of Europe and the European Union, so far, the international organizations that have been most active on the question of human genome modification are, at the global level, the United Nations, through some of its specialized agencies, mainly the United Nations Educational, Scientific and Cultural Organization (UNESCO) and the World Health Organization (WHO), and UN bodies, such as the General Assembly and the Human Rights Council, and, at the regional level, the Organisation for Economic Cooperation and Development (OECD). In the subsections below we will analyse what has been done in each of these fora to address and respond to the opportunities and challenges presented by the possibility to modify the human genome in general and germline cells in specific.

1 *UNESCO*

The United Nations Educational, Scientific and Cultural Organization (UNESCO) is a 'specialized agency' of the United Nations.[26] Presently, it counts 195 members, although the United States and Israel announced their withdrawal on 1 January 2019 due to the admission of Palestine to the organization in 2011.[27] Of all UN specialized agencies it is probably the one with the

---

[26] 'Specialized Agencies are legally independent international organizations with their own rules, membership, organs and financial resources [that] were brought into relationship with the United Nations through negotiated agreements. Some of the agencies existed before the First World War, some were associated with the League of Nations, others were created almost simultaneously with the United Nations and yet others were created by the United Nations itself to meet emerging needs. Given the diversity of their respective fields of action, history and experience, each agency has its own needs and concerns, not to speak of corporate culture'. Chief Executives Board for Coordination, 'Directory of United Nations System Organizations' (Chief Executives Board Secretariat, 2016) www.unsystem.org/members/specialized-agencies accessed 8 November 2018.

[27] See, United States Withdraws from UNESCO, U.S. Department of State, Press Release, 12 October 2017. The decision takes effect on 31 December 2018. US Department of State, 'The United States Withdraws From UNESCO' (2017) www.state.gov/r/pa/prs/ps/2017/10/274748.htm accessed on 28 December 2018. This is not the first time the US turned its back to UNESCO. During the 1970s and 1980s, UNESCO was denounced by the United States and some of its allies as a platform for communists and Third World dictators to attack the West. In 1984, the United States withheld its contributions and withdrew from the organization in protest, followed by the United Kingdom and Singapore in 1985. The UK rejoined in 1997. The United States rejoined in 2003, followed by Singapore in 2007. For the Israeli withdrawal, see, Declaration by UNESCO Director-General Audrey Azoulay on the withdrawal of Israel from the Organization, UNESCO, 29 December 2017. UNESCO, 'Declaration by UNESCO Director-General Audrey Azoulay on the withdrawal of Israel from the Organization' (2017)

broadest mandate, which, in its history, has been both a blessing and a curse. Its stated purpose is 'to contribute to peace and security by promoting collaboration among the nations through education, science and culture in order to further universal respect for justice, for the rule of law and for the human rights and fundamental freedoms'.[28] It pursues these objectives through five major programmes: education, natural sciences, social/human sciences, culture and communication/information.

Probably, UNESCO's most famous activity is the attempt to protect the world's cultural and natural heritage through its World Heritage sites list. However, it is less known that over the past twenty years, it has been active on the question of the human genome and bioethics, too. In 1993, the Secretary-General of UNESCO established the International Bioethics Committee (IBC) to 'follow progress in the life sciences and its applications in order to ensure respect for human dignity and freedom'.[29] The IBC is currently composed of thirty-six persons, mostly specialists in medicine and biology but counting also several legal scholars, and reflecting, as all UN bodies, the whole organization membership according to the principle of 'equitable geographic representation'.[30] Over the years it has adopted three soft law instruments in the field of biotechnology that are crucial for the purposes of this volume.

a UNESCO Universal Declaration on Human Genome and Human Rights (1997)

On 11 November 1997, the members of UNESCO adopted, unanimously but after much deliberation and discussion at the IBC, the Universal Declaration on Human Genome and Human Rights.[31] This declaration forms 'the basis of

https://en.unesco.org/news/declaration-unesco-director-general-audrey-azoulay-withdrawal -israel-organization accessed on 28 December 2018.

[28] Constitution of UNESCO, 4 *UNTS* 275, art. 1.

[29] See, UNESCO, 'International Bioethics Committee (IBC)' (2017) www.unesco.org/new/en/ social-and-human-sciences/themes/bioethics/international-bioethics-committee/ accessed 8 November 2018. On the IBC and its role in the governance of bioethics, see, in general: A Bagheri, J Moreno and S Semplici (eds.), *Global Bioethics: The Impact of the UNESCO International Bioethics Committee* (Springer 2016).

[30] See, UNESCO, 'Members of the International Bioethics Committee' (2019) www.unesco.org /new/en/social-and-human-sciences/themes/bioethics/international-bioethics-committee/me mbers/ accessed 8 November 2018.

[31] UNESCO, Universal Declaration on the Human Genome and Human Rights (adopted at the 29th Session of the General Conference on 11 November 1997) BR/2001/PI/H/1. The 1997 Declaration was subsequently endorsed by the United Nations General Assembly with Resolution AIRES/53/152 on 9 December 1998 at its 53rd session.

"soft law" in the area of human genome governance'.[32] The main objective of this declaration is to preserve the human genome from improper manipulations that may imperil the identity and physical integrity of future generations.[33] The first four articles set out the broadest principles of bioethics relating to human genome. The first one is contained in Article 1: 'The human genome underlies the fundamental unity of all members of the human family, as well as the recognition of their inherent dignity and diversity. In a symbolic sense, it is the heritage of humanity.' The second regards 'dignity', an over-arching but ill-defined concept in international bioethics law: '(a) Everyone has a right to respect for their dignity and for their rights regardless of their genetic characteristics; (b) That dignity makes it imperative not to reduce individuals to their genetic characteristics and to respect their uniqueness and diversity.'[34] The third one is the mutability and individuality of the human genome: '[t]he human genome, which by its nature evolves, is subject to mutations. It contains potentialities that are expressed differently according to each individual's natural and social environment, including the individual's state of health, living conditions, nutrition and education.'[35] Finally, Article 4 proclaims: 'The human genome in its natural state shall not give rise to financial gains.'

Of the four, the most controversial statement is probably that the human genome is the 'heritage of humanity'. The 'common heritage of mankind' (also referred to gender-neutrally as 'common heritage of humanity', or 'common heritage of humankind') is a philosophical concept, an international legal principle and, in the context of the human genome, a biological concept, too.[36] Philosophically, its roots can be traced back to the father of international law, Hugo Grotius, and to the father of philosophical Enlightenment, Immanuel Kant. However, the concept started acquiring normative significance beginning the late 1950s, as humanity developed technology to reach and exploit the resources in spaces that had been hitherto unreachable (i.e. Antarctica, the sea bed of the high sea and outer space).[37] The need to avoid a scramble for those resources in a world locked in a nuclear-armed Cold War

---

[32] C Kuppuswamy, *The International Legal Governance of the Human Genome* (Routledge 2009) 28.

[33] R Andorno, *Principles of International Biolaw: Seeking Common Ground at the Intersection of Bioethics and Human Rights* (Brussels, Bruylant 2013) 14.

[34] UNESCO 1997 Declaration on the Human Genome (n 31), art. 2.

[35] Ibid., art. 3.

[36] Kuppuswamy, *The International Legal Governance of the Human Genome* (n 32) 49.

[37] On the notion of 'common heritage of mankind' in international law, see, in general, R Wolfrum, 'Common Heritage of Mankind' (*Max Planck Encyclopedia of Public International Law*, last updated November 2009).

and the process of decolonization, which created scores of developing countries eager to acquire a share of the resources yet to be reached, helped turn the
old philosophical concept into a legal principle. The principle eventually
found its way into a number of major multilateral treaties governing the so-
called global commons, including the high seas, with the 1982 United Nations
Convention on the Law of the Sea,[38] and the moon and other celestial
bodies.[39]

The principle is said to have multiple dimensions: non-appropriation;
international management; benefit sharing; peaceful use and preservation
for the benefit of future generations.[40] Non-appropriation holds that common
heritage of mankind is *res communis*, common property of mankind. As such,
it cannot be appropriated by anyone, state or individuals. Instead, common
heritage must be managed and exploited jointly, through international agencies that can ensure the benefits are equally and proportionally shared by all
states. Finally, shared spaces and resources cannot be used for non-peaceful
uses and must be managed in such a way as to be able to pass them to the next
generation as they have been received from the previous.

When UNESCO issued its 1997 Declaration, boldly declaring that 'the
human genome [is], [i]n a symbolic sense, … the heritage of humanity', it
made heads turn. Albeit the concept of common heritage of mankind has been
invoked in recent decades for much more than just global commons, such as
the internet, cultural heritage, photosynthesis, the Earth's climate and many
others, pushing its practical and logical limits, its application to the human
genome is far from obvious. Save for a few scholars who have read into that
statement more than it says,[41] it does not seem possible to conclude that
human genome is actually a common property of all humanity, not even
according to the Declaration itself. The reasons are several. First, the
Declaration does not rule out appropriation. Although Article 4 declares that
'[t]he human genome in its natural state shall not give rise to financial gains',
Article 3 recognizes that it is at the same time common to everyone and
individual to each: '[t]he human genome … contains potentialities that are
expressed differently according to each individual's natural and social

---

[38] Convention on the Law of the Sea (published 10 December 1982, entered into force
     16 November 1994) 1833 *UNTS* 397.
[39] Treaty on Principles Governing the Activities of States in the Exploration and Use of Outer
     Space, including the Moon and Other Celestial Bodies (signed 27 January 1967, entered into
     force 10 October 1967) 18 UST 2410.
[40] K Baslar, *The Concept of Common Heritage of Mankind in International Law* (Nijhoff
     1998) 82.
[41] E.g. Kuppuswamy, *The International Legal Governance of the Human Genome* (n 32).

environment, including the individual's state of health, living conditions, nutrition and education'. Second, there is not yet an international regime in place to manage the human genome and ensure benefit sharing. All the declaration does is to call for solidarity and international cooperation in human genetic research,[42] and to ask the IBC to 'contribute to the dissemination of the principles set out in this Declaration and to the further examination of issues raised by their applications and by the evolution of the technologies in question'.[43] Granted, a mere declaration could not create an international regime to manage a common resource, but the fact that a binding legal instrument on the status of the human genome has not been adopted is indicative of states' *opinio juris*.

In any event, the Declaration clearly stops short of declaring human genome as common heritage, with all legal consequences that it entails. It intentionally qualifies the statement by saying 'in a symbolic sense'.[44] Moreover, unless the human genome is considered to be an exception to the other international regimes created so far for the other global commons, it is clear that the status of 'common heritage' entails the creation of mechanisms for management and benefits sharing. That cannot be reconciled with the idea that the human genome is sacred and unmodifiable.

Articles 5 to 8 reassert well-established rights that individual subjects enjoy when participating in biomedical research. They include the right to informed consent,[45] the prohibition of discrimination,[46] confidentiality of genetic data[47] and the right to 'just reparation for any damage sustained as a direct and determining result of an intervention affecting his or her genome'.[48] These articles are followed by a claw back clause, which provides that 'to protect human rights and fundamental freedoms, limitations to the principles of consent and confidentiality may only be prescribed by law, for compelling reasons within the bounds of public international law and the international law of human rights'.[49]

Articles 10 to 12.a set three limits for genetic research: 'No research or research applications concerning the human genome ... should prevail over respect for the human rights, fundamental freedoms and human dignity of

[42]   UNESCO Declaration on the Human Genome (n 31), arts. 17–19.
[43]   Ibid., art. 24.
[44]   Ibid., art. 1.
[45]   Ibid., art. 5.
[46]   Ibid., art. 6.
[47]   Ibid., art. 7.
[48]   Ibid., art. 8.
[49]   Ibid., art. 9.

individuals or, where applicable, of groups of people';[50] 'practices which are contrary to human dignity, such as reproductive cloning of human beings, shall not be permitted';[51] and 'benefits from advances in biology, genetics and medicine, concerning the human genome, shall be made available to all, with due regard for the dignity and human rights of each individual'.[52] Moreover,

> States should respect and promote the practice of solidarity towards indivi-
> duals, families and population groups who are particularly vulnerable to or
> affected by disease or disability of a genetic character. They should foster,
> inter alia, research on the identification, prevention and treatment of geneti-
> cally based and genetically influenced diseases, in particular rare as well as
> endemic diseases which affect large numbers of the world's population.[53]

Articles 12.b to 16 set out the conditions for the conduct of research. First, '[f] reedom of research, which is necessary for the progress of knowledge, is part of freedom of thought. The applications of research, including applications in biology, genetics and medicine, concerning the human genome, shall seek to offer relief from suffering and improve the health of individuals and human-kind as a whole.'[54] Second,

> [t]he responsibilities inherent in the activities of researchers, including
> meticulousness, caution, intellectual honesty and integrity in carrying out
> their research as well as in the presentation and utilization of their findings,
> should be the subject of particular attention in the framework of research on
> the human genome, because of its ethical and social implications. Public and
> private science policy-makers also have particular responsibilities in this
> respect.[55]

Third, 'States should take appropriate measures to foster the intellectual and material conditions favourable to freedom in the conduct of research on the human genome and to consider the ethical, legal, social and economic implications of such research.'[56] Fourth, 'States should take appropriate steps to provide the framework for the free exercise of research on the human genome ... to safeguard respect for human rights, fundamental freedoms and human dignity and to protect public health. They should seek to ensure

[50] Ibid., art. 10.
[51] Ibid., art. 11.
[52] Ibid., art. 12.a.
[53] Ibid., art. 17.
[54] Ibid., art. 12.b.
[55] Ibid., art. 13.
[56] Ibid., art. 14.

that research results are not used for non-peaceful purposes.'[57] And, finally, 'States should recognize the value of promoting ... the establishment of independent, multidisciplinary and pluralist ethics committees to assess the ethical, legal and social issues raised by research on the human genome and its applications.'[58]

States have several duties with regard to fostering international cooperation in genetic research. They should 'make every effort ... to continue fostering the international dissemination of scientific knowledge concerning the human genome, human diversity and genetic research and, in that regard, to foster scientific and cultural co-operation, particularly between industrialized and developing countries'.[59] Article 19 continues:

'(a) In the framework of international co-operation with developing countries, states should seek to encourage measures enabling: (i) assessment of the risks and benefits pertaining to research on the human genome to be carried out and abuse to be prevented; (ii) the capacity of developing countries to carry out research on human biology and genetics, taking into consideration their specific problems, to be developed and strengthened; (iii) developing countries to benefit from the achievements of scientific and technological research so that their use in favour of economic and social progress can be to the benefit of all; (iv) the free exchange of scientific knowledge and information in the areas of biology, genetics and medicine to be promoted.

Finally, and crucially for the purposes of this book, according to Article 24, the International Bioethics Committee of UNESCO should make recommendations and give advice concerning the follow-up of this declaration, including regarding 'the identification of practices that could be contrary to human dignity, *such as germ-line interventions*'.[60]

b UNESCO International Declaration on Human Genetic Data (2003)

In the early 2000s, the Human Genome Project breakthrough and the growing number of national genetic banking projects were the spur for the adoption of the UNESCO International Declaration on Human Genetic Data. This instrument is a fascinating example of the interplay between scientific advancements and the development of a legal framework to regulate them. On 15 February 2001, the Human Genome Project announced it had

---

[57] Ibid., art. 15.
[58] Ibid., art. 16.
[59] Ibid., art. 18.
[60] Ibid., art. 24 (italics added).

completed sequencing 90 percent of all three billion base pairs in the human genome.[61] UNESCO had been following these developments. The IBC had already considered the issues created by the collection of human genetic data and had produced a report entitled 'Confidentiality and Genetic Data' in June 2000. In May 2001, the Director-General of UNESCO announced he had asked the IBC to examine the possibility of drafting an international instrument on human genetic data. In May 2002, the IBC issued a second report, entitled 'Human Genetic Data: Preliminary Study of the IBC on their Collection, Processing, Storage and Use'.[62] On 14 April 2003, the National Human Genome Research Institute, the US Department of Energy and their partners in the International Human Genome Sequencing Consortium announced the successful completion of the Human Genome Project. On 16 October 2003, UNESCO member states adopted unanimously the International Declaration on Human Genetic Data.[63]

Article 1 sets the goal of the Declaration as

> to ensure the respect of human dignity and protection of human rights and fundamental freedoms in the collection, processing, use and storage of human genetic data, human proteomic data and of the biological samples from which they are derived ... in keeping with the requirements of equality, justice and solidarity, while giving due consideration to freedom of thought and expression, including freedom of research; to set out the principles which should guide States in the formulation of their legislation and their policies on these issues; and to form the basis for guidelines of good practices in these areas for the institutions and individuals concerned.[64]

The Declaration defines 'human genetic data' as 'information about heritable characteristics of individuals obtained by analysis of nucleic acids or by other scientific analysis'[65] and 'human proteomic data' as 'information pertaining to an individual's proteins including their expression, modification and interaction'.[66]

---

[61] International Human Genome Sequencing Consortium, Initial Sequencing and Analysis of the Human Genome (2001) 409 *Nature* 860, 860–921.

[62] UNESCO, *International Bioethics Committee, Human Genetic Data: Preliminary Study by the IBC on its Collection, Processing, Storage and Use* (SHS-503/01/CIB-8/3 (Rev.2), 15 May 2002).

[63] UNESCO, International Declaration on Human Genetic Data (adopted at the 32nd Session of the General Conference on 16 October 2003).

[64] Ibid., art. 1.a.

[65] Ibid., art. 2.i.

[66] Ibid., art. 2.ii.

Recognizing that '[e]ach individual has a characteristic genetic makeup' and that, '[n]evertheless, a person's identity should not be reduced to genetic characteristics, since it involves complex educational, environmental and personal factors and emotional, social, spiritual and cultural bonds with others and implies a dimension of freedom',[67] the Declaration calls for collecting, treating, using and storing genetic data using transparent and ethically acceptable procedures.[68] As a general principle, '[a]ny collection, processing, use and storage of human genetic data, human proteomic data and biological samples shall be consistent with the international law of human rights'.[69] Under Article 5, 'human genetic data' and 'human proteomic data' may be

> collected, processed, used and stored only for the purposes of: (i) diagnosis and health care, including screening and predictive testing; (ii) medical and other scientific research, including epidemiological, especially population-based genetic studies, as well as anthropological or archaeological studies ... ; (iii) forensic medicine and civil, criminal and other legal proceedings, ... ; (iv) or any other purpose consistent with the Universal Declaration on the Human Genome and Human Rights and the international law of human rights.

At the data collection stage, the Declaration emphasizes 'prior, free, informed and express consent, without inducement by financial or other personal gain' of the person providing the data.[70] Data collected for one purpose should not be used for a different purpose that is incompatible with the original consent.[71] At the processing stage, the Declaration recommends that genetic data linked to an identifiable person not be disclosed nor made accessible to third parties – in particular, employers, insurance companies, educational institutions or families – except for 'an important public interest reason in cases restrictively provided for by domestic law that is consistent with the international law of human rights'.[72] Moreover, '[t[he provisions of this Declaration apply ... except in the investigation, detection and prosecution of criminal offences and in parentage testing that are subject to domestic law that is consistent with the international law of human rights'.[73]

Finally, states 'should regulate ... the cross-border flow of human genetic data, human proteomic data and biological samples so as to foster

---

[67] Ibid., art. 3.
[68] Ibid., art. 6.
[69] Ibid., art. 1.b.
[70] Ibid., art. 8.
[71] Ibid., art. 16.
[72] Ibid., art. 16.a.
[73] Ibid., art. 1.c.

international medical and scientific cooperation and ensure fair access to these data'.[74] Also, '[s]tates should make every effort ... to continue fostering the international dissemination of scientific knowledge concerning human genetic data and human proteomic data and, in that regard, to foster scientific and cultural cooperation, particularly between industrialized and developing countries'.[75] Scientists are encouraged to 'endeavour to establish cooperative relationships, based on mutual respect with regard to scientific and ethical matters and ... should encourage the free circulation of human genetic data and human proteomic data in order to foster the sharing of scientific knowledge ... To this end, they should also endeavour to publish in due course the results of their research'.[76] 'Benefits resulting from the use of human genetic data, human proteomic data or biological samples collected for medical and scientific research should be shared with the society as a whole and the international community.'[77]

c UNESCO Universal Declaration on Bioethics and Human Rights (2005)

The third relevant UNESCO 'soft law' instrument is the Universal Declaration on Bioethics and Human Rights.[78] It was adopted by acclamation by UNESCO member states on 10 January 2005. This was the end of a process that began with an invitation by the 2001 General Conference to the UNESCO International Bioethics Committee to report on the possibility of elaborating a universal instrument on bioethics.[79] Of the triad, the 2005 Declaration is the broadest in scope. It 'addresses ethical issues related to medicine, life sciences and associated technologies as applied to human beings, taking into account their social, legal and environmental dimensions'.[80] In the words of UNESCO,

> its originality lies in the fact that it goes much further than the various professional codes of ethics concerned. It entails reflection on societal changes and even on global balances brought about by scientific and

---

[74] Ibid., art. 18.a.
[75] Ibid., art. 18.b.
[76] Ibid., art. 18.c.
[77] Ibid., art. 19.a.
[78] UNESCO, Universal Declaration on Bioethics and Human Rights (adopted by the General Conference on 19 October 2005).
[79] J F Martin, 'The National Bioethics Committees and the Universal Declaration on Bioethics and Human Rights: Their Potential and Optimal Functioning' in A Bagheri, J Moreno and S Semplici (eds.), *Global Bioethics: The Impact of the UNESCO International Bioethics Committee* (Springer 2016) 125–136.
[80] Universal Declaration on Bioethics and Human Rights (n 78), art. 1.1.

technological developments. To the already difficult question posed by life sciences – how far can we go? – other queries must be added concerning the relationship between ethics, science and freedom.[81]

Notably, although the Declaration is addressed to states, '[a]s appropriate and relevant, it also provides guidance to decisions or practices of individuals, groups, communities, institutions and corporations, public and private'.[82]

For the first time in the history of bioethics, the international community compiled a list of fundamental principles of bioethics within a single text and expressed its desire to respect and apply them.[83] These include respect for 'human dignity, human rights and fundamental freedoms';[84] giving priority to the 'interests and welfare of the individual ... over the sole interest of science or society';[85] maximization of benefits and minimization of harm deriving from research;[86] autonomy of individuals to make decisions and responsibility for those decisions;[87] prior, free and informed consent of those subject to research;[88] respect for human vulnerability and personal integrity;[89] respect for privacy and confidentiality;[90] respect for the fundamental equality of all human beings in dignity and rights;[91] prohibition of discrimination and stigmatization;[92] respect for cultural diversity and pluralism;[93] solidarity among human beings and international cooperation;[94] sharing of benefits;[95] promotion of health and social development;[96] and protection of future generations[97] and of the environment, the biosphere and biodiversity.[98]

[81] Ibid.
[82] Ibid., art. 1.2.
[83] On the importance of the Universal Declaration on Bioethics and Human Rights, see, in general, R Andorno, 'Global Bioethics at UNESCO: In Defense of the Universal Declaration on Bioethics and Human Rights' (2007) 33:3 Journal of Medical Ethics 150, 150 –154.
[84] Universal Declaration on Bioethics and Human Rights (n 78), art. 3.1.
[85] Ibid., art. 3.2.
[86] Ibid., art. 4.
[87] Ibid., art. 5.
[88] Ibid., art. 6.
[89] Ibid., art. 8.
[90] Ibid., art. 9.
[91] Ibid., art. 10.
[92] Ibid., art. 11.
[93] Ibid., art. 12.
[94] Ibid., art. 13.
[95] Ibid., art. 15.
[96] Ibid., art. 14.
[97] Ibid., art. 16.
[98] Ibid., art. 17.

To fulfil these principles, the Declaration calls on states to promote '[p]rofes-sionalism, honesty, integrity and transparency in decision-making ... [and] ... to use the best available scientific knowledge and methodology in addressing and periodically reviewing bioethical issues',[99] dialogue between 'persons and profes-sionals concerned and society as a whole'[100] and '[o]pportunities for informed pluralistic public debate, seeking the expression of all relevant opinions, should be promoted'.[101] States are also urged to create ethics committees to assess the relevant ethical, legal, scientific and social issues related to research projects involving human beings, provide advice on ethical problems in clinical settings, and assess scientific and technological developments;[102] to promote 'appropriate assessment and adequate management of risk related to medicine, life sciences and associated technologies';[103] to facilitate 'transnational practices';[104] and to foster 'international dissemination of scientific information and encourage the free flow and sharing of scientific and technological knowledge'.[105]

d Overall Assessment of the Work of UNESCO and Future Developments

With its three declarations, UNESCO attempted to provide a general frame-work with global reach, which sets out the fundamental principles and rights that must be respected in the process of regulating scientific advancements, in the field of biomedicine. While the effort should be appreciated, in the end these instruments, being not binding, fail to provide the regulatory framework that the policy impetus intended to have. By and large, they simply echo already well-established international instruments on bioethics, such as the World Medical Association's Declaration of Helsinki, which was adopted back in 1964 and has been updated several times since,[106] and the 2002 Ethical Guidelines for Biomedical Research Involving Human Subjects of the Council for International Organizations of Medical Sciences,[107] and the best practices already followed by most developed states of the world.

99　Ibid., art. 18.1.
100　Ibid., art. 18.2.
101　Ibid., art. 18.3.
102　Ibid., art. 19.
103　Ibid., art. 20.
104　Ibid., art. 21.
105　Ibid., art. 24.1.
106　World Medical Association, *Declaration of Helsinki: Recommendations Guiding Physicians in Biomedical Research Involving Subjects* (as amended through 2013).
107　Council for International Organizations of Medical Sciences, *International Ethical Guidelines for Biomedical Research Involving Human Subjects* (CIOMS, Geneva, 2002).

However, breaking new ground and originality are hardly the point of international law-making efforts. In a community made of sovereign states, progress can be achieved only gradually, by broadening support of these efforts to as many states as possible. It means most international instruments, and certainly those of a global scope, are just the minimum common denominator that is conceivably achievable between almost 200 sovereign states. Often, universal declarations are scorned for their non-binding nature and vague language, but that is inevitable given the nature of the international community. These legal instruments are intentionally non-binding for a reason. Norm-making through 'soft law' instruments as opposed to through 'hard law' ones, such as treaties, which create binding legal obligations, permits states to take on commitments they otherwise would not have taken. Furthermore, soft law instruments present the advantage of permitting countries to gradually become familiar with the proposed standards before they are confronted with the adoption of enforceable rules or with the development of a binding instrument. As we saw, soft law instruments can, and often do, morph into hard law over time, in the form of customary law, as long as they are followed.[108]

As we will see throughout this volume, vague or lacking definitions are not just a problem of international bioethics instruments. National laws, and, disturbingly, even criminal norms, can often be as vague and undetermined. Again, it is not the result of oversight but rather a deliberate choice. Except for very technical terms, lawmakers typically prefer not to define precisely most of the words they use. Rather, they tend to leave that task to common understanding and, ultimately, to courts' interpretation. Flexibility, especially in a field fast developing like bio-medical science, is a virtue. In the specific case of UNESCO, vagueness is unavoidable because it is impossible to reach a global agreement on the precise meaning of terms such as 'dignity', 'autonomy', 'justice', 'benefit', 'harm' or 'solidarity', terms that have a long philosophical history and are, to some extent, conditioned by cultural factors. Universal principles must be contextualized before they can be applied in a meaningful sense at the national level, but contextualization without a global framework of reference opens the door to diverging standards and goes against the very fundamental idea undergirding all international law: the belief that humanity is one, even if it is divided in several sovereign states.

As we saw, the more recent UNESCO declaration directly relevant to the topic of the book was issued in 2005. At the speed at which genetic engineering develops, this instrument can be hardly said to be current. Since then, it has fallen to the IBC to continue monitoring development and to issue reflections

---

[108] See, in this Chapter, Section I.2.

on what states and UNESCO ought to do. In October 2015, the IBC issued a report entitled 'Updating Its Reflection on the Human Genome and Human Rights',[109] taking into account its previous recommendations on the matter[110] and the three UNESCO declarations. In this report, the IBC called upon states to agree on a moratorium on genome engineering of the human germline, at least as long as the safety and efficacy of the procedures are not adequately proven as treatments, and to produce an international legally binding instrument to ban human cloning for reproductive purposes;[111] to renounce the possibility of acting alone in relation to engineering the human genome and accept to cooperate on establishing a shared, global standard for this purpose, building on the principles set out in the Universal Declaration on the Human Genome and Human Rights and the Universal Declaration on Bioethics and Human Rights;[112] and to encourage the adoption of rules, procedures and solutions, which can be as non-controversial as possible, especially with regard to the issues of modifying the human genome and producing and destroying human embryos.[113]

The IBC also underlined the need to adopt legislative and other measures to ensure that quality-assured information be provided with regard to direct-to-consumer tests, including non-medical tests, in order to mitigate risks and avoid misuse;[114] to organize healthcare systems so that the new opportunities offered by precision/personalized medicine can be shared with society as a whole, without becoming a new source of inequality and discrimination;[115] to develop a trustworthy form of governance for biobanks and biobank secrecy and harmonize the corresponding rules at the international level;[116] and to ensure that new possibilities of genetic screening and in particular non-invasive prenatal testing comply with both the right to autonomous choices and the principles of non-discrimination and non-stigmatization and respect for every human being in her or his uniqueness.[117] Finally, the IBC called on

---

[109] UNESCO, *International Bioethics Committee: Updating Its Reflection on the Human Genome and Human Rights* (SHS/YES/IBC-22/15/2 REV.2, 2 October 2015).

[110] For a list of International Bioethics Committee's advices and recommendations, see, UNESCO, 'Reports and Advices of the International Bioethics Committee' (2017) www.unesco.org/new/en/social-and-human-sciences/themes/bioethics/international-bioethics-committee/reports-and-advices/ accessed 8 November 2018.

[111] UNESCO, *International Bioethics Committee, Updating Its Reflections* (n 109), para. 118.

[112] Ibid., para 116.

[113] Ibid., para 113.

[114] Ibid., paras 120–121.

[115] Ibid., para 122.

[116] Ibid., para 123.

[117] Ibid., para 125.

states and UNESCO to consider revising the three declarations, emphasizing that the cogency of principles remains untouched but some applications could need updating.[118]

The IBC concluded with a plea:

> We are human because of the interplay of many biological, historical, cultural determinants, which preserve the feeling of our fundamental unity and nourish the richness of our diversity. The international community, States and governments, scientists, actors of civil society and individuals are called upon to consider the human genome as one of the premises of freedom itself and not simply as raw material to manipulate at leisure. At the same time, considering that scientific advancements in this field are likely to offer unprecedented tools against diseases, it is crucial to acknowledge that these opportunities should never become the privilege of few. What is heritage of humanity entails sharing both of responsibilities and benefits.[119]

2 WHO

The World Health Organization (WHO) is another specialized agency of the United Nations whose mission is 'the attainment by all peoples of the highest possible level of health'.[120] This goal is pursued through three 'core functions': (1) 'normative', including drafting and adopting treaties, regulations, and other non-binding legal standards and recommendations; (2) 'directing and coordinating', including launching focused programmes, such as those on poverty and health, essential medicine and specific diseases; and (3) 'research and technical cooperation', including disease eradication and coordinating response to health emergencies.[121]

The WHO is mostly known for its work on the latter two functions. Compared to many other UN agencies, the WHO has been a relatively less active law-maker.[122] Historically, it has chosen to eschew the legal approach in favour of developing international guidelines of practice for specific health issues.[123] So far, only three international legal instruments have been adopted

[118]   Ibid., para 127.
[119]   Ibid., para 128.
[120]   WHO Constitution, 14 UNTS 185, art. 1.
[121]   Ibid., art. 2. See, in general, G L Burci and C-H Vignes, *The World Health Organization* (Kluwer 2004).
[122]   See, in general, L Gostin and D Sridhar, 'Global Health and the Law' (2014) 370 *New England Journal of Medicine* 1732, 1732–1740.
[123]   Kuppuswamy, *The International Legal Governance of the Human Genome* (n 32), 40–41.

under its aegis: the Framework Convention on Tobacco Control,[124] the International Health Regulations[125] and the Pandemic Influenza Preparedness Framework.[126]

The WHO's reluctance to engage in norm-setting and to make genetics and genomics a priority, preferring instead to focus on minimizing public health risks and expanding healthcare coverage,[127] explains why there are no WHO standards applicable to genetic research, in general, and human genome modification, in particular. The WHO's contribution to the governance of genomic research has been mostly through 'directing and coordinating' and encouraging 'research and technical cooperation'. For instance, the WHO's Human Genomics in Global Health Initiative aims to provide information and raise awareness within the health sector, governments and the wider public on the health challenges and opportunities of human genomics, and to share information and to develop innovative approaches in the field of human genetics and genomics.[128] It builds on the work of the former Human Genetics Programme and of the Initiative on Genomics & Public Health, acting under the responsibility of the Department of Service Delivery and Safety and working across the Organization, with WHO Collaborating Centres, NGOs and other international organizations active in this field.[129]

Until recently, the WHO had left the driver's seat on the question of the governance of human genome manipulation to UNESCO, leading some observers to accuse UNESCO of having exceeded its mandate and trespassing on WHO turf.[130] However, it is hard to see how the charge of having exceeded its mandate can stand when one considers how broad UNESCO's mandate is to begin with. Furthermore, within the United Nations, there are hardly other agencies that could claim the same level of experience at the intersection of

---

[124]   WHO Framework Convention on Tobacco Control (opened for signature 16 June 2003, entered into force 27 February 2005) 2302 UNTS 166.

[125]   International Health Regulations (2005), 79 UNTS 2509.

[126]   World Health Organization, *Pandemic Influenza Preparedness Framework for the Sharing of Influenza Viruses and Access to Vaccines and Other Benefits* (World Health Organization, Geneva, 2011).

[127]   See, WHO, 'WHO Priorities' hwww.who.int/dg/priorities/en/ accessed 14 November 2018.

[128]   See, WHO, 'Human Genomics in Global Health' www.who.int/genomics/en/ accessed 8 November 2018.

[129]   See, WHO, 'About WHO's Human Genomics in Global Health Initiative' www.who.int/genomics/about/en/ accessed 8 November 2018.

[130]   W Landman and U Schüklenk, 'UNESCO "Declares" Universals on Bioethics and Human Rights: Many Unexpected Universal Truths Unearthed by UN Body' (2005) 5:3 *Developing World Bioethics* iii–vi; JR Williams, 'UNESCO's Proposed Declaration on Bioethics and Human Rights: A Bland Compromise' (2005) 5:3 *Developing World Bioethics* 210, 210–215.

sciences, ethics and human rights. UNESCO is the only UN agency specialized in sciences (both natural and human sciences) and has served for decades as a forum for philosophical discussion on cross-disciplinary issues. As it has been aptly said,

> a conflict of competence between two or more UN agencies interested in this matter would be as absurd as a dispute between a philosopher and a doctor over the 'ownership' of bioethics. Of course, bioethics does not belong in exclusivity to any of them. As it is by its very nature an interdisciplinary specialty, all related professions (and likewise, all related UN bodies) have the right – and the duty – to make their specific contribution to this emerging and complex domain."[131]

At the same time, as we also know, overlapping mandates and competences of international organizations increase the likelihood of fragmentation of international law, opening the door for conflicting understandings of how the problems that arise in the field of human genome engineering and biomedicine more generally should be regulated.

As we said earlier, the WHO played a limited role until recently. In December 2018, just days before this book went to the press, the WHO's Director-General announced a plan to put together an expert panel looking at international standard for human genome germline modification. 'We will work with member states', Director-General Tetros stated, 'to do everything we can to make sure of all issues – be it ethical, social, safety – before any manipulation is done'.[132] We will see what comes out of this, although it is reasonable to assume that any such effort will take into account existing soft law standards in the field of international bioethics law that has been developed by UNESCO but with a focus on health-oriented issues and concerns.

As we saw, overlap between different branches of international law, such as between international bioethics law and international human rights law, is far from exceptional. It actually strengthens the whole construct of international human rights because the repetition of the same principles over and over helps solidifying norms into customary international law. UNESCO's and WHO's standard-setting activities operate at different levels. While UNESCO tends to produce general normative frameworks

---

[131]   R Andorno, 'Global Bioethics at UNESCO (n 83), 152.
[132]   S Nebehay, 'WHO Looks at Standards in "Uncharted Water" of Gene Editing' (Reuters, 2018) www.reuters.com/article/us-china-health-who/who-looks-at-standards-in-uncharted-water-of-gene-editing-idUSKBN1O227Q accessed 13 December 2018.

of a predominantly philosophical and legal nature, the WHO's guidelines are usually more technical and focused on specific health-related issues.[133] However, although there is undoubtedly overlap between bioethics law and human rights law, there are also differences between the two branches when it comes to determining the range of human rights at stake in the field of biomedicine. We will revert to this later on.[134]

3 OECD

The Organisation for Economic Cooperation and Development (OECD) is an intergovernmental organization whose aim is to stimulate economic progress and world trade. It comprises thirty-six high-income states that are committed to democracy and the market economy, a group that has sometimes been dubbed a 'club of rich countries'.[135] Of the states surveyed in this book, only the People's Republic of China and Singapore are not members of it.[136] It provides a platform to compare policy experiences, seek answers to common problems, identify good practices and coordinate domestic and international policies of its members.

The OECD features a Working Party on Biotechnology, Nanotechnology and Converging Technologies that has focused on policy issues in emerging technology fields related to bio, nano and converging technologies, including gene editing.[137] As far as norm-setting is concerned, so far the Working Party on Biotechnology has developed guidelines on human biobanks and genetic research databases;[138] quality assurance of molecular genetic testing offered in a clinical context;[139] licensing of intellectual property rights that relate to genetic inventions used for the purpose of human healthcare;[140] and

---

[133] Ibid.

[134] See, in this Chapter, Section III.5.

[135] *The Economist* explains, 'What is the OECD?' (*The Economist*, 6 July 2017) www.economist.com /the-economist-explains/2017/07/05/what-is-the-oecd accessed 8 November 2018.

[136] OECD, 'List of OECD Member Countries – Ratification of the Convention on the OECD' (OECD, 19 July 2018) www.oecd.org/about/membersandpartners/list-oecd-member-countries.htm accessed 19 July 2018.

[137] OECD, 'Emerging Technologies' www.oecd.org/sti/emerging-tech/ accessed 8 November 2018.

[138] OECD, 'Guidelines for Human Biobanks and Genetic Research Databases (HBGRDs)' www.oecd.org/sti/emerging-tech/guidelines-for-human-biobanks-and-genetic-research-data bases.htm accessed 8 November 2018.

[139] OECD, 'OECD Guidelines for Quality Assurance in Genetic Testing' www.oecd.org/sti/emer ging-tech/oecdguidelinesforqualityassuranceingenetictesting.htm accessed 8 November 2018.

[140] OECD, 'Guidelines for the Licensing of Genetic Inventions' www.oecd.org/sti/emerging-tech/guidelinesforthelicensingofgeneticinventions.htm accessed 8 November 2018.

policy papers on pharmacogenetics,[141] and on biomarkers and targeted therapies.[142]

In 2015, the OECD Biotechnology Working Party launched the 'Project on Gene Editing' to 'produce a forum conducive to evidence-based discussion across countries on the many issues of shared concern . . . [and] to help guide policy at the national and international levels and promote – where appropriate – cooperative governance approaches'.[143] So far, the Project has produced a few working papers, including one on the governance of gene editing and advanced therapies.[144]

4 Civil Society

Finally, since the onset of the 'CRISPR revolution', several dozens of nongovernmental organizations across the globe have issued 'statements', 'views', 'recommendations' and policy papers dealing specifically with germline genome engineering only. Carolyn Brokowski, of the Yale School of Medicine, reviewed sixty-one ethics reports and statements, crafted by more than fifty countries and organizations between 2015 and early 2018.[145] The organizations include learned societies (e.g. the American and European Societies of Human Genetics, the European Society of Human Reproduction and Embryology and the International Society for Stem Cell Research); bioethics organizations (e.g. the Nuffield Council on Bioethics (UK), the Danish

---

[141] OECD, 'Pharmacogenetics: Opportunities and Challenges for Health Innovation' www.oecd.org/sti/emerging-tech/pharmacogeneticsopportunitiesandchallengesforhealthinnovation.htm accessed 8 November 2018.

[142] OECD, 'Biomarkers and Targeted Therapies' www.oecd.org/sti/emerging-tech/biomarkersandtargetedtherapies.htm accessed 8 November 2018.

[143] See, OECD, BNCT, 'Project on Gene Editing (Innovation Policy Platform)' www.innovationpolicyplatform.org/project-gene-editing-oecd-bnct accessed 19 July 2018.

[144] H Garden and D Winickoff, 'Gene Editing for Advanced Therapies: Governance, Policy and Society' (OECD Science, Technology and Industry Working Papers, OECD Publishing, Paris, 2018).

[145] For a survey of 'ethical statements' on the matter, see, in general, C Brokowski, 'Do CRISPR Germline Ethics Statements Cut It?' (n 25), 115–125. Weeks before this book was being finalized, the Second International Summit of Genome Editing took place in Hong Kong (27–29 November 2018). The meeting, sponsored by the Academy of Science of Hong Kong, the UK Royal Society, the US National Academy of Sciences and the Academy of Medicine, followed a meeting held in 2015 in Washington, DC, to discuss the science, application, ethics and governance of human genome editing. The Organizing Committee issued a statement on 29 November 2018. See, The National Academies of Sciences, Engineering, and Medicine, *Statement by the Organizing Committee of the Second International Summit on Human Genome Editing* (29 November 2018) www8.nationalacademies.org/onpinews/newsitem.aspx?RecordID=11282018b accessed 30 November 2018.

Council on Ethics, and the International Bioethics Committee of UNESCO); industry groups and organizations (e.g. the Biotechnology Innovation Organization and various genome-editing biotech companies); and political groups (e.g. the 2015 White House).[146] Most reports and statements were produced by organizations from Europe and the United States, although groups from Canada, New Zealand, Japan, China, Australia, Latin America and other international conglomerates also contributed.[147]

In her review, Brokowski notes that these statements vary considerably in both length and depth of analysis, from succinct, direct and practical to expansive, indeterminate, nuanced and philosophical. Positions range widely, too, but they can be clustered in groups. Overall, a majority of statements surveyed (54 per cent) expressly considered germline editing impermissible at the current time.[148] A further 11 per cent also consider germline editing impermissible currently, but are expressly open to the possibility of allowing it under certain conditions.[149] In 30 per cent of cases, the position is not expressly addressed or is ambiguous.[150] Only 5 per cent of the reports state an openness to further exploration.[151]

Overall, a large majority seem reluctant to proceed with heritable germline editing unless and until more were known about safety, risks, benefits, and efficacy and a broad societal consensus was achieved. Various categories of risk seem to outweigh any potential benefits, for now. Some favour a form of moratorium, ranging from broadly prohibiting 'gene editing of human embryos or gametes which would result in the modification of the human genome' to more narrowly prohibiting 'attempts to apply nuclear genome editing of the human germ line in clinical practice'.[152] A common concern is that editing might pose technical/mechanical obstacles, leading indefinitely to safety risks in the modified organism and future progeny, including inaccurate editing (off- and on-target effects), incomplete editing (mosaicism), efficiency challenges (success rate) and interference from unexpected and/or poorly understood factors (e.g. epigenetic, immune and environmental events; pleiotropy; and penetrance) resulting in unintended consequences.[153] Other concerns included the potential return of eugenics, the misuse of this technology for human

---

[146] Brokowski, table 1, 117–119.
[147] Ibid.
[148] Ibid., figure 3, 122.
[149] Ibid.
[150] Ibid.
[151] Ibid.
[152] Ibid., 116.
[153] Ibid.

enhancement goals and the exacerbation of social inequalities, along with a purported lack of compelling medical rationale justifying such interventions.[154] Additionally, difficulties with obtaining actual informed consent, given the complexity surrounding the status of the human embryo and the potential effects lasting into numerous future generations, were highlighted.[155] Many also seem to believe that national and international laws already prohibit such modifications.[156]

Brokowski's survey of statements does not track whether there has been a shift of positions over time. It could not because of the novelty of CRISPR and the fact that it covers only the last three years. However, as science advances, uncertain if not downright prohibitory legal frameworks notwithstanding, it is likely statements will gradually shift position as well, tracking public opinion sentiments. Indeed, as the German Ethics Council noted, there seems to have been a subtle, though important, shift in opinion about the permissibility of heritable genome editing: from 'not allowed as long as the risks have not been clarified' to 'allowed if the risks can be assessed more reliably'.[157]

As Brokowski noted, from a bioethics or legal perspective, many of these reports are limited.[158] Some offer conclusions but lack significant support for them. At times, purported justifications for limitations lie on shaky logical and ethical foundations, and, although many of these statements call for public engagement and open debates about risks, costs and benefits, few offer concrete ideas on how to organize those debates. Despite their value in raising questions and generating dialogue, it is unlikely that any single ethics report or position statement, now or in the future, could address all critical issues raised by heritable genome-editing technology. Yet, even if one were to take an extreme view of international law, one that holds that it is only the official view of states, as articulated by their governments, that matters to determine *opinio juris*, it would be a serious mistake to dismiss them as irrelevant. Because of the high degree of technical and scientific complexity of human genome modification, governments would be hard pressed to justify departing from the recommendations of the learned societies in their country, and the consensus across boundaries.

---

[154]   Ibid.
[155]   Ibid.
[156]   Ibid.
[157]   Deutscher Ethikrat, *Germline Intervention in the Human Embryo: German Ethics Council Calls for Global Political Debate and International Regulation: Ad Hoc Recommendation* (Berlin, 29 September 2017) 3.
[158]   Brokowski, 'Do CRISPR Germline Ethics Statements Cut It?' (n 25), 116.

III APPLICABLE INTERNATIONAL HUMAN RIGHTS STANDARDS

While human rights are often acknowledged in 'soft law' bioethical instruments, such as the UNESCO declarations we just discussed, international human rights norms have not featured prominently in bioethical analyses, perhaps with the exception of the right to health.[159] As a result, what a human rights approach to biomedical research, and germline engineering in particular, entails is still rather unclear. One of the goals of this book is to tackle this problem and advance our understanding of how germline engineering and human rights intersect in this international framework.

We believe it is essential to take a fresh look at the complex relationship between the international bioethics instruments and international human rights law. Even if UNESCO stresses respect for human rights, its instruments are developed in parallel rather than in integration with international human rights law. The so-called right to science and the rights of science in particular are referenced to but not fully incorporated in the UNESCO declarations and other international bioethics instruments. As we saw in the previous section, the contributions of the WHO and the OECD are still very limited. Although their work may contribute to the international protection of some human rights, none of them have an explicit and general mandate to advance human rights standard-setting in this field.

Our objective is not to give a comprehensive account of all international human rights that are touched by the scientific progress made on human germline genome modification.[160] The list of human rights at play as a result of the recent advances made in the field of human germline modification is long, certainly too long for this book. It includes both civil and political rights (e.g. right to life; right to bodily integrity; right to privacy; right to academic freedom) and economic, social and cultural rights (e.g. right to health; right to a family; right to benefit from progress in science and technology), and their corresponding duties. It includes special protections of particularly vulnerable

---

[159] See, e.g., R Andorno, *Principles of International Biolaw: Seeking Common Ground at the Intersection of Bioethics and Human Rights* (Bruylant, 2013); S Holm, *The Law and Ethics of Medical Research: International Bioethics and Human Rights* (Cavendish Publishing, 2005).

[160] Albeit they are certainly relevant, we will not discuss here the right to health, reproductive rights and the right to integrity. These are complex rights that are touched by activities aimed at modifying the genome of human germline cells but cannot be given here the treatment they deserve. We suggest those interested in them to look further in current literature. E Riedel, 'Health, Right to, International Protection' (*Max Planck Encyclopedia of Public International Law*, last updated April 2011); J Gebhard and D Trimiño Mora, 'Reproductive Rights, International Regulation' (*Max Planck Encyclopedia of Public International Law*, last updated August 2013).

groups (e.g. children and women). Moreover, a comprehensive account of all the rights at play will need to consider both the global and regional regimes of human rights protection. Rather we focus on a set of rights that we believe have been set aside and ignored in the international efforts made so far to further governance of human germline modification, but which in our view should be brought to the centre of the discussion: the so-called right to science and the so-called rights of science.

1 The Right to Science and the Rights of Science: Origin and Development

The expression 'right to science' is commonly used to indicate one specific human right: the right to enjoy the benefits of scientific and technological progress and its applications.[161] However, this is a cursory reading of what right to science means under international human rights law. In this section, we wish to develop a more detailed account of this right in the light of the human rights debates on the meaning and scope of this right. In addition, we believe that there is a need to distinguish the right to science from what we define as the 'rights of science'. We use the term 'rights of science' to indicate the set of rights and corresponding duties that concern scientific research and technological development, such as freedom of expression, academic freedom and the right to seek and disseminate knowledge, the right to associate, the right to work, the right to protection of the moral and material interests resulting from inventors' work, the duty states have to encourage scientific research and to facilitate cross-border cooperation and the like. While the right to science focuses on the right to benefit from scientific progress and its applications, the rights of science focus on the rights of scientists that enable them to conduct research, including in the field of biomedicine.

The roots of the rights of science run deep, arguably all the way to the early 1600s, with Francis Bacon and Galileo Galilei, and intertwine with other better-known rights, such as the right to education and freedom of expression. Although the right to science is a considerably more recent idea, it is still one with lineage as old as any other internationally recognized human rights. Indeed, it was recognized first in the American Declaration of Human

---

[161] E.g. L Shaver, 'The Right to Science : Ensuring that Everyone Benefits from Scientific and Technological Progress' (2015) 4 *European Journal of Human Rights* 411, 411–430; A Chapman, 'Towards an Understanding of the Right to Enjoy the Benefits of Scientific Progress and Its Applications' (2009) 8:1 *Journal of Human Rights* 1, 1–36; M Mancisidor, 'Is There Such a Thing as a Human Right to Science in International Law?' (2015) 4:1 *ESIL Reflections* 1, 1–6.

Rights, adopted at the Ninth International Conference of American States in Bogotá, Colombia, on 2 May 1948.[162] The American Declaration preceded and inspired the proclamation of the same right in the Universal Declaration of Human Rights, adopted by the General Assembly of the United Nations on 10 December 1948.[163]

Article XIII of the American Declaration of Human Rights recites:

> Every person has the right to take part in the cultural life of the community, to enjoy the arts, and to participate in the benefits that result from intellectual progress, especially scientific discoveries. He likewise has the right to the protection of his moral and material interests as regards his inventions or any literary, scientific or artistic works of which he is the author.

Echoing the American Declaration, Article 27 of the Universal Declaration of Human Rights provides that

> '(1) Everyone has the right freely to participate in the cultural life of the community, to enjoy the arts and to share in scientific advancement and its benefits. (2) Everyone has the right to the protection of the moral and material interests resulting from any scientific, literary or artistic production of which he is the author.'

As we discussed earlier, recognition of the non-binding nature of mere declarations led the international community to start negotiations of a binding international human right treaty, but the division between East and West led to the adoption, instead, of twin treaties: one focused on civil and political rights (ICCPR) and the second on economic, social and cultural rights (ICESCR).[164] Some 'rights of science' were included in the list of civil and political rights (e.g. freedom of expression, academic freedom and the right to seek and disseminate knowledge, the right to associate), while others were included in the list of economic, social and cultural rights. These include the right to work, the right to protection of the moral and material interests, the duty states have to encourage scientific research and to facilitate cross-border cooperation and the like, and also the right to health and the right to a family, which are key when discussing reproductive rights. The 'right to science' was included in the list of socio-economic rights, bundled together

---

[162] American Declaration of the Rights and Duties of Man (adopted by the Ninth International Conference of American States, 1948) OAS Res XXX, reprinted in Basic Documents Pertaining to Human Rights in the Inter-American System, OEA/Ser L V/II.82 Doc 6 Rev 1, at 17 (1992).

[163] See, in this Chapter, Section I.3.

[164] Ibid.

with a broader right called the 'right to culture'. Article 15 of the International Covenant on Economic, Social and Cultural Rights reads:

1. The States Parties to the present Covenant recognize the right of everyone: (a) To take part in cultural life; (b) To enjoy the benefits of scientific progress and its applications; (c) To benefit from the protection of the moral and material interests resulting from any scientific, literary or artistic production of which he is the author.

2. The steps to be taken by the States Parties to the present Covenant to achieve the full realisation of this right shall include those necessary for the conservation, the development and the diffusion of science and culture.

3. The States Parties to the present Covenant undertake to respect the freedom indispensable for scientific research and creative activity.

4. The States Parties to the present Covenant recognise the benefits to be derived from the encouragement and development of international contacts and co-operation in the scientific and cultural fields.

As it was discussed, the basic structure of the UN human rights regime, with a declaration and two separate instruments, each dedicated to a family of rights, was echoed in various regions across the globe.[165] As a consequence, we can find the 'right to science' and the 'rights of science' also in various regional legal instruments. For instance, in the Americas, besides in the American Declaration,[166] one can find elements of these rights in several articles of the Charter of the Organization of American States (1948),[167] and in the Additional Protocol to the American Convention on Human Rights in the Area of Economic, Social and Cultural Rights ('Protocol of San Salvador'), which, echoing the Covenant, requires states to recognize the right of everyone 'to enjoy the benefits of scientific and technological progress'[168] and the duty of states to 'extend among themselves the benefits of science and technology by encouraging the exchange and utilization of scientific and technological knowledge'.[169]

---

[165] Ibid.
[166] American Declaration (n 164), art. XIII.
[167] Charter of the Organization of American States (1948), 119 UNTS 3, arts. 17, 30, 34.i, 38, 45, 47 and 51.
[168] ICESCR, art. 14.1.b.
[169] Additional Protocol to the American Convention on Human Rights in the area of Economic, Social, and Cultural Rights (Protocol of San Salvador), OAS Treaty Series No 69, art. 38.

In Africa, the Constitutive Act of the African Union identifies scientific and technical cooperation as essential for meeting its goals,[170] and the Protocol on the Rights of Women in Africa of the African Charter on Human and Peoples' Rights requires states to take specific measures to promote education and training for women, particularly in the fields of science and technology.[171]

In the Arab world, the Arab Charter on Human Rights recognizes the right of everyone 'to take part in cultural life and to enjoy the benefits of scientific progress and its application', together with the obligations of states to 'respect the freedom of scientific research and creative activity, . . . ensure the protection of moral and material interests resulting from scientific, literary and artistic production . . . enhance cooperation at all levels, with the full participation of intellectuals and inventors and their organizations, in order to develop and implement recreational, cultural, artistic and scientific programs'.[172]

Finally, in Southeast Asia, the ASEAN Human Rights Declaration provides that every person has 'the right, individually or in association with others, to freely take part in cultural life, to enjoy the arts and the benefits of scientific progress and its applications and to benefit from the protection of the moral and material interests resulting from any scientific, literary or appropriate artistic production of which one is the author'.[173]

Oddly, in Europe, there is no explicit reference to the right to science either in the European Convention on Human Rights[174] or in the European Social Charter,[175] the two most important human rights treaties in Europe. This is one of the great mysteries of international human rights law. However, at least for what concerns the European Union, this lacuna was partially filled in 2000 with the adoption of the Charter of Fundamental Rights of the European Union, which provides that scientific research shall be 'free of constraint'.[176]

---

[170] Constitutive Act of the African Union (adopted 11 July 2000, entered into force May 26, 2001) OAU Doc. CAB/LEG/23.15, art. 3.m.

[171] Maputo Protocol on the Rights of Women in Africa of the African Charter on Human and Peoples' Rights (signed 11 July 2003, entered into force 25 November 2005) CAB/LEG/66.6, art. 12. 2.b.

[172] Arab Charter on Human Rights (adopted 15 September 1994, entered into force 15 March 2008), reprinted in 12 Int'l Hum Rts Rep 893 (2005), art. 42.

[173] Association of Southeast Asian Nations Human Rights Declaration (adopted at the 21st ASEAN Summit, Phnom Penh, Cambodia on 18 November 2012), art. 32.

[174] [European] Convention for the Protection of Human Rights and Fundamental Freedoms, ETS No. 5.

[175] European Social Charter (revised), ETS No. 163.

[176] Charter of Fundamental Rights of the European Union (proclaimed by the European Parliament on 7 December 2000 and entered into force in adapted wording with the date of the entry into force of the Lisbon Treaty on 7 December 2009) OJ C 326 (TEU) 391–407.

2 From the Vanishing Point of International Human Rights Law to Front and Centre

Although the right to science has been recognized under international law since 1948, until recently international, regional and national bodies, as well as human rights activists and scholars, have paid little attention to it. Writing in 1952, at the dawn of international human rights, Hersch Lauterpacht wrote 'if economic, social and cultural rights lie at the vanishing point of international human rights law . . . then the question of the right to enjoy the benefits of scientific and technological progress and its applications lies at the vanishing point of economic, social and cultural rights'.[177] Science and human rights have long had an uneasy relationship. When science is mentioned, it is more often as a threat to human rights than as a tool to enhance them and protect them. For instance, although the final act of the 1993 Vienna World Conference on Human Rights recognizes that 'everyone has the right to enjoy the benefits of scientific progress and its applications', it also adds immediately after that 'certain advances, notably in the biomedical and life sciences as well as in information technology, may have potentially adverse consequences for the integrity, dignity and human rights of the individual, and calls for international cooperation to ensure that human rights and dignity are fully respected in this area of universal concern'.[178]

The result of this diffidence and neglect is that our understanding of the normative content of the right to science – that is, what exactly are states' obligations – is not yet entirely settled. However, in the past two decades the right to science and the rights of science have gained a more prominent position in human rights debates in international fora, and progress towards a more complete understanding of these rights has been tangible.

At the global level, three developments are particularly significant. The first one, in 2009, is the adoption of the Venice Statement on the Right to Enjoy the Benefits of Scientific Progress and Its Applications ('Venice Statement'), drafted under the auspices of UNESCO.[179] The second one, also in 2009, is the appointment by the Human Rights Council of an Independent Expert in the field of Cultural Rights, whose mandate also includes the right to

---

[177] H Lauterpacht, 'The Problem of the Revision of the Law of War' (1952) 39 *British Yearbook Int'l L* 139.

[178] B Boutros-Ghali, *World Conference on Human Rights: the Vienna Declaration and Programme of Action* (New York, United Nations, Dept of Public Information, June 1993) I.11, para 3.

[179] UNESCO, *Venice Statement on the Right to Enjoy the Benefits of Scientific Progress and Its Applications* (2009).

science.[180] The third one took place in 2015 and is the mandate given by the Committee on Economic, Social and Cultural Rights, the expert body in charge of supervising implementation of the Covenant on Economic, Social and Cultural Rights, to two of its members, Mikel Mancisidor and Rodrigo Uprimny, to draft a 'general comment' on the right to science. The eventual adoption by the Committee of the general comment, perhaps in 2019, will be the crowning moment of a decade-long process of normative development.

The Venice Statement is the outcome of a 2009 meeting sponsored by UNESCO aiming at 'clarifying the normative content of the right to enjoy the benefits of scientific progress and its applications and generating a discussion among all relevant stakeholders with a view to enhance the implementation of this right'.[181] The Venice Statement makes two significant contributions. The first is spelling out the three duties that states parties to the Covenant on Economic, Social and Cultural Rights have, namely the duty to 'respect', to 'protect' and to 'fulfil'. 'Respecting' means guaranteeing the freedoms that are necessary to do science (e.g. autonomy, freedom of speech, freedom to assemble in professional societies and to collaborate, and to ensure science is not used to interfere with enjoyment of other human rights and freedoms).[182] 'Protecting' means ensuring that science is not done by infringing upon the rights of anybody (e.g. research subjects, vulnerable populations and so forth).[183] 'Fulfilling' calls for a variety of strategies, including monitoring harms arising from science, enhancing public engagement in decision-making about science and technology, ensuring access to benefits of scientific progress on a non-discriminatory basis and developing science curricula at all levels of schooling.[184] The second contribution made by the statement is pointing out that it is also incumbent upon non-governmental actors (e.g. scientific societies, for-profit entities, civil society) to contribute to the realization of the right to science.[185]

The second significant development at the global level is the United Nations Human Rights Council's decision to give an Independent Expert a special mandate on cultural rights, including the right to science. The first appointee was Pakistani sociologist Farida Shaheed, and the current one is the Algerian-American law professor Karima Bennoune. In 2012, Shaheed

[180] UN Human Rights Council (43rd Meeting) Resolution 10/23 (26 March 2009).
[181] Venice Statement (n 181), para 2.
[182] Ibid., para 14.
[183] Ibid., para 15.
[184] Ibid., para 16.
[185] Ibid., paras 25–27. The Statement touches also upon the issue of privatization of science and how it could conflict with the right to science. Ibid., para 5.

released a report titled 'The Right to Enjoy the Benefits of Scientific Progress and its Applications'.[186] The report discusses the normative content, state obligations and limitations of the right to science. With regard to the normative content, the report makes four contributions. First, it connects the right to science to the right to participate freely in the cultural life of the community as recognized by article 15 of the ICESCR. Article 15 entails the right to contribute to science (as knowledge producers) and enjoy opportunities to participate in decisions about science (as citizens). The report further maintains that the right should be enjoyed without discrimination.[187] Second, it stresses the importance of freedom of research as a prerequisite of the enjoyment of the right to science. In fact, the ability to 'continuously engage in critical thinking about themselves and the world they inhabit, and ... the opportunity and wherewithal to interrogate, investigate and contribute new knowledge with ideas, expressions and innovative applications, regardless of frontiers' are a prerequisite for implementing both rights.[188] Third, it connects the right to science to the concept of 'human dignity' to the extent that the right protects people's 'ability to aspire – namely, to conceive of a better future that is not only desirable but attainable'.[189] Aspirations, the Special Rapporteur noted, 'embody people's conceptions of elements deemed essential for a life with dignity."[190] Fourth, it identifies links to other rights. In some cases, the right to science is enjoyed in conjunction with other rights, such as the right to seek information, to take part in the conduct of public affairs, to self-determination, to development and to make informed decisions.[191] The right to science is also a prerequisite for the realization of other rights, namely the right to food, health, water, housing, education and a clean and healthy environment.[192]

The second part of the Report focuses on the normative content and related obligations of states. In it, the Special Rapporteur proposes a list of objectives that states must guarantee: access by all without discrimination, freedom of scientific research and opportunities for all to contribute to the scientific enterprise, individual and collective participation in decision-making and an environment that enables knowledge production and exchange.[193]

---

[186] F Shaheed, 'Report of the Special Rapporteur in the Field of Cultural Rights: The right to enjoy the benefits of scientific progress and its applications' (Presented at the twentieth session of the Human Rights Council, 14 May 2012) A/HRC/20/26.
[187] Ibid., s III.b.1.
[188] Ibid., para 18.
[189] Ibid., para 20.
[190] Ibid.
[191] Ibid., paras 21–22.
[192] Ibid., para 23.
[193] Ibid., s III.B.

The third and last section of the Report discusses the limitations of the right to science. The Special Rapporteur pointed out that limitations certainly arise from the very same body of human rights law and thus it must promote the general welfare and be proportionate to the objective.[194] The regulation and protection of research subjects provides an example of a justifiable limitation of the right to science.[195] The prohibition against subjecting a person without his free consent to medical or scientific experimentation is especially important.[196] More controversially, the Rapporteur also cited the precautionary principle as an important guide for science and technology policies in the absence of scientific consensus, arguing that caution the avoidance of steps are required in case an action or policy might cause severe or irreversible harm to the public or the environment.[197]

Finally, the third major development took place in 2015. The Committee on Economic, Social and Cultural Rights, the expert body in charge of supervising implementation of the Covenant on Economic, Social and Cultural Rights, gave the mandate to two of its members, Mikel Mancisidor and Rodrigo Uprimny, to draft a 'general comment' on the right to science. The appointment of the two co-rapporteurs is a significant step because of the authority 'general comments' on human rights treaties have. Besides 'assisting the States parties in fulfilling their reporting obligations',[198] general comments are commonly considered to be the official interpretation of a right on the part of the United Nations.[199] The Committee is expected to adopt the general comment on the right to science by the end of 2019. While a draft is not yet available to the public, remarks made by the co-rapporteurs at various meetings have already provided sufficient insights on the direction the comment will take. The final text is expected to build upon all that had been established so far, starting from the Venice Statement, and solidify it into some clear guidance for states as to the contours of this right. We will further explore the normative content of the right to science and its implications for human germline modification in the conclusions of this book.

---

[194] Ibid., para 49.

[195] Ibid., para 51.

[196] ICCPR (n 17) art. 7: 'No one shall be subjected without his free consent to medical or scientific experimentation.'

[197] Report of the Special Rapporteur in the Field of Cultural Rights (n 186) para 50.

[198] Committee on Economic, Social and Cultural Rights, 'Report on the Third Session, Supplement No. 4' (1989) UN Doc E/1989/22, Annex III 'Introduction: the purpose of general comments'.

[199] L Grover and H Keller, 'General Comments of the Human Rights Committee and their Legitimacy' in L Grover, G Ulfstein and H Keller (eds.), *UN Human Rights Treaty Bodies: Law and Legitimacy* (Cambridge, UK, Cambridge University Press, 2012) 116–198.

3 *Limits to the Right to Science and the Rights of Science*

Few international human rights can never be derogated and have no exceptions (e.g. the right not to be arbitrarily deprived of one's life, the right not to be subject to torture, cruel, inhuman or degrading treatment, the right not to be enslaved). Most are qualified, admit exceptions, and/or can be suspended legally, for various reasons and in various ways. The 'right to science' and the 'rights of science' are some of those. What exceptions and derogations are admissible in the case of each right depend on the wording of the specific legal instrument in which they have been codified.

The Universal Declaration of Human Rights contains two main 'claw back' clauses. The first one is Article 29, which says:

> (2) In the exercise of his rights and freedoms, everyone shall be subject only to such limitations as are determined by law solely for the purpose of securing due recognition and respect for the rights and freedoms of others and of meeting the just requirements of morality, public order and the general welfare in a democratic society; (3) These rights and freedoms may in no case be exercised contrary to the purposes and principles of the United Nations.

The second is Article 30, which says: 'Nothing in this Declaration may be interpreted as implying for any State, group or person any right to engage in any activity or to perform any act aimed at the destruction of any of the rights and freedoms set forth herein.' These are very broad provisions that over the decades have been used to justify all kinds of measures restrictive of freedoms, but they have also been used to protect democracy and human rights from their own excesses.

When states started negotiating the two covenants, they realized more precise limits were needed. Echoing Article 30 of the Universal Declaration, Article 5 of both the ICESCR and the ICCPR recite: '1. Nothing in the present Covenant may be interpreted as implying for any State, group or person any right to engage in any activity or to perform any act aimed at the destruction of any of the rights or freedoms recognized herein, or at their limitation to a greater extent than is provided for in the present Covenant.' Both also add that no restriction upon or derogation from any of the fundamental human rights recognized or existing in any country by virtue of law, conventions, regulations or custom shall be admitted on the pretext that the covenants do not recognize those rights or that they recognize them to a lesser extent.[200]

---

[200]   ICESCR (n 18) art. 5.2; ICCPR (n 17) art. 5.2.

Besides these common provisions, the ICCPR and the ICESCR then take a different approach to acceptable limitations to rights, largely due to the different nature of the rights protected in each instrument. The ICCPR contains a list of rights that can never be derogated.[201] This includes the prohibition to subject a person to medical or scientific experimentation without consent.[202] The others can be suspended '[i]n time of public emergency which threatens the life of the nation and the existence of which is officially proclaimed',[203] or can be limited to 'protect public safety, order, health, or morals or the fundamental rights and freedoms of others'. These include the right to liberty of movement and freedom to choose residence and leave any country, including the own;[204] the right to a public trial;[205] the right to manifest one's religion or beliefs and the rights of parents to ensure the religious and moral education of their children in conformity with their own convictions;[206] the right of peaceful assembly;[207] the right to freedom of association;[208] and, of particular importance for scientists, the right to hold opinions without interference, to freedom of expression, including the freedom to seek, receive and impart information and ideas of all kinds, regardless of frontiers, through any media.[209]

The ICESCR steers clear of 'public safety, order, health, or morals' arguments to justify limitations. Instead, rights under this Covenant 'may be subject . . . only to such limitations as are determined by law' and 'only in so far as this may be compatible with the nature of these rights and solely for the purpose of promoting the general welfare in a democratic society'.[210] The distinction in the acceptable limitations to the rights protected in the two covenants is paramount because the 'right to science' and several of the 'rights of science' (e.g. the right to work, the right to protection of the moral and material interests, the duty states have to encourage scientific research and to facilitate cross-border cooperation and the like), the right to health and reproductive rights, which includes the right to choose when, how and how many children to have, are protected in the ICESCR, not the ICCPR. These rights can be limited only 'for the purpose of promoting the general

---

[201] ICCPR (n 17) art. 4.2.
[202] Ibid., art. 7.
[203] Ibid., art. 4.1.
[204] Ibid., art. 12.
[205] Ibid., art. 14.1.
[206] Ibid., art. 18.
[207] Ibid., art. 21.
[208] Ibid., art. 22.
[209] Ibid., art. 19.
[210] ICESCR (n 18) art. 4.

welfare in a democratic society', not for 'public safety, order, health, or morals'. Of course, one could argue that the two are the same. Certain limitations that, facially, are taken to protect 'public safety, order, health, or morals' might also be 'necessary to promote the general welfare in a democratic society', but the distinction is important and the burden of justifying restrictions to rights, to the international community and to their citizens is on governments. Restrictions based on morality need to have broad support to be consistent with the goal of the promotion of welfare in a democratic society. Restrictions based on health must be supported with an explanation of how the balancing between the right to health of the individual and the right to health of the many has been achieved. Public safety can be invoked only if there is a concrete risk, scientifically proven, not just a speculative one.

4 *Dignity*

The duty to protect 'human dignity' features prominently in the already discussed soft law instruments relevant to human genome modification which form part of bioethics law, above all, in the three UNESCO declarations.[211] In this context, dignity is invoked particularly when pointing to certain practices 'that may pose dangers to the integrity and dignity of the individual'[212] and to justify restrictions, moratoria or bans. For instance, the preamble of UNESCO's 1997 Universal Declaration on Human Genome and Human Rights refers to dignity by recalling it is mentioned in the preamble of the UNESCO Constitution and in the preamble of the Universal Declaration of Human Rights.[213] It also recognizes that 'research on the human genome and the resulting applications open up vast prospects for progress in improving the health of individuals and of humankind as a whole, but emphasizing that such research should fully respect human dignity'.[214] Then, it starts with a section entitled 'Human Dignity and the Human Genome', which proclaims: 'The human genome underlies the fundamental unity of all members of the human family, as well as the recognition of their inherent dignity and diversity',[215] and '[e]veryone has a right to respect for their dignity and for their rights regardless of their genetic characteristics'.[216]

---

[211]   See, in this chapter, Section II.1.
[212]   UN Commission on Human Rights, *Commission on Human Rights Resolution 2003/69: Human Rights and Bioethics* (E/CN.4/RES/2003/69, 25 April 2003) Preamble, para 4.
[213]   1997 Universal Declaration on Human Genome and Human Rights (n 31), first and fourth preambular paragraphs.
[214]   Ibid., six preambular paragraph.
[215]   Ibid., art. 1.
[216]   Ibid., art. 2.a.

Paragraph b of Article 2 of the same declaration provides more insight into the meaning of 'dignity': 'dignity makes it imperative not to reduce individuals to their genetic characteristics and to respect their uniqueness and diversity'. The same concept is in essence repeated in Article 6,[217] 10,[218] 11,[219] 12.a,[220] 15[221] and 21.[222] Notably, the 1997 Declaration seems to single out 'germline interventions' as an example of a possible violation of human dignity. Under Article 24, the International Bioethics Committee of UNESCO is asked to 'give advice concerning the follow-up of this Declaration, in particular regarding the identification of *practices that could be contrary to human dignity, such as germ-line interventions*'.[223]

Also the 2003 International Declaration on Human Genetic Data uses the expression 'human rights, fundamental freedoms or human dignity' repeatedly to indicate the object of the duty of protection that states have. 'Dignity' is found in the preamble, and it is referenced in the description of the aim,[224] prohibition of use of genetic data to infringe human dignity,[225] and prohibition of acts contrary to dignity.[226]

Likewise, the 2005 Universal Declaration on Bioethics and Human Rights refers to 'dignity' repeatedly. Besides the preamble, which echoes those of the two previous declarations, the stated aims of the declaration are, inter alia, '(c) to

---

[217] 'No one shall be subjected to discrimination based on genetic characteristics that is intended to infringe or has the effect of infringing human rights, fundamental freedoms and human dignity.' Ibid., art. 6.

[218] 'No research or research applications concerning the human genome, in particular in the fields of biology, genetics and medicine, should prevail over respect for the human rights, fundamental freedoms and human dignity of individuals or, where applicable, of groups of people.' Ibid., art. 10.

[219] 'Practices which are contrary to human dignity, such as reproductive cloning of human beings, shall not be permitted.' Ibid., art. 11.

[220] 'Benefits from advances in biology, genetics and medicine, concerning the human genome, shall be made available to all, with due regard for the dignity and human rights of each individual.' Ibid., art. 12.a.

[221] 'States should take appropriate steps to provide the framework for the free exercise of research on the human genome with due regard for the principles ... to safeguard respect for ... human dignity.' Ibid., art. 15.

[222] 'States should take appropriate measures to encourage other forms of research, training and information dissemination conducive to raising the awareness of society and all of its members of their responsibilities regarding the fundamental issues relating to the defence of human dignity.' Ibid., art. 21.

[223] Emphasis added.

[224] UNESCO, International Declaration on Human Genetic Data (n 63). Ensuring 'the respect of human dignity and protection of human rights and fundamental freedoms in the collection, processing, use and storage of human genetic data', art. 1.a.

[225] Ibid., art. 7.a.

[226] Ibid., art. 27.

promote respect for human dignity and protect human rights ... ; and (d) to recognize the importance of freedom of scientific research and the benefits derived from scientific and technological developments, while stressing the need for such research and developments ... to respect human dignity, human rights and fundamental freedoms'.[227] Article 3 declares at paragraph 1 that '[h]uman dignity, human rights and fundamental freedoms are to be fully respected', and paragraph 2 contains the corollary of the principle of human dignity: people should not simply become instruments for the benefit of science, because science is not an absolute but only a means at the service of the human person. The duty to respect human dignity is echoed in Article 10,[228] as well as in the prohibitions of discrimination,[229] and of acts contrary to human rights, fundamental freedoms and human dignity.[230] Finally, although the declaration restates the importance of giving due regard to the importance of 'cultural diversity and pluralism', those cannot be invoked 'to infringe upon human dignity, human rights and fundamental freedoms'.[231]

The Council of Europe included the word 'dignity' in the full title of the Oviedo Convention: 'Convention for the Protection of Human Rights and Dignity of the Human Being with regard to the Application of Biology and Medicine'.[232] The Oviedo Convention envisages a very broad protection of human dignity. Article 1 postulates generally that 'parties ... shall protect the dignity and identity of all human beings'. Dignity is also mentioned in the preamble, where it says: 'the misuse of biology and medicine may lead to acts endangering human dignity'.

Dignity is also invoked in general international instruments.[233] The concept was first included in the preamble to the Charter of the United Nations, in 1945.[234] The second paragraph 'reaffirms faith ... in the dignity and worth of the human person'. This reference to dignity was taken up in the 'International

---

[227]   UNESCO, Universal Declaration on Bioethics and Human Rights (n 78) art. 2.
[228]   'The fundamental equality of all human beings in dignity and rights is to be respected so that they are treated justly and equitably.' Ibid., art. 10.
[229]   Ibid., art. 11.
[230]   Ibid., art. 28.
[231]   Ibid., art. 12.
[232]   Convention for the Protection of Human Rights and Dignity of the Human Being with Regard to the Application of Biology and Medicine: Convention on Human Rights and Biomedicine (opened for signatures on 4 April 1997, entered into force 12 January 1999) ETS No. 164 (Oviedo Convention). On the Oviedo Convention, see, in this book, Part 3, Section II, Chapter 6.
[233]   See, in general, D Kretzmer and E Klein (eds.), *The Concept of Human Dignity in Human Rights Discourse* (Kluwer, The Hague, 2002).
[234]   United Nations, Charter of the United Nations (entered into force 24 October 1945) 1 UNTS XVI.

Bill of Rights'. The preamble and Article 1 of the Universal Declaration of Human Rights proclaim: '[a]ll human beings are born free and equal in dignity and rights'.[235] Dignity is further mentioned in Articles 22 and 23, which both proclaim certain social rights. The preambles to the ICCPR[236] and ICESCR[237] affirm that the human rights guaranteed by the covenants 'derive from the inherent dignity of the human person'. The United Nations Millennium Declaration of the United Nations General Assembly mentions the principle of dignity several times.[238] In particular, human dignity is recognized as a core value of the UN system.[239]

Although, all these references to human dignity are not immediately legally binding, as they are either part of the non-binding preamble to a generally binding treaty or they are expressed in a resolution of the UN General Assembly, which has no directly binding effect, nearly all major general human rights instruments mention human dignity in their operative part, for example, how persons deprived of liberty should be treated,[240] the aims of the right to education,[241] what is required by the right to privacy,[242] the limits to forced labour as a penalty[243] and the motive for prohibiting exploitation and degrading treatment.[244] The only exception is the European Convention for the Protection of Human Rights and Fundamental Freedoms, where 'dignity' is nowhere to be found.[245] However, this gap has been filled by the European Court of Human Rights, which has interpreted certain provisions of the Convention in the light of the concept of human dignity.[246] Moreover, the concept was elevated to

---

[235]   UDHR (n 11).
[236]   ICCPR (n 17).
[237]   ICESCR (n 18).
[238]   United Nations Millennium Declaration (published 18 September 2000, UNGA Res 55/2) paras I.2, I.6, and VI.
[239]   Ibid.
[240]   E.g. ICCPR (n 17) art. 10.1; American Convention on Human Rights (adopted 21 November 1969, entered into force 18 July 1978) 1144 UNTS 123, art. 5.2.
[241]   ICESCR (n 18) art. 13.1.
[242]   American Convention on Human Rights (n 240) art. 11.1.
[243]   Ibid., art. 6.2.
[244]   African [Banjul] Charter on Human and Peoples' Rights (adopted 27 June 1981, entered into force 21 October 1986) 1520 UNTS 217, art. 5.
[245]   Convention for the Protection of Human Rights and Fundamental Freedoms, ETS No. 5 (European Convention on Human Rights, as amended) (ECHR). On the ECHR, see, in this book Part 3, Section II, Chapter 6. See, in general, B Maurer and others, *Le principe de respect de la dignité humaine et la convention européenne des droits de l'homme* (Documentation Française, Science et technique de la démocratie, 1999).
[246]   J-P Costa, 'Human Dignity in the Jurisprudence of the European Court of Human Rights' in C McCrudden (ed.), *Understanding Human Dignity* (Oxford University Press, 2014) 393–402; A Oehling de los Reyes, 'The Human Dignity Concept in the Case Law of the European Court of Human Rights' in L Gordillo (ed.), *Constitutionalism of European*

keystone of the whole European human rights edifice by the Charter of Fundamental Rights of the European Union, which solemnly starts by saying: 'Human dignity is inviolable. It must be respected and protected.'[247]

All in all, the term 'dignity' seems to be used in general human rights instruments to encapsulate two different concepts.[248] On the one hand, it is used in a 'formal sense', justifying the existence of human rights and decoupling them, not formally, but philosophically, from the requirement of state consent. Individuals have human rights because they have dignity that is inherent to them, not because states have entered into treaties recognizing them.[249] On the other hand, it is also used in a substantive sense as a legal guarantee assuring the respect of human beings and protecting them against humiliation and degradation.[250] It also plays a 'founding function' for international human rights law,[251] demonstrated by several preambles to international human rights instruments, such as the one of the UDHR, which invokes the 'recognition of inherent dignity . . . the foundation of freedom, justice and peace in the world'.[252]

Supranational Courts: Recent Developments and Challenges (Thomson Reuters, 2015) 21–32. For a view on how international courts other than the European Court have interpreted dignity, see, C McCrudden, 'Human Dignity and Judicial Interpretation of Human Rights' (2008) 19:4 *European Journal of International Law* 655, 655–724; P Carozza, 'Human Dignity and Judicial Interpretation of Human Rights: A Reply' (2008) 19:4 *European Journal of International Law* 931, 931–944.

[247] Charter of the Fundamental Rights of the European Union [2000] OJ C 364/1, art. 1. On the Charter of Fundamental Rights of the EU, see, in this volume, Part 3, Section II, Chapter 6.

[248] N Petersen, 'Human Dignity, International Protection' (*Max Planck Encyclopedia of Public International Law*, last updated October 2012), paras 19 and 37; JA Frowein, 'Human Dignity in International Law' in D Kretzmer and E Klein (eds.) *The Concept of Human Dignity in Human Rights Discourse* (Kluwer, The Hague, 2002) 121–132.

[249] 'This is particularly the case for art. 5 of the African Charter on Human and Peoples' Rights, but also for the preamble of Protocol No 13 to the European Convention on Human Rights concerning the Abolition of the Death Penalty and the jurisprudence of the European Court of Human Rights.' Petersen (n 248), para 19.

[250] 'On the other hand, it is also used in a substantive sense. This is particularly obvious for art. 11.1 of the American Convention on Human Rights. In this context, dignity shall guarantee the respect of the human being and protect him against humiliation and degradation.' ibid. On the foundational role played by dignity, see, also, P Capps, *Human Dignity and the Foundations of International Law* (Hart, Oxford, 2009); A Gewirth, 'Human Dignity as the Basis of Rights' in MJ Meyer and WA Parent (eds.) *The Constitution of Rights: Human Dignity and American Values* (Cornell University Press, Ithaca, 1992) 10–28; O Schachter, 'Human Dignity as a Normative Concept' (1983) 77 *American Journal of International Law* 848, 848–854.

[251] K Dicke, 'The Founding Function of Human Dignity in the Universal Declaration of Human Rights' in D Kretzmer and E Klein (eds.), *The Concept of Human Dignity in Human Rights Discourse* (Kluwer, The Hague, 2002) 111–120.

[252] UDHR (n 11), first preambular paragraph.

International bioethical law instruments seem to emphasize the importance of human dignity in a more powerful way than traditional human
rights law.[253] Indeed, in contrast to the background role assigned to human
dignity in international human rights instruments, international bioethical
law puts it at the foreground as the ultimate rationale for the norms relating
to this discipline.[254] In this context, dignity is not understood as an independent legal guarantee. It is rather an argument justifying the elaboration of
special regulations, restrictive measures and, indeed, bans, in the field of
bioethics.[255]

The use of the 'dignity' in relation to reproductive human cloning is
a perfect illustration. The 1998 World Health Organization Resolution on
Ethical, Scientific and Social Implications of Cloning in Human Health
recalls in its preamble 'its condemnation of human cloning for reproductive
purposes as contrary to human dignity'.[256] State representatives referred to
human dignity during the negotiations on the International Convention
against the Reproductive Cloning of Human Beings to justify a prohibition
of cloning.[257] However, they could not reach any consensus on the forms of
cloning that were to be included in the ban. The compromise formula found
for the UN General Assembly Resolution 59/280 on Human Cloning is the
prohibition of 'all forms of human cloning *inasmuch as they are incompatible
with human dignity*'.[258]

Some go as far as arguing that 'dignity' is part of customary international law or
constitutes a 'general principle of international law'.[259] Others argue that it is the
legal foundation of international human rights law.[260] Yet, some dismiss it as

[253] R Andorno, 'Global Bioethics at UNESCO: UNESCO' (n 83),152–153.
[254] R Andorno, 'La notion de dignité humaine est-elle superflue en bioéthique?' (2005) 16 *Revue
       générale de droit médical* 95, 95–102; J-S Gordon, 'Human Dignity, Human Rights, and
       Global Bioethics' in W Teays, J-S Gordon, A Dundes Renteln (eds.) *Global Bioethics and
       Human Rights: Contemporary Issues* (Rowman & Littlefield, 2014) 68–91; D Beyleveld and
       R Brownsword, *Human Dignity in Bioethics and Biolaw* (Oxford, Oxford University Press,
       2002); C Foster, *Human Dignity in Bioethics and Law* (Hart, Oxford, 2011).
[255] Petersen, 'Human Dignity, International Protection' (n 248), para 23.
[256] World Health Organization, *Resolution on the Ethical, Scientific and Social Implications of
       Cloning in Human Health* (WHA51/1998/REC/1) Preamble, para 1.
[257] United Nations General Assembly, *Press Release: Ethical Issues Stressed as Legal Committee
       Continues – Debate on two Draft Texts on Human Cloning* (UN Doc. GA/L/3258, 22 October
       2004).
[258] UN General Assembly Resolution 59/280 on Human Cloning (A/RES/59/280, 8 March 2005)
       para b. Emphasis added.
[259] Petersen, 'Human Dignity, International Protection' (n 248), para 6.
[260] J Waldron, 'Is Dignity the Foundation of Human Rights?' (Public Law & Legal Theory
       Research Paper Series Working Paper No 12–73, January 2013).

a useless notion;[261] one that does not have an autonomous value since, in international human rights law, the term is always used in combination with other guarantees and prohibitions.[262] As a matter of fact, its importance notwithstanding, no legal instrument contains an effective general guarantee of human dignity and none defines it. Not even the German Constitution, which enshrines the principle that '[t]he dignity of man is inviolable' in Article 1.1, one of the few articles that cannot be changed, defines it.[263] As it has been said, it seems that it is this very nature of the concept of dignity 'that has allowed, on the one hand, human rights to receive such international acceptance as a theoretical enterprise and, on the other hand, has led the concept to be constantly challenged by different cultures worldwide'.[264] Indeed, despite the centrality of 'dignity' to international law and public discourse, there does not seem to be yet an equally universal understanding of its meaning.

Defining 'dignity' is an enduring problem of philosophy. The Roman philosopher Cicero held *dignitas* (dignity) to be the distinctive characteristic of humans compared to animals.[265] However, dignity was not common to all human beings nor inherent, but rather a feature that could be gained or lost and that was influenced by social rank and authority. Medieval Christian theology held humans had inherent dignity because they were a creature of, and the image of, God (*imago dei*).[266] That is still part of the official doctrine of the Catholic Church.[267] During Renaissance, philosophy tried to free dignity

---

[261] E.g. R Macklin, 'Dignity is a Useless Concept: It means no more than respect for persons or their autonomy' (2003) 327:7429 *British Medical Journal* 1419, 1419–1420.

[262] On the uses and abuses of the concept of dignity in law, see, in general: E Hilgendorf, 'The Abuse of Human Dignity – Difficulties in Using the Human Dignity Topos Taking the Bio-Ethics Debate as an Example' in E Hilgendorf and M Kremnitzer (eds.), *Human Dignity and Criminal Law* (Würzburg Conference on Human Dignity, Human Rights and Criminal Law in Israel and Germany, 20–22 July 2015, Duncker & Humblot, 2018) 39–60; J Weinrib, 'Human Dignity and its Critics' in G Jacobsohn and M Schor (eds.), *Comparative Constitutional Theory* (Edward Elgar Pbl, 2018) 167–186; R Andorno, 'The Paradoxical Notion of Human Dignity' (2001) 78 *Rivista Internazionale di Filosofia del Diritto* 151–168.

[263] 'Die Würde des Menschen ist unantastbar. Sie zu achten und zu schützen ist Verpflichtung aller staatlichen Gewalt.' (Human dignity shall be inviolable. To respect and protect it shall be the duty of all state authority.) Grundgesetz für die Bundesrepublik Deutschland 1949.

[264] PA Rodriguez, 'Human Dignity as an Essentially Contested Concept' (2015) 23:2 *Cambridge Review of International Affairs* 3.

[265] M T Cicero, 'De Officiis' (Book I.XXX, Walter Miller trans, Heinemann 1928) paras 5–9, 106–113.

[266] S Moyn, *Christian Human Rights* (University of Pennsylvania Press, 2015); T Lowenthal, 'The Role of Dignity in Human Rights Theory: Constituent or Teleological' (2015) 18 *Trinity College Law Review* 56, 56–83.

[267] Vatican, *Catechism of the Catholic Church*, Part Three (Life in Christ), Section One (Man's Vocation Life in the Spirit), Chapter One (The Dignity of the Human Person), art. 1 (Man: The Image of God).

from these metaphysical restrictions. In the Oration on the Dignity of Man (*Oratio de hominis dignitate*), Giovanni Pico della Mirandola explained that it is the characteristic of human dignity that every human being may decide freely about their way of living.[268] Immanuel Kant further elaborated the idea of 'dignity as autonomy' in the Metaphysics of Morals (*Die Metaphysik der Sitten*), attributing dignity to persons who are capable of reasoning and thus capable of autonomy and morality.[269]

In more recent times, Alan Gewirth, while sharing Kant's view that rights arise from dignity, focuses on the positive obligations that dignity imposes on humans, such as the moral requirement not only to avoid harming others but to actively assist one another in achieving and maintaining a state of well-being.[270] For Jeremy Waldron, the concept of dignity is both a principle of morality and a principle of law. Drawing from the insights about dignity primarily from law, he explains that the use of *human* dignity in constitutional and human rights law 'can be understood as the attribution of a high legal rank or status to every human being'.[271]

5 A Synthesis of Human Rights Principles Applicable to the Scientific Enterprise

As we discussed in Section II, there is not yet a binding international legal instrument dedicated to the regulation of human germline modification from bench to bedside. A reason for that is that there is no international consensus as to what a global regulatory framework for human germline engineering should look like. However, the discovery of CRISPR has spurred a new global

---

[268]   G Pico della Mirandola, *Oratio de hominis dignitate*, para 20, 117 (Francesco Borghesi and others (eds.), Cambridge University Press, 2012, 1486); C McCrudden, 'Human Dignity and Judicial Interpretation of Human Rights' (2008) 19 *European Journal of International Law* 655, 659.

[269]   I Kant, *Metaphysics of Morals*, The Doctrine of Virtue (Mary Gregor trans, Cambridge University Press, 1991, 1797) ss 11 and 38, 230–231, 255; C McCrudden, 'Human Dignity and Judicial Interpretation of Human Rights' (n 268), 659–660.

[270]   A Gewirth, *Self-Fulfillment* (1998) 85, 127, and 180; A Gewirth, *The Community of Rights* (1996) 31–44; A Gewirth, *Reason and Morality* (1978) 134–136, 209–210. On Gewirth's view of dignity, see, D Beyleveld, 'Human Dignity and Human Rights in Alan Gewirth's Moral Philosophy' in M Düwell and others (eds.), *The Cambridge Handbook of Human Dignity: Interdisciplinary Perspectives* (Cambridge University Press, 2014) 230–239.

[271]   J Waldron, 'Dignity, Rank, and Rights: The Tanner Lectures on Human Values' (Berkeley, University of California, 21–23 April 2009) 250.

conversation on this issue, as evidenced by the numerous ethics statements that have been published since 2015. This conversation could result in a consensus for a more solid global regulatory framework to emerge. While penning a shared framework for human germline engineering requires time, '[f]ortunately, many of the institutions needed to regulate [germline engineering already] exist'.[272] Among these 'institutions' there is the corpus of international human rights law that has been developed since the end of the Second World War. We believe that international human rights law must to be at the centre of the global conversation towards the development of a shared framework for human germline modification; the whole of it, not just some select pieces and bits. International bioethics instruments, as developed mainly by UNESCO, provide a narrow and inadequate account of the range of human rights that must be taken into account in this global conversation. While that branch of international law gives attention to some human rights, such as the right to free and informed consent or the right not to be discriminated against, it neglects many others, including the right to health, reproductive rights and the so-called right to science and the rights of science. Moreover, while these international human rights are contained in treaties that states parties are legally bound to respect, protect and fulfil, international bioethics law is almost entirely made of soft law legal instruments, whose legal significance is considerably less clear.

We believe the 'right to science' and the 'rights of science' are critically important as they protect the basic interests not only of scientists to continue advancing scientific research in this field, but also the interests of beneficiaries of the scientific progress made, including its applications, not least to prevent and cure genetic diseases. A more thorough analysis of how these rights come into play and can contribute to the emerging international regulatory framework will be presented in the conclusions of the book.[273] Here, we want to highlight five key principles that we see as emergent from the integration of international bioethics law with international human rights law, particularly the 'right to science' and the 'rights of science', and foundational to an international framework for human germline engineering. These principles are: freedom of research; benefit sharing; solidarity; respect for dignity; and the obligation to respect and to protect the rights and individual freedoms of others. It is important to note that, in respecting these principles, states must provide an enabling environment (i.e. laws, regulations, funding) 'necessary

---

[272] N H Evitt, S Mascharak and R B Altman, 'Human Germline CRISPR-Cas Modification: Toward a Regulatory Framework' (2015) 15 *American Journal of Bioethics* 25–29.
[273] Part 4, Chapter 22.

for the conservation, the development and the diffusion of science' and 'recognize the benefits of international contacts and co-operation in the scientific field'.[274] Also, the three UNESCO declarations reference international cooperation in various articles, often stressing the need to involve developing countries.[275]

I FREEDOM OF RESEARCH The first principle that international instruments recognize is *freedom of research*. The ICESCR requires states parties to 'respect the freedom indispensable for scientific research'.[276] The three UNESCO declarations also call for respect of the freedom of research. The 1997 Universal Declaration on Human Genome and Human Rights states that freedom of research 'is necessary for the progress of knowledge' and 'is part of freedom of thought'.[277] The 2003 International Declaration on Human Genetic Data states that due consideration must be given to 'freedom of thought and expression, including freedom of research'.[278] The 2005 Declaration recognizes 'freedom of science and research' as the basis of 'scientific and technological developments'[279] and 'the importance of freedom of scientific research and the benefits derived from scientific and technological developments'.[280] The 2015 report of the UNESCO International Bioethics Committee, entitled 'Updating Its Reflection on the Human Genome and Human Rights', states that 'freedom of research and freedom of individuals should not be inhibited by too many strict regulations' and that 'basic research ... is not possible without the freedom of researchers'.[281]

Freedom of research is both individual and collective, negative and positive. Individually, freedom entails the right of 'everyone' to participate in the scientific enterprise. The pronoun 'everyone' includes scientists, tissue

---

[274] ICESCR (n 18) art. 15.4.
[275] The 2005 UNESCO Universal Declaration on Bioethics and Human Rights (n 78) requires states to 'foster international dissemination of scientific information and encourage the free flow and sharing of scientific and technological knowledge, [...] promote cultural and scientific cooperation and enter into bilateral and multilateral agreements enabling developing countries to build up their capacity to participate in generating and sharing scientific knowledge, the related know-how and the benefits thereof'. Art. 3.1.
[276] ICESCR (n 18) art. 15.3.
[277] 1997 UNESCO Universal Declaration on Human Genome and Human Rights (n 31) art. 12(b).
[278] 2003 UNESCO International Declaration on Human Genetic Data (n 63) art. 1.
[279] 2005 UNESCO Universal Declaration on Bioethics and Human Rights (n 78) Preamble.
[280] Ibid., art. 2.
[281] UNESCO, International Bioethics Committee, Updating Its Reflections (n 109) paras 10 and 29.

donors and patients. This right must be protected along with other freedoms that scientists enjoy, including the right to intellectual property, to participate in learned societies and travel, and to academic freedom. Collectively, it is the right of scientists to govern the scientific enterprise, the right to 'self-regulation', as well as the right to an environment that enables 'the conservation, the development, and the diffusion of science and culture' (i.e. right to policies that support science, to research funding and infrastructure), as codified in Article 15.2 of the ICESCR.

Self-regulation, which comprises customs, principles, norms and institutions that scientists have developed over centuries as a means to regulate their work, is paramount for two reasons. First, it is constitutive of the scientific enterprise. Scientists have set the epistemological parameters of what constitutes science and what knowledge is scientific. Any individual who claims to 'do science' must comply with the norms and institutions of the scientific enterprise. Scientific self-regulation sets the normative parameters of what constitutes science. Second, self-regulation ensures that science is carried out responsibly.[282] Self-regulation includes adherence to the scientific method, timely communication and publication, refinement of results through replication and extension of the original work, peer review, data sharing, authorship, and training and supervision of associates and students. These norms apply by virtue of membership in the scientific community and require maintaining the integrity of the research process. Science cannot work without them. Ultimately, self-regulation is integral to freedom of research, if this is understood as a negative freedom that demands that governments not interfere with the internal workings of science.

II BENEFIT SHARING The second principle is *benefit sharing*. It is codified in the ICESCR, which requires states parties to 'recognize the right of everyone to enjoy the benefits of scientific progress and its applications'.[283] Benefit sharing must be understood as the right both to the creation of benefits and to access the benefits. In the first sense, the right of 'everyone to enjoy the benefits of scientific progress and its applications' means that governments have an obligation to use scientific knowledge in ways that are beneficial to everyone. In particular, there is a duty to transform basic or foundational knowledge into applications, whenever possible and unless there are legally

---

[282] See, in general, Institute of Medicine, National Academy of Sciences and National Academy of Engineering, *Responsible Science, Volume I: Ensuring the Integrity of the Research Process* (The National Academies Press 1992).

[283] ICESCR (n 18) art. 15.1.

valid reasons not to do so. The 2003 UNESCO Declaration embraces this concept of 'benefit sharing' when it provides that governments must ensure 'access to medical care, provision of new diagnostics, facilities for new treatments or drugs stemming from the research', and 'support for health services'.[284] In the second sense, 'benefit sharing' means that benefits must be enjoyed by everyone, without any discrimination. The UNESCO 1997 Declaration captures this idea by stating that 'benefits from advances in biology, genetics and medicine, concerning the human genome, shall be made available to all'. Like freedom of research, benefit sharing is both a negative and a positive right. Governments must not prohibit the translation of scientific knowledge into applications (negative) and must take steps to ensure that scientific knowledge is translated into applications (positive).[285]

III SOLIDARITY The third principle is *solidarity*. While featuring prominently in the UNESCO declarations, this principle is not expressly mentioned in the ICESCR.[286] This is mostly likely due to historical reasons. Solidarity was not part of the human rights jargon when the ICESCR was adopted. Its first use is credited to Karel Vasak, who used the term 'solidarity rights' in an article written for the UNESCO Courier in 1977 to name the so-called third-generation human rights, which include the rights to development, to peace, to a healthy environment, to share in the exploitation of the common heritage of mankind, to communication and humanitarian assistance.[287] Even today the principle of solidarity remains ill-defined. Often, solidarity is linked to social justice. Discussing it in the context of germline engineering, Debra Mathews reads solidarity as a minimal obligation not to contribute to inequality and social division.[288] Solidarity conveys the moral responsibility that arises from shared human vulnerabilities and entails contributing to institutions that protect vulnerable individuals. This meaning of solidarity finds support in the ICESCR and the principle of non-discrimination. Under Article 2, the states parties 'undertake to guarantee that the rights enunciated in the present Covenant will be exercised without discrimination of any kind as to race, colour, sex, language, religion, political or other opinion, national or social origin, property, birth or other status'. Solidarity may also convey the need to spread risks and benefits evenly. This means inclusion of vulnerable

---

[284] 2003 UNESCO International Declaration on Human Genetic Data (n 63) art. 19.
[285] ICESCR (n 18) art. 15.2.
[286] However, the ICESCR contains several obligations regarding international cooperation. E.g., art. 15.4, on scientific cooperation.
[287] K Vasak, 'A 30-year struggle' (UNESCO Courier, November 1977) 29.
[288] D J H Mathews, 'Solidarity in the Age of CRISPR' (2018) 1 *The CRISPR Journal* 261.

populations. It can be also mean 'intergenerational equity', which demands that states take into account the rights of future generations when undertaking activities that may affect them. Mulvihill and his colleagues of the International Human Genome Organization Committee of Ethics, Law, and Society take solidarity in a different direction, closer to benefit sharing. According to these authors, the principle of solidarity requires to 'recognize the opportunities to share benefits as a public good' and the need for 'a continued broad debate, including issues of benefit sharing versus private commercialisation'.[289] In this sense, the normative dimensions of solidarity and benefit sharing overlap.

IV RESPECT FOR HUMAN DIGNITY The fourth principle is *respect for human dignity*. As we saw, reference is included in the preamble of the ICESCR and repeated several times in the UNESCO declarations.[290] However, as it was explained, 'human dignity' is a contested concept, one over which there has never been and will probably never be a consensus.[291] Its application to germline modifications depends on which view of 'dignity' one espouses. If one follows Christian theology, modifying the human genome can be construed as a departure from the creator's blueprint, the *imago dei*, and, therefore, an assault on dignity.[292] This view emphasizes the absolute worth of each human life (starting from conception), the sanctity of life. It also upholds the absolute character of the prohibition on scientific exploitation of human embryos, sometimes expressed in terms of the need to protect 'human dignity'.[293] This position implies not only that modifying the human genome can be an assault on dignity; it is also associated with the position that scientists must not interfere in the process from conception to birth of a new human being regardless of the purpose for such interference.

However, if the meaning of dignity is limited to *human* dignity and not extended to early stages of human life, and is understood as personal autonomy, as Pico della Mirandola and Kant, and their followers to the present day did, or if 'dignity' is understood not only as a right but also as a duty, one to assist others in achieving and maintaining a state of well-being, then one could argue that human germline genome modification, if used to cure genetic diseases, actually fulfils human dignity.

---

[289] J Mulvihill and others, 'Ethical Issues of CRISPR Technology and Gene Editing Through the Lens of Solidarity' (2017) 122:1 *British Medical Bulletin* 17, 17–29.

[290] See, in this chapter, Section III.4.

[291] See, in this chapter, Section III.4

[292] Ibid. (n 266).

[293] Waldron, 'Dignity, Rank and Rights' (n 271) 222–224.

In international law, neither view prevails over the other. There is no indication that a majority of states espouses either position. Construing 'human dignity' by embracing only one of the two views would not be an accurate reflection of the law as it is (*de lege lata*).

V OBLIGATION TO RESPECT AND TO PROTECT THE RIGHTS AND INDIVIDUAL FREEDOMS OF OTHERS The fifth principle is that everyone has an *obligation to respect and to protect the rights and individual freedoms of others*. This principle is codified in the ICESCR, whose Article 5 provides: 'Nothing in the present Covenant may be interpreted as implying for any State, group or person any right to engage in any activity or to perform any act aimed at the destruction of any of the rights or freedoms recognized herein, or at their limitation to a greater extent than is provided for in the present Covenant.' As we saw earlier, the same is also found in the Universal Declaration of Human Rights and the Covenant on Civil and Political Rights.[294] It is one of the core principles of international human rights law, together with the prohibition of discrimination. Thus, unsurprisingly, the principle is found also in all three UNESCO declarations. For instance, the 1997 Declaration provides: 'No research or research applications concerning the human genome, in particular in the fields of biology, genetics and medicine, should prevail over respect for the human rights, fundamental freedoms and human dignity of individuals or, where applicable, of groups of people.'[295]

The rules protecting human subjects in research certainly fall with this category. They include a preliminary assessment of the risks and benefits before research can begin, research oversight to ensure that rules are followed and that criteria are satisfied, and the requirement of obtaining the free and adequate informed consent of any human being participating in research. Research oversight is usually set up at the institutional level, with peers and administrators approving and overseeing research, or at the national level, with institutional regulators assessing safety and benefit of research and therapies. In some jurisdictions, both levels are present. Clearly, research oversight must always be consistent with the principle of freedom of research.

VI OTHER LIMITATIONS The rights recognized by the ICESCR are not absolute. As we discussed, the Covenant contemplates limitations on them.[296] According to Article 4, limitations are compatible with the

---

[294] See, in this chapter, Section III.3.
[295] 1997 UNESCO Universal Declaration on Human Genome and Human Rights (n 31) art. 10.
[296] See, in this chapter, Section III.3.

Covenant only when 'determined by law' and 'for the purpose of promoting the general welfare in a democratic society'.[297] In the context of the right to science, Article 4 balances freedom of research and benefit sharing, on the one hand, and permissible limitations, on the other. If limitations cannot gut the core content of the Covenant's rights, then they must be reconcilable with freedom of research and benefit sharing. By restricting the purposes for which limitations may be imposed and the manner in which that may be done, the ICESCR ensures that the rights are still protected. Moreover, the burden of proof that freedom of research and benefit sharing must be limited, and the extent of these limitations, is on governments. It is not the duty of scientists to show what society could gain from research: as a default, they enjoy freedom of research.

IV CONCLUSIONS

In this chapter, we reviewed the key elements of the larger international and transnational framework within which the national legal regimes of human germline genome modifications exist. The pillars of this framework are the Universal Declaration of Human Rights and the two Covenants, which collectively form the so-called International Bill of Rights. The most important provisions of the 'International Bill of Rights' that are relevant to human germline genome modification are the so-called right to science and the rights of science. In addition to the provisions of the 'International Bill of Rights', the framework includes non-binding soft law instruments, adopted by international and intergovernmental organizations, and 'statements', 'views', 'recommendations' and policy papers issued by non-governmental organizations. Among these, the three declarations adopted by UNESCO are paramount in articulating some of the principles of the emerging international framework of regulating human germline engineering. The UNESCO declarations and other non-binding soft law instruments form a body of law called 'international bioethics law'. We argue that international bioethics law and its instruments provide a narrow and inadequate description of the range of human rights that must be taken into account in the conversation on the regulation of germline engineering. These instruments must be integrated with the broader international human rights law corpus. When the two bodies of law are integrated, five key principles emerge as foundations of the emerging regulatory framework:

---

[297]   ICESCR (n 18) art. 4.

freedom of research; benefit sharing; solidarity; respect for dignity; and the obligation to respect and to protect the rights and individual freedoms of others. We will come back to discussing the emerging international framework with its core five principles in the final chapter of this book, after Part III where our colleagues will discuss the regulation of human germline genome modification in eighteen chosen states. There, in the final chapter, we will present the basic features of a human rights framework for human germline engineering.

THE REGULATION OF GENOME MODIFICATION AT THE NATIONAL LEVEL

PART I

NORTH AMERICA

3

The Regulation of Human Germline Genome Modification in Canada

Erika Kleiderman

I INTRODUCTION

Being a former British colony, Canada's parliamentary system stems from the British tradition. Canada is a federation and a parliamentary democracy; thereby the Parliament of Canada is made up of three distinct elements: the Crown, the Senate, and the House of Commons.[1] As a federation, the enactment of laws and associated responsibility is distributed across the federal, provincial (ten provinces), and territorial (three territories) governments.[2]

Canada is also a constitutional monarchy in that the Queen acts as the political head of state (i.e., all acts of the government are carried out in her name).[3] The Canadian government is comprised of three branches: (1) the executive branch, (2) the legislative branch, and (3) the judicial branch. The executive branch includes the Monarch, the Prime Minister, and the Cabinet, and is responsible for implementing laws and budget appropriations. The legislative branch includes the Parliament of Canada (i.e., House of Commons and Senate) and is responsible for adopting laws. The judicial branch is made up of judges and the courts, and is responsible for interpreting and applying the law. The Supreme Court of Canada (established in 1875) is Canada's final court of appeal.[4]

For health-related matters, the Constitution Act, 1867, imposes a shared jurisdiction whereby both levels of government are able to enact legislation

---

[1]   Constitution Act 1867 (UK), c 3, s 17.
[2]   R Marleau and C Montpetit (eds.), *House of Commons Procedure and Practice* (House of Commons, January 2000).
[3]   Ibid.
[4]   Office of the United Nations High Commissioner for Human Rights (OHCHR) "Core Document Forming Part of the Reports of States Parties – Canada" (January 12, 1998) UN Doc E/HR1/CORE/1/Add.91, para 80 (Core Document Forming Part of the Reports of States Parties).

pertaining to the subject. According to the division of powers (Sections 91 and
92 of the Constitution), the federal authority aims to reduce health risks arising
from a variety of products.[5] The provincial authority oversees the governance
and delivery of healthcare, including hospital services and medical professions
in the province.[6] As well, the federal level of government provides the provinces with funds to enable their healthcare insurance programs and to
support research and health programming.[7] Finally, it is of note that with
the increase in use of biological materials for therapeutic purposes (e.g., cell
and gene therapies), the distinction between products and services, as well as
provincial and federal jurisdictions, breaks down.[8] This is something that must
be kept in mind as technology evolves.

At present, Canada's approach to human germline modification remains
generally conservative, at times unclear, and largely prohibitive from both
a research and clinical perspective.[9] To date, no research using human germline modification has been undertaken in Canada. This chapter provides an
overview of the Canadian legal and policy landscape surrounding human
germline modification. It will begin by laying the groundwork for the subsequent assessment of specific legal provisions and policies governing the modification of the human germline from basic research to clinical applications.
Finally, it will reflect on the challenges and future possibilities for human
germline modification in Canada.

II THE REGULATORY ENVIRONMENT

1 *Constitution*

No provisions within the Constitution of Canada (Constitution Act, 1867)
specifically refer to research or clinical applications involving human germline genome modification. However, Canada's Constitution, and in particular
Sections 91 and 92, lays out the division of powers between the provinces and
the federal government. The Parliament of Canada, the federal legislature,
has the exclusive right to enact and oversee criminal law.[10] As we will see, it is
currently through federal legislative power by way of criminal law that assisted

---

5   T Lemmens and others (eds.), *Regulating Creation: The Law, Ethics, and Policy of Assisted
    Human Reproduction* (University of Toronto Press, 2017).
6   Ibid.
7   Ibid.
8   Ibid.
9   R Isasi et al., "Editing Policy to Fit the Genome?" (2016) 351:6271 *Science* 337.
10 Constitution Act 1867 (UK), c 3, s 91.27.

human reproduction is governed in Canada. Generally, the provinces have jurisdiction over healthcare and social services, including "the establishment, maintenance, and management of hospitals";[11] "property and civil rights in the province";[12] and "all matters of a merely local or private nature in the province."[13]

With regard to the acceptance and implementation of international human rights treaties in Canada, all three branches (executive, legislative, and judicial) share responsibility.[14] The federal government of Canada, typically following consultations with the provinces and territories, has the authority to ratify and become party to treaties on behalf of Canada.[15] Treaties in force do not automatically acquire force of law in the country.[16] To give them effect in domestic law, incorporation by legislation is required.[17] However, courts have called upon international treaties to which Canada is a party, but has not incorporated, for the purpose of interpreting and applying existing legislation,[18] which has occurred notably in the case of the International Covenant on Economic, Social and Cultural Rights (ICESCR).[19] Hence, the implementation of treaties in Canada is "met through a combination of laws, policies and programs of the federal, provincial and territorial governments," as well as court decisions relating to the interpretation and application of such treaties.[20]

2 Laws

In 1989, following almost a decade of rising ethical, legal, and social concerns surrounding the first baby conceived using in vitro fertilization (IVF) and the emergence of the field of new reproductive technologies (NRT), the Canadian federal government established the Royal Commission on New Reproductive Technologies. It was tasked with both studying public

---

[11] Ibid., s 92.7.

[12] Ibid., s 92.13.

[13] Ibid., s 92.16.

[14] Core Document Forming Part of the Reports of States Parties, para 90.

[15] Ibid., para 136.

[16] Ibid., para 137.

[17] N Duval Hesler, "L'influence du droit international sur la Cour d'appel du Québec" (2013) 54:1 *Les Cahiers de Droit* 177.

[18] 2009 QCCS 3213 *Dumont c. Québec (Procureur général)* [2009], para 127.

[19] International Covenant on Economic, Social and Cultural Rights (adopted December 16, 1966, entered into force January 3, 1976) 993 UNTS 3 (ICESCR).

[20] Canada, *Canada's Fourth Report on the United Nations' International Covenant on Economic, Social and Cultural Rights* (Ontario: Canadian Heritage, Ottawa, 2004), para 138.

perceptions and attitudes in this emerging field and exploring the implications of new and potential future developments regarding NRT, eventually producing seventeen volumes of research.[21] This included
considerations about "genetic manipulation and therapeutic interventions
to correct genetic anomalies in the context of human reproduction and to
make recommendations in the public interest with respect to them."[22]
The final report of the Royal Commission, entitled "Proceed with Care,"
was published in November 1993. The Report recommended "statutory
prohibitions of several activities and the creation of a regulatory framework to ensure that [new reproductive technologies] are provided in
a safe, ethical and accountable way."[23] In the aftermath of the Report,
a voluntary moratorium on nine practices, including human germline
modification, was enacted.[24] The series of public deliberations that took
place between 1993 and 2004 focused primarily on the meaning of
human dignity, the vulnerability of women and children, as well as
dystopian and utopian visions of science. Controversy surrounded not
only NRT but also the creation of embryos and their subsequent use in
research. The moral status of embryos, their possible commercialization,
the creation of embryos for IVF, and the fate of surplus embryos from
IVF treatments generated fear among the public. These uncertainties laid
the groundwork for the adoption, a decade later, of the 2004 Assisted
Human Reproduction Act (AHRA), a federal act that regulates assisted
human reproduction and related research.[25] It outlines prohibitions
against reproductive and research activities (e.g., human germline modification) and specifies the offense and associated sanctions for contravention of its provisions. The AHRA is a federal law with criminal
sanctions. As such, the AHRA applies across the country, thereby circum-
scribing human germline modification within the application of Article
15.1.b of the ICESCR.

---

[21] S Norris and M Tiedemann, "Legal Status at the Federal Level of Assisted Human
       Reproduction in Canada" (Background Paper, Library of Parliament, Ottawa, September 6,
       2011) http://publications.gc.ca/collections/collection_2012/bdp-lop/bp/2011–82-eng.pdf
       accessed April 30, 2018.
[22] Canada, Privy Council Office, Royal Commission on New Reproductive Technologies,
       *Proceed with Care – Final Report of the Royal Commission on New Reproductive
       Technologies* (Privy Council Office, Ottawa, 1993), 921.
[23] Norris and Tiedemann, "Legal Status at the Federal Level of Assisted Human Reproduction
       in Canada" (n 21) 2.
[24] Ibid.
[25] Assisted Human Reproduction Act SC 2004, cl 2 (AHRA).

Not long after its adoption, the province of Quebec challenged the validity of certain provisions of the AHRA claiming that the federal legislature had encroached on provincial legislative authority.[26] In 2010, the Supreme Court of Canada deemed that certain sections of the AHRA, "namely regarding the creation of a federal licensing and control system,"[27] were indeed ultra vires and unconstitutional. Ruling in favor of Quebec,[28] the Court concluded that, according to the Constitution, it is up to the provinces to regulate and oversee IVF clinics and healthcare services.[29] The portions of the law containing criminal offenses were, however, intra vires and remain valid today.

This was not a case of first impression. The Supreme Court of Canada had dealt with the issue of what constitutes a valid criminal law objective in 1949 in the *Margarine Reference* case.[30] According to the Court's opinion in this case, such a valid objective requires (1) a prohibition, (2) a penal sanction, and (3) the advancement of a legitimate public purpose.[31] Examples of legitimate public purposes include "public peace, order, security, health, morality."[32] In 2010, in deciding the case brought by Quebec against the AHRA, the Supreme Court referred to the holding in the *Margarine Reference* case and concluded that to be constitutionally valid in matters of health under federal jurisdiction, a criminal statute serves a valid public purpose when it is "directed at a legitimate public health evil."[33] Furthermore, in 1982, the Law Reform Commission of Canada had stated that criminal law should be used solely for "conduct which is culpable, seriously harmful, and generally conceived of as deserving of punishment,"[34] and that it should be "an instrument of last resort."[35] This

---

[26] *Attorney General of Quebec* v *Attorney General of Canada*, 2008 QCCA 1167 (*Quebec v Canada* 2008); Norris and Tiedemann, "Legal Status at the Federal Level of Assisted Human Reproduction in Canada" (n 21).

[27] BM Knoppers and others, "Human Genome Editing: Ethical and Policy Considerations" (Policy Brief, Genome Quebec Innovation Centre, Centre of Genomics and Policy and McGill University, 2018) www.genomequebec.com/DATA/PUBLICATION/34_en~v~Human_Genome_Editing_-_Policy_Brief.pdf accessed April 30, 2018, at 4.

[28] *Reference Re Assisted Human Reproduction Act*, 2010 SCC 61 (*Re AHRA* 2010).

[29] Ibid.

[30] *Reference Re Validity of Section 5(a) of the Dairy Industry Act*, 1949 SCR 1.

[31] Ibid., 49.

[32] Ibid., 50.

[33] *Re AHRA* 2010 (n 28) para 56.

[34] Canada, Law Reform Commission of Canada, *The Criminal Law in Canadian Society* (Ottawa: Government of Canada Publications, August 1982) http://publications.gc.ca/collections/collection_2017/jus/J2-38-1982-eng.pdf accessed May 7, 2018, at 4.

[35] Ibid., 44.

report further supported the notion that criminal prohibitions in the area of health are exceptional. Indeed, it could be argued that such prohibitions are unsuitable instruments to regulate science and research on reproductive technologies as they foreclose public debate, lack flexibility, and exhibit slow responsiveness to evolving societal perceptions to technology.[36]

Having successfully challenged the validity of certain provisions contained in the AHRA, Quebec passed its own legislation on the research and clinical activities related to human reproduction – the Loi sur les activités cliniques et de recherche en matière de procréation assistée (Act Respecting Clinical and Research Activities Related to Procreation)[37] – making it the only Canadian province or territory to have done so. The jurisdictional reach of this Act is clearly limited to Quebec. It does not refer to human germline modification per se but rather outlines the requirements for conducting high-quality, safe, and ethical activities relating to assisted reproduction services within accredited centers.[38]

Finally, in October 2016, Health Canada, the federal government department responsible for the AHRA, opened a public consultation with the aim of developing regulations to strengthen the Act beginning with a specific focus on sections concerning reimbursement for gamete donation and surrogacy. Therefore, there seems to be a willingness to reconsider and clarify certain provisions to better align with current societal attitudes. It is possible that other prohibited activities under the AHRA may also be reopened in the future (e.g., human germline modification).

3 Government Regulations

The Canadian Institutes of Health Research (CIHR),[39] Natural Sciences and Engineering Research Council of Canada (NSERC),[40] and Social Sciences and Humanities Research Council of Canada (SSHRC)[41] are the three main

---

[36] Knoppers and others, "Human Genome Editing" (n 27); BM Knoppers and others, "Human Gene Editing: Revisiting Canadian Policy" (2017) 2:3 *npj Regenerative Medicine* www.ncbi.nlm.nih.gov/pmc/articles/PMC5677987/ accessed April 30, 2018; T Caulfield, "Bill C-13 The Assisted Human Reproduction Act: Examining the Arguments Against a Regulatory Approach" (2002) 11:1 *Health Law Rev* 20.

[37] *Loi sur les activités cliniques et de recherche en matière de procréation assistée* (Québec), LRQ 2010, c A-5.01.

[38] Ibid. See, in general, Knoppers and others, "Human Genome Editing" (n 27).

[39] Canadian Institutes of Health Research (CIHR) www.cihr-irsc.gc.ca/e/193.html accessed April 30, 2018.

[40] Natural Sciences and Engineering Research Council of Canada (NSERC) www.nserc-crsng.gc.ca/index_eng.asp accessed April 30, 2018.

[41] Social Sciences and Humanities Research Council (SSHRC) www.sshrc-crsh.gc.ca/home-accueil-eng.aspx accessed April 30, 2018.

research-funding agencies in Canada. Jointly, they adopted the Tri-Council Policy Statement: Ethical Conduct for Research Involving Humans (TCPS 2),[42] a policy statement that provides ethical guidance and acts as a benchmark for the ethical conduct of research involving humans. As a condition of funding by any of the three agencies, researchers conducting research under the auspices of an eligible institution and its affiliates (e.g., Canadian universities and affiliated hospitals) are expected to comply with the TCPS 2.[43] It was initially released in 1998 with subsequent revisions leading to the 2014 version that is currently used. The TCPS 2 defines "germline alteration" as involving "changes that could be transmitted to future generations."[44] It goes on to define gene alteration as "the transfer of genes into cells to induce an altered capacity of the cell" and notes that it "is irreversible – [as] the cell and its descendants are forever altered and introduced changes cannot be removed."[45] The TCPS 2 addresses gene transfer research involving the alteration of human germ cells and reiterates the prohibition on human germline modification outlined in the AHRA.[46] Hence, the research and clinical applications of human germline modification are addressed by both the AHRA and the TCPS 2.

In 2016, the CIHR's Standing Committee on Ethics published a document entitled "Human Germline Gene Editing: Points to Consider from a Canadian Perspective."[47] The goal of the document was to outline issues about human germline modification and to inform future discussions about the applications of gene editing technologies in human embryos within the Canadian context.[48] It avoided taking a position on how best to proceed (or not) with human germline modification in Canada, while providing an overview of the current Canadian regulatory landscape, the key considerations for potential research and clinical applications, as well as presenting

---

[42] Canadian Institutes of Health Research, Natural Sciences and Engineering Research Council of Canada, Social Sciences and Humanities Research Council of Canada, *Tri-Council Policy Statement: Ethical Conduct for Research Involving Humans* (December 2014) (TCPS 2).

[43] Panel on Research Ethics (PRE), "Scope" www.pre.ethics.gc.ca/eng/policy-politique/interpretations/scope-portee/ accessed August 6, 2018.

[44] TCPS 2 (n 42) 198.

[45] Ibid.

[46] Ibid., c 13, s G.

[47] Canadian Institutes of Health Research (CIHR), *Human Germline Gene Editing: Points to Consider from a Canadian Perspective* (Canadian Institutes of Health Research, 2016) www.cihr-irsc.gc.ca/e/documents/sce_human_germline_gene_editing_en.pdf accessed May 7, 2018.

[48] Ibid.

alternatives to criminal prohibitions as an approach toward regulating human germline modification.

III OVERSIGHT BODIES AND FUNDING FOR RESEARCH ON HUMAN GERMLINE MODIFICATION

Currently in Canada, there is no governmental body specifically responsible for overseeing human germline modification research, nor is there a source of public or private funding available for such research, as it is currently prohibited under the AHRA. The Stem Cell Oversight Committee (SCOC),[49] a standing committee of CIHR's Governing Council, is, however, responsible for reviewing applications for research on human pluripotent stem cells – including embryonic stem cells, induced pluripotent stem cells, and embryonic germ cells – to ensure conformity with the TCPS 2.[50] The SCOC also plays an advisory role to the CIHR Governing Council regarding the scientific, ethical, legal, and social implications of human stem cell research and its possible clinical applications.[51] However, its current role does not expand to include oversight of research involving human germline modification or regulatory responsibilities such as licensing. Prior to the Quebec challenge, the Assisted Human Reproduction Agency was the body responsible for licensing-controlled activities and implementing the AHRA.[52] But the existence of the Agency was short, as the sections of the AHRA governing this body were repealed following the 2010 Supreme Court ruling, which abolished the Agency altogether.

As mentioned above, the three main research-funding agencies in Canada are: CIHR, NSERC, and SSHRC. In addition, Genome Canada, a nonprofit organization funded by the government of Canada, provides large-scale investments for the development of genomics-based technologies and support for research on the ethical, environmental, economic, legal, and social aspects of genomics.[53] Although public funding opportunities for genomic, biomedical, and natural science research are readily available in Canada, none of these funding bodies specifically encourage or focus on funding for human

[49] Stem Cell Oversight Committee, "Terms of Reference – Stem Cell Oversight Committee" (*Canadian Institutes of Health and Research*, December 19, 2014) www.cihr-irsc.gc.ca/e/20410.html accessed April 30, 2018.
[50] TCPS 2 (n 42) c 12, s F.
[51] See, in general, Knoppers and others, "Human Genome Editing" (n 27).
[52] Norris and Tiedemann "Legal Status at the Federal Level of Assisted Human Reproduction in Canada" (n 21).
[53] Genome Canada www.genomecanada.ca/ accessed April 30, 2018.

germline modification research. It is, however, possible to submit a research protocol proposing the use of human somatic genome modification (as this is permitted in Canada) to any of the above agencies' call for funding, and if successful, the funded research will require research ethics board approval.

IV SUBSTANTIVE PROVISIONS

1 *Assisted Human Reproduction Act*

In Canada, human germline modification is criminally prohibited by the AHRA. Section 5.1.f of the AHRA specifically prohibits any alteration of "the genome of a cell of a human being or in vitro embryo such that the alteration is *capable of being transmitted to descendants*."[54] Anyone found guilty of such a criminal offense is liable to a fine up to CAD $500,000, imprisonment up to 10 years, or both.[55]

There is a lack of clarity surrounding this provision, notably regarding the interpretation of "capable of being transmitted to descendants."[56] Does it apply to both research and reproductive applications of the technology (i.e., irrespective of intent), or does it apply solely to reproductive uses of human germline modification? On its face, the AHRA would appear to target both the research and reproductive applications of assisted human reproduction. One of the key principles recognized by the Act is that:

> the benefits of assisted human reproductive technologies *and related research for individuals, for families and for society in general* can be most effectively secured by taking appropriate measures for the protection and promotion of human health, safety, dignity and rights in the use of these technologies *and in related research*.[57]

The reference to "related research" suggests that the application of the AHRA and its criminal prohibitions extends to research activities as well.

Yet, according to a guidance document emitted by Health Canada in 2014, Section 5.1.f of the AHRA intends to prevent "people from using genetic technologies to alter the DNA of embryos *before transferring them*

---

[54]   *AHRA* (n 25) (emphasis added by the author).
[55]   Ibid., s 60.
[56]   Knoppers and others, "Human Gene Editing" (n 36); U Ogbogu and others, "Research on Human Embryos and Reproductive Materials: Revisiting Canadian Law and Policy" (2018) 13:3 *Healthcare Policy* 10; Z Master and P Bedford, "CRISPR Gene Editing Should Be Allowed in Canada, but Under What Circumstances" (2017) 40:2 *Journal of Obstetrics and Gynaecology Canada* 224.
[57]   *AHRA* (n 25) s 2.b (emphasis added by the author).

to a uterus."[58] It further mentions that the provision also bans "embryo 'gene therapy' that is intended to remove a disease-causing gene."[59] In other words, it could be argued that according to Health Canada, the AHRA prohibits only the clinical and reproductive applications of human germline modification, and not research endeavors, if implantation of the modified gametes or embryos is not foreseen.

.

2 *Basic Research Using Germline Modification in Human Embryos*

Although the AHRA seems to prohibit basic research using germline modification on human embryos in Canada, the above-mentioned 2014 Health Canada guidance document seems to allow basic research using human germline modification in human embryos (if not to be implanted). Irrespective of the correct interpretation, the fact is that no such research is currently being undertaken, as the criminal prohibition outlined in Section 5.1.f of the AHRA has a deterrent effect on researchers. Understandably, no one seems to be willing to test the interpretation out and risk being sanctioned. Instead, initiatives by researchers who wish to continue to expand and better understand early embryo development, genetic disease, and non-implantation and miscarriage have been limited to calling for further clarification from Health Canada on the matter.[60]

However, the use of human embryos for research purposes more generally is permitted in Canada, provided they are leftover from IVF and have not been created specifically for research purposes. The AHRA prohibits the creation of embryos for "any purpose other than creating a human being or improving or providing instruction in assisted procedures"[61] and acting contrary to this prohibition is subject to a fine up to CAD $500,000, imprisonment up to 10 years, or both.[62] Therefore, an embryo may only be created for the purpose of establishing a pregnancy or for testing and training purposes within a lab, after

---

[58]   Health Canada, "Prohibitions Related to Scientific Research and Clinical Applications" (*Government of Canada*, January 19, 2014) www.canada.ca/en/health-canada/services/drugs-health-products/biologics-radiopharmaceuticals-genetic-therapies/legislation-guidelines/assisted-human-reproduction/prohibitions-scientific-research-clinical-applications.html accessed May 7, 2018 (emphasis added by the author).

[59]   Ibid.

[60]   T Blackwell, "End Canada's Criminal Ban on Contentious CRISPR Gene-Editing Research, Major Science Group Urges" *National Post* (Toronto, November 8, 2017) http://nationalpost.com/health/end-canadas-criminal-ban-on-contentious-crispr-gene-editing-research-major-science-group-urges accessed May 16, 2018.

[61]   *AHRA* (n 25) s 5.1.b.

[62]   Ibid., s 60.

which the embryo would be destroyed or could be used in research (only if no longer needed for reproductive purposes and consent has been duly obtained from the gamete donors).[63] The AHRA defines "human embryo" as "a human organism during the first 56 days of its development following fertilization or creation, excluding any time during which its development has been suspended, and includes any cell derived from such an organism that is used for the purpose of creating a human being."[64]

The TCPS 2 further stipulates that embryos no longer needed for reproductive purposes can be donated for use in research.[65] Embryo and gamete donors (if they are different individuals) will be informed of the available options regarding the use of their embryos and their consent will be sought prior to any future use. As such, the TCPS 2 allows research involving embryos in two situations: (1) embryos created and intended for implantation[66] and (2) leftover embryos from IVF treatments,[67] provided that certain criteria are met. In the first situation, where embryos are created with the intent of achieving a pregnancy, the research must: (a) aim at benefiting the embryo; (b) not include the care of the woman or subsequent fetus; (c) provide close monitoring of the safety and comfort of the woman and the safety of the embryo; and (d) be carried out only after consent is obtained from the gamete donors.[68] In the second situation, where embryos are left over from IVF treatments, the following criteria must be met: (a) ova and sperm from which they are formed were obtained in accordance with Article 12.7 of the TCPS 2; (b) consent was obtained from the gamete donors; (c) embryos exposed to manipulations not specific to their ongoing normal development will not be transferred for continuing pregnancy; and (d) research will not go beyond fourteen days after their formation by combination of the gametes (the so-called fourteen-day rule).[69]

The deliberate creation of excess embryos is prohibited by Article 12.14 of the TCPS 2 according to which "researchers shall not ask, encourage, induce or coerce members of the health care team to generate more embryos than necessary for the optimum chance of reproductive success." This act would be consistent with the AHRA prohibition to create embryos for research purposes.[70]

[63] Health Canada, "Prohibitions Related to Scientific Research and Clinical Applications" (n 58).
[64] *AHRA* (n 25) s 3.
[65] TCPS 2 (n 42) art. 12.11.
[66] Ibid., art. 12.7.
[67] Ibid., art. 12.8.
[68] Ibid., art. 12.7.
[69] Ibid., art. 12.8.
[70] *AHRA* (n 25) s 5.1.b.

3 *Preclinical Research Using Germline Modification Technologies in Animals*

As Health Canada is responsible for helping to maintain and improve the health of Canadians, one of its requirements, prior to conducting any studies in humans, is to review findings from prior animal studies. More specifically, Health Canada states that

> animal studies used to support safety and/or effectiveness in humans should be presented. These studies should be undertaken using good laboratory practices. The objectives, methodology, results, analysis and manufacturer's conclusions should be covered by reports submitted.[71]

This requirement ensures that researchers and sponsors meet the safety obligations for the protection of Canadians as set out by Health Canada.[72]

Created in 1968, the Canadian Council on Animal Care (CCAC), a nonprofit organization, is the primary regulatory body for managing research on animals in Canada through education, development of guidelines, and assessment of laboratories. Its guidelines, which are regularly updated to respond to emerging new technologies and concerns, outline the institutional requirements and standards surrounding the oversight of animal care and ethics.[73] The CCAC's policy statement entitled *"Ethics of Animal Investigation"* provides that animals may be used for research, training, or testing purposes "only if it promises to contribute to understanding of fundamental biological principles, or to the development of knowledge that can reasonably be expected to benefit humans or animals" and "only if the researcher's best efforts to find an alternative have failed."[74] The ethical basis upon which the *Guide to the Care and Use of Experimental Animals*[75] rests relies on adherence to the so-called "Three Rs" tenet of humane

---

[71] Canada, Minister of Health, Health Products and Food Branch, "Guidance Document: Guidance on Supporting Evidence to be provided for New and Amended License Applications for Class III and Class IV Medical Devices, not including In Vitro Diagnostic Devices (IVDDs)" (Ottawa, Minister of Public Works and Government Services Canada 2012) www.canada.ca/content/dam/hc-sc/migration/hc-sc/dhp-mps/alt_formats/pdf/md-im/applic-demande/guide-ld/md_gd_data_im_ld_donnees_ciii_civ-eng.pdf accessed May 10, 2018, at 31.

[72] I Murnaghan, "Animal Testing in Canada" (*About Animal Testing*, 2018) www.aboutanimaltesting.co.uk/animal-testing-canada.html accessed May 10, 2018.

[73] Canadian Council on Animal Care (CCAC) www.ccac.ca/en/standards/guidelines/ accessed May 10, 2018.

[74] CCAC, "Policy Statement: Ethics of Animal Investigation" (October 1989) www.ccac.ca/Documents/Standards/Policies/Ethics_of_animal_investigation.pdf accessed May 10, 2018, at 1.

[75] ED Olfert and others (eds.), *Guide to the Care and Use of Experimental Animals* (1st vol. 2nd ed., Canadian Council on Animal Care, 1993) www.ccac.ca/Documents/Standards/Guidelines/Experimental_Animals_Vol1.pdf accessed May 10, 2018.

experimental technique: replacement, refinement, and reduction.[76] According to the "Three Rs" tenet, "animals may only be used if there is no alternative experimental approach to obtain the required information" (replacement); "the use of animals for biomedical science must be based on the appropriateness of the model for the scientific goal of the study, and not on other factors such as economics or the availability of animals within the institution" (refinement); and "the fewest animals appropriate to provide valid information and statistical significance should be used [. . .] and the numbers of animals maintained should not exceed the number that an institution can successfully house and care for" (reduction).[77] When animals are used, they must be provided with an environment (physical and social) that is conducive to their welfare, which is a key consideration for animal research.[78] It is of note that Canada does not have a formal ban on using chimpanzees or other great apes for research purposes, as they fall under the general guidelines put forward by the CCAC.[79]

In 1997, the CCAC released the *"Guidelines on: Transgenic Animals"*[80] to assist Animal Care Committees (ACC) in evaluating the ethical creation and use of genetically modified animals. It defines a transgenic animal as "an animal in which there has been a deliberate modification of the genome – the material responsible for inherited characteristics – in contrast to spontaneous mutation."[81] In order to encompass technological advances in the field that allowed researchers to go beyond the use of transgenesis, new guidelines are in preparation. A draft of these guidelines, entitled *"Guidelines on Genetically Engineered Animals Used in Science,"* employs "genetically engineered" as a broader term that captures the range of genetically altered animals.[82] To clarify what is meant by genetically modified animal, the CCAC provides the following definition:

---

[76] WMS Russell and RL Burch, *The Principles of Humane Experimental Technique* (London: Methuen and Co. Ltd, 1959).

[77] Canadian Council on Animal Care, *Guidelines on: Procurement of Animals Used in Science* (CCAC, 2007) 5 www.ccac.ca/Documents/Standards/Guidelines/Procurement.pdf accessed May 10, 2018.

[78] Ibid.

[79] Canadian Broadcasting Corporation, "Safe Haven for Chimps: Canada's Position on Chimpanzees in Research" (*The Nature of Things with David Suzuki*, 11 June 2017) www.cbc.ca/natureofthings/features/using-chimps-for-research-in-canada accessed July 4, 2018.

[80] Canadian Council on Animal Care, "Guidelines on: Transgenic Animals" (February 5, 1997) www.ccac.ca/Documents/Standards/Guidelines/Transgenic_Animals.pdf accessed May 10, 2018.

[81] Ibid., 1.

[82] EH Ormandy and others, "Genetic Engineering of Animals: Ethical Issues, Including Welfare Concerns" (2011) 52:5 *The Canadian Veterinary Journal* 544.

an animal that has had a change in its nuclear or mitochondrial DNA (addition, deletion, or substitution of some part of the animal's genetic material or insertion of foreign DNA) achieved through a deliberate human technological intervention.[83]

During the process of discussing and developing these guidelines, certain ethical and welfare concerns arose. Notably, concerns about the invasiveness of the interventions used; the need for large numbers of animals; the possibility for unanticipated welfare concerns; and the need to establish ethical limits to genetic engineering of animals.[84] Thus, although Environment Canada and Health Canada impose restrictions on how genetically engineered animals that have been put down should be disposed of, the CCAC does not prohibit the use of human germline modification in the context of preclinical research on animals.

Furthermore, the CCAC is responsible for certifying institutions that conduct animal-based work as promoting Good Animal Practice (GAP), following an assessment of their animal ethics and care programs and based on compliance with the standards put forth by the CCAC.[85] To ensure respect for guidelines and animal welfare, the CCAC requires each institution conducting research on animals to be monitored by an ACC. Each Committee is composed of individuals with varying expertise (e.g., researchers, veterinarians, community members, technical staff, etc.) and who have a responsibility to review protocols in an ethical and scientific manner.[86] However, as this is a power held by the provinces, each province and territory is also responsible for developing their own laws on animal care.

4 Clinical Research and Applications Using Germline Modification Technologies in Humans

As per Section 5.1.f of the AHRA, any alteration of the genome that is "capable of being transmitted to descendants" is considered a criminal offense. Therefore, clinical research and future clinical applications of human

---

[83] Ibid., 546.

[84] Ibid.

[85] Canadian Council on Animal Care, "CCAC Policy Statement on: The Certification of Animal Ethics and Care Programs" (January 2016) www.ccac.ca/Documents/Standards/Poli cies/Certification-of-animal-ethics-and-care-programs.pdf accessed May 10, 2018.

[86] Canadian Council on Animal Care, "CCAC Guidelines on: Animal Use Protocol Review (1997)" (September 29, 1996) www.ccac.ca/Documents/Standards/Guidelines/Protocol_Revi ew.pdf accessed May 10, 2018.

germline modification are prohibited in Canada. Section 60 of the AHRA outlines the offense and punishment for contravening Section 5 (i.e., a fine up to CAD $500,000, imprisonment up to ten years, or both).

The human genome is defined in the AHRA as "the totality of the deoxyribonucleic acid sequence of a particular cell."[87] As such, the prohibition on human germline modification also applies to mitochondrial replacement therapy (MRT), which aims to prevent the transmission of heritable mitochondrial disorders by replacing any mutated mitochondrial DNA (mtDNA) in the woman's oocytes (eggs) with that from a healthy donor.[88] These disorders are maternally inherited and can have serious implications for the health of the child. Pathogenic mutations in this genome are associated with a broad spectrum of diseases, ranging from severe metabolic disorders that are fatal in early infancy to late-onset neurodegenerative conditions.[89] MRT typically "involves transferring the nucleus from an oocyte that contains pathogenic mitochondrial DNA to an enucleated donor oocyte that contains only normal mitochondrial genomes. The reconstituted oocyte then is fertilized *in vitro*, and the embryo is implanted."[90] To date, research on the safety and efficacy of the technology has been conducted in both animal models (mice and nonhuman primates) and human oocytes (in accordance with the fourteen-day rule). Findings from these studies indicate a level of safety and efficiency with the technique[91] that is sufficient to warrant carefully supervised clinical trials in some jurisdictions, such as the United Kingdom.[92] However, as mentioned above, in Canada clinical trials are not possible due to the applicability of the criminal ban on human germline alterations to MRT, which makes it illegal to "knowingly

---

[87] *AHRA* (n 25) s 3.

[88] Health Canada, "Prohibitions Related to Scientific Research and Clinical Applications" (n 58).

[89] BM Knoppers and others, "Mitochondrial Replacement Therapy: The Road to the Clinic in Canada" (2017) 39:10 *Journal of Obstetrics and Gynaecology Canada* 916.

[90] Ibid., 916.

[91] M Tachibana and others, "Mitochondrial Gene Replacement in Primate Offspring and Embryonic Stem Cells" (2009) 461:7262 *Nature* 367; M Tachibana and others, "Towards Germline Gene Therapy of Inherited Mitochondrial Disease" (2013) 493:7434 *Nature* 627; E Kang and others, "Mitochondrial Replacement in Human Oocytes Carrying Pathogenic Mitochondrial DNA Mutations" 540:7632 *Nature* 270.

[92] T Klopstock and others, "Mitochondrial Replacement Approaches: Challenges for Clinical Implementation" (2016) 8:126 *Genome Medicine*; I Sample, "First UK Licence to Create Three-Person Baby Granted by Fertility Regulator" (*The Guardian*, March 16, 2017) www.theguardian.com/science/2017/mar/16/first-licence-to-create-three-person-baby-granted-by-uk-fertility-regulator accessed May 7, 2018.

create embryos that have nuclear DNA from two people and mtDNA from
a third person" (i.e., three-parent baby).[93]

Another application for mitochondria within the context of assisted human
reproduction, which may escape regulation in Canada,[94] is called autologous
germline mitochondrial energy transfer (AUGMENT). It is a relatively new
procedure, developed by OvaScience, an American company. AUGMENT is
used to improve egg health and thus the rate of successful pregnancies.[95] The
treatment works by adding "energy-producing mitochondria from a patient's
own egg precursor (EggPC^{SM}) cells, immature egg cells found in the protective ovarian lining, [. . .] to the patient's mature eggs to supplement the existing
mitochondria."[96] This is not prohibited under the AHRA as modifications are
not being made to the germline; it is considered to be a rejuvenation of sorts.[97]
AUGMENT has been shown to be effective in demonstrating "increases [of]
clinical pregnancy rates per initiated cycles by threefold."[98] However, the early
clinical applications of the treatment indicated that there is a need for more
data and room for improvement with regard to current assisted human reproduction programs. Only time will tell the real impact of AUGMENT as its use
is still quite limited.[99]

AUGMENT, which has never been offered in the United States, has been
made available by a handful of fertility centers in Canada, Japan, and the
United Arab Emirates.[100] These centers offer AUGMENT as a treatment for
women who have had ongoing difficulty becoming pregnant. One of these,
was the Toronto Center for Assisted Reproductive Technologies (TCART)
Fertility Partners (now TRIO), located in Ontario, Canada. In 2015, the first

93   Health Canada, "Prohibitions Related to Scientific Research and Clinical
     Applications" (n 58).
94   A Motluck, "IVF Booster Offered in Canada but not in the US" (2015) 187:3 *Canadian
     Medical Association Journal* 89.
95   European Society of Human Reproduction and Embryology, "Energizing Eggs with a Patient's
     own Mitochondria Offers No Benefit in Assisted Reproduction" (3 July 2018) *EurekAlert!*
     www.eurekalert.org/pub_releases/2018–07/esoh-eewo62818.php accessed December 31, 2018.
96   Cision, "New AUGMENT(SM) Fertility Treatment Now Available in Canada" (*Cision
     Canada*, December 18, 2014) www.newswire.ca/news-releases/new-augmentsm-fertility-
     treatment-now-available-in-canada-516801991.html accessed May 14, 2018.
97   K Weintraub, "Turmoil at Troubled Fertility Company OvaScience" (December 29, 2016)
     MIT Technology Review www.technologyreview.com/s/603274/turmoil-at-troubled-fertility-
     company-ovascience/ accessed July 5, 2018.
98   DC Woods and JL Tilly, "Autologous Germline Mitochondrial Energy Transfer
     (AUGMENT) in Human Assisted Reproduction" (2015) 33:6 *Seminars in Reproductive
     Medicine* 410, 417.
99   Ibid.; European Society of Human Reproduction and Embryology (n 95).
100 Weintraub "Turmoil at Troubled Fertility Company OvaScience" (n 97).

baby born using this technique was born in Canada to a couple who had undergone the treatment at TRIO.

However, in November 2017, a class action lawsuit was filed in Massachusetts against OvaScience Inc: by all common shareholders in the company. The basis of the class action suit rests on allegations that the company

> made materially false and misleading statements and/or failed to disclose that: (1) the science behind AUGMENT was untested and in doubt; (2) the patients that had received OvaScience's AUGMENT procedure in 2014 did not achieve a pregnancy success rate that was significantly higher than the rate achieved without the Company's AUGMENT procedure; (3) the Company had not chosen to undertake its studies outside of the United States, but was forced to as it did not want to meet stringent and expensive federal regulations; and (4) the Company was far from being profitable, or even approaching profitability.[101]

Finally, although not the focus of this book, it is worth mentioning that, unlike the modification of human germ cells (i.e., sperm and ova – leading to heritable genetic changes), the modification of somatic cells (i.e. genetic changes that are not heritable) is permitted in Canada. The TCPS 2 defines somatic cells as "any body cell other than gametes (egg or sperm)."[102] As existing regulations and guidance are not specific to somatic modifications, such changes typically fall into the category of cell and gene therapies more broadly and are governed as "biological drugs" under Health Canada's Food and Drugs Act and the Food and Drug Regulations.[103] Currently, human somatic genome modification is being used in both research and clinical trials. Despite this, additional guidance regarding its safety, traceability, and quality may be beneficial, notably as the number of somatic gene therapy clinical trials increases.[104] Further discussion will also be needed surrounding the distinctions and unique ethical, legal, and social challenges that arise in the context of such research with pediatric versus adult populations

---

[101] Bernstein Liebhard LLP, "Bernstein Liebhard LLP Announces That a Class Action Has Been Filed Against OvaScience, Inc." (*Cision Canada*, November 28, 2017) www.prnewswire.com /news-releases/bernstein-liebhard-llp-announces-that-a-class-action-has-been-filed-against-ovascience-inc-300562971.html accessed May 14, 2018.

[102] TCPS 2 (n 42) at 210.

[103] Food and Drugs Act, R.S.C., 1985, c. F-27, Schedule D; Food and Drug Regulations, C.R.C., 2009, c 870, Divisions 1A, 2, 4, 5 and 8.

[104] M Hawryluk, "Gene Therapy Development on the Rise but Costs Remain an Issue" (*Association of Health Care Journalists*, April 13, 2018) https://healthjournalism.org/blog/201 8/04/gene-therapy-development-on-the-rise-but-costs-remain-an-issue/ accessed May 7, 2018; Knoppers, "Human Genome Editing" (n 27).

(e.g., informed consent, best interests, decision-making, risk-benefit ratio, quality of life).[105]

V CURRENT PERSPECTIVES AND FUTURE POSSIBILITIES

Fear is the natural response to emerging disruptive technologies as there are many unknowns. As such, the initial reaction adopted by the 2004 AHRA to prohibit rather than regulate the technology, although not optimal, is no surprise. However, the lack of reassessment of the social landscape over time as the technology evolves leaves citizens with laws and policies that no longer represent today's reality.[106] To allow further deliberation and perhaps use of human germline modification in Canada, Health Canada must further clarify the interpretation and scope of Section 5.1.f of the AHRA. Although its 2014 guidance document appears to make clear that the ban is solely of the reproductive applications of the technology, Canadian researchers remain unwilling to apply human germline modification in their laboratories for fear of being prosecuted under the rather draconian criminal law.[107]

Basic research must be outside the scope of the application of the AHRA, as it is distinct from the clinical applications of assisted human reproduction. Since the enactment of the AHRA, the research community has criticized the use of criminal law to regulate and oversee science and reproductive research. Strict prohibitions and restrictions on research applications lack the necessary flexibility and nuance to evolve with science.[108] This would allow for basic and preclinical research using human germline modification (i.e., as long as there is no implantation/intention to create pregnancy) in Canada.

Public engagement is another area that is often cited as important to moving human germline modification forward. There is a need to periodically assess societal attitudes toward emerging technologies to ensure that the legal and regulatory frameworks in place reflect current opinions. Several studies have been conducted to capture public perceptions of human germline modification – in the United States, Australia, the United Kingdom, and Japan – with most demonstrating acceptance of the technology for preventive or therapeutic purposes but disagreement with regard to nontherapeutic

[105]   Knoppers and others, "Human Genome Editing" (n 27).
[106]   Ibid.
[107]   *AHRA* (n 25) s 60.
[108]   See, in general, Lemmens, *Regulating Creation* (n 5); Knoppers and others, "Human Gene Editing" (n36); Isasi "Editing Policy to Fit the Genome?" (n 9).

applications (e.g., enhancement).[109] There has also been a call for a global observatory to encourage broad societal conversation and facilitate consensus-building through the necessary "cosmopolitan conversation."[110] Some researchers have even proposed the need to rethink our approach to public engagement to incorporate organizations that have a declared interest in rethinking human germline modification, such as the World Health Organization, patient advocacy groups, research centers, disability rights and environmental groups, and pharmaceutical companies and agricultural institutions.[111] This will result in a consortium that aims to connect the public to the ongoing debates among science and policy experts, and to connect researchers and policymakers to various other stakeholders.[112]

Finally, as required by both the Universal Declaration of Human Rights[113] and the International Covenant on Economic, Social and Cultural Rights,[114] which Canada ratified,[115] Canada must support the right of all citizens "to enjoy the benefits of scientific progress and its applications." For the most part, this right has lain dormant, only recently being awakened.[116] The obligation to enable access to scientific informa-tion and to promote the diffusion of science is also an integral part of this right,[117] along with the obligation to "respect the freedom indispensable for

---

[109] T McCaughey and others, "A Global Social Media Survey of Attitudes to Human Genome Editing" (2016) 18:5 *Cell Stem Cell* 569; DA Scheufele and others, "U.S. Attitudes on Human Genome Editing" (2017) 357:6351 *Science* 553; G Gaskell and others, "Public Views on Gene Editing and Its Uses" (2017) 35:11 *Nature Biotechnology* 1021; Hopkins van Mil: Creating Connections Ltd (HVM), *Potential Uses for Genetic Technologies: Dialogue and Engagement Research Conducted on Behalf of the Royal Society: Findings Report* (Commissioned by the Royal Society, Finding Reports, December 2017) https://royalsociety.org/~/media/policy/pro jects/gene-tech/genetic-technologies-public-dialogue-hvm-full-report.pdf accessed April 30, 2018; M Uchiyama and others, "Survey on the Perception of Germline Genome Editing Among the General Public in Japan" (2018) 63 *Journal of Human Genetics* 745.

[110] S Jasanoff and JB Hurlburt, "A Global Observatory for Gene Editing" (2018) 555 *Nature* 435; K Saha and others, "Building Capacity for a Global Genome Editing Observatory: Institutional Design" (August 2018) 36:8 *Trends in Biotechnology* 741; C Brokowski, "Do CRISPR Germline Ethics Statements Cut It?" (April 1, 2018) 1:2 *The CRISPR Journal* 115.

[111] S Burall, "Rethink Public Engagement for Gene Editing" (March 22, 2018) 555 *Nature* 438.

[112] Ibid.

[113] Universal Declaration of Human Rights (adopted December 10, 1948 UNGA Res 217 A(III)) (UDHR) art. 27.

[114] *ICESCR, op. cit.*, art. 15.1.b.

[115] Canada acceded on 19 May 1976: United Nations Treaty Collection, "Depositary" https://tr eaties.un.org/pages/ViewDetails.aspx?src=IND&mtdsg_no=IV-3&chapter=4&clang=_en accessed May 7, 2018.

[116] R Yotova and BM Knoppers, "The Right to Benefit from Science and Big Data" (2018) *European Journal of International Law* (publication pending).

[117] Ibid.; *ICESCR* (n 19) art. 15.2.

scientific research."[118] This implies that Canada must aim to facilitate, rather than hinder, basic and preclinical research (with rigorous oversight) using human germline modification so that knowledge and a deeper understanding can be fostered. All the while, Canada must also ensure that measures are in place to protect individuals from the possible negative implications of science.[119] Therefore, the application of this right, in a responsible and transparent manner, would enable Canadian researchers to remain engaged with the international research community, while promoting scientific freedom as a core tenet of liberal democracies and ensuring that governments provide access to new technologies.[120] Furthermore, Canada must support the "right of everyone to the enjoyment of the highest attainable standard of physical and mental health,"[121] which highlights the promise of human genome modification (both somatic and germline) to potentially prevent, treat, and cure serious, life-threatening diseases. As stated by Eric T. Juengst, adopting a human rights approach to human germline modification policy would "reorient our conversation from policing science to governing society and would shift our focus from avoiding risks to protecting opportunities."[122]

---

[118]   *ICESCR* (n 19) art. 15.3.
[119]   See, in general, Yotova and Knoppers (n 116).
[120]   Knoppers and others, "Human Genome Editing" (n 27) at 5.
[121]   *ICESCR* (n 19) art. 12.
[122]   ET Juengst, "Crowdsourcing the Moral Limits of Human Gene Editing?" (2018) 47:3 *Hastings Center Report* 15, 21.

4

The Regulation of Human Germline Genome Modification in the United States

Kerry Lynn Macintosh

In human germline genome modification (germline editing), scientists use CRISPR/Cas9, base editors, and other molecular tools to alter the genomes of human gametes and/or embryos.[1] Some edit human embryos in an effort to correct genetic mutations that cause disease.[2] At present, their work is restricted to the laboratory, but their long-term goal is to help parents have healthy children.[3] Theoretically, germline editing can also be used to enhance children, but applications are speculative at present. Genes are complex, our knowledge of them is limited, and environmental factors also contribute to human traits.[4]

So far, the United States has not adopted a comprehensive legal regime to manage germline editing. However, as this chapter explains, the US Congress has prevented a key regulatory agency from receiving applications to conduct clinical trials, thereby ensuring that scientists cannot attempt to create children with modified genomes in America for the time being. Basic lab research on human gametes and embryos remains legal in most states, but federal funds are not available to support research in which human embryos are created for

---

[1] This chapter does not address 'mitochondrial replacement therapy' – that is, techniques that substitute healthy mitochondria for the mutated mitochondria that some women carry in their eggs. Ewen Callaway, 'The Power of Three' (2014) 509 *Nature* 414, 417. Mitochondrial replacement therapy involves a heritable genetic modification: female children born through it can pass the substituted mitochondria to descendants through their eggs. K Ludlow, 'The Policy and Regulatory Context of U.S., U.K., and Australian Responses to Mitochondrial Donation Governance' (2018) 58 *Jurimetrics* 247, 255. Thus, mitochondrial replacement therapy cannot be performed legally in the United States due to the de facto moratorium discussed in Section I.1.b.I G Cohen and E Adashi, 'The FDA Is Prohibited from Going Germline' (2018) 353 *Science* 545, 546.

[2] For example, scientists working with cloned human embryos have corrected a mutation that causes the blood disorder known as beta thalassemia. P Liang and others, 'Correction of β-thalassemia Mutant by Base Editor in Human Embryos' (2017) 8 *Protein & Cell* 811.

[3] Ibid.

[4] For a discussion, see KL Macintosh, *Enhanced Beings: Human Germline Modification and the Law* (Cambridge University Press 2018) 16–19.

research, destroyed, discarded, or subjected to risk of serious harm or death. Federal and state proposals to outlaw germline editing may emerge as technology matures, particularly if children with modified genomes are born abroad.

I THE REGULATORY ENVIRONMENT

Article 15 of the International Covenant on Economic, Social, and Cultural Rights acknowledges the right of everyone '[t]o enjoy the benefits of scientific progress and its applications'.[5] It requires parties '[t]o respect the freedom indispensable for scientific research and creative activity'[6] and to undertake such steps as may be 'necessary for the conservation, the development, and the diffusion of science'.[7] Although the United States signed the Covenant in 1977, it never ratified it and is not bound to implement it.[8] Thus, this section will not discuss whether US law is consistent with Article 15 of the Covenant. Instead, it will focus on the regulatory framework, definitions, oversight bodies, and funding provisions that are relevant specifically to the United States.

1 Regulatory Framework

The United States has two systems of law: federal and state. The US Constitution and legislation enacted by the US Congress apply throughout the entire nation. In addition, each of the fifty states has its own constitution and legislature. A specific state constitution or act applies only within that state. According to the Supremacy Clause of the US Constitution, federal law prevails over state law in case of conflict.[9]

This section discusses legal authorities that affect germline editing directly or restrict it indirectly (Table 4.1).[10] It begins with the US Constitution, turns

---

[5] International Covenant on Economic, Social, and Cultural Rights (adopted 16 December 1966, entered into force 3 January 1976) 993 UNTS 3 (ICESCR) art. 15.1.b. See also, Universal Declaration of Human Rights (adopted 10 December 1948) UNGA Res 217 A(III) (UDHR), art. 27 (non-binding declaration asserting right 'to share in scientific advancement and its benefits').

[6] ICESCR, art. 15.3.

[7] Ibid., art. 15.2.

[8] G MacNaughton and M McGill, 'Economic and Social Rights in the United States: Implementation without Ratification' (2012) 4 *N.E. U. L.J.* 365, 366–69.

[9] U.S. Constitution (U.S. Const.), art. VI.2.

[10] General statutes and regulations that govern laboratories, medical practice, or assisted reproduction will not be discussed. For a more comprehensive discussion of human genome editing, including germline editing, see National Academy of Sciences and National Academy of Sciences, Engineering, and Medicine, *Human Genome Editing: Science, Ethics, and Governance* (2017) (NAS Report, *Human Genome Editing*).

TABLE 4.1 *United States: federal constitution, legislation, regulations, guidance, and funding policies*

Name	Citation	Notes
US Constitution	First, Fifth, and Fourteenth Amendments	May limit the ability of Congress and state legislatures to ban germline editing
Federal Food, Drug, and Cosmetic Act	21 U.S.C. s 301 et seq.	Regulates drugs and medical devices
Public Health Service Act	42 U.S.C. s 201 et seq.	Regulates biological products
Application of Current Statutory Authorities to Human Somatic Cell Therapy Products and Gene Therapy Products	58 Fed. Reg. 53,248 (14 October 1993)	Explains that human somatic cell therapy products and gene therapy products are biological products and/or drugs within FDA jurisdiction
Consolidated Appropriations Act, 2018	115 Pub. L. No. 141, Div. A, Title VII, s 734, 132 Stat. 348, 389.	Restrains FDA from receiving applications for clinical trials
Guidance for Industry #187	https://wayback.archive-it.org/7993/20170 1100 5939/http://www.fda.gov/down loads/AnimalVeterinary/GuidanceCo mplianceEnforcement/GuidanceforIn dustry/UCM113903.pdf	Explains regulation of genetically engineered animals with heritable recombinant DNA constructs. Proposed revision covers animals with intentionally altered genomic DNA
Animal Welfare Act	7 U.S.C. s 2131 et seq.	Seeks to secure welfare of experimental animals
Public Health Service Policy on Humane Care and Use of Laboratory Animals	https://olaw.nih.gov/policies-laws/phs-policy.htm	Applies to institutions that receive Public Health Service funds
Common Rule	45 C.F.R. s 46.101 et seq.	Regulates research that DHHS conducts or funds. Adopted by many other federal departments and agencies
Additional Protections for Pregnant Women, Human Fetuses and Neonates Involved in Research	45 C.F.R. s 46.201 et seq.	Regulates research that DHHS conducts or funds
Additional Protections for Children Involved as Subjects in Research	45 C.F.R. s 46.401 et seq.	Regulates research that DHHS conducts or funds

(continued)

TABLE 4.1 (continued)

Name	Citation	Notes
Statement on NIH Funding of Research Using Gene-editing Technologies in Human Embryos	www.nih.gov/about/director/04292015_sta tement_gene_editing_technologies .htm	Declares that federal funds are not available for editing of human embryos
Dickey-Wicker Amendment (included in Consolidated Appropriations Act, 2018)	115 Pub. L. No. 141, Div. H, Title V, s 508, 132 Stat. 348, 764.	Denies federal funds for research that creates, destroys, discards, or harms human embryos
National Institutes of Health, NIH Guidelines for Research Involving Recombinant or Synthetic Nucleic Acid Molecules (April 2016)	https://osp.od.nih.gov/wp-content/upload s/2013/06/NIH_Guidelines.pdf	Regulates NIH-funded research involving recombinant or synthetic nucleic acid molecules. Revisions proposed

next to federal legislation and regulations, moves on to state laws, and closes with a look at reports, policies, and guidelines.

a The US Constitution

The US Constitution was written in 1787, more than 230 years ago,[11] and is difficult to amend.[12] Thus, it should come as no surprise that it is silent on the issues raised by modern biomedicine. However, certain rights can be read into the existing provisions of the Constitution through judicial interpretation. Although to date there has not been any major case touching directly upon the issues discussed in this book, the US Supreme Court might be called upon to rule on the constitutionality of future laws that Congress or the states might pass to regulate the field. The key question is whether the Constitution could conceivably be interpreted to confer a right to engage in scientific research and/or benefit from scientific and technological progress.

Although one could imagine several hypothetical scenarios, given the overall political climate in the country, two are most likely to be relevant. First, suppose that Congress or a state legislature passes a law that bans germline editing, including basic research. The US Constitution does not expressly protect a right to conduct scientific research.[13] However, legal scholars have attempted to locate such a right within provisions that secure other rights. For example, the First Amendment provides that Congress shall not enact laws that abridge freedom of speech.[14] The late John Robertson argued that scientific research is protected under the First Amendment as a precondition to dissemination of scientific information,[15]

---

[11] History.com (ed.), 'U.S. Constitution Ratified' (History.com, 30 December 2018) www .history.com/this-day-in-history/u-s-constitution-ratified accessed 12 October 2018.

[12] E Posner, 'The U.S. Constitution Is Impossible to Amend' (Slate, 5 May 2014) www.slate.com /articles/news_and_politics/view_from_chicago/2014/05/amending_the_constitution_is_ much_too_hard_blame_the_founders.html accessed 12 October 2018.

[13] The Constitution provides that Congress has the power '[t]o promote the Progress of Science and Useful Arts, by securing for limited Times to Authors and Inventors the exclusive Right to their respective Writings and Discoveries'. U.S. Const., art. I.8. Although this provision reflects a desire to further science, it does not articulate a right to engage in experiments.

[14] Ibid., First Amendment. The Fourteenth Amendment, which provides that a state cannot deprive any person of liberty without due process, incorporates the First Amendment's guarantee of free speech. *Gitlow* v. *New York* (1925) 268 U.S. 652, 666.

[15] J Robertson, 'The Scientist's Right to Research: A Constitutional Analysis' (1977) 51 *S. Cal. L. Rev.* 1203, 1217–18.

while others have asserted that it is protected as an integral part of the scientific method [16]

The US Supreme Court has not yet resolved the question of whether the First Amendment secures a right to conduct scientific research,[17] nor has it identified a standard of review for evaluating laws that infringe upon that right.[18] If scientists use the aforementioned academic theories to challenge a ban on germline editing, the Court must come down on one side or the other. If it determines that the First Amendment protects scientific research, it might rule in favour of the scientists and invalidate the ban. Alternatively, it could hold that the First Amendment does not protect scientific research – an unfortunate outcome that would invite further legislative encroachments into the scientific realm.[19]

To be sure, scientists can also argue that federal and state laws that ban germline editing deprive them of liberty without due process in violation of the Fifth and Fourteenth Amendments, respectively.[20] Yet, although due process has a substantive dimension, the Supreme Court probably will not recognize scientific research as a *fundamental* liberty.[21] Instead, it will uphold a ban if it is rationally related to a legitimate government interest. For example, Congress or a state legislature could persuade the Supreme Court that a total ban on all germline editing, including basic research, is a logical way to prevent reproductive applications that are unsafe or objectionable on other grounds.[22]

Second, suppose that Congress or a state legislature enacts a law that bans germline editing in human reproduction. In other words, it prohibits men and women from using modified gametes or embryos to initiate pregnancies and have children. The Supreme Court has declared that substantive due process includes a right to privacy. Procreation is nestled within this right.[23] So far, the Supreme Court has not discussed whether the Constitution protects

---

[16] RG Spece Jr. and J Weinzierl, 'First Amendment Protection of Experimentation: A Critical Review and Tentative Synthesis/Reconstruction of the Literature' (1998) 8 S. Cal. Interdisc. L. J. 185, 213–15.

[17] Macintosh, *Enhanced Beings* (n 4), 158.

[18] See Spece & Weinzierl, 'First Amendment Protection of Experimentation' (n 16), 219–20 (discussing possible standards of review, including intermediate and strict scrutiny).

[19] Macintosh, *Enhanced Beings* (n 4), 159.

[20] U.S. Const., Fifth Amendment and Fourteenth Amendment, s 1.

[21] Robertson, 'The Scientist's Right to Research' (n 15), 1212–14.

[22] Macintosh, *Enhanced Beings* (n 4), 158.

[23] *Carey* v. *Population Services Int'l* (1977), 431 U.S. 678, 684–85 (*Carey* v. *Population Services Int'l*).

procreation via assisted reproductive technologies, but a federal district court has held that the right to privacy secures access to such technologies.[24]

Although most men and women can exercise their procreative liberty without germline editing, for some carriers of genetic mutations resorting to this technology might be the only way to conceive a healthy offspring. In their case, a ban on germline editing imposes a significant burden on exercise of their privacy right, thereby triggering strict scrutiny – a higher standard of review that forces a legislative body to prove that the law is narrowly tailored to further a compelling government interest.[25] Although a ban on reproductive uses of germline editing could be justified on the ground that it is necessary to ensure the safety of mothers and children, should the technology become reasonably safe, carriers of genetic mutations might be able to persuade the Supreme Court that a ban is a violation of their constitutional right to procreate.[26]

b Federal Legislation

The US Congress has not yet enacted specific legislation to control genomic editing in animals or humans. Specifically, no federal statute directly prohibits the creation of an animal with an edited genome, a human gamete or embryo with an edited genome, or even a human being with an edited genome. In the absence of such legislation, the agency with primary regulatory authority is the Federal Drug Administration (FDA). It draws its power to regulate drugs and medical devices from the Federal Food, Drug, and Cosmetic Act and its authority over biological products from the Public Health Service Act.[27]

I HUMANS In the United States, one cannot market new drugs, medical devices, or biological products for use in humans without obtaining approval or licensure pursuant to an arduous process that includes submission of data from clinical trials in humans.[28] Before conducting a clinical trial, one must submit to the FDA an 'investigational new drug' (IND) application requesting exemption.[29] The FDA scrutinizes such applications to ensure the safety and autonomy of the human subjects involved.[30]

---

[24] *Lifchez* v. *Hartigan* (N.D. Ill. 1990) 735 F. Supp. 1361, 1377.

[25] *Carey* v. *Population Services Int'l*, 686.

[26] For a more detailed analysis, see Macintosh, *Enhanced Beings* (n 4), 159–61.

[27] Richard A. Merrill and Bryan J. Rose, 'FDA Regulation of Human Cloning: Usurpation or Statesmanship?' (2001) 15 *Harv. J.L. & Tech.* 85, 107–8.

[28] Ibid., 108–9.

[29] Ibid., 109. The IND application process applies to biological products as well as drugs. Ibid., 108, note 118.

[30] Ibid., 109.

Turning to germline editing, the foregoing regulatory regime does not apply to research on human gametes and embryos.[31] However, it does pertain to germline editing in reproduction. To explain, the FDA has long claimed that human somatic cell therapy products and gene therapy products qualify as biological products and/or drugs within its jurisdiction.[32] In 2015, after news broke that scientists had edited the genomes of human embryos, the National Institutes of Health (NIH) declared that it would not fund such experiments and added:

> The Public Health Service Act and the Federal Food, Drug, and Cosmetic Act give the FDA the authority to regulate cell and gene therapy products as biological products and/or drugs, which would include oversight of human germline modification. During development, biological products may be used in humans only if an investigational new drug application is in effect (21 CFR Part 312).[33]

In other words, the NIH named the FDA as the agency responsible for regulating clinical trials of germline editing in humans. The statement indicated that the FDA had jurisdiction because biological products and/or drugs were involved.

Shortly after the NIH issued its statement, Congress enacted the Consolidated Appropriations Act, 2016. The Act funded the Department of Health and Human Services (DHHS), which includes the FDA. It provided:

> None of the funds made available by this Act may be used to notify a sponsor or otherwise acknowledge receipt of a submission for an exemption for investigational use of a drug or biological product under section 505(i) of the Federal Food, Drug, and Cosmetic Act (21 U.S.C. 355(i)) or section 351(a)(3) of the Public Health Service Act (42 U.S.C. 262(a)(3)) in research in which a human embryo is intentionally created or modified to include a heritable genetic modification. Any such submission shall be deemed to have not been received by the Secretary, and the exemption may not go into effect.[34]

---

[31] *See* NAS Report, *Human Genome Editing* (n 10), 42 (stating that preclinical research on human gametes and embryos is subject to state regulation and funding restrictions, while clinical trials are subject to FDA regulation).

[32] Food and Drug Administration, *Application of Current Statutory Authorities to Human Somatic Cell Therapy Products and Gene Therapy Products* (58 Fed. Reg. 53248, 1993) (Application of Current Statutory Authorities).

[33] FS Collins, *Statement on NIH Funding of Research Using Gene-editing Technologies in Human Embryos* (National Institutes of Health, 29 April 2015) www.nih.gov/about/director/04292015_statement_gene_editing_technologies.htm accessed 12 October 2018 (Statement on NIH Funding of Research).

[34] Consolidated Appropriations Act 2016, 114 Pub. L. No. 113, Div. A, Title VII, s 749, 129 Stat. 2242, 2283.

By disabling the FDA from carrying out its mission, Congress imposed a moratorium de facto on germline editing in human reproduction. It included the same provision in appropriations bills for 2017 and 2018,[35] suggesting that the moratorium may continue indefinitely.

To illustrate, suppose a scientist modifies human gametes (which could be used later to create embryos) or modifies existing embryos. Her goal is to conduct a clinical trial in which men and women reproduce using these gametes or embryos. However, by virtue of this provision in the Consolidated Appropriations Act, the FDA cannot receive an IND application, leaving the scientist no legal way to obtain an exemption and conduct the clinical trial.[36]

II NON-HUMAN ANIMALS Before employing germline editing in human reproduction, scientists may wish to experiment by using germline editing to create non-human animals. Such preclinical research in non-human animals is subject to federal legislation.

The Federal Food, Drug, and Cosmetic Act prohibits introduction of a new animal drug into interstate commerce unless the FDA approves it.[37] A person who wishes to conduct a clinical investigation of a new animal drug can request an exemption pursuant to FDA regulations.[38] The FDA considers the 'recombinant DNA construct in a genetically engineered animal' to be a new animal drug and views development of that animal as a clinical investigation necessitating an exemption.[39] Fortunately for researchers, the FDA has decided not to enforce these requirements for '[genetically engineered] animals of non-food species that are raised and used in contained and controlled conditions such as [genetically engineered] laboratory animals used in research institutions'.[40] A proposed revision of its guidance document refers to animals with intentionally altered genomic DNA, rather than genetically engineered animals, and continues the policy of non-enforcement for lab animals used at research institutions.[41]

---

[35] Consolidated Appropriations Act 2017, 115 Pub. L. No. 31, Div. A, Title, VII, s 736, 131 Stat.135, 173; Consolidated Appropriations Act 2018, 115 Pub. L. No. 141, Div. A, Title VII, s 734, 132 Stat. 348, 389.

[36] Macintosh, *Enhanced Beings* (n 4), 124–25.

[37] See 21 U.S.C. ss 321(v), 331(a), 351(a)(5), and 360b(a)(1)(A).

[38] Ibid., s 360b(j) (authorizing FDA to issue regulations exempting new animal drugs intended for investigational use); 21 C.F.R. s 511.1(b) (detailing the rules for establishing the exemption).

[39] Food and Drug Administration, *Guidance for Industry #187, Regulation of Genetically Engineered Animals Containing Heritable Recombinant DNA Constructs* (June 2015) 6, 9.

[40] Ibid., 7.

[41] Food and Drug Administration, *Guidance for Industry #187, Regulation of Intentionally Altered Genomic DNA in Animals, Draft Guidance* (January 2017) 8.

Also, the Animal Welfare Act[42] regulates research conducted on dogs, cats, monkeys, guinea pigs, hamsters, and rabbits but does not protect certain other animals, including birds, rats, and mice bred for research.[43] The Animal and Plant Health Inspection Service, which is part of the Department of Agriculture, ensures compliance with the Animal Welfare Act.[44] Institutions that take funds from the Public Health Service must comply with its Policy on Humane Care and Use of Laboratory Animals,[45] including the establishment of an Institutional Animal Care and Use Committee to review proposed experiments.[46]

c Federal Regulations

The so-called Common Rule is a body of DHHS regulations that establishes a framework for research involving human subjects. Multiple other federal departments and agencies incorporate it into their own regulations or comply with it due to executive order.[47] The Rule has recently been revised, with a general compliance date scheduled for 21 January 2019.[48] Key elements include Institutional Review Board (IRB) evaluation of research proposals (including a risk-benefit calculus) and procurement of informed consent from human subjects.[49] Related regulations provide additional protections

---

[42] 7 U.S.C. ss 2131–2159.

[43] Ibid., s 2132(g).

[44] NAS Report, *Human Genome Editing* (n 10), 45.

[45] Office of Laboratory Animal Welfare, *PHS Policy on Humane Care and Use of Laboratory Animals* (National Institutes of Health, Revised 2015) https://olaw.nih.gov/policies-laws/phs-policy.htm accessed 30 September 2018 (PHS Policy on Humane Care and Use of Laboratory Animals).

[46] National Research Council of the National Academies, *Science, Medicine, and Animals* (The National Academies Press, 2004) 31.

[47] Macintosh, *Enhanced Beings* (n 4), 127. For a list of other departments and agencies with citations, see Office of Human Research Protections, *Federal Policy for the Protection of Human Subjects ('Common Rule')* (U.S. Department of Health & Human Services, 18 March 2016) www.hhs.gov/ohrp/regulations-and-policy/regulations/common-rule/index.html accessed 22 August 2018.

[48] The revised Common Rule took effect on 19 July 2018. Regulated entities must comply with the pre-revision Common Rule until 21 January 2019 but have the option to implement certain burden-reducing provisions during the six-month delay. For more details, see Federal Policy for the Protection of Human Subjects: Six Month Delay of the General Compliance Date of Revisions While Allowing the Use of Three Burden-Reducing Provisions During the Delay Period (19 June 2018) 83 Fed. Reg. 28497. Like the original Common Rule, the revised Common Rule will be codified at 45 C.F.R. section 46.101 through section 46.124. Citations in this chapter are to the revised sections as published in the Federal Policy for the Protection of Human Subjects (19 January 2017) 82 Fed. Reg. 7149.

[49] For a discussion of IRBs and informed consent, see NAS Report, *Human Genome Editing* (n 10), 45–48.

for research subjects who are pregnant women, fetuses, and neonates[50] or children.[51]

The Common Rule applies to research conducted, funded, or otherwise regulated by a federal department or agency that adopts it.[52] As Section I.4 will explain, scientists can receive federal funds to support research in which human gametes are edited, but not research in which human embryos are edited – a limitation that makes it seem as if the Common Rule has little relevance here. However, this appearance is deceiving. Many private research institutions apply the Rule to *all* research conducted there, whatever the funding source.[53]

Still, the Common Rule has a major loophole: independent scientists and fertility clinics that refuse federal funds need not comply with it.[54] For example, a wealthy scientist with a privately funded laboratory could edit human embryos to her heart's content without minding the Common Rule and its restrictions. More significantly, a fertility doctor could transfer a modified gamete or embryo to a woman to initiate a pregnancy, and the Common Rule and related regulations would not constrain his activities.

d State Legislation

State legislatures have not yet enacted laws aimed directly at germline editing. However, many states restrict research on human embryos, which should give pause to scientists who plan to combine edited gametes to create human embryos or to edit human embryos. For instance, in Louisiana, an *in vitro* fertilized ovum cannot be cultured for research.[55] In Minnesota and South Dakota, human embryos may not be subjected to experiments that could kill or harm them, as genome editing certainly could.[56] Arizona makes it a felony to disaggregate a human embryo to create a pluripotent stem cell line, a law that could ensnare a researcher who creates a stem cell line from an edited

---

[50] Additional Protections for Pregnant Women, Human Fetuses and Neonates Involved in Research, 45 C.F.R. ss 46.201–207.

[51] Additional Protections for Children Involved as Subjects in Research, 45 C.F.R. ss 46.401–409.

[52] 45 C.F.R. s 46.101(a); see also 45 C.F.R. s 46.201 (providing that regulations that protect human subjects who are pregnant women, fetuses, and neonates apply to research conducted or supported by DHHS); and 45 C.F.R. s 46.401 (providing that regulations that protect human subjects who are children apply to research conducted or supported by DHHS).

[53] NAS Report, *Human Genome Editing* (n 10), 40, note 6.

[54] Macintosh, *Enhanced Beings* (n 4), 131–32.

[55] La. Rev. Stat. s 9:122.

[56] Minn. Stat. ss 145.421-.422; S.D. Codified Laws ss 34–14-16 to -20.

embryo.[57] Florida, Maine, New Mexico, North Dakota, Pennsylvania and Rhode Island restrict experimentation on human fetuses or unborn children,[58] and their laws may extend to embryos.[59] Even in states that allow embryo research in general, unexpected barriers can arise. For example, in Massachusetts and Missouri, scientists cannot create an embryo via fertilization for research or stem cell research, respectively.[60]

State laws restricting human embryo research may also limit reproductive uses of edited gametes or embryos, although defences are possible. For example, fertility doctors and patients who employ edited gametes or embryos in the context of *in vitro* fertilization may argue that they are engaged not in research or experimentation but rather in procreation.[61] However, fertility specialists will probably find it safer to offer germline editing services only in states without laws against human embryo research.

Seventeen states of the Union also have laws against human cloning.[62] Most address human cloning specifically, but a few include vague language that could extend to germline editing. For example, in Arizona, it is illegal to create an embryo by any means other than union of human egg with human sperm.[63] Depending on how the term 'create' is read, the law may bar genomic modification of an embryo in the lab or fertility clinic. Illinois and Missouri ban human cloning, defined as the transfer to a woman of an embryo produced by any method other than fertilization of human egg by human sperm.[64] Scientists in Illinois and Missouri are safe as long as they do not transfer embryos to women. However, fertility doctors and their patients who employ germline editing in conjunction with *in vitro* fertilization could be in trouble if a court later decides that the modified embryos were produced by a method other than fertilization.[65]

Lastly, in Arizona and Louisiana, it is illegal to create 'human-animal hybrids', a category that includes human embryos into which non-human

---

57 Ariz. Rev. Stat. ss 36.2311.1, 36.2313.
58 Fla. Stat. s 390.0111(6); Me. Rev. Stat. tit. 22, s 1593; N.M. Stat. s 24-9A-3; N.D. Cent. Code s 14–02.2–01.1; 18 Pa. Cons. Stat. s 3216(a); R.I. Gen. Laws s 11–54-1(a).
59 Macintosh, *Enhanced Beings* (n 4) 132.
60 Mass. Gen. Laws Ch. 111L, s 8(b); Mo. Const. of 1945, art. III, s 38(d).2(2).
61 Macintosh, *Enhanced Beings* (n 4) 133.
62 The states are: Arizona, Arkansas, California, Connecticut, Illinois, Indiana, Iowa, Maryland, Massachusetts, Michigan, Missouri, Montana, New Jersey, North Dakota, Oklahoma, South Dakota, and Virginia. Ibid., 134.
63 Ariz. Rev. Stat. s 36–2312.A
64 410 Ill. Comp. Stat. 110/40; Mo. Const. of 1945, art. III, s 38(d).2(1), .6(2).
65 Macintosh, *Enhanced Beings* (n 4) 134.

cells or cellular components have been introduced.[66] Molecular editing tools often employ enzymes native to bacteria, so the laws may preclude their use to modify human embryos. Tools built with DNA-free sources of these enzymes may provide a workaround for scientists, fertility doctors, and patients.[67]

e Policies and Guidelines

The National Academy of Sciences and National Academy of Medicine are private, non-profit organizations that provide impartial advice on scientific and medical issues.[68] In 2017, they released a report on human genome editing which included a discussion of germline editing.[69] The report recommended that basic lab research – including experiments in which scientists edit the genomes of human gametes and embryos – proceed within the existing regulatory framework for such research.[70] Further, it noted that heritable genomic editing offered potential benefits, particularly for carriers of genetic mutations and their children, but acknowledged that more research was needed before clinical trials could proceed.[71] However, if technical challenges were surmounted, and benefits justified risks, regulators could approve clinical trials designed to prevent a serious disease or condition under specified conditions, including oversight to protect research subjects and their descendants.[72] For the time being, regulators should not approve clinical trials for reasons other than treating or preventing disease or disability.[73] In other words, the report distinguished therapeutic and enhancing uses of germline editing, even as it acknowledged the inherent blurriness of that distinction.[74]

Similarly, in 2017, the American Society of Human Genetics, a private organization of human genetics specialists,[75] opined that research in which human gametes and embryos are edited should be lawful and qualified to receive public funds.[76] However, it concluded that attempts to initiate

---

[66] Ariz. Rev. Stat. ss 36–2311.2(a), 36–2312.B.1; La. Rev. Stat. s 14:89.6.A(1), D(1)(a).

[67] Macintosh, *Enhanced Beings* (n 4) 135.

[68] 'Who We Are' (The National Academies of Sciences, Engineering, and Medicine) www .nationalacademies.org/about/whoweare/index.html accessed 18 August 2018.

[69] NAS Report, *Human Genome Editing* (n 10).

[70] Ibid., 81–82, Recommendation 3.1.

[71] Ibid., 113–16, 134.

[72] Ibid., 134–35, Recommendation 5–1.

[73] Ibid., 159, Recommendation 6–1.

[74] Ibid., 191–92.

[75] For a description of the organization, see 'About the ASHG' (American Society of Human Genetics, 2016) www.ashg.org/about/index.shtml accessed 6 September 2018.

[76] Kelly E. Ormond and others, 'Human Germline Genome Editing' (2017) 101 *Am. J. Human Genetics* 167.

a human pregnancy should not be made at this time. It did not close the door on clinical applications in the future, but indicated such uses would require a compelling medical rationale and an ethical justification.[77] Significantly, the American Society for Reproductive Medicine, an influential organization of doctors, nurses, and other professionals in the fertility industry, endorsed the statement.[78] Its endorsement is a reminder of the importance of industry self-regulation and provides reassurance that reputable fertility doctors and clinics will not offer germline editing prematurely.

In 2016, the International Society for Stem Cell Research (ISSCR), a non-profit organization, issued its *Guidelines for Stem Cell Science and Clinical Translation*.[79] The ISSCR supported research in which the nuclear genomes of human gametes or embryos were edited, but stated that ethical and scientific issues had to be resolved before modified embryos were used in human reproduction.[80] It advocated that such research occurs within its 'embryo research oversight process', a general procedure for all human embryo research detailed in its document.[81]

Finally, in 1994, the NIH released a report describing the conditions under which the federal government might fund human embryo research. This report suggested that funded research should not extend past the fourteenth day after fertilization, when the primitive streak appears in the embryo.[82] Ten years later, in 2004, an advisory body that counselled President George W. Bush on bioethical matters recommended that Congress restrict embryo research to the first ten to fourteen days after fertilization,[83] but Congress did not enact any such legislation. Thus, unlike some other nations, the United States has not codified the fourteen-day rule as a matter of federal law. However, professional associations and institutions often adopt the fourteen-day rule as a guiding principle.[84] Researchers who wish to edit the genomes of human embryos must mind this potential restriction.

---

[77]   Ibid.
[78]   Ibid.
[79]   ISSCR Guidelines Updates Task Force, *Guidelines for Stem Cell Science and Clinical Translation* (International Society for Stem Cell Research, May 2016) www.isscr.org/docs/default-source/all-isscr-guidelines/guidelines-2016/isscr-guidelines-for-stem-cell-research-and-clinical-translation.pdf?sfvrsn=4 accessed 1 October 2018.
[80]   Ibid., 8, Recommendation 2.1.4.
[81]   Ibid., 5–7, Recommendations 2.1.1, 2.1.2, 2.1.3.
[82]   National Institutes of Health, *Report of the Human Embryo Research Panel, Volume I* (September 1994) 67.
[83]   President's Council on Bioethics, *Reproduction and Responsibility: The Regulation of New Biotechnologies* (March 2004) 223.
[84]   Ibid., 141.

2 Definitions

Because the United States has no single source of national legislation on germline editing, identifying key terms and their definitions is not a simple task. The legal authorities cited in Section I.1 define more terms than can be discussed in this brief chapter. Therefore, this section will comment on a few terms that have particular salience.

a Research on Human Gametes and Embryos

First, consider scientists engaged in research on human gametes and embryos. As Section I.1.c explained, most will be obliged to comply with Common Rule protections for human subjects. The Common Rule defines 'human subject' this way:

> Human subject means a living individual about whom an investigator (whether professional or student) conducting research:
> (i) Obtains information or biospecimens through intervention or interaction with the individual, and uses, studies, or analyzes the information or biospecimens; or
> (ii) Obtains, uses, studies, analyzes, or generates identifiable private information or identifiable biospecimens.[85]

Significantly, the definition is limited to a 'living individual'. Human gametes and embryos are not considered to be living individuals but men and women who donate them are. These donors qualify as human subjects when scientists extract gametes from them (an intervention), interview them (an interaction), access medical records linked to names (identifiable private information), or perform research on gametes or embryos associated with names (identifiable biospecimens). However, anonymous donors are not human subjects when scientists do not intervene or interact with them.[86]

Scientists engaged in basic research must also pay attention to the term 'human embryo'. As Section I.4 explains in more detail below, federal funds are not available to support most research on human embryos. The rider that imposes this restriction defines 'human embryo or embryos' to include any organism 'that is derived by fertilization, parthenogenesis, cloning, or any other means from one or more human gametes or human diploid cells'.[87]

---

[85]   45 C.F.R. s 46.102(e)(1).

[86]   Macintosh, *Enhanced Beings* (n 4), 128–29.

[87]   Consolidated Appropriations Act 2018, 115 Pub. L. No. 141, Div. H, Title V, s 508(b), 132 Stat. 348, 764.

Through this broad definition, Congress protects as many kinds of embryos as possible. Scientists cannot sidestep this restriction by working with embryos derived through cloning or other unusual means. Moreover, the rider does not define the term 'organism'. The *Cambridge Dictionary* defines 'organism' as a 'single living plant, animal, or other living thing'.[88] Thus, even fertilized eggs or zygotes that are alive might qualify as embryos.

As noted in Section I.1.d, certain states restrict research on human embryos and/or fetuses. Louisiana refers specifically to the 'in vitro fertilized ovum' that shall not be cultured for research.[89] South Dakota, which bars research that destroys an embryo, defines 'human embryo' as a 'living organism of the species Homo sapiens at the earliest stages of development (including the single-celled stage) that is not located in a woman's body'.[90] Other states adopt definitions broad enough to protect anything from a fertilized egg or zygote to a much more developed entity. For example, Arizona, which prohibits the disaggregation of any human embryo to create pluripotent stem cells, defines 'human embryo' as 'a living organism of the species homo sapiens through the first fifty-six days of its development'.[91] Minnesota protects the 'human conceptus', taken to mean 'any human organism, conceived either in the human body or produced in an artificial environment other than the human body, from fertilization through the first 265 days thereafter'.[92] New Mexico, which precludes research on the 'fetus', describes that entity as 'the product of conception from the time of conception until the expulsion or extraction of the fetus or the opening of the uterine cavity'.[93] Pennsylvania, which prohibits nontherapeutic research on an 'unborn child', defines that term to include 'an individual organism of the species homo sapiens from fertilization until live birth'.[94] Lastly, Rhode Island, which also bars research on the 'fetus', adopts an expansive definition that includes both embryos and neonates.[95]

States differ in their willingness to protect unusual embryos. For example, by adopting definitions that avoid any reference to fertilization, Arizona and South Dakota extend protection to cloned human embryos. In contrast, Massachusetts, which prohibits the creation of human embryos for research

---

[88] 'Organism', Cambridge Dictionary https://dictionary.cambridge.org/us/dictionary/english/organism accessed 29 September 2018.

[89] La. Rev. Stat. s 9:122.

[90] S.D. Codified Laws s 34-14-20.

[91] Ariz. Rev. Stat. s 36-2311.3.

[92] Minn. Stat. ss 145.421-.422(1).

[93] N.M. Stat. s 24-9A.1.G.

[94] 18 Pa. Cons. Stat. s 3203.

[95] R.I. Gen. Laws. s 11-54-01(f).

via fertilization, expressly excludes from protection embryos created for research through somatic cell nuclear transfer (cloning), parthenogenesis, or other asexual methods.[96] Missouri bars the creation of human embryos for stem cell research via fertilization, which it defines as 'the process whereby an egg of a human female and the sperm of a human male form a zygote (i.e., fertilized egg)'.[97] Through these definitions, Massachusetts and Missouri favor one class of human embryo over another and open the door to genomic editing of clones and parthenotes.

b Germline Editing in Human Reproduction

Next, consider scientists, fertility doctors, and others who wish to use germline editing to initiate pregnancies. The NIH has said that the FDA has jurisdiction over clinical trials pursuant to the laws that give it the power to regulate 'drugs' and 'biological products'.[98] Federal law defines these terms as follows:

> The term [']drug['] means (A) articles recognized in the official United States Pharmacopoeia, official Homoeopathic Pharmacopoeia of the United States, or official National Formulary, or any supplement to any of them; and (B) articles intended for use in the diagnosis, cure, mitigation, treatment, or prevention of disease in man or other animals; and (C) articles (other than food) intended to affect the structure or any function of the body of man or other animals; and (D) articles intended for use as a component of any article specified in clause (A), (B), or (C).[99]
>
> The term [']biological product['] means a virus, therapeutic serum, toxin, antitoxin, vaccine, blood, blood component or derivative, allergenic product, protein (except any chemically synthesized polypeptide), or analogous product, or arsphenamine or derivative of arsphenamine (or any other trivalent organic arsenic compound), applicable to the prevention, treatment, or cure of a disease or condition of human beings.[100]

In 1993, the FDA issued a document explaining why human somatic cell therapy products and gene therapy products were 'biological products' and/or 'drugs' within its jurisdiction.[101] Decades have passed since then and technology has advanced. The FDA has yet to explain how and why these defined terms apply when human gametes and embryos are edited and put to

[96] Mass. Gen. Laws. c 111L, s 8(b).
[97] Mo. Const. of 1945, art. III, ss 38(d).2(2) and 6(4).
[98] *Statement on NIH Funding of Research* (n 33).
[99] 21 U.S.C. s 321(g)(1).
[100] 42 U.S.C. s 262(i)(1).
[101] Application of Current Statutory Authorities (n 32).

reproductive use. To aid scientists and clinicians, it should provide an updated account of the basis for its jurisdiction over this sensitive form of human reproduction.

3 Oversight Bodies

The United States has not enacted legislation to control germline editing and thus has not established a new agency to manage its use. As Section I.1.b explained, the FDA does not regulate research performed on human gametes or embryos, but it does control clinical trials and applications in humans. To the extent clinical trials and applications of germline editing involve biological products, the Center for Biologics Evaluation and Research (CBER), a unit within the FDA, is responsible for providing regulatory oversight.[102] However, such oversight is theoretical at the moment, for Congress has barred the FDA from receiving applications for clinical trials of germline editing.

The FDA oversees development of non-human animals with edited genomes. As Section I.1.b noted, currently it is not enforcing its regulations against scientists who create such animals for use in research institutions. The Animal Welfare Act is also relevant, and the Animal and Plant Health Inspection Service, an agency within the Department of Agriculture, oversees compliance with it.

Two other federal bodies are relevant: the DHHS and NIH. As Section I.1.c explained, the DHHS issues regulations that control federally funded research involving human subjects, and as discussed immediately below, the NIH issues guidelines that regulate all recombinant DNA research conducted at institutions that receive NIH funds for any such research.

4 Funding

As mentioned previously, in 2015, the NIH stated that it would not fund research in which scientists edited the genomes of human embryos.[103] In justification, it mentioned the so-called Dickey-Wicker Amendment, a rider Congress has added to appropriations bills since 1996.[104] The most recent iteration of the rider provides:

---

[102] U.S. Food & Drug Administration, 'About the Center for Biologics Evaluation and Research (CBER)' (2017) www.fda.gov/aboutfda/centersoffices/officeofmedicalproductsandtobacco/c ber/ accessed 1 October 2018.

[103] *Statement on NIH Funding of Research* (n 33).

[104] NAS Report, *Human Genome Editing* (n 10) 42, note 10.

(a) None of the funds made available in this Act may be used for—

  (1) the creation of a human embryo or embryos for research purposes; or

  (2) research in which a human embryo or embryos are destroyed, discarded, or knowingly subjected to risk of injury or death greater than that allowed for research on fetuses in utero under 45 CFR 46.204(b) and section 498(b) of the Public Health Service Act (42 U.S.C. 289 g(b)).

(b) For purposes of this section, the term [']human embryo or embryos['] includes any organism, not protected as a human subject under 45 CFR 46 as of the date of the enactment of this Act, that is derived by fertilization, parthenogenesis, cloning, or any other means from one or more human gametes or human diploid cells.[105]

Stated in simpler terms, the Dickey-Wicker Amendment forbids the DHHS from funding projects in which scientists create human embryos for research, or destroy, discard or expose human embryos to risk of serious injury or death. Genome editing subjects human embryos to physical risk. Moreover, once research is complete, edited embryos are discarded. Thus, it is not surprising that the NIH will not fund such research. However, the NIH has not addressed research in which scientists edit the genomes of human gametes, or develop animals with edited genomes, suggesting it might be willing to fund such projects.

Whether research involves humans or non-human animals, the *NIH Guidelines for Research Involving Recombinant or Synthetic Nucleic Acid Molecules*[106] are relevant.[107] Designed to protect lab workers and the environment,[108] the Guidelines establish practices for assembling and handling recombinant nucleic acid molecules, synthetic nucleic acid molecules, and cells, organisms, and viruses that include those molecules.[109] If an

---

[105] Consolidated Appropriations Act 2018, 115 Pub. L. No. 141, Div. H, Title V, s 508, 132 Stat. 348, 764.

[106] National Institutes of Health, *NIH Guidelines for Research Involving Recombinant or Synthetic Nucleic Acid Molecules* (April 2016) https://osp.od.nih.gov/wp-content/uploads/2013/06/NIH_Guidelines.pdf accessed 3 October 2018 (NIH Guidelines).

[107] The NIH has proposed reducing its role in evaluating gene therapy proposals and leaving such matters to the FDA. F Collins and S Gottlieb, 'The Next Phase of Human Gene-Therapy Oversight' (2018) 379 *N. Engl. J. Med.* 1393; National Institutes of Health (NIH) Office of Science Policy (OSP), Recombinant or Synthetic Nucleic Acid Research: Proposed Changes to the NIH Guidelines for Research Involving Recombinant or Synthetic Nucleic Acid Molecules (NIH Guidelines) (17 August 2018) 83 Fed. Reg. 41082.

[108] NAS Report, *Human Genome Editing* (n 10) 39.

[109] *NIH Guidelines* (n 106) s I-A.

institution conducts or sponsors such research, and receives NIH funds to support any of it, the Guidelines apply to all such research conducted there.[110] An institution that is subject to the Guidelines must establish an Institutional Biosafety Committee (IBC) to review and approve such research.[111]

The Guidelines state that the NIH will not consider proposals for 'germ line alteration', defined as 'a specific attempt to introduce genetic changes into the germ (reproductive) cells of an individual, with the aim of changing the set of genes passed on to the individual's offspring'.[112] If a scientist wishes to conduct an experiment in which men and women procreate with edited gametes, the NIH will not consider her proposal. However, if she plans only to edit gametes in the lab, altered genes will not be passed on to offspring, and NIH review is available. Thus, she can comply with the Guidelines, as a recipient of federal funds must. One significant limitation should be noted: because of the Dickey-Wicker Amendment, those who receive NIH funds cannot test the viability of edited gametes by conceiving human embryos with them.[113]

II SUBSTANTIVE PROVISIONS

This section applies the foregoing legal authorities to hypothetical scenarios to illustrate which experiments are permitted, prohibited, or restricted in the United States.

1 *Regulation of Basic Research*

Suppose a scientist wishes to conduct basic research in which she edits the genomes of human gametes and/or embryos. As a part of this research, she might create embryos for the specific purpose of editing them; or, she might test the viability of edited gametes by combining them to create embryos. As Section I.1.b explains, under federal statutory law as it currently stands, such work is permitted as long as she does not attempt to initiate a pregnancy. However, depending on her funding source and research institution, federal regulations may apply.

To elaborate, assume the scientist wants to edit donated human embryos. As Section I.4 noted, the NIH will not fund her research. She may obtain

---

[110]   Ibid., s I-C-1-a-(1).
[111]   Ibid., s IV-B-2 (describing membership, procedures, and functions of the IBC).
[112]   Ibid., 100, Appendix M.
[113]   Macintosh, *Enhanced Beings* (n 4) 127.

private funds to support her research, in which case the Common Rule does not apply to her work directly. However, she still must comply with the procedures adopted by her research institution, and those procedures probably incorporate the Common Rule. Thus, as detailed in Section I.1.c, she must submit a proposal to an IRB and obtain informed consent from donors if their names are linked to the embryos. If her research institution follows the fourteen-day rule for research on human embryos, she cannot continue her work beyond that developmental threshold. Further, many research institutions in the United States receive NIH funds for recombinant DNA research. If the scientist works at such an institution, she must comply with the NIH Guidelines and submit to IBC oversight, as noted in Section I.4.

Alternatively, assume the scientist wants to edit human gametes. The NIH has not clarified whether it will fund such research. If it does, the scientist must comply with the Common Rule and the NIH Guidelines. But, again, even if her research is privately funded, her research institution might have incorporated the Common Rule into its own procedures and/or have taken NIH funds for other recombinant DNA research, thereby obliging her to comply with the Common Rule and/or NIH Guidelines.

The scientist must also observe the laws of the state where she conducts her experiments. Although states have not yet prohibited germline editing as such, one or more could do so at any time. In addition, various laws regarding human embryo research, human cloning, and human-animal hybrids might also apply to and restrict her work. For example, a scientist cannot culture fertilized ova for research in Louisiana. She cannot edit embryos in Minnesota or South Dakota because her work might kill or harm them. She cannot create an embryo via fertilization for research in Massachusetts or stem cell research in Missouri. Other state laws that could restrict germline editing are discussed in Section I.1.d and listed in Table 4.2.

2 Regulation of Preclinical Research

Suppose a scientist wants to use CRISPR/Cas9 or other molecular tools to edit non-human gametes or embryos. She may even plan to use the edited products to create non-human animals with modified genomes. Under federal law, such preclinical research is permitted. As Section I.1.b explained, the FDA considers animals with intentionally altered genomic DNA to be within its jurisdiction but does not intend to enforce its regulations against those who develop and maintain lab animals in research

TABLE 4.2 *United States: selected state legislation (as of November 2018)*

Bans Or Restricts What	States	Citations
Restricts research on human embryos and/ or fetuses	Arizona, Florida, Louisiana, Maine, Massachusetts, Minnesota, Missouri, New Mexico, North Dakota, Pennsylvania, Rhode Island, South Dakota	Ariz. Rev. Stat. ss 36.2311.1, 36.2313; Fla. Stat. s 390.0111(6); La. Rev. Stat. s 9:122; Me. Rev. Stat. tit. 22, s 1593; Mass. Gen. Laws Ch. 111L, s 8(b); Minn. Stat. ss 145.421-.422; Mo. Const. of 1945, art. III, s 38(d).2(2); N.M. Stat. ss 24-9A-1, 24-9A-3; N.D. Cent. Code s 14-02.2-01.1; 18 Pa. Cons. Stat. ss 3203, 3216(a); R.I. Gen. Laws s 11-54-1(a), (f); S.D. Codified Laws ss 34-14-16 to -20.
Bans human cloning; statute may reach germline editing	Arizona, Illinois, Missouri	Ariz. Rev. Stat. s 36-2312.A; 410 Ill. Comp. Stat. 110/40; Mo. Const. of 1945, art. III, s 38 (d).2(1), .6(2).
Bans creation of human-animal hybrids; statute may reach germline editing	Arizona, Louisiana	Ariz. Rev. Stat. ss 36-2311.2(a), 36-2312.B.1; La. Rev. Stat. s 14:89.6.A(1), D(1)(a).

institutions. Thus, if the scientist plans to confine her animals to her lab, she need not comply with burdensome FDA regulations. However, if the Animal Welfare Act applies to the species with which she works, she must mind its restrictions.

Federal funds bring additional regulations into play. For example, suppose the scientist works at an institution that receives funds from the Public Health Service. As Section I.1.b described, under the Policy on Humane Care and Use of Laboratory Animals her institution must establish an Institutional Animal Care and Use Committee to review research proposals, including hers.[114] And as noted in Section I.4, if she works at an institution that receives federal funds for any research subject to the NIH Guidelines, she too must comply with the Guidelines, and an IBC will oversee her research.

---

[114] *PHS Policy on Humane Care and Use of Laboratory Animals* (n 45) 31.

3 *Regulation of Clinical Research*

Suppose a scientist wishes to conduct a clinical trial in which she edits the genomes of human gametes and/or embryos and then puts them to reproductive use. For example, she proposes inseminating women with edited sperm, or transferring edited embryos to their uteri to bring about the birth of human beings with heritable modifications. As explained in Section I.1.b, the United States does not have any federal law that expressly prohibits germline editing. Nevertheless, her clinical trial is not permitted. To proceed, she would first have to submit an IND application to the FDA, but Congress has barred that agency from receiving such applications. If Congress lifts this restriction, and the FDA begins to receive applications, the agency will demand evidence that the proposed clinical trial is safe for human subjects before allowing her to proceed. Given the human and political stakes, the FDA will be in no rush to authorize such a trial.

4 *Regulation of Clinical Applications*

Lastly, suppose a fertility clinic offers *in vitro* fertilization cycles to the public. Carriers of genetic mutations come to the clinic seeking correction of the mutations in their sperm or eggs prior to the cycle, hoping that any embryos created and transferred to uteri will be mutation-free and produce healthy children. Alternatively – but more speculatively – men and women who desire children with enhanced traits ask the clinic to edit their gametes accordingly prior to the cycle. In either case, heritable modifications are involved. Again, the United States does not have a federal law that expressly prohibits the clinic from providing germline editing services. Even so, clinical applications of germline editing are not permitted. Due to the de facto moratorium Congress has imposed, the FDA cannot receive requests for clinical trials, let alone provide the premarket approval that clinical applications require. The FDA may require decades of clinical trials before approving a single clinical application.[115] Arguably, this arbitrary delay in clinical trials denies carriers

---

[115] For example, the FDA approved the first gene therapy clinical trial in 1990. E M Kane, 'Human Genome Editing: An Evolving Regulatory Climate' (2017) 57 *Jurimetrics J*. 301, 307. Twenty-seven years later, in 2017, it approved the first gene therapy for use in the United States: Kymriah, which treats leukaemia in children and young adults. U.S. Food & Drug Administration, 'FDA Approval Brings First Gene Therapy to the United States' (30 August 2017) www.fda.gov/NewsEvents/Newsroom/PressAnnouncements/ucm574058.htm accessed 3 October 2018. It plans to release a new framework for approval of gene therapy products soon. U.S. Food & Drug Administration, 'Remarks by Commissioner Gottlieb to the Alliance

of genetic mutations and their children the ability to 'enjoy the benefits of scientific progress and its applications'[116] within a reasonable time.

III CURRENT PERSPECTIVES AND FUTURE POSSIBILITIES

In the United States today, scientific organizations and lawmakers hold divergent opinions about germline editing. Consider first basic research in which scientists edit the genomes of human gametes and embryos. As noted in Section I.1.e, the American Society of Human Genetics has declared that such research should be legal and scientists who perform it should have access to public funds. However, although scientists in most states can edit human gametes and embryos without breaking the law, the NIH has refused to fund research on human embryos, due in part to the Dickey-Wicker Amendment. This refusal to fund discourages talented scientists from entering the field and making important discoveries.

Consider next clinical trials in which scientists use edited gametes and embryos to help men and women procreate. While acknowledging that germline editing is not ready for reproductive use, the National Academy of Sciences and the National Academy of Medicine have suggested that it may be ethical in the future to conduct clinical trials designed to prevent transmission of a serious genetic disease. By contrast, Congress has slammed the door on clinical trials by refusing to let the FDA receive applications.

Looking to the future, one can safely predict that scientific innovation will continue outside the United States. Somewhere else scientists will perfect CRISPR/Cas9 and other molecular tools and learn the most effective means of modifying human gametes and embryos. Some may use germline editing to help parents have children free of deadly genetic diseases. Indeed, as this book goes to press, germline editing may already have been used in humans, albeit prematurely. A Chinese scientist claims to have edited the CCR5 gene in human embryos, resulting in the birth of twin girls who may be resistant to infection with the human immunodeficiency virus (HIV).[117] Meanwhile, back

---

for Regenerative Medicine's Annual Board Meeting' (22 May 2018) www.fda.gov/NewsEvents/ Speeches/ucm608445.htm accessed 3 October 2018.

[116] International Covenant on Economic, Social and Cultural Rights (n 5) art. 15.1.b.

[117] Associated Press, 'Chinese Researcher Claims to Have Altered Babies' DNA' (Fox News, 25 November 2018) www.foxnews.com/health/chinese-researcher-claims-to-have-altered-babies-dna accessed 25 November 2018.

in the United States, Congress will probably respond to this and other reports of children born abroad in one of three different ways.

First, if the public reacts favourably to the news, Congress may decide to lift its moratorium. A 2018 poll suggests this scenario is realistic: 72 per cent of Americans already approve of altering an 'unborn baby's' genetic traits to treat a serious disease or condition it would otherwise have at birth.[118] Once the FDA can receive IND applications, it can allow clinical trials of germline edits that have been proven safe in other countries; and, if the data from clinical trials are reassuring, it can approve clinical applications. Because correction of genetic mutations is likely to be simpler than engineering polygenic traits such as intelligence, FDA regulation may naturally yield a distinction between therapeutic applications, which have a chance of being approved, and enhancing applications, which do not.[119] However, FDA regulation has its limits. Generally, the agency reviews applications for clinical trials to ensure the safety of human subjects and renders decisions without regard to moral concerns.[120] And critics have raised moral concerns about germline editing, asserting that it expresses hubris, disrespects nature, harms human dignity, and impairs human diversity.[121]

Second, if the public reacts unfavourably to the news of children born abroad, Congress may attempt to enact comprehensive federal legislation to restrict germline editing. According to the same 2018 poll, 65 per cent of Americans oppose the testing of gene editing on human embryos,[122] a statistic that hints at support for proposals to ban all germline editing. Yet, 72 per cent approve of altering an unborn baby's genetic traits to treat a serious disease or condition, while 80 per cent think enhancing the intelligence of an unborn baby goes too far.[123] These results point to support for proposals to ban only certain reproductive uses of germline editing. Reaching a compromise may be difficult. Conservatives will not support legislation that permits germline editing for research but not reproduction,

---

[118] C Funk and M Hefferon, 'Public Views of Gene Editing for Babies Depend on How It Would Be Used' (3 Pew Research Center, 26 July 2018) http://assets.pewresearch.org/wp-content/uploads/sit es/14/2018/07/25173131/PS_2018.07.26_gene-editing_FINAL.pdf accessed 29 September 2018.

[119] Macintosh, *Enhanced Beings* (n 4) 151.

[120] NAS Report, *Human Genome Editing* (n 10) 134 (noting that FDA does not have authority to consider morality when deciding to allow clinical trials).

[121] Ibid., 111–12.

[122] Funk & Hefferon, 'Public Views of Gene Editing for Babies Depend on How It Would Be Used' (n 118) 4.

[123] Ibid., 3.

thereby compelling scientists to destroy or discard embryos when research is complete.[124]

Third, if public responses to children born abroad are mixed, and no compromise is possible, Congress might simply renew its moratorium, year after year, until it becomes a de facto ban on germline editing in human reproduction.[125] Federal legislators may prefer this solution, which allows them to duck the tough policy issues surrounding germline editing. Americans who carry genetic mutations and need germline editing to procreate safely will be out of luck unless they have the financial means to travel to countries where it is available.

Meanwhile, the fifty states will remain free to act on their own. To illustrate, imagine that Congress continues to renew its moratorium. That will not be enough to satisfy conservatives, because human embryos will still be subject to experimentation. States leaning politically to the conservative side of the spectrum might ban germline editing outright to protect embryos, while states of a liberal bent might prohibit only reproductive uses, thereby signalling to scientists and investors that basic research is safe there.[126] Unfortunately, politics has become very divisive in the United States, and the likelihood that the states will adopt a consistent response to this controversial new technology is small indeed.

In any event, a federal or state ban on germline editing may prompt scientists, doctors, and/or prospective parents to bring a challenge under the US Constitution. As noted in Section I.1.a, it is not easy to predict the outcome of cases reaching the US Supreme Court, which adds a further element of uncertainty to an already complex field of law. Optimistically, the Constitution will emerge as an important bulwark against legislative encroachments upon the right to participate in science and enjoy its fruits.

---

[124] *See* KL Macintosh, *Human Cloning: Four Fallacies and Their Legal Consequences* (Cambridge University Press 2013) 181–85 (explaining that Congress failed to ban human cloning due to a fundamental disagreement about the moral status of the human embryo).

[125] R Spivak, G Cohen, E Adashi, 'Moratoria and Innovation in the Reproductive Sciences: Of Pretext, Permanence, Transparency, and Time Limits' (2018) 14 *J. Health & Biomed. L.* 5, 23.

[126] Macintosh, *Enhanced Beings* (n 4) 152.

5

The Regulation of Human Germline Genome Modification in Mexico

María de Jesús Medina Arellano

I INTRODUCTION

In April 2016, Mexico made world headlines when the first child with DNA from three different individuals was born in Guadalajara.[1] The procedure was done by a team of researchers of the New Hope Fertility Center, New York City, led by Dr John Zhang, to prevent the child from inheriting a genetic disease from the mother's mitochondrial DNA. The embryo was created and manipulated in the United States but was implanted and brought to term in Mexico.

This was certainly not the first time in history that Americans travelled south of the border to do things that are either illegal or in the legal grey zone in the United States. Indeed, the current regulatory framework governing research on human embryos and genetic modification in Mexico is considerably less constraining, comprehensive, consistent and enforced compared with the United States and much of the developed world.[2] Mexico lacks specific legislation regulating assisted reproductive technology (ART) and genetic engineering. These issues are instead regulated by the General Health Law (GHL) (*Ley general de salud*), a very broad statute that has not kept up with developments in human embryo and genetic research.

As soon as the mitochondrial replacement procedure became public, Mexican lawmakers scrambled to figure out how to close the loophole that

---

[1] S Reardon, 'Three-Parent Baby Claim Raises Hope – and Ethical Concerns' (*Nature*, 228 September 2016) www.nature.com/news/three-parent-baby-claim-raises-hopes-and-ethical-concerns-1.20698 accessed 16 October 2018; S González Santos, N Stephens, R Dimond, 'Narrating the First "Three-Parent Baby": The Initial Press Reactions from the United Kingdom, the United States, and Mexico' (2018) 40:4 *Science Communication*, 419, 419–441.

[2] See, in general, M Medina Arellano, 'Contested Secularity: Governing Stem Cell Science in Mexico' (2012) 39:3 *Science and Public Policy* 386, 386–402.

made it possible.[3] The rapidly developing field of genetic engineering, with the advent of precise genome modification techniques, adds urgency. So far, Mexican legislators have been concerned mostly with genome modification of plants and animals (GMOs), biosafety and bio-piracy.[4] Also, in 2008, it became the first country in the world to 'nationalize' the genome of its population. The so-called Genomic Sovereignty Amendments to the General Health Law made the genome data of the Mexican population property of Mexico's government, by prohibiting and penalizing data collection and utilization in research without prior government approval. However, things might change soon. At the time of this writing, the Congress of the Union, the bicameral legislature of the federal government of Mexico, is considering amending the General Health Law to regulate more stringently assisted reproduction, surrogacy and ban research on human embryos.[5]

This chapter discusses the current regulatory framework for human germ-line genome modification in Mexico. First, it will review the overall legal system, highlighting its federal structure. Then, it will discuss the federal constitution, laws and regulations. Particular attention will be given to the GHL, to date the key legislation on the matter. Then, I will briefly review state legislation, mostly to highlight inconsistencies with the federal framework. Finally, since Mexico's legislation on the matter discussed in this volume is in a state of flux, I will conclude by advancing some considerations that I believe the legislators should keep in mind while rethinking the overall regulatory regime for interventions on the human genome and assisted reproductive technologies.

II THE REGULATORY FRAMEWORK

1 Overview

The Mexican legal system is a combination of the French civil law tradition and American constitutional and federal paradigms. As the United States of America, the United Mexican States (*Estados Unidos Mexicanos*) is a federal union, comprising 31 federated states and a Federal District containing the

---

[3] See, in general, S Chan, C Palacios-González, M Medina Arellano, 'Mitochondrial Replacement Techniques, Scientific Tourism, and the Global Politics of Science' (2017) 47:5 *Hastings Center Report* 7, 7–9.

[4] S Chan and M Medina Arellano, 'Genome Editing and International Regulatory Challenges: Lessons from Mexico' (2016) 2:3 *Ethics, Medicine and Public Health* 426, 431.

[5] S Reardon, 'Mexico Proposal to Ban Human-Embryo Research Would Stifle Science' (2016) 540:7632 *Nature* 180, 181.

capital of the Union, Mexico City.[6] As all countries with a civil law tradition, the main source of law in Mexico is statutory law. Under Article 133 of the Federal Constitution, '[t]his Constitution, the laws of the Congress of the Union that emanate therefrom, and all treaties that have been made and shall be made in accordance therewith by the President of the Republic, with the approval of the Senate, shall be the supreme law of the whole Union'.[7] In the hierarchy of laws, below the 'supreme law' of the Union sit five major Federal Codes (civil, commercial, criminal, civil procedure and criminal procedure), which are compilations of the basic rules regulating those broad fields.[8] Below these codes there are Federal Laws. In the Mexican legal system, there is a particular kind of law known as 'general laws', which are broad statutes that regulate entire fields of activities. For instance, as we will see, activities modifying the human genome fall under Mexico's General Health Law.[9] Finally, Federal Regulations and Mexican Official Norms (*Normas Oficiales Mexicanas* – NOMs) are complementary and mandatory administrative rules issued by federal authorities, regulating technical issues in specific areas.[10]

Politically, Mexico is a Presidential Republic, with the typical tripartite division of powers of modern democracies: Executive (the President of the Mexican Republic), Legislative (the Congress of the United States, also called the Federal Congress, which consists of the Senate and the House of Representatives) and the Judiciary (Mexican Supreme Court, federal and local courts). This tripartite division of power is also mirrored in each of the federated states. Within the Union, each of the federated entities has their own constitution, laws and regulation. State constitutions and regulations must comply with the Federal Constitution; if any inconsistencies or incompatibilities arise, the Federal Constitution prevails.[11]

---

[6] The United Mexican States are: Aguascalientes, Baja California, Baja California Sur, Chihuahua, Colima, Campeche, Coahuila, Chiapas, Durango, Guerrero, Guanajuato, Hidalgo, Jalisco, Estado de México, Michoacán, Morelos, Nayarit, Nuevo León, Oaxaca, Puebla, Querétaro, Quintana Roo, Sinaloa, San Luis Potosí, Sonora, Tabasco, Tlaxcala, Tamaulipas, Veracruz, Yucatán and Zacatecas. Federal District: Mexico City.

[7] *Constitución Política de los Estados Unidos Mexicanos* (Political Constitution of the Mexican States) (Federal Constitution) [1917, amended last August 2018].

[8] On the hierarchy of sources of the law in the Mexican legal system, see, in general, JA Vargas, *Introduction to Mexico's Legal System* (Legal Studies Research Papers School of Law, University of San Diego, 2008).

[9] See, in this Chapter, Section II.c.i.

[10] 'NOMs' can also be understood as 'Standards'. See, in general, WD Signet, 'Official Mexican Norms and Mexican Normalization: The Ticket to Modernization in an Emerging Economy?' (1997) 29:1/2 *The University of Miami Inter-American Law Review* 253, 253–296.

[11] JM Serna de la Garza, *The Constitution of Mexico, A Contextual Analysis* (Hart Publishing 2013), in particular Chapter '7: The Protection of Human Rights'.

Finally, the Federal Government and the federated entities share competence over the regulation of health, science and technology. Legislators both at the local and federal levels are able to enact regulation on these issues, but their approach to these matters varies greatly: overall, while the federal framework prohibits little, several states give embryos full legal personality at conception (by means of the local constitutions protection, civil and criminal local codes).[12] These considerations make an exhaustive and comprehensive overview of the regulatory framework for human germline genome modification in Mexico a challenge. This chapter will discuss mostly the federal regulatory framework.

2 *The Federal Legal Framework*

a Constitution

The Political Constitution of the United Mexican States (*Constitución Política de los Estados Unidos Mexicanos*) was adopted during the Mexican Revolution of 1917, replacing earlier constitutions. At the time of its adoption, it was a truly remarkable and progressive document, one of the first in the world to spell out not just human rights and freedoms but also to recognize citizens' economic and social rights, serving as a model for the later German Weimar Constitution of 1919 and the Russian Constitution of 1918.[13]

Article 3 establishes that:

> Education provided by the State shall develop harmoniously all human abilities and will stimulate ... respect for human rights and the principle of international solidarity, independence and justice .... I. ... the education provided by the State shall be secular ... II. The guiding principles for state education shall be based on scientific progress and shall fight against ignorance and its effects, servitude, fanaticism and prejudices ... V. In addition ... [it] will support scientific and technological research and will promote strengthening and spreading our culture ...[14]

Article 4 provides for the right to family protection and right to health: 'The law shall protect the organization and development of the family ...

---

[12]   C Palacios-González and M Medina-Arellano, 'Mitochondrial Replacement Techniques and Mexico's Rule of Law: On the Legality of the First Maternal Spindle Transfer Case' (2017) 4:1 *Journal of Law and the Biosciences* 50, 50–69.

[13]   For a brief exploration of the origin and evolution of the Mexican Constitution, see JA Vargas, *Mexican Law for the American Lawyer* (Durham, NC: Carolina Academic Press 2009).

[14]   Federal Constitution, art. 3, ss I, II and V.

Every person has the right to health protection."[15] Remarkably, the right to an education based on science and the right to health were provided already in the original 1917 text.

Although the Mexican Constitution has been amended several times since its adoption, most recently in 2011, it does not specify when the right to life begins and it does not define the embryo or its legal status.[16] While it mentions human dignity, it does not elaborate on the concept.[17] Constitutional jurisprudence has not filled these gaps.

b Treaties and Customary International Law

As already mentioned, under Article 133 of the Federal Constitution, all duly ratified treaties, together with the Constitution and laws approved by the Federal Congress, are the 'supreme law' of Mexico.[18] Remarkably, the 2011 constitutional reform gave international human rights norms the same rank as constitutional norms.[19] Under Article 1 of the amended Constitution,

> all individuals shall be entitled to the human rights granted by this Constitution and the international treaties signed by the Mexican State, as well as to the guarantees for the protection of these rights. The provisions relating to human rights shall be interpreted according to this Constitution and the international treaties on the subject, working in favor of the broader protection of people at all times.

In other words, the Constitution requires laws and norms that affect human rights to be interpreted consistently not only with the Constitution but also with international human rights treaties Mexico has entered into with the aim to guarantee the widest protection of human rights.[20]

It is a remarkable constitutional provision, since it opens the door not only to the direct incorporation of international human rights standards but, potentially, also to decisions of international human rights adjudicative bodies. However, because the reform is recent, there is still considerable

[15] Ibid., arts. 4.1 and 3.
[16] M Medina Arellano, 'The Need for Balancing the Reproductive Rights of Women and the Unborn in the Mexican Courtroom' (2010) 18:3 *Medical Law Review* 427, 427–433.
[17] Federal Constitution, art. 1 (as amended in 2011).
[18] See, in this Chapter, Section II.1.
[19] On the 2011 reform, see, in general, P Salazar Ugarte, JL Caballero Ochoa and LD Vázquez, *La Reforma Constitucional sobre Derechos Humanos: Una Guía Conceptual* (México, Instituto Belisario Domínguez: Senado de la República, LXII Legislatura 2014).
[20] E Ferrer Mac-Gregor, 'El Control Difuso de Convencionalidad en el Estado Constitucional' (2014) 1:11 *Revista Urbe et Ius*, 151, 182.

disagreement over its exact implications. So far, the Mexican Supreme Court has shown certain circumspection when called to rule on the role of international human rights norms in the domestic legal system.[21]

As a matter of customary international law, Mexico is bound by both the Universal Declaration of Human Rights[22] and the American Declaration of Human Rights.[23] Article XIII of the American Declaration of Human Rights recites: 'Every person has the right to ... participate in the benefits that result from intellectual progress, especially scientific discoveries. He likewise has the right to the protection of his moral and material interests as regards his inventions or any literary, scientific or artistic works of which he is the author.' Echoing the American Declaration, Article 27 of the Universal Declaration of Human Rights provides that: '(1) Everyone has the right freely to ... share in scientific advancement and its benefits. (2) Everyone has the right to the protection of the moral and material interests resulting from any scientific, literary or artistic production of which he is the author.'

As to treaties, Mexico acceded to the International Covenant on Economic, Social and Cultural Rights on 23 March 1981.[24] The Covenant contains several provisions that are relevant to human germline genome modification, including Article 15.1.b and 15.2 through 4 ('Right to Science' and 'Rights of Science'), Article 12 (Right to Health) and Article 10 (Right to a Family), when human germline genome modification is done for reproductive purposes. Mexico is also party to the Inter-American Convention on Human Rights,[25] and to its Additional Protocol to the American Convention on Human Rights in the Area of Economic, Social and Cultural Rights ('Protocol of San Salvador'), which, echoing the Covenant, requires states to

[21] I De Paz González, *The Social Rights Jurisprudence in the Inter-American Court of Human Rights* (United Kingdom, Elgar Studies in Human Rights, 2018) 10.

[22] Universal Declaration of Human Rights (adopted 10 December 1948 UNGA Res 217 A(III) (UDHR).

[23] American Declaration of the Rights and Duties of Man, OAS Res. XXX, adopted by the Ninth International Conference of American States (1948), reprinted in Basic Documents Pertaining to Human Rights in the Inter-American System, OEA/Ser L V/II.82 Doc 6 Rev 1, at 17 (1992).

[24] International Covenant on Economic, Social and Cultural Rights (adopted 16 December 1966, entered into force 3 January 1976) 993 UNTS 3 (ICESCR), art. 15.1.b. United Nations Treaty Collection, 'States Party to the International Covenant on Economic, Social and Cultural Rights' https://treaties.un.org/Pages/ViewDetails.aspx?src=IND&mtdsg_no=IV-3&chapter=4&clang=_en accessed 14 September 2018. Mexico has not yet ratified the Optional Protocol to the Covenant on Economic, Social and Cultural rights giving individuals the possibility to file communications with the Committee on Economic, Social and Cultural Rights regarding violations of human rights they allegedly suffered.

[25] American Convention on Human Rights, OAS Treaty Series No. 36, 1144 UNTS 123, reprinted in Basic Documents Pertaining to Human Rights in the Inter-American System, OEA/Ser L V/II.82 Doc 6 Rev 1 at 25 (1992).

recognize everyone's right 'to enjoy the benefits of scientific and technological progress',[26] and the duty of states 'to extend among themselves the benefits of science and technology by encouraging the exchange and utilization of scientific and technological knowledge'.[27]

By virtue of Articles 1 and 133 of the Federal Constitution, the Covenant and the Protocol of San Salvador have the rank of federal constitutional norms within the Mexican legal system. At a minimum, the right to freedom of research should prevent federal and state governments from imposing overly restrictive regulations that inhibit progress of human genetic engineering research. Arguably, the government has also a constitutional obligation to create a legal framework that gives full effect to these rights.

Finally, Mexico has accepted the jurisdiction of the Inter-American Court of Human Rights with jurisdiction over cases involving violations of the Inter-American Convention on Human Rights brought by the Inter-American Commission of Human Rights.[28] Although most of the Court's case law has focused on basic human rights issues, such as extrajudicial killings, conditions of detention and the like, more recently it has made pronouncements on advanced human rights issues, including two cases stemming from Costa Rica's ban of in vitro fertilization (IVF) procedures: *Artavia Murillo* v. *Costa Rica*[29] and *Gómez Murillo* v. *Costa Rica*.[30] In *Artavia Murillo* the Court determined that Costa Rica had violated several provisions of the American Convention on Human Rights by denying couples who could only conceive through IVF access to reproductive health services and reproductive freedom.[31] The Court held that the right to private life includes 'reproductive

---

[26] Additional Protocol to the American Convention on Human Rights in the Area of Economic, Social and Cultural Rights (Protocol of San Salvador) (entered into force 16 November 1999) OAS Treaty Series No 69 (1988) reprinted in Basic Documents Pertaining to Human Rights in the Inter-American System OEA/Ser L V/II.82 Doc 6 Rev 1 at 67 (1992), art. 14.1.b.

[27] Ibid., art. 38.

[28] See OAS, 'American Convention on Human Rights: States Parties Status' (2011) <www.oas.org /en/iachr/mandate/basics/conventionrat.asp> accessed 16 October 2018.

[29] Inter-American Court of Human Rights, *Artavia Murillo et al* v. *Costa Rica* case, Judgment of November 28, 2012, Series C, No. 257, Operative 'Declares' para 1.

[30] Inter-American Court of Human Rights, Case of *Gómez Murillo et al* v. *Costa Rica*, Judgment of November 29, 2016. Series C No. 326. Costa Rica did not comply immediately with the decision on *Artavia Murillo*. An executive decree of the President of Costa Rica legalizing IVF was declared unconstitutional by the Constitutional Court, arguing that this regulation must come from the legislative assembly and not the executive branch. This caused the Inter-American Commission to bring a second case before the Court. However, this time, the Costa Rica government renounced to defend itself on the merits and admitted responsibility, and, soon thereafter, started taking the necessary legislative steps to guarantee that IVF is available in both public and private health spheres in the country.

[31] Ibid., para 277.

autonomy and access to reproductive health services, which includes the right to have access to the medical technology necessary to exercise this right'.[32] This right protects the decision when to become a genetic parent.[33] It also determined that the right to enjoy the benefits of scientific progress includes accessing medical technology necessary to exercise the right to private life and reproductive freedom to found a family.[34] This right requires access to the best healthcare, including assisted reproductive techniques, and prohibits any arbitrary or disproportionate restrictions on accessing this technology.[35] As to the question at what point life begins to trigger Article 4 (Right to Life) of the American Convention, the Court concluded that life begins once the embryo is transferred in the uterus, and not at fertilization.[36]

The *Artavia Murillo* decision, and in particular the dictum that IVF (cryo-preserved) human embryos do not have a right to life, might open the door for permissive regulation of embryo research in Mexico. Indeed, since the 2011 Constitutional reform, when interpreting and applying federal and state laws, Mexican judges must take into consideration not only the norms contained in treaties ratified by Mexico but arguably also the decisions of international adjudicative bodies on how these norms are to be interpreted.[37]

c Federal Laws

To date, Mexico has not adopted specific federal legislation to regulate assisted reproductive technologies, stem cell research or genetic engineering.[38] Despite growing interest in those areas, the regulation of healthcare, genomic research, experimental treatments and the use of human tissues is wanting, with poor enforcement of existing federal and state laws, leaving researchers under a cloud of uncertainty.[39]

---

[32] Ibid., para 146.
[33] Ibid.
[34] Ibid., para 146.
[35] Ibid.
[36] Ibid., para 264.
[37] The binding nature of the decisions of the Inter-American Court of Human Rights was affirmed by the Supreme Court of Justice of Mexico in Case *Varios 912/2010*, session of 14 July 2011, 31. See, in general, E Ferrer Mac-Gregor, 'Interpretación Conforme y Control difuso de Convencionalidad: El Nuevo Paradigma para el Juez Mexicano' (2011) 9:2 *Estudios Constitucionales* 531, 531–622; A Von Bogdandy and others (eds.), *Transformative Constitutionalism in Latin America: The Emergence of a New Ius Commune* (Oxford University Press 2017).
[38] See, in general, M Medina Arellano, 'The Rise of Stem Cell Therapies in Mexico: Inadequate Regulation or Unsuccessful Oversight?' (2010) 2 *Revista RedBioética UNESCO*, 63, 63–78.
[39] See, in general, M Medina-Arellano, 'Stem Cell Regulation in Mexico: Current Debates and Future Challenges', (2011) 5:1 *Studies in Ethics, Law, and Technology*.

I THE GENERAL HEALTH LAW The General Health Law (GHL) (*ley general de salud*) is the key federal law regarding biomedical research and therapy.[40] It is binding not only at the federal level, but it overrides inconsistent state legislation (although state legislation stands until it is declared incompatible by the Supreme Court). It was enacted in 1982 and, although it has been amended several times since, it is a dated legal instrument that is largely failing to stay abreast with developments in biotechnology, old and new. It implements the constitutionally sanctioned right to health and healthcare.[41]

Because of its 'general law' status, the GHL regulates all healthcare activities in the country and contains few provisions specifically relating the issues addressed in this volume. Notably, Article 330 prohibits the transplant of gonads or gonads tissues (i.e. sperm and oocytes), and the use, for any purpose, of embryo tissues or fetal products that are the result of medical abortion. This means that donation of sperm, oocytes and all germline cells, even for reproductive purposes, is prohibited in Mexico. Violations can be sanctioned with a fine from 6,000 to 12,000 times the minimum salary in the relevant geographic area.[42]

The GHL defines 'germ cell', 'embryo' and 'fetus'. Under Article 314 'germ cells' are 'male and female reproductive cells that are capable of giving origin to an embryo', an 'embryo' is 'the product of conception from the moment of it, and until the end of the twelfth gestational week' and a 'fetus' is 'the product of conception from the thirteenth week of gestation until delivery from the mother's womb'. However, the GHL does not define 'conception'. In the Mexican legal system, the term 'conception' is understood to indicate the union of gametes during sexual intercourse.[43] Thus, it is unclear whether embryos created during IVF are included in the notion of 'conception'.

In 2011, Congress added several articles to the GHL in a section entitled 'the Human Genome' (*el genoma humano*).[44] There, the human genome is

---

[40]   *Diario Oficial de la Federación* (Official Gazette of the Federation – DOF), DOF 14 July 2008 (Mexico).

[41]   'Every person has the right to health protection. The law shall determine the bases and terms to access health services and shall establish the competence of the Federation and the Local Governments in regard to sanitation.' Federal Constitution, art. 4. Also, Article 73 of the Federal Constitution establishes the power of Congress to "[e]nact laws on . . . public health". Ibid., art. 73.

[42]   GHL, art. 421.

[43]   See, in general, A Madrazo, 'The Evolution of Mexico City's Abortion Laws: From Public Morality to Women's Autonomy' (2009) 106:3 *International Journal of Gynecology & Obstetrics* 266, 266–269.

[44]   GHL, Fifth Title Bis (Título Quinto Bis), arts. 103.bis-103.bis.7, Official Gazette of the Federation, 16 November 2011.

defined as 'genetic material that characterizes the human species and that contains all the genetic information of the individual, considered as the basis of the fundamental biological unit of the human being and its diversity'.[45] In addition to an overarching prohibition of 'genetic discrimination', the 2011 reform added language on scientific research on human genome, establishing that: 'The scientific research, innovation, technological development and applications of the human genome, will be oriented to the protection of health, prevailing respect for human rights, freedom and dignity of the individual; being subject to the respective regulatory framework.'[46] This is consistent with similar legislation found in most states where research on human genome is done.

However, the 2011 amendments went considerably further. Consistently with the UNESCO Declaration on the Human Genome and Human Rights,[47] Article 103 bis declares: 'The human genome and knowledge about it are the heritage of humanity. The individual genome of each human being belongs to each individual.' In other words, all research on the genome of an individual, from study, sequencing or modification, cannot be done unless the individual in question has given his or her informed consent.[48]

At the same time, the added articles make the genome of each individual a 'public good'. Mexico had already started turning the genome of individuals into a public good in 2008, with the so-called Genomic Sovereignty Amendments to the GHL.[49] These Amendments made the 'Mexican human genome data' property of Mexico's government, by prohibiting and penalizing data collection and utilization of such data in population genomic studies without prior government approval. 'Population genomic studies' are defined as studies 'whose purpose is analysis of one or more genetic markers in unrelated individuals that describe the genomic structure of a given population, identify an ethnic group or identify genes associated with a trait, a disease

[45]   Ibid., art. 103 bis.

[46]   Ibid., art. 103 bis.5.

[47]   UNESCO Office Brasilia, The Universal Declaration on the Human Genome and Human Rights: from theory to practice (2000) BR/2001/PI/H/1.

[48]   GHL, arts. 103 bis.2-103 bis.3.

[49]   Decreto por el que se reforma la fracción V del artículo 100 y el artículo 461, y se adicionan los artículos 317 Bis y 317 Bis 1, todos de la Ley General de Salud. A critical appraisal of the 'Mexican Genomic Sovereignty' can be found in: E Schwartz-Marín and E Restrepo, 'Biocoloniality, Governance and the Protection of "Genetic Identities" in Mexico and Colombia' (2013) 47:5 *Sociology* 993, 993–1010. See A Rojas-Martínez, 'Confidentiality and Data Sharing: Vulnerabilities of Mexican Genomics Sovereignty Act' (2015) 6:3 *J Community Genet* 313, 313–319.

or the response to drugs'.[50] Article 317 imposes various restrictions to transferring tissue that could contain genetic material out of Mexico to perform population genomic studies elsewhere. First, the transfer must be part of a research project approved by the Mexican institution for scientific research in conformity with Article 100 of the GHL, GHL-derived regulations and all applicable provisions. Second, those who want to carry out the study must obtain approval from the Ministry of Health.[51] Sanctions for violations of this article are rigid: four to fifteen years of prison, and a fine and loss of professional licence.[52] To put things in perspective, these are sanctions similar to those given in Mexico for manslaughter during a brawl (*homocidio en riña*).[53] Although the 'Genomic Sovereignty Act' does not consider human genome modification, it is indicative of the overall attitude of the Mexican legislator towards genetics and genetic engineering: conservative and protective to a fault.

In the wake of the 2016 mitochondrial replacement stunt, more amendments, mostly aimed to restrict research, are pending. During the previous legislature (the LXIII, 2015–2018), a draft bill was introduced to the lower house, the Chamber of Deputies, prohibiting the creation of human embryos for any purpose except reproduction and any research with existing IVF cryopreserved human embryos.[54] It had the backing of the National Action Party (PAN) and President Peña Nieto's Institutional Revolutionary Party (PRI).[55] However, it failed to be converted into law before the Chambers were dismissed and elections were called. A new Parliament was sworn in July 2018. It remains to be seen whether it will resume the initiative of the previous legislature or take a different approach.

d Other Federal Governmental Regulations

In Mexico, there are three main providers of healthcare: social security institutions (i.e. IMSS, ISSSTE, ISSFAM, PEMEX); the healthcare scheme for uninsured population (*Seguro Popular*); and the private sector. Each operates under specific secondary regulations implementing the GHL and addressing specific aspects of their work.[56]

---

[50]   GHL, art. 317 bis III.
[51]   Ibid., arts. 317 and 416.
[52]   Ibid., art. 461.
[53]   Federal Criminal Code (*Código Penal Federal*), art. 308.
[54]   Reardon, 'Mexico Proposal to Ban Human-Embryo Research Would Stifle Science' (n 5), 6.
[55]   Ibid.
[56]   E.g. *Reglamento de la Ley General de Salud en Materia de Prestación de Servicios de Atención Médica* (1986) (Regulations on Delivery of Health Services); *Reglamento en Materia de Publicidad*, (Regulations on Publicity) (1986, amended last 2012); *Reglamento en*

In the case of human genome editing particularly noteworthy is the 1987 Regulation on Health Research (*Reglamento de la Ley General de Salud en Materia de Investigación para la Salud*).[57] It includes rules on biomedical research, ARTs (although only very broadly without regulating the details of this practice) and other clinical practices. Overall, like the rest of the Mexican regulatory framework, it has not kept up with biomedical developments.[58] Besides indicating that research on embryos must be conducted for the benefit of that particular embryo, the Regulation does not contain rules regulating embryo research or genetic modification more specifically, including the fate of supernumerary IVF embryos or the donation and cryopreservation of germline cells. It defines 'embryos' and 'fetus' consistently with the GHL.[59] It defines 'assisted fertilization' as 'that related to artificial insemination (homologous or heterologous), including *in vitro fertilization*'.[60] Clearly, these definitions per se do not provide either specific guidance or certainty for the existing ART clinics in the country or for basic research.[61]

Article 56 of the 1987 Regulation limits the scope of research on human embryos to inquiries aiming 'to solve sterility problems that cannot be solved otherwise, respecting the couple's moral, cultural, and social point of view, even if these differ from those of the researcher'.[62] Under Article 55, such research activities are governed by the provisions of the GHL regarding 'donation, transplants, and end of life'.[63]

I THE OFFICIAL MEXICAN STANDARDS (NOMS) Finally, we should mention the Official Mexican Standards (*Normas Oficiales Mexicanas* – NOMs). They are federal technical regulations that complement the general regulations. NOMs

---

Materia de Control Sanitario de la Disposición de Órganos, Tejidos y Cadáveres de Seres Humanos (Regulation on the Sanitary Disposal of Human Organs, Tissues and Corpses) (1985). See, in general, L Motta-Murguia and G Saruwatari-Zavala, 'Mexican Regulation of Biobanks' (2016) 44 *The Journal of Law, Medicine & Ethics* 58, 59.

57 *Reglamento de la Ley General de Salud en Materia de Investigación para la Salud* (Regulation on Health Research).

58 See, in general, D Feinholz, 'Las Investigaciones Biomédicas', in I Brena Sesma and G Teboul (eds.) *Hacia un Instrumento Regional Interamericano sobre la Bioética: Experiencias y Expectativas* (Mexico, IIJ-UNAM, 2009) 233–278.

59 Regulation on Health Research, arts. 40.III and 40.IV.

60 Ibid., art. 40.XI.

61 RM Isasi and others, 'Legal and Ethical Approaches to Stem Cell and Cloning Research: A Comparative Analysis of Policies in Latin America, Asia, and Africa' (2004) 32:4 *The Journal of Law, Medicine & Ethics* 626–640.

62 Regulation on Health Research, art. 56.

63 Ibid., art. 55.

lay down the general characteristics, conditions and quality of products and services.[64] Usually adopted by the Ministry of Economy, matters covered by the GHL are issued by the Ministry of Health.[65] There are more than 180 specific NOMs dealing with health-related matters.[66] For example, there are NOMs on the care of women during pregnancy and prevention of violence against them (NOM-046-SSA-2005),[67] access to clinical records (NOM-046-SSA2-2005)[68] and criteria of clinical research protocols of research on human health (NOM-012-SSA-2012).[69] However, specific NOMs about clinical genetic applications have yet to be adopted. Compliance with NOMs is mandatory since they usually provide for penalties, including fines or cancellation of licenses.

3 The State Legal Framework

The United Mexican States comprises thirty-one states and a Federal District containing the capital of the Union, Mexico City. Each federated entity has its constitution, laws and regulations, often diverging from state to state and with federal legislation, creating a normative kaleidoscope far too complex to be fully summarized here. Overall, about half of the Mexican states have an extremely conservative attitude towards pre-natal life and genetic research, as opposed to the laissez-faire attitude that has distinguished the federal government so far. To complicate things further, both the Federal Government and the states have shared jurisdiction over health, science and technology, and the Supreme Court has been called repeatedly to demarcate the boundaries of the respective powers. For instance, at the moment of this writing, the Supreme Court is considering a case over whether Tabasco's regulation of

---

64 R Gutiérrez-Vega, 'Mexican Official Norms; Concept, Background and Legal Scope' (2016) 79:3 *Revista Médica del Hospital General de México* 115, 115–116. A list of all Mexican NOMs currently in force can be found at www.economia-noms.gob.mx/noms/inicio.do, accessed 24 September 2018.

65 GHL, art. 195.

66 All relevant norms related to the area of health are summarized in D Karam Toumeh and R Placencia Villanueva (eds.) *Compendio de Normas Oficiales Mexicanas Vinculadas con el Derecho a la Protección de la Salud, I & II* (Mexico, CNDH & IMSS 2010).

67 NOM-046-SSA2-2005, *Violencia familiar, sexual y contra las mujeres. Criterios para la prevención y atención* (Family Violence, Sexual Violence, and Violence Against Women: Criteria for Prevention and Treatment) [amended last 24 March 2016].

68 NOM-004-SSA3-2012, *Del expediente clínico* (On the Clinical Records).

69 NOM-012-SSA3-2012, *Que establece los criterios para la ejecución de proyectos de investigación para la salud en seres humanos* (Which Establishes the Criteria for the Implementation of Health Research Projects in Humans).

ART, included in its Civil Code,[70] violates federal prerogatives and the federal constitution.[71]

Eighteen of the thirty-two state constitutions protect life from 'conception'.[72] For instance, the Constitution of Jalisco 'recognizes, protects and underwrites the right to life of every human being, by expressively maintaining that, since fertilization, they are protected under the law and considered to be living human beings for all legal corresponding effects, until their natural death'.[73] Most amendments were adopted in response to the 2007 decriminalization of abortion in the Federal District of Mexico City.[74]

Interestingly, and consistently with the federal move to 'nationalize' the genome, Article 9 of the Constitution of Nayarit, a state home to four main native-American groups, protects the right to benefit from scientific progress in the field of genomic medicine. It recognizes the link between the right to health protection and the right to research and scientific development, and that genomic research must follow the principles of human dignity, autonomy of the will, respect for physical and mental integrity, privacy, confidentiality, non-discrimination and genetic identity.[75]

[70] *Código Civil Tabasco* (Tabasco Civil Code), Title Eight 'Filiation', c I and II 'generalities and paternity', art. 329, and c VI Bis 'Assisted gestation and surrogacy', arts. 380 Bis and 320 Bis 7.
[71] Action of Unconstitutionality 16/2016, Petitioner: General Prosecutor Office in the Mexican Republic. The full text of this action is not publicly available. However, the registration of the case is available at www.scjn.gob.mx/sites/default/files/acuerdos_controversias_constit/documento/2016-11-28/MI_AccInconst-16-2016%20(1)_0.pdf accessed 19 September 2018. It is expected that this action is going to be solved out before the end of 2018.
[72] Baja California, Sonora, Chihuahua, Zacatecas, Tamaulipas, Nayarit, Jalisco, Colima, Aguascalientes, Hidalgo, Chiapas, Quintana Roo, Guanajuato, San Luis Potosí, Querétaro, Yucatán, Veracruz & Michoacán.
[73] Constitución Política del Estado de Jalisco (Political Constitution of the State of Jalisco), art. 4.
[74] See, in general, A Madrazo, 'Narratives of Prenatal Personhood in Abortion Law', in RJ Cook, JN Erdman, and BM Dickens (eds.) *Abortion Law in Transnational Perspective: Cases and Controversies*, (USA, University of Pennsylvania Press 2014) 327–346.
[75] Constitución Política del Estado de Nayarit (Political Constitution of the State of Nayarit), art. 9: 'Todo individuo tiene derecho a beneficiarse del progreso científico en el área de la medicina genómica, por tanto, el estado reconoce el vínculo existente entre el derecho a la protección de la salud y el derecho a la investigación y al desarrollo científico. En la investigación en el área de las ciencias genómicas, deberán prevalecer los principios de dignidad humana, autonomía de la voluntad, respeto por la integridad física y psíquica, intimidad, confidencialidad, no discriminación e identidad genética; por tanto, queda prohibida cualquier práctica que atente contra estos principios, contra los derechos humanos o contra cualquier instrumento internacional que regule las ciencias genómicas. Todo individuo tiene derecho a conocer la información genómica personal y sus vínculos biológicos de parentesco, para tal efecto, la ley determinará los límites y modalidades mediante las pruebas científicas correspondientes.'

As to state statutory law and codes, notably the Criminal Code of the Federal District prohibits the fertilization of eggs for any purpose other than reproduction.[76] This precludes the creation of embryos specifically for research, which may be important in gene-editing work, but not the use of supernumerary embryos. In the Federal District, gamete can be used in research as long as the donor consented to this use.[77] Still, it prohibits manipulation of human genes that 'alter the genotype' for purposes other than eliminating or improving disease. The statute does not provide much more guidance than that, but one can reasonably conclude that genome modifications with a therapeutic goal are not prohibited.[78] This also raises the issue of whether research of how genome modifications can be used for therapeutic purposes is lawful. Finally, genetic engineering for 'illicit ends' is prohibited. However, the statute does not define which ends are illicit.[79] In light of the vagueness of the current legislation, it may be feasible to infer that germline engineering for clinical purposes is allowed since it will be in the benefit of the offspring, which can be interpreted as 'licit end'. This opens the door to a progressive interpretation of the law (constitutional principles of pro-person, progressiveness, universality, interdependence and indivisibility):[80] any action that is not unequivocally prohibited and that could be of benefit of the embryo (future offspring) is permitted, especially if it is in line with human rights protected by the Federal Constitution and international treaties (such as the right to health and the right to benefit from science advancements).

III OVERSIGHT BODIES FOR RESEARCH ON HUMAN GERMLINE GENOME MODIFICATION

Under the General Health Law and its various implementing regulations, the authorities in charge of health and medical-related services include the President of Mexico, the General Health Council (*Consejo General de Salud*), the Federal Ministry of Health (*Secretario de Salud*) and state governments. For brevity's sake, we discuss only federal oversight bodies.

The Ministry of Health, through the Federal Commission for the Protection against Sanitary Risks (*Comisión Federal para la Protección contra Riesgos Sanitarios* – COFEPRIS), has the broadest jurisdiction regarding the

---

[76]   Código Penal del DF (Criminal Code of the Federal District) (2002), art. 154.
[77]   Ibid., art. 149.
[78]   Ibid., art. 154.
[79]   Ibid., art. 154.III.
[80]   Federal Constitution, art. 1.

control and supervision of genome manipulation activities. COFEPRIS is the Mexican equivalent of the US Food and Drug Administration (FDA).[81] It is responsible for enforcing clinical research rules in research and treatment settings.[82] Significantly, it has the authority to control and oversee clinical trials and novel therapeutic activities involving human subjects and to monitor the development and advertising of new drugs, medicines and therapies entering the Mexican market. All research protocols, clinical trials and experimental medical treatments involving human beings or their tissues for which there is not enough scientific evidence to prove their preventive, therapeutic and regenerative efficiency must obtain previous approval by COFEPRIS before being conducted and remain under its supervision.[83] That being said, in Mexico therapeutic and research uses of tissues and cells are only vaguely regulated. There are no specific guidelines, provisions or principles to be followed when carrying out the storage, use or transplantation of biomaterials.

Importantly, the GHL[84] establishes that the National Commission of Bioethics in Mexico (*Comisión Nacional de Bioética* – CONBIOÉTICA) will be responsible for the establishment, registration and oversight of the Research Ethics Committees (REC) and Bioethics Hospital Committees, in all public and private healthcare, health-research and research centres across the country.[85] Broadly, the mandate of this commission includes the identification and promotion of ethical practices concerning life sciences, biomedical research and emerging biotechnologies.[86] The CONBIOÉTICA has operated as a governmental advisory figure and has influenced the national bioethics debates, as was originally intended.[87] CONBIOÉTICA works in

[81] Among the issues under COFEPRIS jurisdiction are the monitoring of environmental risks, publicity on health and supplies, sanitary surveillance on food and connected aliments, assessment of pharmaceutical products and so forth. See, in general, O Gómez Dantés and others, 'Health System in Mexico', (2011) *Salud Pública de México* 53 Supp 2 S229.

[82] GHL, art. 17.

[83] *Reglamento de la Comisión Federal para la Protección de Riesgos Sanitarios* (Regulation of the Federal Commission for the Protection against Sanitary Risks) (2004), art. 14.VIII. Mexico, 'Reglamento de la comisión federal para la protección contra riesgos sanitarios (The Internal Regulations of COFEPRIS)' (2004) <www.diputados.gob.mx/LeyesBiblio/regla/29.PDF> accessed 31 July 2018.

[84] GHL, arts. 41 bis, 98, 316.

[85] See, in general, M Ruíz De Chavez, A Orozco and G Olaiz, 'The Past, Present, and Future of Mexico's National Bioethics Commission' (2017) 47 *Hasting Center Report* S31–34.

[86] For a deeper review of the creation and functions of this commission, see I Luengas, D Feinholz and GM Soberón, 'National Bioethics Commission: Its Mandate and Approach' (2007) 2 *Bioethical Debate* 43.

[87] See G Jiménez-Sánchez, CF Lara-Álvarez and A Arellano-Méndez, 'A Survey of the Development of Mexican Bioethics: Genomic Medicine as One of Its Greatest Challenges' in L Pessini and others (eds.) *Ibero-American Bioethics* (Springer 2010) 159–73.

coordination with COFEPRIS and has established guiding principles to create, register and oversee these committees.[88]

Under the GHL, COFEPRIS, in coordination with the National Transplant Centre (*Centro Nacional de Transplantes* – CENATRA), exerts control and vigilance over the disposal and transplant of tissues and cells of human beings.[89] COFEPRIS has the exclusive statutory competence to oversee the inspection, approval and authorization of activities concerning the use, storage and transplant of cells and their components (including human embryos or embryonic cell lines), although it lacks specific guidelines to exercise this competence.[90] However, its actual oversight capacity over the whole territory of Mexico is questionable since there are approximately 25,000 facilities registered to provide health services, approximately 4,500 of which are hospitals (1,200 corresponding to public institutions and 3,200 to the private sector).[91] Alas, although the GHL prohibits the commercial use of tissues and cells,[92] it is well known that Mexico has a thriving market of tissues and cells, especially those devoted to stem cell therapies.[93]

IV FUNDING

Overall, Mexico spends very little on research and development. When measured by gross domestic product (GDP), Mexico is the world's eleventh largest economy, and it is growing at a remarkable rate of at least 3.5 per cent each year.[94] However, Mexico spends only 0.5 per cent of its GDP on scientific research and development, which is the second lowest percentage among OECD states after Chile. That is less than economically troubled and smaller nations such as Greece.[95] When measured by patent production and inventive

---

[88] *Guía Nacional para la Integración y el Funcionamiento de los Comités de Ética en Investigación* (National Guidelines on Establishing and Operation of Research Ethics Committees) (2016).

[89] GHL, art. 342 bis. See Mexico, 'National Transplant Centre' (CENATRA) www .cenatra.salud.gob.mx accessed 31 July 2018. CENATRA administers the National System and Registry of Transplants, which oversees and records all allogeneic transplants carried out in the country, as well as a list of people waiting for organ transplants and donors.

[90] Ibid., arts. 340 and 341.

[91] See, in general, J Frenk and O Gómez-Dantés, 'Health System in Mexico' (2017) *Health Care Systems and Policies*, 1, 1–11.

[92] Ibid., art. 327.

[93] See, in general, I Berger and others, 'Global Distribution of Businesses Marketing Stem Cell-based Interventions' (2016) 19:2 *Cell Stem Cell* 158, 158–162.

[94] E Vance, 'Why Can't Mexico Make Science Pay Off?'(October 2013) *American Science* 67.

[95] In 2013, Greece spent 0.81 per cent of its GDP on research and development. UNESCO Institute for Statistics, 'Research and Development Expenditure (% of GDP)'

capacity, Mexico ranks last among OECD countries. Although politicians at the federal and state level stress the importance of science in Mexico's development, they rarely follow through.[96] Support for science does not translate in the Mexican budget. Congress seems not to be particularly interested in scientific research.[97]

Article 97 of the GHL establishes that it is the Ministry of Public Education's obligation, in coordination with the Ministry of Health and with the participation of the National Council of Science and Technology (*Consejo Nacional de Ciencia y Tecnología* – CONACyT), to set the policy for advancing scientific and technological research for health.[98] The Science and Technology Law establishes the National System of Researchers (*Sistema Nacional de Investigadores* – SNI) and the overall framework for funding scientific research.[99] Most health research in Mexico is carried out by SNI members in national public institutes of health services and hospitals.[100] The bulk of publicly funded healthcare research is carried out in the social security institute, including the National Institutes of Health (*Institutos Nacionales de Salud* – NIH).[101]

In 2004, the federal government created the National Institute for Genomic Medicine (*Instituto Nacional de Medicina Genómica* – INMEGEN).[102] INMEGEN's immediate goal is to create personalized diagnoses and medicines for the Mexican population, based on genetic data. The larger and more ambitious goal is to eliminate the dependence on foreign technology and innovations, by generating our own.[103] However, INMEGEN cannot fund activities related to human cloning, embryos or stem cell research,[104] and, in

(The World Bank, 2018) http://data.worldbank.org/indicator/GB.XPD.RSDV.GD.ZS accessed 28 September 2018.

[96] See C Rosen, 'Rebuilding Mexico's Science and Technology Capacity' (2011) *Science Development Network*, para 3, 23–31.

[97] Ibid., para 12.

[98] GHL, art. 97.

[99] *Ley General de Ciencia y Tecnología* (2015) (General Law on Science and Technology).

[100] E Martínez-Martínez and others, 'Health Research Funding in Mexico: The Need for a Long-Term Agenda' in SH Vermun (ed.) 7:12 *PLoS ONE*, e51195.

[101] The NIHs are governed by the National Institutes of Health Law (*Ley de los Institutos Nacionales de Salud*, 2000).

[102] Mexico, 'Secretaría de Salud' www.inmegen.gob.mx/ accessed 16 October 2018. See, in general, M Muñoz De Alba Medrano, 'Reflexiones En Torno Al Derecho Genómico', in *Aspectos Sobre La Regulación Del Genoma Humano En México* (Mexico, IIJ-UNAM 2002) 191–209.

[103] On these goals, see, in general, G Jimenez-Sanchez, 'Developing a Platform for Genomic Medicine in Mexico', (2003) 300:5617 Science 295, 295–296.

[104] Art. 3, s I, Statute of INMEGEN, www.inmegen.gob.mx/media/filer_public/9c/a3/9ca3942a-26b2-42d0-bc56-2699db4c1c6a/estatuto_organico_2017.pdf accessed 3 November 2018. Also

any event, it has paltry resources. Between. 2016 and 2017, it operated on a budget of $46,288,000 Mexican pesos (less than $2,312,000 USD).[105]

As in the case of many countries, where public funds are not available, the gap can be filled by private money. In 2010, Mexican billionaire Carlos Slim, for many years considered to be the richest man in the world, invested $65 million in this field. In conjunction with the INMEGEN, a new project was created, known as the 'Slim Initiative in Genomic Medicine'.[106] The aim is to secure independence from foreign biotechnology, as well as to position Mexico as a regional leader and serious global competitor in this scientific arena. However, given the size of the task ahead, that amount of money can be considered, at best, only seed grant.

V THE SPECIFIC REGULATORY ENVIRONMENT

Let us now look at how Mexico regulates human germline genome modification at every step of the translational pipeline. Overall, the highly fragmented national regulatory system creates uncertainty for scientists conducting research on embryos, hampering research, even if the scant existing regulatory framework is rarely enforced, as the scandal of the child born in Guadalajara in 2016 proved.

1 Basic Research Using Germline Modification in Human Embryos

Research protocols and clinical trials must be approved by the relevant local research committees.[107] Article 98 of the GHL requires all health institutes (*instituciones de salud*) to establish: (I) an Investigation Committee (*Comité de Investigación*); (II) an Investigation Ethics Committee (*Comité de Ética en Investigación*) in case research on humans is carried out; and (III) a Biosafety Committee (*Comité de Bioseguridad*), when genetic engineering techniques

---

see on the reasons for these prohibitions as political compromise in order to carry out genomic research in Mexico: G Jimenez-Sanchez G and others, 'Genomic Medicine in Mexico: Initial Steps and the Road Ahead' (2008) 18:8 *Genome Research* 1191, 1191–1198; B Séguin and others, 'Genomics, Public Health and Developing Countries: The Case of the Mexican National Institute of Genomic Medicine (INMEGEN)', (October 2008) *Nat Rev Genet*, S5-9.

[105] INMEGEN 2017 Annual Report. Available at: www.inmegen.gob.mx/media/filer_public/ba/87/ba8793ba-5fce-4395-b783-3f64eebd3a76/informe_anual_autoevaluacion_2017_final_pagina_web.pdf, accessed 28 September 2018.

[106] See www.carlosslim.com/preg_resp_slim_genoma_ing.html, accessed 31 July 2018.

[107] See, in general, F Santiago Rodríguez, 'Governing ethical clinical research in developing countries: exploring the case of Mexico' (2010) 37:1 *Science and Public Policy* 587, 587–596.

are used. Moreover, under Article 41 bis, all medical-care establishments of the public or private sector of the national health system are required to set up a Hospital Bioethical Committee (*Comité Hospitalario de Bioética*). The Hospital Bioethical Committees and the Investigation Ethics Committees are interdisciplinary, gender-balanced and composed of medical personnel in various fields and of people with training in bioethics from professions in psychology, nursing, social work, sociology, anthropology, philosophy or law.[108] As mentioned, the CONBIOÉTICA, which works in coordination with COFEPRIS, has established guiding principles to create, register and oversee these committees.[109]

As we saw, Article 330 of the GHL prohibits the transplant of gonads (i.e. organs that produces gametes, such as a testis or ovary) or gonads tissues (i.e. sperm, oocytes and all germline cells) and the use, for any purpose, of embryo tissues or fetal products that are the result of medical abortion. This means that donation of sperm, oocytes and all germline cells, even for reproductive purposes, is prohibited in Mexico. Also, by virtue of second transitory article of the Tissue Regulation,[110] couples cannot donate embryos either to another couple or scientific health research. Because of that, between 2010 and 2016 it was reported that 657 embryos were kept in cryopreservation *sine die* in the National Institute of Perinatology (INPeR).[111]

Second, under Article 56 of the Regulation on Health Research, ART research can only be done to address infertility problems. However, in Mexico supernumerary IVF embryos are often used for basic science research in private and public health research settings.[112] In 2012, public and private professionals working in the field of assisted reproduction, from thirty-four centres, adopted a National Consensus document.[113] It expressed the need to create specific and permissive rules in the assisted reproduction field and envisaged the possibility of donating or transferring non-used gametes and

---

[108] GHL, art. 41 bis.

[109] *Guía Nacional para la Integración y el Funcionamiento de los Comités de Ética en Investigación* (National Guidelines on Establishing and Operation of Research Ethics Committees) (2016).

[110] On a detailed explanation of this prohibition, see L Motta-Murguia and G Sauwatari-Zavala, 'Mexican Regulation of Biobanks' (2016) 44 *The Journal of Law, Medicine & Ethics*, 58, 58–67.

[111] RV Esparza-Pérez, 'The Infra-value of Human Rights in the Context of Assisted Regulation in México' (2017) 153:5 *Gaceta Médica de México* 570, 574.

[112] See MJ López-Rioja and others, 'Estudio Genético Preimplantación para Aneuploidias: Resultados de la Transición entre Diferentes Tecnologías' 86:2 *Ginecol Obstet Mex* 96, 96–107.

[113] Ginecología y Obstetricia de México, 'Consenso Nacional Mexicano de Reprodución Asistida' (2012) https://ginecologiayobstetricia.org.mx/secciones/articulos-originales-numero 83/consenso-nacional-mexicano-de-reproduccion-asistida/ accessed 3 November, 2018.

embryos to other couples for reproductive use or using them for scientific research.[114] In any event, although the GHL is silent about the creation of embryos or gametes for research, the Criminal Code of the Federal District permits the creation of embryos only for reproductive purposes, thus prohibiting the creation of embryos for research purposes.[115]

However, as we saw, several states and federated entities have more stringent rules when it comes to research on human embryos. For instance, the Federal District Criminal Code prohibits manipulation of human genes 'so as to alter the genotype' for any purpose other than eliminating or improving disease, but it is not clear what is meant by this.[116] It forbids any procedure of genetic engineering or human genetic manipulation for 'illicit ends' but fails to describe which kind of ends would be licit.[117] Interestingly, it is rumoured that Zhang's team decided to carry out the mitochondrial replacement therapy in the New Hope Clinic located in Jalisco, rather than in the one located in Mexico City, to bypass the more strict Federal District regulation.[118]

2 Preclinical Research Using Germline Modification Technologies in Animals

All clinical research activities require sound data from basic science research and preclinical research before moving to clinical trials involving human beings. Preclinical research using germline modification technologies in non-human animals is permitted but subject to the safeguards established by the Law on Biosafety of Genetically Modified Organisms[119] and the Official Mexican Standard 'Technical Specification for the Production, Care and Handling of Laboratory Animals' (NOM-062-ZOO-1999).[120] These regulations seek to ensure that research involving animals before clinical trials minimizes the number of non-human animals used and is carried out with

---

[114] A broader explanation of every aspect of this consensus can be found in general, A Kably-Ambe and others, 'Consenso Nacional Mexicano de Reproducción Asistida' (2012) 5:2 *Rev Mex Reprod* 68, 68–113.

[115] Federal District Criminal Code, art. 154.

[116] Ibid., art. 154, s 1.

[117] Ibid.

[118] C Palacios-González and M Medina-Arellano, 'Author's Response to Peer Commentaries: Mexico's Rule of Law and MRT's' (2017) *Journal of Law and the Biosciences* 623, 627.

[119] *Ley de Bioseguridad de los Organismos Genéticamente Modificados* (2015).

[120] NOM-062-ZOO-1999, *Especificaciones técnicas para la producción, cuidado y uso de los animales de laboratorio* (Technical Specifications for the Production, Control and Use of Laboratory Animals).

all due care when experiments include non-human primates.[121] The local Bioethics Committees approve of preclinical research protocols, and oversee their implementation, focusing on minimizing the suffering of non-human animals being used in scientific research.

However, the regulatory framework is largely permissive. There are no restrictive norms regarding animal experimentation. In general, in Mexico, little attention is paid to the ethics of preclinical research on non-human subjects, although there is an emerging academic literature focusing on non-human research protection.[122]

3 Clinical Trials and Clinical Application of Human Germline Genome Modification

Clinical trials are regulated by the Health Research Regulation, as well as NOM-012-SSA3-2012, which establishes the criteria for the implementation of health research on humans.[123] The National Registry of Clinical Trials (*Registro Nacional de Estudios Clínicos*, RNEC) keeps records of all research protocols and clinical trials authorized and conducted in Mexico. In theory, it can be accessed by anyone. In practice, it is most of the time offline and not accessible.

To date, COFEPRIS has authorized 369 healthcare establishments to handle organs, tissues and cells, of which 206 are private and 103 in the public healthcare sector.[124] It has authorized the use in clinical trials of germinal cells for assisted reproductive purposes, as well as various kinds of stem cells (i.e. hematopoietic endothelial, endothelial, autologous mesenchymal, neural, autologous myoblast, mesenchymal derived from placentae and hematopoietic derived from bone marrow, umbilical cord blood and the bloodstream). In all cases, it authorized their use exclusively for research

[121]   TA Fregoso Aguilar and E Guarneros Bañuelos, 'Bioethics in the Use of Experimental Animals', in JA Morales-Gonzalez and E Aguilar Nájera, *Reflections on Bioethics* (IntechOpen, Mexico 2018).

[122]   See B Vanda Cantón, 'Bioética y aspectos jurídicos en la relación con los animales no humanos', in S Chan and others (eds.) *Bioética y Bioderecho: Reflexiones Clásicas y Nuevos Desafíos* (Mexico, IIJ-UNAM, 2018) 383–410.

[123]   MC López-Pacheco and others, 'Normatividad que rige la investigación clínica en seres humanos y requisitos que debe cumplir un centro de investigación para participar en un estudio clínico en México' (2016) 37:3 *Acta Pediatr Mex* 175, 175–182.

[124]   Data obtained from COFEPRIS on 5 January 2017, through the Mexican government's national platform for transparency (*Plataforma Nacional de Transparencia* PNT), of the National Institute for Access to Information, and Access to Information (*Instituto Nacional de Accesso a la Información*): 'Plataforma Nacional de Transparencia' (PNT 2007) www .infomex.org.mx/gobiernofederal/home.action accessed 29 September 2018.

purposes, not therapeutic, and not aiming at the modification of the human genome.[125]

Yet, it is well documented that COFEPRIS is poorly financed and empowered.[126] Limited financial budgets, infrastructure and human resources inhibit its capacity to enforce the confusing biomedical legal provisions and to monitor medical research activities. For example, COFEPRIS does not follow up on the success or failure of clinical trials. The expectation is that researchers would make results publicly available, but COFEPRIS is not equipped to follow up on that.

It is worth noting that the lack of specific legislation in this area is a barrier for any oversight body, but also to scientific progress, since it leaves clinicians and scientists in a climate of uncertainty, unable to pursue their research or any clinical experimentation with confidence. However, as already mentioned, since genetic engineering on IVF human embryos is not prohibited when it is carried out for the benefit of the embryo or fetus, it could be inferred that it is permitted. In addition, if genetic modification of the human germline leads to a benefit to the embryo/fetus, it can be said that clinical application can be allowed. This conclusion assumes a purposive or progressive interpretation of the statute, which is justified by the fact that articles 3 and 4 of the Federal Constitution recognize the right to health and the right to benefit from the advances of science.

Again, the case of the 2016 mitochondrial replacement procedure in Guadalajara illustrates the chasm that exists in Mexico between the letter of the law and practice.[127] After the operation, Dr Zhang published a paper in which he stated that the procedure was part of an approved clinical research protocol.[128] A public official request, filed with COFEPRIS to find out more about the procedure, was answered by simply stating that the New Hope Fertility Clinic had applied for, and obtained, a license to carry out ART and that since there was no specific regulation regarding ART, the Clinic had

---

[125] Ibid.
[126] Santiago-Rodríguez has conducted extensive work on the deficiencies of COFEPRIS in oversight clinical trials in Mexico. Some of that investigation can be found, in general, F Santiago-Rodríguez, 'Facing the Trial of Internationalizing Clinical Research to Developing Countries: Evidence from Mexico', in W Dolfsma, G Duysters, and I Costa (eds.) *Multinationals and Emerging Economies: The Quest for Innovation and Sustainability* (Edward Elgar Publishing 2009) 58–74.
[127] On the 2016 mitochondrial operation in Guadalajara, see, in general, M García Barragan López and M Medina-Arellano, *Investigación en Genética, Bioética y Laicidad* (Mexico, IIJ-UNAM 2019) (publication pending).
[128] J Zhang and others, 'Live Birth Derived from Oocyte Spindle Transfer to Prevent Mitochondrial Disease' (2017) 34:4 *Reprod Biomed Online*, 361, 361–368.

not committed any violation.[129] However, it is remarkable that the Clinic registered its Research Ethics Committee before the CONBIOÉTICA, in December 2016, while the announcement of the 'first three parents' baby' born in Mexico was in September 2016, three months before registration. To date, there has been no investigation by COFEPRIS of the Jalisco's New Hope Fertility Clinic or of the doctors who participated in the procedure.

VI CURRENT PERSPECTIVES AND FUTURE POSSIBILITIES

All in all, Mexico's regulatory framework for genome modification, whether that be of somatic or germinal cells, is deficient. It is not the result of a deliberate choice but rather the consequence of outdated legal instruments that the legislator and regulatory agencies have failed to keep in sync with technological developments. By and large, the GHL and its secondary regulations (i.e. the Health Research Regulation and the Tissue Regulations) are of little use to research on human embryos and human germline genome modification, since these activities are inherently different from all other scientific and therapeutic activities. They require more specific regulations and effective enforcement. Adequate mechanisms of compliance combined with more targeted regulatory provisions are necessary if Mexico is to invest seriously in this innovative health research setting.

Mexico urgently needs a national regulatory regime for ART, human embryo research and genetic engineering. What this regime will look like remains to be seen. After the constitutional reform of 2011, Mexico decided to make international human rights instruments an integral part of its legal system. When it comes to scientific research, the Congress should keep in mind the obligations it has under the Covenant on Economic, Social and Cultural Rights to ensure everyone's rights to health and to benefit from progress in science and technology, and Mexico's duty to let scientists work without arbitrary and overly burdensome interference. Once more, the Mexican Congress has a chance to be a trailblazer, as it was when it passed the 1917 Constitution. Hopefully, when discussing the next round of amendments to the GHL, it will seize the opportunity and conquer its worst anti-scientific instincts.

---

[129] The official request number 02317017- Exp. 248/2017 was formulated through the Mexican government's National Platform for Transparency (*Plataforma Nacional de Transparencia*).

6

The Regulation of Human Germline Genome Modification in Europe

Jessica Almqvist and Cesare P.R. Romano

I INTRODUCTION

One of the distinctive traits of Europe is the scope and breath of international cooperation and integration projects that states of the 'Old Continent' have developed since the end of the Second World War. Devastated by two consecutive continent-wide wars during the first half of the twentieth century – conflicts that eventually spread out to engulf the whole globe – during the second half of the century European nations embarked on an ambitious project of integration and transfer of sovereignty to shared supranational institutions to avoid future wars. The two main pillars of this 'European Project' are the European Union (EU) and the Council of Europe (CoE). The CoE was founded in 1949 by 10 western European states to uphold human rights, democracy and the rule of law in Europe. Over the next 70 years, it expanded to include, nowadays, 47 states. In 1951, six of those core European states (France, Germany, Italy, Belgium, the Netherlands and Luxembourg) started a limited attempt to integrate the market of the resources over which Germany and France had fought repeatedly since industrialization: coal and steel. Later, it expanded to pooling the resource of the future – atomic energy – and, eventually, to establish a customs union and integrate markets in general to ensure free circulation of persons, goods, services and capital. Eventually, the three European Communities became one, and then, in the 1990s, the European Community morphed into the present European Union: a quasi-federal project that has been given by its member states considerable powers to regulate all aspects of their economic and social life, and that is increasingly acting as one vis-à-vis the rest of the world.[1]

---

[1] Euratom remains an entity distinct from the European Union, but it is governed by the same EU institutions.

Although the CoE and the European Union are separate and distinct organizations, with somewhat different goals, they have much in common and overlap. Besides having a remarkably similar flag (12 yellow stars on a background of just a different shade of blue), all 28 members of the European Union are members of the CoE. Indeed, in the past, membership to the latter has been considered a prerequisite for accession to the former. Their authority and legal instruments (e.g. treaties, directives, regulations) affect the way in which members operate within the national boundaries, between them and with the rest of the world.

Historically, European nations have played a key role in humanity's scientific and technological progress, and this carries on into the twenty-first century. In aggregate, they continue influencing the direction and speed of scientific progress worldwide, directly and indirectly. In 2016, the gross domestic expenditure on research and development of the combined 28 EU members stood at 303 billion euros.[2] While that was just two-thirds of the same expenditure of the United States, it was almost 50% higher than China's, more than double the expenditure of Japan and more than five times higher than South Korea.[3] European states, through the European Union, influence the direction of research globally, because research done by European research institutions often involves non-European researchers as co-investigators. The European Union has international agreements for scientific and technological cooperation with 20 countries. These create a framework for joint projects, sharing of facilities, staff exchanges or the organization of specific events. Also, EU research funding is accessible to non-EU scientists. For instance, 13 non-EU states (including Norway, Israel and Switzerland) have 'Associated Country' status and contribute to the budgets of 'Framework Programmes for Research and Technological Development' proportionally to their GDP.[4] Their scientists have access to funding through EU Framework Programmes.

---

[2] Gross domestic expenditure on R&D (GERD) includes expenditure on research and development by business enterprises, higher education institutions, as well as government and private non-profit organizations. Eurostat Statistics Explained, 'R & D Expenditure' (March 2018) https://ec.europa .eu/eurostat/statistics-explained/index.php/R_%26_D_expenditure accessed 24 September 2018. However, measured by proportion of the gross domestic product (Research and Development Intensity), the 28 EU members combined rank was well below the corresponding ratios recorded in Japan (3.29 per cent, 2015 data) and the United States (2.79 per cent, 2015 data), as has been the case for a lengthy period of time. In 2015, R&D intensity in China surpassed that of the EU-28, with Chinese R&D expenditure equivalent to 2.07 per cent of GDP. Ibid.

[3] Ibid.

[4] The EU 'Framework Programmes for Research and Technological Development' are funding programmes created by the European Union/European Commission to support and foster research in the European Research Area. The specific objectives and actions vary between

Finally, as we will see, so far Europe is the only region of the globe to have regulatory frameworks for biomedical research. This consideration alone justifies a discussion of Europe separate and distinct from the one that we will have in the following chapters of the national regulatory framework of some selected EU and CoE members. Thus, in this chapter we discuss first the CoE. After a brief introduction of its history, goals and structure, we will discuss the Convention for the Protection of Human Rights and Dignity of the Human Being with regard to the Application of Biology and Medicine (better known as the 'Oviedo Convention'), the first, and to date still the only, multilateral treaty entirely devoted to biomedicine and its human rights aspects. Then, we will turn to the European Union. Again, after a brief introduction, we will discuss the specific EU legislation that affects research on human embryos, germline cells and their genetic modification. We will follow it 'from the bench to bedside', highlighting contradictions, gaps and issues. Our conclusions, drafted in a historical time of uncertainty over the 'European Project', suggest a way forward.

II THE COUNCIL OF EUROPE

1 Introduction

The Council of Europe (in French, *Conseil de l'Europe*) is the continent's oldest political organization.[5] It was founded in 1949, in the aftermath of the Second World War, by 10 western European states, to uphold human rights, democracy and the rule of law in the continent.[6] It was headquartered in Strasbourg, Alsace, a European region that had been bitterly fought over by France and Germany at least since 1870. Over the next 70 years, the CoE expanded its membership to include, nowadays, nearly all European states: 47 states, from Iceland and Portugal to the west, to Russia and Turkey to the east. Twenty-eight of these are also members of the European Union. Moreover, a number of non-European states (i.e. Australia, Canada, Japan, Mexico and

---

periods. So far, there have been eight 'Framework Programmes' (abbreviated FP1 to FP8). The Focus of FP8, also known as Horizon 2020, is innovation. See, in general, European Commission, 'Research and Innovation' https://ec.europa.eu/info/research-and-innovation_en accessed 24 September 2018. On Horizon 2020, see below, Section III.2.d.iii.

[5] On the Council of Europe, see, in general, S Schmahl and M Breuer (eds.), *The Council of Europe: Its Law and Policies* (Oxford University Press 2017).

[6] Belgium, Denmark, France, Ireland, Italy, Luxembourg, the Netherlands, Norway, Sweden and the United Kingdom.

the United States) and other entities (e.g. the Holy See and the European Union) participate in its works as 'Observers'.

The CoE's statutory bodies are: (i) the Council of Ministers, the decision-making body comprising the foreign ministers of all member states or their permanent diplomatic representatives in Strasbourg; (ii) the Parliamentary Assembly, composed of 324 national politicians representing the parliaments of the CoE's 47 member states; and (iii) the Secretariat. Given the topic of this volume, another CoE body worth of mention is the Committee on Bioethics (DH-BIO). It was created in 2012, following a reorganization of intergovernmental bodies at the CoE.[7] This Committee meets twice a year, consisting of delegations of the 47 member states with expertise in the various aspects of bioethics. It reports to the Council of Ministers and it is assisted by a permanent secretariat, the Bioethics Unit, acting under the Directorate General Human Rights and Rule of Law of the Council of Europe.

Unlike the European Union, the Council of Europe does not have the power to create norms that are binding for its members. What it does, instead, is provide a forum for the discussion and adoption of treaties in the fields of its competence (i.e. human rights, democracy and rule of law) that members are subsequently encouraged to ratify.[8] The most famous of such treaties is certainly the European Convention on Human Rights and Fundamental Freedoms (ECHR).[9] The ECHR was adopted in 1950 and is the linchpin of the European human rights regime. Over the years, it has been modified and upgraded by a series of protocols (16 to date), which have expanded the list of rights and modified the oversight mechanisms.[10] Besides the Statute of the Council of Europe, which is the organization's constitutive treaty, the ECHR is the only other treaty that all Council members must ratify, and its protocols enter into force only after they have been ratified by all members. We will discuss it in more detail further below.[11]

The ECHR, both in its original form and after revisions brought about by the protocols, focuses mostly on civil and political freedoms. Many economic, social and cultural rights are not included in it but are rather protected under

---

[7]   The DH-BIO has taken over the responsibilities of the Steering Committee on Bioethics (CDBI) for the tasks assigned by the Oviedo Convention, as well as for the intergovernmental work on the protection of human rights in the field of biomedicine.

[8]   Also states with Observer Status, as well as the European Union, can become parties to certain CoE treaties, if they wish to do so.

[9]   [European] Convention for the Protection of Human Rights and Fundamental Freedoms, ETS No 5 (European Convention on Human Rights, as amended) (ECHR).

[10] Protocols to the ECHR enter into force only when they have been ratified by all CoE member states.

[11] See, in this chapter, Section II.4.

a second CoE treaty called the European Social Charter.[12] For the purposes of this volume, it is worth mentioning that the 'right to science' is not mentioned either in the ECHR or in the European Social Charter, despite the fact that at the time of their drafting other international human rights instruments had already declared it. The 'right to health' is only mentioned in the European Social Charter, not in the ECHR, despite subsequent protocols to the ECHR adding other social and cultural rights to it, such as the right to education.

The ECHR was the first major treaty adopted under the aegis of the CoE. Since then, more than 200 treaties and protocols have been adopted. Among those, one in particular is relevant for the question of the regulation of human germline genome editing: the Convention for the Protection of Human Rights and Dignity of the Human Being with regard to the Application of Biology and Medicine, adopted by the Council of Ministers of the CoE on 4 April 1997 in Oviedo, Spain.[13]

2 The Oviedo Convention

a Background and Overview

The 'Oviedo Convention' is the first, and to date still the only, multilateral treaty entirely devoted to biomedicine and its human rights aspects, not just in the CoE but in the world.[14] Although the CoE had actually been involved in addressing bioethical issues since the 1980s, the drafting of the Oviedo Convention started in 1992 and lasted through 1996, an aeon ago as far as research on genetics is concerned, before many of the discoveries that have revolutionized biomedicine and genetics during the past 20 years were made.

In 1985, the Committee of Ministers created the Ad Hoc Committee of Experts on Bioethics (CAHBI), working under its direct authority, and

---

[12]   European Social Charter (revised), ETS No 163.

[13]   Convention for the Protection of Human Rights and Dignity of the Human Being with Regard to the Application of Biology and Medicine: Convention on Human Rights and Biomedicine (opened for signatures on 4 April 1997, entered into force 12 January 1999) ETS No 164 (Oviedo Convention).

[14]   On the Oviedo Convention, see, in general: R Uerpmann-Wittzack, 'Convention on Human Rights and Biomedicine' in Stefanie Schmahl and Marten Breuer (eds.), *The Council of Europe: Its Law and Policies* (Oxford University Press 2017) 572–588; H Gros Espiell, J Michaud and G Teboul (eds.), *Convention sur les droits de l'homme et la biomédecine: analyses et commentaires* (Paris: Economica 2010); Council of Europe, *Biomedicine and Human Rights: the Oviedo Convention and its Additional Protocols* (Council of Europe 2009); R Andorno, 'First Steps in the Development of an International Biolaw' in C Gastmans and others (eds.), *New Pathways for European Bioethics* (Intersentia 2007) 121–138.

entrusted it with the intergovernmental activities of the CoE in the field of bioethics. In 1992, the CAHBI became the Steering Committee on Bioethics (CDBI). The CDBI set up a Working Party responsible for drafting a 'Convention on Biomedicine'.[15] In 1994, the CDBI adopted a first draft, which was released by the Council of Ministers for public consultation. The draft received considerable criticism and was consequently thoroughly revised. In 1996, the CDBI submitted a final draft to the Council of Ministers, which adopted it and opened it for signature.

The Oviedo Convention was conceived from the very beginning as a 'framework treaty', a binding international legal instrument but one that contains only broad, general principles, which are intended to be developed subsequently, internationally by additional protocols on specific issues and nationally by specific legislation. It was also one that CoE member states were free to decide to ratify or ignore. To fill the framework with specific content, to date, the CoE has adopted and opened for signature and ratification four additional protocols: on the Prohibition of Cloning of Human Beings (1998);[16] on Transplantation (2002);[17] on Biomedical Research (2004);[18] and on Genetic Testing for Health Purposes (2008).[19] We will revert to these later.

Article 1 of the Oviedo Convention leaves it to each ratifying state to 'take in its internal laws the necessary measures to give effect to the provisions' of the Convention. Some provisions are regarded as self-executing, such as those related to some individual rights – for example, the 'right to information', the requirement of 'informed consent' and the prohibition of non-discrimination.[20] Also some prohibitory norms established by the Convention, such as the prohibition of creation of embryos for research, might have direct application

---

[15]   The Working Party was originally chaired by Dr Michael Abrams (UK) and, after his untimely death, by Mr Salvatore Puglisi (Italy). Explanatory Report, para. 5.

[16]   Additional Protocol on the Prohibition of Cloning Human Beings (adopted by the Committee of Ministers on 6 November 1997, entered into force on 1 March 2001) ETS No 164. To date, it has been ratified by 24 states.

[17]   Additional Protocol concerning Transplantation of Organs and Tissues of Human Origin (adopted by the Committee of Ministers on 8 November 2001, entered into force on 1 May 2006) ETS No 186. To date, it has been ratified by 15 states.

[18]   Additional Protocol on Biomedical Research (adopted by the Committee of Ministers on 30 June 2004, entered into force on 1 September 2007) ETS No 195. To date, it has been ratified by 12 states.

[19]   Additional Protocol concerning Genetic Testing for Health Purposes (adopted by the Committee of Ministers on 7 May 2008, opened for signature on 27 November 2008, entered into force on 1 July 2018) ETS No 203. To date, it has been ratified by five states.

[20]   Convention for the Protection of Human Rights and Dignity of the Human Being with Regard to the Application of Biology and Medicine: Convention on Human Rights and Biomedicine (Explanatory Report), para. 20. Available at: https://rm.coe.int/168ooccde5 accessed on 8 December 2018.

in some member states, depending on their constitutional and legal system. However, it is for each state to adopt the necessary domestic legal instruments to give effect to the Convention and establish sanctions for its violation,[21] and it is for national courts to enforce the rights.[22]

In other words, the Oviedo Convention just established minimum common standards. States that ratify it cannot adopt a lower level of protection of human rights in the biomedical field when they decide to legislate on bioethics.[23] And those who do not agree with these standards, for whatever reason, are free not to ratify it. Of course, when the Convention was negotiated and drafted, a concerted effort was made to find as wide as possible common ground between European states, if not all, at least the major ones, even if many fundamentally disagreed on how to approach the most ethically divisive issues relating to biomedicine. The difficulty of reaching agreement explains why the Convention lacks any definition of terms, and why many of its provisions are very general.

b The Rights and Duties Contained in the Oviedo Convention

The Oviedo Convention consists of a preamble and 28 articles, organized into 14 chapters. The general norms are contained in Chapter I (Articles 1 to 4): Purpose; Primacy of Human Being; Equitable Access to Healthcare; Professional Standards. Chapters II to VII set up substantive provisions relating to specific bioethical issues, such as: consent; right to information and right not to be informed; protection of persons undergoing research; principles regulating organs and tissue removal; prohibition of financial gain; as well as two issues particularly relevant for the present discussion: interventions on the human genome (Article 13) and research on embryos in vitro (Article 18). Finally, Chapters VIII to XIV include the procedural norms, treaty organs and final clauses.

---

[21] Oviedo Convention, art. 25.

[22] Ibid., art. 23.

[23] Restrictions are allowed only if 'prescribed by law and are necessary in a democratic society in the interest of public safety, for the prevention of crime, for the protection of public health and for the protection of rights and freedoms of others' (Oviedo Convention, art. 26.1). No restriction can be put on the rights contained in art. 11 (non-discrimination), 13 (intervention on human genome), art. 14 (prohibition of sex selection), art. 16 (rights of persons undergoing research), art. 17 (protection of persons not able to consent); arts. 19 and 20 (removal of organs and tissue from living donors for transplantation purposes), and art. 21 (prohibition of financial gain). However, art. 18 (prohibition of creation of embryos for research) is not one of the non-derogable rights.

The notion of 'human dignity' is clearly the bedrock of the Oviedo Convention. It is enshrined in its full title: 'Convention for the Protection of Human Rights and Dignity of the Human Being with regard to the Application of Biology and Medicine'. The Preamble refers three times to this concept: the first, when it recognizes 'the importance of ensuring the dignity of the human being'; the second, when it recalls that 'the misuse of biology and medicine may lead to acts endangering human dignity'; the third, when it expresses the resolution of taking the necessary measures 'to safeguard human dignity and the fundamental rights and freedoms of the individual with regard to the application of biology and medicine'. Finally, according to Article 1, the Convention aims to 'protect the dignity and identity of all human beings and guarantee everyone, without discrimination, respect for their integrity and other rights and fundamental freedoms with regard to the application of biology and medicine'.

The fact that Article 1 mentions both 'everyone' and 'human beings' is not due to poor drafting. Again, it was a deliberate choice to bypass disagreement between member states on the legal status of the human embryo and whether and at what stage of development legal personality is attached.[24] According to the Explanatory Report of the Oviedo Convention, 'it was a generally accepted principle that human dignity and the identity of the human being had to be respected as soon as life began' but without clarifying when that occurs.[25] Thus, the drafters deliberately used simultaneously two different expressions – 'everyone' (in French *toute personne*) and 'human being' (in French *être humain*) – to refer to the subject of the protection granted by the Convention, without defining either concept nor specifying whether they are synonymous.

As a direct corollary of the idea of human dignity, Article 2 assigns the highest priority to the interests and welfare of the 'human being', whose respect 'shall prevail over the sole interest of society or science'. As the Explanatory Report of the Convention says, '[p]riority is given to the [interests of the human being], which must in principle take precedence over [the interests of science or society] in the event of a conflict between them. One of the important fields of application of this principle concerns [scientific] research'.[26] Also, '[t]he whole Convention, the aim of which is to protect

---

[24] Steering Committee on Bioethics (CDBI), *Preparatory Work on the Convention for the Protection of Human Rights and Dignity of the Human Being with Regard to the Application of Biology and Medicine* (ETS No 164, CDBI/INF (2000) 1, Council of Europe, 2000) 10–13.

[25] Explanatory Report (n 20), paras. 18–19; Steering Committee on Bioethics (CDBI), *Preparatory Work* (n 24), 10–13.

[26] Explanatory Report (n 20) para. 21.

human rights and dignity, is inspired by the principle of the primacy of the human being, and all its articles must be interpreted in this light'.[27] Similar provisions are also found in the Declaration of Helsinki on Biomedical Research[28] and in the UNESCO Universal Declaration on the Human Genome and Human Rights.[29]

Be that as it may, the practical consequences of giving priority to the interests and welfare of the human being over the interests of society or science are unclear. First of all, it builds a straw man out of the interests of society and science. It suggests that the interests of science and the interests of human beings are in opposition with one another and that there is a need to protect humans against scientific research and its applications. Granted, there have been egregious cases where humans have been forced or manipulated to participate in experiments against their will or without having been informed about the risks involved. The prohibition against such conducts by the scientific community is firmly established in international human rights law. For instance, the International Covenant on Civil and Political Rights (ICCPR) establishes that subjecting persons without his or her free consent to medical or scientific experimentation is prohibited.[30] In international bioethics law, this prohibition translates into a fundamental right of everyone to free and informed consent, and specifically to the right to be able to freely give or refuse any intervention involving their person regardless of purpose, including research, and special protection to those persons who are unable to give free and informed consent, from being used as means to achieve scientific progress.[31] However, rare and egregious cases of transgressions of these limits by scientists notwithstanding should not detract us from the fact that the general objective of biomedical research is to develop knowledge for the diagnosis, treatment and prevention of disease and to improve human health. This is, after all, the purpose of all medical activity and research, and the work of scientists, over the centuries, has improved and continues improving the

---

[27] Ibid., para. 22.a.
[28] World Medical Association, *Declaration of Helsinki: Recommendations Guiding Physicians in Biomedical Research Involving Subjects* (as amended through 2013); World Medical Association, *WMA Declaration of Helsinki – Ethical Principles for Medical Research Involving Human Subjects* (as amended through 2013).
[29] UNESCO Universal Declaration on the Human Genome and Human Rights (adopted by UNGA Res 152) A/RES/53/152.
[30] International Covenant on Civil and Political Rights (ICCPR) (adopted 19 December 1966, *entered into force* 23 March 1976) 999 U.N.T.S. 171, art. 7.
[31] Oviedo Convention, art. 5. See also Explanatory Report (n 20), paras. 34–40, and *VC v Slovakia*, App No 18968/07, European Court of Human Rights (Judgment of 8 November 2011).

human condition by all metrics. If that is taken into consideration, the interests of science and the interests of human beings may not only be compatible but actually reinforce each other. The latter is the rationale for the former. This understanding of the relationship between scientific progress and human interests is central in international human rights law. Both the Universal Declaration of Human Rights and the International Covenant on Economic, Social and Cultural Rights (ICESCR) uphold the right to science and the rights of science as international human rights that are compatible with other rights, including civil, political, economic and social rights, recognized in international law.[32]

Of course, the 'right to science' and the 'rights of science' are not absolute human rights.[33] They can be limited but only in so far as their limitations may be compatible with the nature of these rights and solely for the purpose of promoting the general welfare in a democratic society.[34] However, precisely because of their status as *human* rights, the protection of the right to science and the rights of science cannot be automatically subordinated to the protection of other rights. Nevertheless, the Oviedo Convention seems to consider these rights as secondary to all other rights. The fact that the Oviedo Convention itself, in the Preamble, says that the drafters bore in mind the Covenant on Economic, Social and Cultural Rights and the Universal Declaration adds to the perplexity.

For the purposes of this volume, Chapter IV and Chapter V are the most salient ones. Chapter IV (Articles 11–14) is entitled 'Human Genome'. The Explanatory Report of the Oviedo Convention shows that the drafters had a particular understanding of genetic testing and gene therapy, one that reflects the state of knowledge of the time, but one that is becoming outdated: '[g]enetic testing consists of medical examinations aimed at detecting or ruling out the presence of hereditary illnesses or predisposition to such illnesses in a person by directly or indirectly analysing their genetic heritage (chromosomes, genes)'.[35] As also stated in the same report:

> The aim of gene therapy is to correct changes to the human genetic heritage which may result in hereditary diseases. The difference between gene therapy and the analysis of the genome lies in the fact that the latter does not modify

---

[32] International Covenant on Economic, Social and Cultural Rights (ICESCR) (adopted 16 December 1966, entered into force 3 January 1976) 993 UNTS 3, art. 15.1.b; Universal Declaration of Human Rights (adopted 10 December 1948 UNGA Res 217 A(III) (UDHR), art. 27.

[33] On the 'right to science' and the 'rights of science', see, in this book, Chapter 2.

[34] See, in this book, Chapter 2 and Chapter 22.

[35] Explanatory Report (n 20) para. 72.

the genetic heritage but simply studies its structure and its relationship with the symptoms of the illness. In theory, there are two distinct forms of gene therapy. Somatic gene therapy aims to correct the genetic defects in the somatic cells and to produce an effect restricted to the person treated. Were it possible to undertake gene therapy on germ cells, the disease of the person who has provided the cells would not be cured, as the correction would be carried out on the cells whose sole function is to transmit genetic information to future generations.[36]

Article 11 contains a generic and uncontroversial prohibition of unfair discrimination on grounds of 'genetic heritage'. Article 12 restricts 'predictive genetic test', that is to say '[t]ests which are predictive of genetic diseases or which serve either to identify the subject as a carrier of a gene responsible for a disease or to detect a genetic predisposition or susceptibility to a disease', to 'health purposes or for scientific research linked to health purposes, and subject to appropriate genetic counselling'. While it did not rule out preimplantation genetic diagnostics (PGD) per se, it did not specify that it could be used in the context of artificial reproductive technology. The ambiguity made it so that, in the following years, some states outlawed PGD (e.g. Italy and Germany), while others allowed it.[37] Also, developments in genetic engineering have rapidly put in question the wisdom of this provision and whether it does actually protect fundamental human rights, at least as long as it is worded as it is. For instance, in recent years, a whole new industry has emerged that offers genetic test kits that allow finding out information about an individual ancestry. This testing does not have medical purposes (or medical purposes might be incidental), can be obtained without genetic counselling and has become very popular. This development generated the adoption in 2008 of the Additional Protocol to the Convention on Human Rights and Biomedicine, concerning Genetic Testing for Health Purposes that we will discuss below.[38]

Article 13, entitled 'Interventions on the Human Genome', provides: 'An intervention seeking to modify the human genome may only be undertaken for preventive, diagnostic or therapeutic purposes and only if its aim is not to introduce any modification in the genome of any descendants.' Again, the Explanatory Report sheds some light on the purpose of this norm.

---

[36] Ibid., para. 73.

[37] On Germany's and Italy's bans of PDG, see, in general, B Bock Von Wülfingen, 'Contested Change: How Germany Came to Allow PGD' (2016) 3 *Reproductive Biomedicine & Society Online* 60, 60–67; A Boggio and G Corbellini, 'Regulating Assisted Reproduction in Italy: A 5-year Assessment' (2009) 12.2 *Human Fertility* 81–88.

[38] See, in this chapter, Section II.3.

The progress of science, in particular in knowledge of the human genome and its application, has raised very positive perspectives, but also questions and even great fears. Whilst developments in this field may lead to great benefit for humanity, misuse of these developments may endanger not only the individual but the species itself. The ultimate fear is of intentional modification of the human genome so as to produce individuals or entire groups endowed with particular characteristics and required qualities.[39]

To address these still-speculative fears, the drafters clarified that Article 13 establishes that 'any intervention which aims to modify the human genome must be carried out for preventive, diagnostic or therapeutic purposes'. Moreover, 'interventions aimed at modifying genetic characteristics not related to a disease or to an ailment are prohibited'.[40] They left the door open to somatic cell gene therapy, at that time at the research stage, but only as long as done in compliance with Chapter V (Articles 15 through 18).[41] This includes not only the uncontroversial requirements of protection of persons undergoing research but also the prohibition of the creation of embryos solely for research, which later on in history created numerous problems to stem cell research. What is categorically prohibited are

> [i]nterventions seeking to introduce any modification in the genome of any descendants ... [I]n particular genetic modifications of spermatozoa or ova for fertilization are not allowed. Medical research aiming to introduce genetic modifications in spermatozoa or ova which are not for procreation is only permissible if carried out *in vitro* with the approval of the appropriate ethical or regulatory body.[42]

Clearly, the drafters approached the 'human genome' as a single public good, one that needs special protection.[43] They did not pause to consider that within the 'human genome' there are considerable variations, both between populations and down to the individual level. They also did not pause to consider, or decided to avoid, the intricate question of what is 'normal' human

---

[39] Explanatory Report (n 20) para. 89.

[40] Ibid., para. 90.

[41] Idem. They excluded from the norm unwanted side effects on the germ cell line, too. Oviedo Convention, Explanatory Report (n 20) para. 92.

[42] Ibid., para. 91.

[43] This is the approach followed also by the UNESCO Universal Declaration on the Human Genome and Human Rights, which declares in Article 1: 'The human genome underlies the fundamental unity of all members of the human family, as well as the recognition of their inherent dignity and diversity. In a symbolic sense, it is the heritage of humanity.' See, in general, J Buttigieg, *The Human Genome as Common Heritage of Mankind* (Ibidem Verlag 2018).

genome and what a genetic defect is, and what implications that distinction has, and the ethics of prohibiting interventions that limit the capacity to treat inheritable genetic diseases. Perhaps they did not intend to protect the human genome per se, but rather the embryo, and, to enhance its protection, added this befuddling provision.

The legislative history of Article 13 reveals that the drafters struggled mightily finding the right balance between protection of the human genome and not blocking science. During an early meeting, in 1992, several experts who had been summoned to provide the Working Party with scientific advice were reportedly in favour of prohibiting interventions on the germ cell line. They felt it necessary, given the then state of scientific knowledge, to prohibit such interventions considering the unpredictability of their side effects and effects on subsequent generations.[44] However,

> other participants felt that the option should nevertheless be left open and that it might be possible to authorize germ cell therapy, although the intervention would need to carry a certain number of guarantees which were not available at the present stage of scientific knowledge. If on the other hand such therapy proved its worth and reliability, these experts might be able to accept it under certain conditions.[45]

Alternative language was proposed. The Working Party eventually chose to keep the language prohibiting germ cell therapy but 'agreed unanimously to specify that the provision would need to be reviewed within a certain time (e.g. five years after the entry into force of the Convention) having regard to the current progress in knowledge'.[46] Regrettably, the provision requesting revision of Article 13 after five years never made it to the final text. What made it was instead a general and optional process to amend the Convention, through a public debate.[47]

The Steering Committee on Bioethics debated at length also whether research on germ cell lines was to be allowed.[48] Alternative language considered included: 'Any intervention with the aim of modifying the genetic characteristics in the germ cell line is prohibited' and '[a]ny intervention with the aim of modifying genetic characteristics transmissible to descendants of persons is prohibited'.[49] However, the final text more convolutely provides:

[44] CDBI, Preparatory Work (n 24) CORED 14–16/12/92, 63.
[45] Ibid.
[46] Ibid.
[47] Oviedo Convention, art. 28 and 32.
[48] See, e.g., CDBI, Preparatory Work (n 24) CDBI 20–22/11/95, 66.
[49] Ibid.

'An intervention seeking to modify the human genome may only be under-taken for preventive, diagnostic or therapeutic purposes and only if its aim is not to introduce any modification in the genome of any descendants'.[50]

As Iñigo de Miguel Beriain and Carlos Casabona note in this volume, the Oviedo Convention does not specify what 'human genome' or 'descendant' are nor clarifies what 'aimed at' in Article 13 means.[51] Acutely, they pointed out that the human genome changes only when a new gene is added to the already vast and diverse pool of the human genome. However, human germline editing is limited to replacing a pathologic gene with its healthy expression. In such a limited intervention, one that nowadays is technically possible thanks to the advent of the CRISPR family of technologies, nothing 'new' is added to the human genome pool. If gene editing is done to prevent or correct genetic mutations – themselves a threat to the integrity and future of human identity – and no new genetic material is introduced, then it is difficult to see how germline genetic editing could be regarded as an assault on human dignity, one that the drafters of the Oviedo Convention intended to prohibit.[52]

Also, because the Convention does not define the term 'descendant', it is not clear whether a mere embryo, as opposed to a fetus or even a newborn, might be considered a 'descendant' and whose genetic manipulation might be prohibited. That hinges on the legal definition and consequent status of the embryo, a question that the drafters of the Oviedo Convention deliberately left unaddressed to avoid interjecting themselves in a cultural and scientific contentious debate dividing European nations.

Chapter V (Articles 15–18) is entitled 'Scientific Research'. Article 15 contains a general statement whereby: 'Scientific research in the field of biology and medicine shall be carried out freely'. It also adds 'subject to the provisions of the Convention' and other unspecified 'legal provisions ensuring the protection of the human being'.[53] Articles 16 and 17 regard the protection of persons subject to research. Article 18, entitled 'Research on Embryos *In Vitro*', is another article of the Oviedo Convention that is key for the purposes of this volume. It recites: '(1) Where the law allows research on embryos *in vitro*, it

---

[50]   Oviedo Convention, art. 13.

[51]   See, in this book, Chapter 13.

[52]   I De Miguel Beriain, 'Should Human Germline Editing Be Allowed? Some Suggestions on the Basis of the Existing Regulatory Framework' (2018) *Bioethics* https://doi.org/10.1111/bioe .12492 accessed 12 October 2018; A Nordberg and others, 'Cutting Edges and Weaving Threads in the Gene Editing (Я)evolution: Reconciling Scientific Progress with Legal, Ethical, and Social Concerns' (2018) *Journal of Law and the Biosciences* 1–49, 26.

[53]   The Explanatory Report and the Preparatory Work fail to shed any light on what these 'other legislative instruments' are.

shall ensure adequate protection of the embryo. (2) The creation of human embryos for research purposes is prohibited.'

As in the case of much of the rest of the Convention, the drafters tried to strike a delicate balance between opposing views about the status of human embryos. It leaves to member states to legally define the 'embryo' and to decide whether to allow or ban research on embryos that are in excess after in vitro fertilization (IVF) and in lab. All it demands is that when states allow research on embryos in vitro, they give embryos 'adequate protection'. Likewise, the Convention does not explain what 'adequate protection' means and how that is compatible with their use as research material. At the same time, it draws the line at the creation of embryos ad hoc for research purposes, a provision that in some states was invoked to block altogether scientific research involving human embryos and in others to create considerable difficulty.

While most of the Convention avoided drastic legal innovations and tended to simply repeat what was already stated at that time in many national legislations, Article 18.2 was the only new innovative norm and, unsurprisingly, was hotly debated during the drafting.[54] The Working Party could not decide on whether the article should be included. To avoid stalling the drafting of the whole Convention, it passed the hot potato to the Steering Committee on Bioethics.[55] At the Committee, several delegations proposed to leave the matter to a separate protocol, but the idea was eventually abandoned when it became clear that such a protocol would have little chance of success.[56] After a discussion, the Committee voted 11 to 6 (2 abstentions) to include Article 18.2 in the Convention.[57] Discussions continued for four more years on how to word it, with several votes, many of them very close.[58]

Despite all, the final wording of the Convention ended up being unacceptable to many states, but for opposite reasons. Germany, because of its history and domestic politics that give southern regions, mostly catholic, a strong voice, tends to have very strong ethical and legal oppositions to any research involving human embryos.[59] During the drafting, the German representatives argued that the Oviedo Convention was too liberal, in particular on embryo experimentation, and, thus, incompatible with the Embryo Protection Act it

---

[54] V Lúcia Raposo, 'The Convention of Human Rights and Biomedicine Revisited: Critical Assessment' (2016) 20:8 *International Journal of Human Rights* 1277–1294, 1279; R Ashcroft, 'Could Human Rights Supersede Bioethics?' (2010) 10:4 *Human Rights Law Review* 639–660, 657.

[55] CDBI, Preparatory Work (n 24) CORED 9-12/11/92, 81.

[56] Ibid., BDBI 24-27/11/92, 81.

[57] Ibid.

[58] Ibid., 82–88.

[59] On Germany, see in this book, Chapter 8.

had adopted in 1990.[60] To this date, it has not ratified the Oviedo Convention and it is unlikely to do so. Italy, Ireland, Poland and Austria, countries where the Catholic Church has considerable influence, voiced similar concerns and are still on the fence regarding ratification.[61] On the other hand, the United Kingdom,[62] but also Belgium[63] and Netherlands,[64] which allow some of the conduct prohibited by the Oviedo Convention, could not accept the prohibition of creating embryos for research purposes. None of them has, so far, ratified the Oviedo Convention, nor do they seem likely to do so, as long as the text remains what it is.

Both sides overplayed their position at the negotiating table during the drafting of the Convention, resulting in an outcome that satisfies few. The Oviedo Convention to date has been ratified by just 29 out of 47 CoE members. Of the states surveyed in this book, only Spain, France and Switzerland have done so. The odd result is that many of the states who played a key role in the drafting of the Convention (e.g. the United Kingdom, Germany and Italy) are not party to it, and therefore not bound by it, while those who had little say in its making are the ones bound by it.[65]

As long as Articles 13 and 18 are worded as they are, it is unlikely the number of ratifications will grow. The Convention provides for an amendment procedure. Any party to the Convention, the Committee on Bioethics and the Committee of Ministers can propose amendments. The Committee on Bioethics is to discuss the amendment, vote by two-thirds majority on it and then forward it to the Committee of Ministers for approval.[66] This process must also take into account Article 28 of the Convention, which calls for an 'appropriate public discussion in the light, in particular, of relevant medical, social, economic, ethical and legal implications'.

---

[60]   Embryonenschutzgesetz vom 13. Dezember 1990 (BGBl. I S. 2746), das zuletzt durch Artikel 1 des Gesetzes vom 21. November 2011 (BGBl. I S. 2228) geändert worden ist [Act for the Protection of Embryos (The Embryo Protection Act), Federal Law Gazette I 2746 (December 13, 1990), Article 1 amended in Federal Law Gazette I 2228 (November 21, 2011). The human embryo is also protected under the German Constitution (*Grundgesetz*). The Constitution states that 'human dignity is inviolable' and that 'everyone has the right to life and inviolability of his person' (art. 1.1). Nonetheless, it also states that freedom to pursue science and research is protected (art. 5.3). Basic Law for the Federal Republic of Germany, as last amended 23 December 2014.

[61]   On Italy, see, in this book, Chapter 12.

[62]   On the UK, see, in this book, Chapter 7.

[63]   On Belgium, see, in this book, Chapter 9.

[64]   On the Netherlands, see, in this book, Chapter 11.

[65]   V Bellver Capella, 'Los Diez Primeros Años del Convenio Europeo sobre Derechos Humanos y Biomedicina: Reflexiones y Valoración' (2008) XIX: 3 Cuadernos de Bioética, 401–421, 405.

[66]   Oviedo Convention, art. 32.6.

In 2015, the Committee on Social Affairs, Health and Sustainable Development of the Parliamentary Assembly of the CoE noted that while Article 13 prohibits interventions on the human genome that are not for preventive, diagnostic or therapeutic purposes and are inheritable, 'this Convention has not yet been ratified by all Council of Europe member States and even those that may have interpret the limits of this prohibition differently'.[67] Thus, the Committee asked the Parliamentary Assembly 'to study the health, ethical, and human rights risks and challenges related to the [gene-editing] techniques' use and regulation with a view to making the appropriate recommendations to the Committee of Ministers on possible action to be taken to provide a common framework for the use of these technologies'.[68]

In November 2017, the same Committee issued a report, entitled 'The Use of New Genetic Technologies in Human Beings', recommending the Parliamentary Assembly to recommend the Council of Ministers to adopt a five-step plan that includes: (i) urging member States which have not yet ratified the Oviedo Convention to do so without further delay, or, as a minimum, to put in place a national ban on establishing a pregnancy with germline cells or human embryos having undergone intentional genome editing; (ii) fostering a broad and informed public debate; (iii) instructing the Council of Europe Committee on Bioethics (DH-BIO) to assess the attendant ethical and legal challenges; (iv) developing a common regulatory and legal framework; and (v) recommending that member States, on the basis of the other steps, develop a clear national position on the practical use of new genetic technologies, setting the limits and promoting good practices.[69] The Parliamentary Assembly adopted the recommendation as its own almost verbatim shortly thereafter and passed it on to the Council of Ministers.[70]

Yet, it is hard to see how this could remove the blocks that have prevented the Convention from gathering, if not ratification by all CoE members, at least support from the major states. It is obvious that ratification of the Oviedo Convention must be the last step in the plan, not the first, and must be reached only after the Convention has been amended. One possible way out of the

[67] Parliamentary Assembly, Committee on Social Affairs, Health and Sustainable Development, 30 November 2015, Doc 13927.

[68] Ibid.

[69] Parliamentary Assembly, Committee on Social Affairs, Health and Sustainable Development, 24 May 2017, Doc 14328.

[70] Parliamentary Assembly, Recommendation 2115 (2017) (The use of new genetic technologies in human beings).

impasse would be to delete from the Convention at least Articles 13 and 18. These two articles could become the object of much more detailed regulation in a separate protocol or two, as other controversial biomedical issues such as end-of-life decisions. This way states that want to give embryos and germline cells high level of protection could go ahead and ratify them, while states that are happy with the status quo could finally ratify the framework Convention. This would ensure that the citizens of this second group of states enjoy, as a matter of international law, the rights that all other articles of the Oviedo Convention describe. This would be a step forward towards the adoption of a truly single bioethics law in Europe.[71] However, for political reasons, it is unlikely this pragmatic solution will be adopted. 'Prohibitionist' states are more concerned about preventing conduct in the territory of other fellow European states than about the Convention preventing them from adopting higher standards of protection of the human embryo for activities taking place within their jurisdiction.

At the time of writing this book, the Committee on Bioethics has yet to present a detailed analysis that could provide some further insights. In 2015, it issued a general statement concerning the call for an in-depth analysis of the potential risks of genome editing and for international and regional debate on its implications for human beings. It recognized the potential of new genome-editing technologies, such as CRISPR-Cas9, for research to understand the causes of diseases and for future treatment as well as to improve health. However, it also expressed concern about the application of genome-editing technologies to human gametes or embryos in the light of the many ethical, social and safety issues, particularly from any modification of the human genome, which could be passed on to future generations. It then held that the ethical and legal challenges raised by these emerging genome-editing technologies are better addressed in the light of the principles laid down in the Oviedo Convention.[72] Recently, in December 2018, following the second International Summit on Human Genome Editing and the announcement of the birth of two babies in China following germline genome modification, the Committee reiterated its 2015 statement, stressing that 'ethics and human rights must guide any use of genome editing technologies' and that the Oviedo Convention provides a unique reference framework to that end.[73]

---

[71] R Andorno, *Principles of International Biolaw: Seeking Common Ground at the Intersection of Bioethics and Human Rights* (Brussels: Bruylant 2013).

[72] Committee on Bioethics, *Statement on Genome Editing Technologies* (Council of Europe, Strasbourg, 2015), Doc. DH-BIO/INF (2015) 13 FINAL.

[73] Newsroom, 'Statement by the Council of Europe Committee on Bioethics: "Ethics and Human Rights Must Guide Any Use of Genome Editing Technologies in Human Beings"' (Council of Europe, Strasbourg, 30 November 2018) www.coe.int/en/web/portal/-/-ethics-and-

Alas, keep referring to the Oviedo Convention as the gold standard is not going to do much to improve it.

3 *The Additional Protocols to the Oviedo Convention*

As it was mentioned, to fill the framework with specific content, to date, the CoE has adopted and opened for signature and ratification four additional protocols: on the Prohibition of Cloning of Human Beings (1998);[74] on Transplantation (2002);[75] on Biomedical Research (2004);[76] and on Genetic Testing for Medical Purposes (2008).[77]

The Additional Protocol to the Convention on Human Rights and Biomedicine on the Prohibition of Cloning Human Beings was adopted in 1998 and entered into force in 2001. So far, it has been ratified by 24 states.[78] It is the first and only binding international legal instrument on this issue. It prohibits 'any intervention seeking to create a human being genetically identical to another human being alive or dead'.[79] While it does not define 'human being', by 'human being genetically identical' it means the creation of a 'human being sharing with another the same nuclear gene set'.[80] Thus, it does not apply to the cloning of cells and tissue for research and therapeutic purposes. No exemption from this prohibition (e.g. for reasons of public safety, prevention of crime, protection of public health or the protection of the rights and freedoms of others) is admissible.

The Additional Protocol on Transplantation was adopted in 2001 and entered into force in 2006. So far, it has been ratified by 15 states. It contains general principles and specific provisions regarding the transplantation of organs and tissues of human origin for therapeutic purposes. Among the general principles there are equitable access to transplantation services for patients; transparent rules for organ allocation; health and safety standards; the prohibition of financial gain by donors; and the need for donors, recipients,

---

human-rights-must-guide-any-use-of-genome-editing-technologies-in-human-beings-, accessed 21 December 2018.

[74] (n 16).
[75] (n 17).
[76] (n 18).
[77] (n 19).
[78] Ratification of the Oviedo Convention is a prerequisite for ratification of its protocols. Thus, those states that have not ratified the Oviedo Convention have not ratified its additional protocols either.
[79] Additional Protocol to the Convention on Human Rights and Biomedicine on the Prohibition of Cloning Human Beings, art. 1.1.
[80] Ibid., art. 1.2.

health professionals and the public to be properly informed. The specific provisions cover the removal of organs from living and deceased persons; the use made of the organs and tissues removed; confidentiality; sanctions and compensation.

The Additional Protocol on Biomedical Research was adopted in 2004 and entered into force in 2007. To date, it has been ratified by 12 states. It builds on the principles embodied in the Oviedo Convention, to protect human rights and dignity in the specific field of biomedical research. Its purpose is to define and safeguard fundamental rights in biomedical research, in particular of those participating in research. The Protocol covers the full range of biomedical research activities involving interventions on human beings. It restates the fundamental principles guiding research involving human beings, such as the free, informed, express, specific, and documented consent of the person(s) participating. It addresses issues such as risks and benefits of research, consent, protection of persons not able to consent to research, scientific quality, independent examination of research by an ethics committee, confidentiality and the right to information, undue influence, safety and duty of care.

Finally, the Protocol on Genetic Testing for Health Purposes was adopted in 2008 and entered into force on 1 July 2018. So far, it has been ratified by five states, which is the threshold for entry into force. This protocol sets down principles relating inter alia to the quality of genetic services, prior information and consent and genetic counselling. It lays down general rules on the conduct of genetic tests, and, for the first time at the international level, deals with genetic tests directly accessible to the public. It specifies the conditions in which tests may be carried out on persons not able to consent. Also covered are the protection of private life and the right to information collected through genetic testing.

4 *The European Convention on Human Rights*

As it was said, the European Convention on Human Rights and Fundamental Freedoms (European Convention) is the linchpin of the European human rights regime.[81] Neither the European Convention nor its protocols, which extended the list of rights, mention the 'right to health', the 'right to science' or the 'rights of science'. The Convention, both in its original form and after revisions brought about by the protocols, focuses mostly on civil and political freedoms. However, it contains some articles that over the years have been used by advocates and patients to address issues raised by biotechnology,

---

[81]   See, in this chapter, Section II.1.

artificial reproductive technology and the question of the legal status of the human embryo. These include Article 2 (Right to Life),[82] Article 8 (Right to Respect for Private and Family Life)[83] and Article 14 (Prohibition of Discrimination).[84] As in the case of every human rights treaty, the rights contained in these articles are formulated very broadly, leaving much room for interpretation. However, interpretation and reading between the lines of the Convention cannot go as far as inventing rights that states did not intend to recognize.

The European Court of Human Rights is the international court with jurisdiction over alleged violations of the human rights contained in the Convention of natural and legal persons committed by any of the 47 members of the CoE within their jurisdiction. Composed of 47 judges – one for each CoE member state, and divided into four chambers and a Grand Chamber – it issues binding decisions that can be enforced by national courts.[85] By all metrics, it is the most important and effective of all international human rights adjudicative bodies.[86] While a full discussion of the Court is certainly beyond the scope of this book, we should mention here a few cases that are relevant for biomedicine and artificial reproductive technology.

In 2012, in *Costa and Pavan v. Italy*, the Court found the prohibition of PGD contained in the version of Law 40/2004 then in force in Italy to be a violation of Article 8 of the European Convention on Human Rights because the applicants' desire to resort to artificial reproductive technology and

---

[82] European Convention on Human Rights, Article 2: '1. Everyone's right to life shall be protected by law. No one shall be deprived of his life intentionally save in the execution of a sentence of a court following his conviction of a crime for which this penalty is provided by law. 2. Deprivation of life shall not be regarded as inflicted in contravention of this Article when it results from the use of force which is no more than absolutely necessary: (a) in defense of any person from unlawful violence; (b) in order to effect a lawful arrest or to prevent the escape of a person lawfully detained; (c) in action lawfully taken for the purpose of quelling a riot or insurrection.'

[83] Ibid., art. 8: '1. Everyone has the right to respect for his private and family life, his home and his correspondence. 2. There shall be no interference by a public authority with the exercise of this right except such as is in accordance with the law and is necessary in a democratic society in the interests of national security, public safety or the economic well-being of the country, for the prevention of disorder or crime, for the protection of health or morals, or for the protection of the rights and freedoms of others.'

[84] Ibid., art. 14: 'The enjoyment of the rights and freedoms set forth in this Convention shall be secured without discrimination on any ground such as sex, race, color, language, religion, political or other opinion, national or social origin, association with a national minority, property, birth or other status.'

[85] See, in general, E Lambert Abdelgawad, *The Execution of Judgments of the European Court of Human Rights* (2nd ed., 2008).

[86] On the European Court of Human Rights, see, in general: D John Harris and others, *Law of the European Convention on Human Rights* (4th ed., 2014).

embryo screening to have a child not affected by a genetic disorder of which they were healthy carriers was an expression of their private and family life.[87]

A second case, again against Italy, was decided in 2015. *Parrillo v. Italy* was about a couple whose IVF surplus embryos had been cryopreserved. After the death of her partner, Ms Parrillo decided to donate the embryos to scientific research. However, Law 40/2004 prohibits the use of human embryos for anything other than reproduction.[88] Ms Parrillo alleged that the prohibition violated her right to respect for private life (Article 8),[89] as well as of her right to private property (Article 1 of Protocol No. 1),[90] and freedom of expression (Article 10). Noting that the embryos contain the genetic material of the applicant and thus represent a constituent part of her identity, the Court's Grand Chamber concluded that Ms Parrillo's ability to exercise a choice regarding the fate of the embryos concerned an intimate aspect of her personal life and was related to her right to self-determination.[91] Therefore, the prohibition to donate embryos to scientific research interfered with Ms Parrillo's right to private life.[92]

However, the right to a private life is not an absolute right. It can be limited. Thus, the Court next considered whether this interference was 'in accordance with the law' as required by Article 8.2 of the Convention, which recites: 'There shall be no interference by a public authority with the exercise of this right except such as is in accordance with the law and is necessary in a democratic society in the interests of national security, public safety or the economic well-being of the country, for the prevention of disorder or crime, for the protection of health or morals, or for the protection of the rights and freedoms of others'.[93] While the Court recognized that the aim pursued by

---

[87] *Costa and Pavan* v. *Italy*, Application No 54270/10, European Court of Human Rights (Judgment of 28 August 2012) para. 57.

[88] Law No 40 of 19 February 2004 'Rules on Medically Assisted Procreation' (In Italian, *Norme in materia di procreazione medicalmente assistita*) art. 13. For a rough translation in English, see European Institute of Bioethics, 'Rules on Medically Assisted Procreation' (Italian Parliament, 2004) www.ieb-eib.org/en/pdf/loi-pma-italie-english.pdf accessed 27 February 2017.

[89] European Convention on Human Rights, art. 8.

[90] Article 1.1 of Protocol No 1 (Protection of Property) to the European Convention on Human Rights states: 'Every natural or legal person is entitled to the peaceful enjoyment of his possessions. No one shall be deprived of his possessions except in the public interest and subject to the conditions provided for by law and by the general principles of international law.' Protocol to the Convention for the Protection of Human Rights and Fundamental Freedoms (18 May 1954) ETS No 9, art. 1.1.

[91] *Parrillo* v. *Italy*, Application No 46470/11, European Court of Human Rights (Grand Chamber Judgment of 27 August 2015), paras. 158–159.

[92] Ibid., para. 161.

[93] Ibid.

Italy (the protection of the embryo's potential for life) could be justified by the aim of 'protecting morals and the rights and freedoms of others', it also stressed that this did not imply any assessment by the Court as to whether the word 'others' extended to human embryos.[94] Eventually, the Court decided that, given the lack of a European consensus on the matter, Italy was to be given a wide margin of appreciation,[95] and did not find a violation of Article 8.2 of the European Convention.[96] As to the claims that Article 1 of Protocol No. 1 and Article 10 had been violated, the Grand Chamber found them inadmissible because an embryo cannot be considered property in the economic and pecuniary sense of that article,[97] and the right to freedom of expression in this case was not vested in the applicant directly, but rather in researchers and scientists.[98]

Three more relevant cases are *Evans* v. *United Kingdom, Dickson* v. *United Kingdom* and *S.H. and Others* v. *Austria*. They all concern various aspects of artificial reproductive technology. In *Evans*, the Court found that the UK laws allowing withdrawal of consent to use cryopreserved embryos by one of the partners had not violated the Convention.[99] *Dickson* was about a couple who could not resort to IVF because the husband was in detention serving a 15-year sentence for murder, and the Court ruled in their favour.[100] Artificial insemination was the applicants' only realistic hope to conceive a child. The Grand Chamber observed that, while the inability to beget a child might be a consequence of imprisonment, it was not an inevitable one, since giving access to artificial insemination facilities would not have involved any security issues or imposed any significant administrative or financial demands on the state. Accordingly, the Court held that there had been a violation of Article 8 of the Convention, as a fair balance had not been struck between the competing public and private interests.[101]

S.H. and Others v. *Austria* was brought by two couples who challenged Austria's ban of heterologous artificial insemination, claiming violations of

---

[94] Ibid., para. 167.
[95] Ibid., paras. 174–176.
[96] Ibid., para.197.
[97] Ibid., para. 215.
[98] *Parrillo* v. *Italy*, Application No 46470/11, European Court of Human Rights (Decision on Admissibility of 28 May 2003).
[99] *Evans* v. *United Kingdom*, Application No 6339/05, European Court of Human Rights (Grand Chamber Judgment of 10 April 2007).
[100] *Dickson* v. *United Kingdom*, Application No 44362/04, European Court of Human Rights (Judgment of 4 December 2007).
[101] Ibid., para. 82.

Article 8 and Article 14.[102] The Court found a violation of Article 14 of the Convention in conjunction with Article 8. However, on appeal, in 2010, the Court's Grand Chamber concluded that there had been no violation of Article 8 per se because the Austrian legislature did not exceed the margin of appreciation afforded to it at the relevant time, either in respect of the prohibition of egg donation for the purposes of artificial procreation or in respect of the prohibition of sperm donation for IVF.[103] In fact, although there was a clear trend across Europe in favour of allowing gamete donation for IVF, there was not yet a consensus on the matter nor settled legal principles.[104]

All in all, when confronted with matters raising bioethical contentious or difficult questions, the European Court of Human Rights has shown significant willingness to defer to states and to allow them sometimes a wide margin of appreciation, as long as the laws or actions in question do not appear to be discriminatory or arbitrary and strike a fair balance between the competing interests and values at play.

5   Other CoE-Relevant Treaties

Finally, to conclude the overview of CoE's legal instruments relevant for a discussion of human genome germline modification one must mention the Council's data protection framework, in particular the Convention for the Protection of Individuals with regard to Automatic Processing of Personal Data[105] and the European Convention for the Protection of Vertebrate Animals used for Experimental and other Scientific Purposes.[106]

As the name of the former suggests, the first is a treaty aiming to protect the right to privacy of individuals, taking account of the increasing flow across frontiers of personal data undergoing automatic processing. When it was adopted, in 1981, it was the first treaty in the world of its kind. In addition to providing guarantees in relation to the collection and processing of personal data, it outlaws the processing of 'sensitive' data on a person's race, politics, health, religion, sexual life, criminal record, etc., in the absence of proper legal safeguards. The Convention also enshrines the individual's right to know

---

[102] *SH and others* v. *Austria*, Application No 57813/00, European Court of Human Rights (Judgment of 1 April 2010).

[103] Ibid., para. 115.

[104] Ibid., para. 96.

[105] Convention for the Protection of Individuals with Regard to Automatic Processing of Personal Data, CETS 108.

[106] European Convention for the Protection of Vertebrate Animals used for Experimental and other Scientific Purposes, CETS 123.

that information is stored on him or her and, if necessary, to have it corrected. The Convention also imposes some restrictions on transborder flows of personal data to states where legal regulation does not provide equivalent protection. The data protection Convention was updated and improved in Amendments approved by the Committee of Ministers, in Strasbourg, on 15 June 1999, to take into account the advent of the Internet and the expansion of data processing capacities.[107] The principles of transparency, proportionality, accountability, data minimization, privacy by design, etc., are now acknowledged as key elements of the protection mechanism and have been integrated in the modernized instrument.[108] All members of the Council of Europe have ratified the treaty, as well as six non-members (Mauritius, Senegal, Tunisia, Uruguay, Cabo Verde and Mexico).[109]

The European Convention for the Protection of Vertebrate Animals used for Experimental and other Scientific Purposes was adopted by the CoE in 1986 primarily to reduce both the number of experiments and the number of animals used for research.[110] To this end, it contains a number of principles that guide the national policies of those states that have ratified it, such as refraining from experimenting on animals except where there is no alternative; seeking alternative methods; selecting animals to experiment on the basis of clearly established quantitative criteria and well caring for and sparing avoidable suffering whenever possible.

III THE EUROPEAN UNION

1 *Introduction*

The European Union was established in 1992 by the Treaty of Maastricht to 'continue the process of creating an ever-closer union among the peoples of Europe',[111] a process that had started in the aftermath of the Second World

---

[107] 'Amendments approved by the Committee of Ministers, in Strasbourg, on 15 June 1999, ETS No 108' (Council of Europe, 1999) https://rm.coe.int/CoERMPublicCommonSearchServic es/DisplayDCTMContent?documentId=09000016800c2b8 accessed 10 December 2018.

[108] See CoE, 'The Modernised Convention 108: Novelties in a Nutshell' https://rm.coe.int/168 08accf8 accessed 5 November 2018.

[109] See CoE, 'Chart of Signatures and Ratifications of Treaty 108' (2018) www.coe.int/en/web/conven tions/full-list/-/conventions/treaty/108/signatures?p_auth=maZNKP38 accessed 5 November 2018.

[110] To date, it has been ratified only by 22 CoE members, including the European Union. See CoE, 'Chart of Signatures and Ratifications of Treaty 123' (2018) www.coe.int/en/web/conventions/full-list/-/conventions/treaty/123/signatures?p_auth=7Jfxuyy8 accessed 5 November 2018.

[111] Preamble of the Treaty on European Union (signed in Maastricht on 7 February 1992) OJ C 191, 1–112.

War with the creation of the three European Communities: the European Coal and Steel Community,[112] the European Economic Community (EEC)[113] and the European Atomic Energy Community.[114] Over the next four decades, this limited project of regional economic integration broadened and deepened to include political issues, such as foreign and security policy, justice and home affairs, social policy, consumer protection, industry, environment, public health and safety, and many others.

Membership expanded, too. Starting with the founding six states (Belgium, France, West Germany, Italy, Luxembourg and the Netherlands), the Communities expanded first, in 1973, to the north-east (the United Kingdom, Ireland and Denmark), then, in the 1980s, to the south (Portugal, Spain, and Greece), then, in the 1990s to the north and east (Sweden, Finland, Austria and East Germany, as a result of German reunification), and finally in the 2000s to the east and south, when 11 states from Central and Eastern Europe and the Balkans, as well as Cyprus and Malta, became part of the Union.[115] Nowadays, the European Union is composed of 28 member states. While in the near future the UK might leave it due to the outcome of the 'Brexit referendum',[116] others, particularly in the Balkans, might join.[117]

The expansion and deepening of the European integration process happened in stages, each marked by a new treaty. At the beginning of the 2000s, its members looked ready to leap forward towards the creation of a European federal state. Negotiations to modify EU institutions began in 2001, resulting in the adoption of a European Constitution, which would have repealed the existing European

---

[112]   ESCS Treaty or Paris Treaty: Treaty Establishing the European Coal and Steel Community (entered into force 18 April 1951, no longer in force and end of validity 23 July 2002) 261 UNTS 140.

[113]   EEC Treaty or Treaty of Rome: Treaty Establishing the European Economic Community (entered into force 25 March 1957) 298 UNTS 3, 4 Eur. Y.B. 412

[114]   EAEC Treaty or EURATOM Treaty: Treaty Establishing the European Atomic Energy Community (entered into force 25 March 1957) 298 UNTS 259.

[115]   Bulgaria (2007), Croatia (2013), Czech Republic (2004), Cyprus (2004), Estonia (2004), Hungary (2004), Latvia (2004), Lithuania (2004), Malta (2004), Poland (2004), Romania (2007), Slovakia (2004) and Slovenia (2004).

[116]   However, at the time of this writing it is not yet clear whether the UK will actually exit. On 10 December 2018 the Court of Justice of the European Union ruled that the UK could unilaterally suspend the exit process, if it wishes to do so. See Case C-621/18, Request for a preliminary ruling under Article 267 TFEU from the Court of Session, Inner House, First Division (Scotland, United Kingdom), made by decision of 3 October 2018, received at the Court on the same day, in the proceedings, Judgment of the European Court of Justice, 10 December 2018.

[117]   As of 2018, there are five candidates for future EU membership: Turkey, Macedonia, Montenegro, Albania and Serbia. Other potential candidates for future EU membership are Kosovo and Bosnia Herzegovina.

treaties. However, when it was put to the vote of the citizens of each member state, it was rejected after the electors in France and the Netherlands voted against it. After a period of reflection, member states agreed to amend instead the existing treaties, salvaging a number of the reforms that had been envisaged in the botched constitution. The result was the so-called Reform Treaty, adopted in Lisbon in 2007 and entered into force in 2009.[118]

The most important changes brought about by the Treaty of Lisbon are the abandonment of unanimity for qualified majority to take decisions in the major policy areas; a more powerful European Parliament; the conferral of legal personality to the European Union, distinct and separate from that of its member states; the creation of a President of the European Council and a High Representative of the Union for Foreign Affairs and Security Policy; and the adoption of the Union's 'Bill of Rights', called the Charter of Fundamental Rights of the European Union (EU Charter).

As to governance, the European Union's main institutions are the Council of the European Union, the European Parliament, the European Commission and the Court of Justice of the European Union. The Council is composed of the 28 member states, represented by their ministers, who meet to discuss, amend, adopt laws, and coordinate policies.[119] The European Parliament is directly elected by the EU citizens every five years and is composed of 751 members.[120] It shares legislative and budgetary powers with the Council and has certain exclusive scrutiny and appointments powers. The European Commission is the European Union's executive arm.[121] It initiates the law-making process by proposing new EU legislation and is responsible for enforcing EU law and policies. Finally, the Court of Justice of the European Union (CJEU, or Court of Justice) ensures that the Union's laws are interpreted and applied the same in all member states.[122] It also settles disputes between national governments and EU institutions and can in some circumstances adjudicate claims brought by legal or natural persons against EU

[118] Treaty of Lisbon amending the Treaty on European Union and the Treaty establishing the European Community (signed at Lisbon, 13 December 2007) OJ C 306, 1–271.

[119] The Council of the European Union is not to be confused with the European Council, which are quarterly summits where EU leaders meet to set the broad direction of EU policy-making. 'The Council of the European Union' www.consilium.europa.eu/en/council-eu/ accessed 9 October 2018; 'The European Council' www.consilium.europa.eu/en/european-council/ accessed 9 October 2018.

[120] 'The European Parliament' www.europarl.europa.eu/portal/en accessed 9 October 2018.

[121] 'The European Commission' https://ec.europa.eu/commission/index_en accessed 9 October 2018.

[122] 'The Court of Justice of the European Union' https://curia.europa.eu/jcms/jcms/j_6/en/ accessed 9 October 2018.

institutions. To ensure harmony in the interpretation of EU law, national courts can refer cases to the Court of Justice for preliminary rulings that are binding both for the court that requested it and also for all member states.

2 The Regulatory Environment

a EU Primary Law

Although the European Union failed to give itself a federal constitution, the triad made of the EU Charter on Fundamental Rights[123] and the two constitutive treaties – the Treaty of the European Union (TEU, originally the Treaty of Maastricht)[124] and the Treaty of the Functioning of the European Union (TFEU, originally the Treaty of Rome, which created the European Economic Community)[125] – is the Union's 'primary law'. They are the de facto 'constitution of the European Union', in the sense that all EU law-making activities must find their legality in them and must be compatible with them.[126]

The TEU and TFEU set out the goals of the Union and the principles on which it is based, as well as its organs, powers, composition and competences. The EU Charter of Fundamental Rights contains a list of fundamental rights everyone within the jurisdiction of the European Union enjoys. When first proclaimed in 2001, the EU Charter was not legally binding for the member states. However, following the Lisbon Treaty, the Charter was attached to, and given the same legal value as, the constitutive treaties.[127] However, unlike the European Convention of Human Rights, those who must ensure respect for the rights included in the EU Charter are not primarily states but the

---

[123] Consolidated Version of the Treaty of the European Union, with the amendments introduced by the Treaty of Lisbon (signed on 13 December 2007 and entered into force on 7 December 2009) OJ C 326, 13–390.

[124] Charter of Fundamental Rights of the European Union (proclaimed by the European Parliament on 7 December 2000 and entered into force in adapted wording with the date of the entry into force of the Lisbon Treaty on 7 December 2009) OJ C 326, 26 October 2012 (TEU) 391–407.

[125] Consolidated version of the Treaty on the Functioning of the European Union, with the amendments introduced by the Treaty of Lisbon (signed on 13 December 2007 and entered into force on 7 December 2009) OJ C 326 (TFEU) 47–390.

[126] Since the entry into force of the Lisbon Treaty in 2009, the EU Charter has the same legal value of the two constitutive treaties and must also be respected by the Union as a matter of primary or 'constitutional' law. TEU, art. 6.1.

[127] Ibid. See, also, N Foster, *Foster's EU Law* (6th ed., Oxford, Oxford University Press 2017) 106–107.

European Union itself.[128] Under the Charter, the European Union must act and legislate consistently with the Charter and courts will strike down legislation adopted by the European Union's institutions that contravenes it. Member states must comply with it only in so far as they are implementing EU law.[129]

The EU Charter was drafted in the early 2000s. Because of its novelty, to date it is the only enunciation of human rights of a general nature to consider how the rapid scientific developments in biology and medicine may affect fundamental human rights. The most important right is included in Title I related to dignity rights and incorporates some of the basic rights and principles included in the Oviedo Convention.[130] Notably, Article 3 of the Charter, entitled 'Right to the Integrity of the Person', establishes that in the fields of biology and medicine the 'free and informed consent of the person concerned' must be respected.[131] The same article also sets forth three prohibitions: the prohibition of 'eugenic practices, in particular those aiming at the selection of persons',[132] of 'making the human body and its parts as such a source of financial gain'[133] and of 'reproductive cloning of human

---

[128]   EU Charter on Fundamental Rights, art. 51.1: 'The provisions of this Charter are addressed to the institutions, bodies, offices and agencies of the Union with due regard for the principle of subsidiarity and to the Member States only when they are implementing Union law. They shall therefore respect the rights, observe the principles and promote the application thereof in accordance with their respective powers and respecting the limits of the powers of the Union as conferred on it in the Treaties.' Charter of the Fundamental Rights of the European Union [2000] OJ C 364/1.

[129]   A Ward, 'Article 51 – Field of Application' in S Peers and others (eds.), *The EU Charter of Fundamental Rights: A Commentary* (Oxford, Beck/Hart Publishers 2014) 1456–1497.

[130]   The provisions of the Oviedo Convention of special importance are Article 5 of the Oviedo Convention related to free and informed consent and Article 21 prohibiting using the human body for financial gain as well as Additional Protocol to this Convention related to the prohibition against human cloning. See S Michalowski, Article 3 in S Peers and others (eds.), *EU Charter of Fundamental Rights: A Commentary* (Oxford, Beck/Hart Publishers 2014) 39–102, 44.

[131]   Ibid., art. 3.2.a. That this right includes the right of donors has been confirmed in a Court of Justice judgment of 9 October 2001 in Case C-377/98 *Netherlands* v. *European Parliament and Council* [2001] ECR-I 7079, at grounds 70, 78 to 80. To some extent, this right finds a basis in Article 7 of the International Covenant of Civil and Political Rights (n 30) according to which 'no one shall be subject without his free consent to medical or scientific experimentation'. Insofar as it applies to reproductive health, Article 3 finds some support in Article 16 of the UN Convention on the Elimination of Discrimination against Women, which establishes the rights of women to 'decide freely and responsibly on the number and spacing of their children and to have access to the education, information and means to enable them to exercise these rights'. Convention on the Elimination of All Forms of Discrimination Against Women (adopted on 18 December 1979) 1249 UNTS 13.

[132]   EU Charter on Fundamental Rights, art. 3.2.b.

[133]   Ibid., art. 3.2.c.

beings'.[134] The EU Charter does not define 'eugenic practices'. However, the Explanations of the Charter indicate that the drafters intended to prohibit those practices that constitute a crime under the Rome Statute of the International Criminal Court,[135] such as the organization and implementation of selection programmes for sterilization, forced pregnancy and compulsory ethnic marriage, among others.[136] The prohibition of human cloning is limited, too, since it only concerns reproductive cloning of human beings, but no other forms of cloning.[137]

Whether Article 3 of the EU Charter prohibits human germline genome modification is unclear.[138] For a start, the EU Charter does not mention it. Although Article 1 of the Charter declares 'human dignity is inviolable ... [i]t must be respected and protected', in the absence of a definition of 'dignity' in the Charter or its Explanations, it is not evident that human genetic engineering would per se amount to a violation of the obligation to respect human dignity. Also, although Article 3.1 declares '[e]veryone has the right to respect for his or her physical and mental integrity', it is not evident how human germline genome modification might violate someone's physical and mental integrity either. Moreover, it is unclear whether the right to the physical integrity of persons extends also to future generations and under what conditions. At the same time, there are other articles in the Charter that protect the right to conduct scientific research, such as Article 13, according to which 'the arts and scientific research shall be free of constraint' and that 'academic freedom shall be respected'. Although the Explanations stress that freedom of scientific research is not absolute and that its exercise must be compatible with the obligation to respect and protect human dignity and may be subject to the limitations authorized by Article 10 of the ECHR (freedom of expression), so far neither the legislative bodies of the European Union nor the judicial ones have clarified where the balance should be struck.[139]

---

[134] Ibid., art. 3.2.d.
[135] Rome Statute of the International Criminal Court (entered into force 1 July 2002) 2187 UNTS 90.
[136] Explanations Relating to the Charter of Fundamental Rights (2007) OJ C 303/2, 17–35 (Explanations) 18.
[137] Ibid., 18. According to the Explanations, the prohibition against the reproductive cloning of human beings 'neither authorises nor prohibits other forms of cloning. Thus it does not in any way prevent the legislature from prohibiting other forms of cloning,' in line with the Oviedo Convention.
[138] For example, it is not mentioned in the analysis of Article 3 of the EU Charter of Fundamental Rights by Sabine Michalowski, 'Article 3' in S Peers and others (eds.), *EU Charter of Fundamental Rights: A Commentary* (Oxford, Beck/Hart Publishers 2014) 39–102.
[139] Ibid., 17 and 22.

Finally, it should be noted that the EU Charter recognizes the 'right to health', which is included in Title IV related to solidarity rights. According to Article 35: 'Everyone has the right of access to preventive health care and the right to benefit from medical treatment under the conditions established by national laws and practices. A high level of human health protection shall be ensured in the definition and implementation of all the Union's policies and activities.' The formulation of this article has been inspired by international human rights treaties, including Article 3 of the Oviedo Convention, according to which 'Parties, taking into account health needs and available resources, shall take appropriate measures with a view to providing, within their jurisdiction, equitable access to health care of appropriate equality' as well as the European Social Charter.[140] According to the explanatory notes of the EU Charter, Article 35 is based on Articles 11 and 13 of the European Social Charter.[141] The objectives are to improve public health, prevent physical and mental illness and diseases, and obviate sources of danger to physical and mental health. The EU involvement on matters of public health covers 'the fight against major health scourges, by promoting research into their causes, their transmission and their prevention, as well as health information and education, and monitoring, early warning of and combating serious cross-border threats to health'.[142] However, the fact that, according to EU law, Article 35 establishes 'principles' instead of 'rights' means that more detailed legislation must be adopted for any health-related rights to be judicially enforceable in the courts.[143] Whether European Union and national law blocking research on human germline modification, research that could lead to eliminating entirely many hereditary genetic diseases, could be seen as contrary to the objective of attaining a 'high level of human health protection' remains to be seen.

---

[140] Articles 11 and 13 of the European Social Charter. See C A Young, 'Fundamental Rights and EU Health Law and Policy' in T K Hervey, C A Young, and L E Bishop, *Research Handbook on EU Health Law and Policy* (Cheltenham, UK, Edward Elgar Publishing 2017) 82–108, 90–91.

[141] Explanations relating to the Charter of Fundamental Rights of the European Union, 27.

[142] TFEU, art. 168.1.

[143] On the distinction between 'rights' and 'principles', see Explanations relating to the Charter of Fundamental Rights of the European Union, 35, which clarifies Article 52.5 of this charter; K Leanerts, 'Exploring the Limits of the EU Charter of Fundamental Rights' (2012) 8 *European Constitutional Law Review* 375, 400, and the analysis of the Advocate General Cruz Villalón in his Opinion AMS v. CGT ECLI:EU:C:491, [2013] paras. 43–80.

b International Law and the European Union

According to the Treaty of Lisbon the European Union has legal personality distinct and separate from that of its member states.[144] The fact that the Union has legal personality means it has international obligations of its own.[145] The European Union is a member of international organizations alongside some or all of its members (e.g. the World Trade Organization (WTO)) and has the power to conclude treaties where it has competence to act, either explicitly or implicitly.[146] International agreements entered into by the European Union are regarded as an integral part of the EU legal order and may under certain circumstances be directly effective (self-executing). However, to date, much of the case law of the Court of Justice in this field concerns the legal effects of WTO agreements,[147] and does not centre on international human rights instruments. Moreover, although the European Union is party to several treaties, including the CoE's Convention for the Protection of Vertebrate Animals Used for Experimental and Other Scientific Purposes,[148] it has not yet ratified the European Convention on Human Rights, even though it is required to do so by its own constitutive treaties.[149] Neither has it ratified the European Convention on Human Rights and Biomedicine (Oviedo Convention).[150] Also, the European Union is not party to the ICESCR, which is a treaty only open for signature to states.[151]

That the European Union is not a party to these human rights treaties does not affect the human rights obligations by its member states, which are, all of them, parties to the Covenant and the European Convention on Human Rights.[152] However, the fact that all EU states must respect the rights and

---

[144] TEU, art. 47. See, however, pre-Lisbon case law of the Court of Justice according to which the European Community was granted international legal personality, e.g. Case 22/70 *Commission v. Council* (AETR/ERTA) (1971) ECR 263.

[145] TFEU, art. 216.2.

[146] TFEU art. 216.1; TEU art. 37.

[147] P Craig and G de Búrca, *EU Law: Text, Cases and Materials* (6th ed., Oxford, Oxford University Press 2015) 362.

[148] European Convention for the Protection of Vertebrate Animals used for Experimental and Other Scientific Purposes, 18 March 1986, ETS No 123. The EU ratified it on 30 April 1998. It entered into force for the EU on 1 November 1998.

[149] TEU, art. 6.2.

[150] Nine out of 28 EU states have not ratified the Oviedo Convention, among them, the United Kingdom.

[151] ICESCR, art. 26.1. The EU has not ratified any international human rights treaties with the exception of the Convention on the Rights of Persons with Disabilities (2006), which it ratified on 23 December 2010. Convention on the Rights of Persons with Disabilities (entered into force 13 December 2006) 2515 UNTS 3.

[152] TEU, art. 6.2.

freedoms established in these legal instruments by virtue of their own ratification is not sufficient to make them binding on the European Union itself, even though the European Union still needs to take account of them 'through interpretation in the light of the Member States' obligations due to the principle of sincere cooperation'.[153] Thus, even if the Union itself is not formally bound to give effect to, for example, the right to science or the right to health as laid down in the Covenant, its institutions must consider these rights, to the extent that they are regarded as general principles of the Union's law and 'result from the constitutional traditions common to the Member States'.[154]

In sum, there are three formal sources of EU human rights law: the EU Charter, the European Convention on Human Rights and the general principles of EU law. Out of these, the EU Charter is the most important one, as also manifested in the case law of the Court of Justice. The general principles are a body of legal principles, including human rights, which were articulated and developed by the Court of Justice before the EU Charter was drafted. General principles are a more ambiguous source said to derive from national constitutional traditions, from the ECHR, and from other international treaties signed by the member states.[155] Even if the definition of general principles included in Article 6.3 of the TEU makes no explicit reference to other international human rights instruments besides the ECHR as a source of EU human rights law, the Court of Justice occasionally cites such instruments,[156] including the Covenant on Economic, Social and Cultural Rights,[157] even if the European Union is not itself party to these instruments. However, importantly, it has rejected reliance on the Oviedo Convention since the European Union has not signed it and since, as the Court of Justice pronounced in 2010, 'of the Member States, only a small majority of them have actually ratified the Convention'.[158]

---

[153]   TEU, art. 4.3. See also K S Ziegler, 'The Relationship between EU Law and International Law' in D Patterson and A Södersten (eds.), *A Companion to European Union Law and International Law* (Malden, John Wiley & Sons 2016) 42–61.

[154]   TEU, art. 6.3.

[155]   Craig and de Búrca, *EU Law* (n 147) 380.

[156]   Ibid., 385. For example, the Court of Justice has cited the UN Convention on the Rights of the Child in Case C-540/03 *European Parliament v. Council* (2006) ECR I-5769, (57) and the UN Convention on the Rights of Persons with Disabilities (which the EU ratified in 2010) in Case C-354/13 *FOA v. Kommunernes Landsforening (Kaltoft)* EU:C:C:2014:2463 (53).

[157]   See, e.g., Case C-5/12, *Marc Betriu Montull v. Instituto Nacional de la Seguridad Social* ECLI:EU:C:2013:571 and Case C-73/08 *Nicolas Bressol and Others and Céline Chaverot and Others v Gouvernement de la Communauté française* ECLI:EU:C:2010:181.

[158]   See judgment of the Court of Justice on 3 June 2010 in Case C-237/09 *État belge v. Nathalie De Fruytier* ECLI:EU:C:2010:316.

Moreover, it is unclear whether the sources of EU human rights law include customary international law,[159] including the Universal Declaration of Human Rights, which proclaims the right of everyone to share in scientific advancements and its benefits,[160] Even if the Court of Justice sometimes cites the Universal Declaration of Human Rights in its judgments, it is when the Declaration refers to so-called general principles of EU law, such as the principle of non-discrimination, in preliminary rulings.[161] Beyond this, the case law of the Court of Justice reveals reluctance to discuss international human rights law as a matter of customary international law, and instead prefers to read them in the EU Charter and the European Convention on Human Rights.[162] Indeed, even if many provisions of the EU Charter 'are themselves based on international human rights instruments, as the explanatory notes to the Charter indicate, those international instruments and the courts or bodies established to interpret them have not yet – apart from the European Convention on Human Rights and Court of Human Rights – been treated as influential or persuasive authority in the interpretation by the ECJ of Charter provisions'.[163]

Against this background, it is unlikely the 'right to science' and the 'rights of science' contained in the Covenant on Economic, Social and Cultural Rights and the Universal Declaration of Human Rights could be successfully invoked before the Court of Justice and other EU institutions. The Court of Justice's emphasis on the autonomy of the EU legal order in relation to international law in its recent case law, particularly when rejecting the draft Agreement on Accession of the European Union to the ECHR, indicates a reluctance to rely on and subordinate itself to international human rights instruments and mechanisms.[164] Moreover, should the right to science one day be included in the EU Charter, it will probably find its place alongside the right to health and other solidarity rights that require the adoption of specific legislation by

[159] S Besson, 'General Principles and Customary Law in the EU Legal Order', in S Vogenauer and S Weatherill (eds.), *General Principles of Law. European and Comparative Perspectives* (Hart Publishing 2017) 105–129.

[160] UDHR, art. 27.1.

[161] See, e.g., C-144/04 *Mangold* ECLI:EU:C:2005:709 and C-411/05 *Palacios de la Villa* ECLI: EU:C:2007:604. But note that in Case C-135/08 *Rottmann* ECLI:EU:C:2010:104, the Court of Justice of the European Union cites Article 15 of the Universal Declaration of Human Rights (related to the right to nationality).

[162] Cases C-402 and 415/05 P *Kadi & Al Barakaat International Foundation* v. *Council and Commission (Kadi I)* (2008) ECR I-6351, (308); and Case 584/10 P *Commission* v. *Kadi (Kadi II)* EU:C:2013:518.

[163] Craig and de Búrca, *EU Law* (n 147) 387.

[164] Opinion 2/13 on EU Accession to the ECHR EU:C:2014:2454 (192).

member states or the European Union itself before the right can be judicially enforced before courts.[165]

c Relevant Regulations and Directives

The key legal instruments of the European Union are 'regulations' and 'directives'. A 'regulation' is specific legislation that, once it has entered into force, is self-executing and is immediately and directly applicable across the Union.[166] EU member states are not obliged to transform EU regulations into national regulations except for when required by the regulation itself or required to ensure its effectiveness.[167] Conversely, a 'directive' just sets out goals that all EU states must achieve, leaving it up to each of these states to devise and adopt their own laws on how to reach them.[168]

In the case of 'directives' EU states are given some discretion and time to define the measures needed to give effect to them, while they have no such leeway when it comes to 'regulations'.[169] EU directives are not directly applicable in the sense of being automatic and general in application, and affording rights without further implementation. However, there are directives that give rise to directly enforceable rights in specific circumstances, such as when a directive has not been implemented at all or has been implemented incorrectly.[170]

In either case, EU member states have the primary responsibility to enforce EU regulations and directives within their jurisdictions. The national courts of EU member states have an important function in enforcing EU law. Regulations have 'direct effect' in the sense of being enforceable in national courts provided the provisions are clear and precise, as well as unconditional,

---

[165] For the distinction between 'rights' and 'principles', see EU Charter, art. 52.5. Several *rights* recognized in the ICESCR are understood as *principles* in EU law. In practice, this means that these rights will be justiciable only following implementing acts of the European Union or the member states and only in relation to interpretation or rulings on the legality of such acts. See Craig and de Búrca, *EU Law* (n 147) 398–399.

[166] TFEU, art. 288: 'A regulation shall have general application. It shall be binding in its entirety and directly applicable in all member states.'

[167] Case 39/72 *Commission* v. *Italy* ECLI:EU:C:1973:13; and Case 128/78 *Commission* v. *United Kingdom* ECLI:EU:C:1979:32.

[168] Other legal acts of the EU include 'decisions' that are binding on those to whom it is addressed and is directly applicable. By contrast, neither 'recommendations' nor 'opinions', while defined as legal acts, have any legal consequences since they are not binding.

[169] TFEU, art. 288: 'A directive shall be binding as to the results to be achieved, upon each member state to which it is addressed, but shall leave to the national authorities the choice of form and methods.'

[170] N Foster, *Foster on EU Law* (6th ed., Oxford, Oxford University Press 2017) 120–121.

and do not require implementing measures by the member states or the European Union itself that give them discretion.[171] Such regulations have 'vertical direct effect', meaning individuals can invoke the rights recognized in them against EU member states, and 'horizontal direct effect', meaning individuals can invoke them against other individuals. Also directives can give rise to direct effect provided their provisions meet the mentioned criteria concerning clarity, precision, etc.[172] and the time limit for implementing the directive has expired.[173] However, unlike regulations, directives only have vertical, not horizontal, effect.[174]

EU law-making procedures are notoriously complex even if the Lisbon Treaty rationalized them, at least to some extent. There are now essentially three ways in which laws are made in the European Union of which the most important one is the 'ordinary legislative procedure', the normal method for making EU legislation.[175] The legislative process is initiated by a Commission proposal.[176] Under the 'ordinary legislative procedure', EU legislation, such as a regulation or a directive, is adopted jointly by the Parliament and the Council.[177]

Whether and to what degree the European Union can legislate on a given issue or process varies and depends on its competences in specific fields. The EU law-making competences in different areas are not necessarily exclusive but often shared with EU states (e.g. in the case of internal market, research,

---

[171]   Case 26/62 *Van Gend Loos* ECLI:EU:C:1963:1

[172]   Case 9/70 *Grad* ECLI:EU:C:1970:78; and Case 41/74 *Van Dyjn* v. *Home Office*, ECLI:EU: C:1974:133.

[173]   Case 148/78 *Public Ministero* v. *Ratti* ECLI:EU:C:1979:110. Also see Case C-144/04 *Mangold* ECLI:EU:C:2005:709, according to which the time limit is not applicable if unimportant.

[174]   Case 152/84 *Marshall* v. *Southhampton Area Health Authority* ECLI:EU:C:1986:84.

[175]   The 'ordinary legislative procedure' is detailed in art. 294 of the TFEU. The other two law-making procedures are the 'special legislative procedure' (TFEU, art. 289.2–4) and the 'consent procedure'. The consent procedure means that the European Parliament's consent is required by the Council concerning membership applications to the European Union, the Union's membership of international agreements and organizations, and association agreements with third countries (TEU, arts. 49 and 50; TFEU, arts: 218 and 217). For EU law-making procedures, see Foster, *Foster on EU Law* (n 170) 132–136.

[176]   But note TFEU, art. 289.4: 'In the specific cases provided for by the Treaties, legislative acts may be adopted on the initiative of a group of Member States or of the European Parliament, on a recommendation from the European Central Bank or at the request of the Court of Justice or the European Investment Bank.'

[177]   TFEU, art. 289.1: 'The ordinary legislative procedure shall consist in the joint adoption by the European Parliament and the Council of a regulation, directive or decision on a proposal from the Commission. This procedure is defined in Article 294.' As defined in Article 294 of the TFEU, the co-decision procedure is central to the Community's law-making procedures. It is based on the principle of parity and means that neither institution (European Parliament or Council) may adopt legislation without the other's assent.

and common safety concerns in public health matters),[178] or supplementary to state competences (e.g. in the case of public health).[179] This complex arrangement reflects the centrality of shared competence areas within the European Union and the question when it may act in these areas. The general thrust is that EU action should be taken as openly and as closely as possible to the citizens, and must respect the 'subsidiarity principle', which states that the Union acts 'only if and insofar as the objectives of the proposed action cannot be sufficiently achieved by the Member States, either at the central level or at regional or local level, but can rather, by reason of the scale or effects of the proposed action, be better achieved at the Union level'.[180] Also relevant is the proportionality principle, according to which the European Union shall act only when deemed necessary.[181] However, the practical significance of these principles remains unclear. This renders the areas falling under 'shared competences' especially complex.

Most EU legislation related to human germline genome modification is the result of EU exercise of 'shared competence' in the area of the 'internal market'.[182] Here the European Union is granted subsidiary law-making power to provide for measures to complete the internal market and for the harmonization of laws affecting the establishment and functioning of the internal market.[183] More specifically, the Union 'shall adopt measures with the aim of establishing or ensuring the functioning of the internal market ... [which] shall comprise an area without internal frontiers in which the free movement of good, persons, services and capital is ensured'.[184] To achieve this objective, '[t]he European Parliament and the Council shall, acting in accordance with the ordinary legislative procedure and after consulting the Economic and Social Committee, adopt the measures for the approximation of the provisions laid down by law, regulation or administrative action in Member States'.[185] The Commission, when making proposals on measures concerning 'health, safety, environmental protection and consumer protection, will take as a base a high level of protection, taking account in particular of any new development based on scientific facts. Within their respective powers, the European Parliament and the Council will also seek to achieve this objective.'[186]

[178] TFEU, art. 4.2.
[179] Ibid., art. 6.
[180] Ibid., art. 5.
[181] Ibid., art. 5.2.
[182] Ibid., art. 4.2.a.
[183] Ibid., arts. 114 and 115.
[184] Ibid., art. 26.
[185] Ibid., art. 114.1.
[186] Ibid., art. 114.3.

Other relevant EU legislation has been adopted on the basis of 'common safety concerns in public health matters'.[0] Article 168.4 of the TFEU gives the European Parliament and the Council a specific mandate to adopt: (a) 'measures setting high standards of quality and safety of organs and substances of human origin'; (b) 'measures in the veterinary and phyto-sanitary fields which have as their direct objective the protection of public health'; and (c) 'measures setting high standards of quality and safety for medicinal products and devices for medical use'.

The EU efforts to achieve the established objectives related to the internal market as well as health and safety concerns have generated several directives and regulations that are relevant for the topic in this book. The question of human germline genome modification arose for the first time in the context of the consideration of legislation to harmonize the protection of biotechnological inventions across the Union to facilitate investment in biotechnology. The action was justified by the Union's shared competence in the field of internal market, specifically to harmonize national laws in the field of intellectual property.[188] It resulted in the adoption in 1998 of Directive 98/44/EC on the Legal Protection of Biotechnological Inventions (Biotech Directive).[189] Although ethical and human rights considerations had not put the question on the EU agenda, they played a major role in the deliberations, delaying the adoption of the Directive for almost ten years. In the end, the Biotech Directive incorporated a well-established rule found in patent law of all developed countries and in international treaties, according to which inventions are not patentable if they are contrary to public order or morality. The Biotech Directive considers processes for cloning of human beings, processes for modifying the germline genetic identity of human beings and uses of human embryos for industrial or commercial purposes not patentable.[190]

---

[187]   TEU, art. 4.

[188]   Case 377/98 *Netherlands* v. *Parliament* and *Council* ECLI:EU:C:2001:523. This case was an unsuccessful application for annulment of the Biotech Directive on the ground that it had been incorrectly adopted on the basis of Article 100a of the EC Treaty (art. 95 following the entry into force of the Amsterdam Treaty and art. 114 of the TFEU).

[189]   Directive 98/44/EC of the European Parliament and of the Council of 6 July 1998 on the legal protection of biotechnological inventions, OJ L 213, 13–21. This directive was adopted on the basis of Article 100a of the EC Treaty (Article 95 following the entry into force of the Treaty of Amsterdam and Article 114 following the entry into force of the Lisbon Treaty).

[190]   See Article 6 of the Biotech Directive. The same exceptions are included in the Implementation Regulations of the European Patent Convention of 1977 as revised in 2000. According to Rule 28 concerning the implementation of article 53(a) of this Convention: 'European patents shall not be granted in respect of biotechnological inventions which, in particular, concern the following: (a) processes for cloning human beings; (b)

Following the adoption of the Biotech Directive, the European Union has adopted several other directives and regulations that are relevant for human germline modifications, including:

(i) Directive 2001/20/EC on the Approximation of the Laws, Regulations and Administrative Provisions of the Member States Relating to the Implementation of Good Clinical Practice in the Conduct of Clinical Trials on Medicinal Products for Human Use ('Clinical Trials Directive');[191]

(ii) Regulation 536/2014 of the European Parliament and of the Council of 16 April 2004 on Clinical Trials on Medicinal Products for Human Use, repealing Directive 2001/20/EC ('Clinical Trials Regulation');[192]

(iii) Directive 2004/23 of the European Parliament and of the Council of 31 March 2004 on Setting Standards of Quality and Safety for the Donation, Procurement, Testing, Processing, Preservation, Storage and Distribution of Human Tissues and Cells ('Human Tissues and Cells Directive');[193]

(iv) Regulation 1394/2007 of the European Parliament and the Council of 13 November 2007 on Advanced Therapy Medicinal Products, amending Directive 2001/83/EC and Regulation (EC) No 726/2004 ('Advanced Therapy Regulation');[194] and

processes for modifying the germ line genetic identity of human beings; and (c) uses of human embryos for industrial or commercial purposes.'

[191] Directive 2001/20/EC of the European Parliament and of the Council of 4 April 2001 on the approximation of the laws, regulations and administrative provisions of the member states relating to the implementation of good clinical practice in the conduct of clinical trials on medicinal products for human use, OJ L 121, 34–44.

[192] Regulation (EU) No 536/2014 of the European Parliament and of the Council of 16 April 2014 on clinical trials on medicinal products for human use, and repealing Directive 2001/20/EC Text with EEA relevance, OJ L 158, 27.5.2014, 1–76.

[193] Directive 2004/23/EC of the European Parliament and of the Council of 31 March 2004 on setting standards of quality and safety for the donation, procurement, testing, processing, preservation, storage and distribution of human tissues and cells, OJ L 102, 48–58. See also Commission Directive 2006/86/EC of 24 October 2006 implementing Directive 2004/23/EC of the European Parliament and of the Council as regards traceability requirements, notification of serious adverse reactions and events and certain technical requirements for the coding, processing, preservation, storage and distribution of human tissues and cells.

[194] Regulation (EC) No 1394/2007 of the European Parliament and of the Council of 13 November 2007 on advanced therapy medicinal products and amending Directive 2001/83/EC and Regulation (EC) No 726/2004 (Text with EEA relevance), OJ L 324, 121–137.

(v) Directive 2010/63/EU of the European Parliament and of the Council of 22 September 2010 on the Protection of Animals Used for Scientific Purposes ('Animal Protection Directive').[195]

Among them, the most important ones are the first two: the 2001 Clinical Trials Directive and the 2014 Clinical Trials Regulation, which will replace the former in 2019. The aim of the Clinical Trials Directive is to harmonize laws related to clinical trials of medicinal products across the European Union. Although it permits clinical trials involving medical products for 'gene therapy, somatic cell therapy, including xenogenic cell therapy and all medicinal products containing genetically modified organisms', it imposes a ban on gene therapy trials 'which can result in modifications to the subject's germ line genetic identity'.[196] The Clinical Trials Regulation was adopted to make it easier to conduct multicentre clinical trials involving several EU states and to improve transparency of information so as to avoid unnecessary duplication of trials.[197] It reaffirms the ban of gene therapy trials which can result in modifications to the subject's germline genetic identity.[198] Additionally, it requires EU states to specify penalties applicable to infringements of its rules and to 'take all measures necessary to ensure that they are implemented'.[199]

Finally, the European Union shares competences with member states in the field of research.[200] According to primary EU law, 'in the areas of research, technological development and space, the Union shall have competence to carry out activities, in particular to define and implement programs; however, the exercise of that competence shall not result in member states being prevented from exercising theirs'.[201] In this regard, the European Parliament and the Council, acting in conformity with the ordinary legislative procedure after consulting the Economic and Social Committee, shall adopt multiannual 'Framework Programmes' setting out the objectives to be achieved in the fields of research and technological development, including the amount of the EU budget allocated to Union financial participation in these

---

[195] Directive 2010/63/EU of the European Parliament and of the Council of 22 September 2010 on the protection of animals used for scientific purposes Text with EEA relevance, OJ L 276, 33–79.

[196] Clinical Trials Directive, art. 9.6.

[197] This will be achieved through the creation of a single EU clinical trial portal and database by the European Medicines Agency that will also manage it.

[198] Clinical Trials Regulation, art. 90.

[199] Ibid., art. 94.

[200] TEU, art. 4.

[201] TFEU, art. 4.3.

programmes.[202] These are the so-called Framework Programmes for Research and Technological Development created by the Union to support and foster research in what has been coined the 'European Research Area'.[203] The specific objectives and actions of these programmes vary between funding periods. So far, there have been eight 'Framework Programmes' (FP). The focus of FP8, also known as Horizon 2020, is innovation, delivering economic growth faster and delivering solutions to end users.[204] The budget of FP8 (2014–2020) is about 77 billion euros. The Commission has proposed to increase it to 97.6 billion euros for the follow-up programme (FP9 – Horizon Europe) that will run from 2021 to 2027.

Horizon 2020 was established by EU Regulation 1291/2013.[205] It is the result of the EU exercise of its competences in the field of research to achieve the objective of ensuring that the conditions for the competitiveness of the Union's industry exists through the adoption of specific measures, excluding the harmonization of national laws.[206] Article 19 of this regulation specifies the fields of research not eligible for funding. Among them there is 'research activity intended to modify the genetic heritage of human beings which could make such changes heritable', while exempting 'research relating to

---

[202] TFEU, art. 182.1.

[203] TFEU, art. 179.1. In line with this article: 'The Union shall have the objective of strengthening its scientific and technological bases by achieving a European research area in which researchers, scientific knowledge and technology circulate freely, and encouraging it to become more competitive, including in its industry, while promoting all the research activities deemed necessary by virtue of other Chapters of the Treaties.'

[204] In the first three years of Horizon 2020's implementation, the main beneficiaries of the programme have been higher education and research organizations, which together received 64.9% of the funding, the private sector receiving 27.7% and public authorities and other types of organizations 7.3%. European Commission, 'Key Findings from the Horizon 2020 Interim Evaluation' (European Commission, 2017) 4 https://ec.europa.eu/research/evaluations/pdf/brochure_interim_evaluation_horizon_2020_key_findings.pdf accessed 12 October 2018.

[205] Regulation (EU) No 1291/2013 of the European Parliament and of the Council of 11 December 2013 establishing Horizon 2020 – the Framework Programme for Research and Innovation (2014–2020) and repealing Decision No 1982/2006/EC Text with EEA relevance, OJ L 347 (Horizon 2020 Regulation) 104–173. This regulation was adopted on the basis of Articles 173.3 and 182.1 of the TFEU. The first article refers to policies and activities to achieve the Union's commitments. Its adoption was motivated by the Union's commitment to foster 'better exploitation of the industrial potential of policies of innovation, research and technological development'. The latter article states that the European Parliament and the Council, acting in accordance with the ordinary legislative procedure after consulting the Economic and Social Committee, shall adopt a 'multiannual framework programme, setting out all the activities of the Union'. The framework shall establish the scientific and technological objectives to be achieved in the field of research and technological development.

[206] TFEU, art. 173.3.

cancer treatment of the gonads', which can be financed. Also excluded from EU funding is 'research activity aiming at human cloning for reproductive purposes' and 'research activities intended to create human embryos solely for the purpose of research or for the purpose of stem cell procurement, including by means of somatic cell nuclear transfer'. Finally, while research on human stem cells may be financed, depending both on the contents of the scientific proposal and on the legal framework of the member states involved, 'no funding shall be granted for research activities that are prohibited in all the Member States'. Furthermore, 'no activity shall be funded in a member state where such activity is forbidden'.[207]

Overall, although the European Union is famous – or infamous, its detractors would say – for regulating issues down to the smallest detail, when it comes to biotechnology, its directives and regulations are surprisingly vague. For instance, the Human Tissues and Cells Directive does not define 'gametes', 'eggs', 'sperm', 'adult stem cells' and 'embryonic stem cells', although they are its main object. It does not univocally define an 'embryo' either.[208] Although 'human embryos' are mentioned several times in the Biotech Directive and the Horizon 2020 Regulation, neither defines them.[209] EU law does not specify what 'activities which can result in modifications of germ line genetic identity'[210] or what activities 'intended to modify the genetic heritage of human beings which could make such changes heritable'[211] are. It does not clarify when a genetic modification becomes inheritable. Also, albeit EU law bans general research involving human germline genome modification, the ban is not formulated identically in various pieces of legislation. For instance, whereas the Biotech Directive concerns

---

[207]   Horizon 2020 Regulation, art. 19.4.
[208]   In 2016, the EU Expert Group on the Development and Implications of Patent Law in the field of Biotechnology and Genetic Engineering discussed the meaning of human embryo. In its report, it posed the question whether a further clarification of human embryo is needed as new technologies become available, like the artificial creation of human germ cells that could lead to the artificial creation of human embryos as entities. It concluded that if these entities are inherently capable of developing into a human being, they, too, must be considered human embryos. Therefore, for the purpose of the Biotech Directive, an embryo produced by means of artificial germ cells should be treated in the same way as natural embryo produced by the fusion of an oocyte and a sperm cell and that no further clarification is needed. Final Report of the Expert Group on the Development and Implications of Patent Law in the Field of Biotechnology and Genetic Engineering (E02973), 17 May 2016, 145.
[209]   See Article 6.2.c of the Biotech Directive concerning the unpatentability of uses of human embryos for industrial or commercial purposes, and Article 19 of the Horizon 2020 Regulation concerning the non-eligibility for EU funding of research activity limited to create human embryos solely for the purpose of research or for the purpose of stem cell procurement.
[210]   Clinical Trials Regulation, art. 90.
[211]   Horizon 2020 Regulation, art. 19.

modifications of 'the genetic heritage of human beings',[212] the Clinical Trials Directive/Regulation refers to modifications to 'the subject's germ line identity'.[213] The language changes somewhat again in the Horizon 2020 Regulation, which talks about modifications of 'the genetic heritage of human beings which could make such changes heritable'.[214]

The vagueness of EU legislation in this area may be explained in the light of the subsidiarity principle that governs and limits EU efforts to harmonize national laws in this area. In areas where the EU has shared competence, intervention by the Union is permissible only when the objectives of an action cannot be sufficiently achieved by the member states, but can better be achieved at the Union level, 'by reason of the scale and effects of the proposed action'.[215] At the same time, as will be discussed in the next section, the case law of the CJEU seems to reveal the importance of providing an EU definition of 'embryos'.

All in all, the lack of clarity is perplexing if one considers that the European Union prides itself on its transparent legislative process, one where the public has ample opportunities to weigh in. If these vague provisions confuse experts, it is hard to see how the wider public could be able to debate their merits.[216] Also, the vagueness of EU legislation on the matter is even more striking if one compares it with the very detailed and sophisticated EU legal regime in place for genetically modified plants and animals (genetically modified organisms – GMO).[217]

d Decisions of the Court of Justice of the European Union

The Court of Justice of the European Union is the principal judicial organ of the European Union and its mission is to ensure that the law is observed in the interpretation and application of the treaties of the European Union and secondary legislation.[218] To this end, it reviews the legality of actions taken by

---

[212] Clinical Trials Directive, art. 9.6.

[213] Ibid., art. 90.

[214] Horizon 2020 Regulation, art. 19.

[215] TEU, art. 5.c.

[216] See, e.g., EU Expert Group on Ethics, Science and Technology, Statement on Gene Editing (2016).

[217] On the regulation of GMOs in the EU, see M Weimer, *Risk Regulation in the Internal Market: Lessons from Agricultural Biotechnology* (Oxford University Press 2019).

[218] On the CJEU, see, in general, T Horsley, *The Court of Justice of the European Union as an Institutional Actor: Judicial Lawmaking and its Limits* (Cambridge University Press 2018), and SK Schmidt, *The European Court of Justice and the Policy Process: The Shadow of Case Law* (Oxford University Press 2018).

the EU's institutions, enforces compliance by member states with their obligations, in co-operation with the national judiciary of the member states, and interprets EU law. The CJEU also resolves legal disputes between national governments and EU institutions, and may take action against EU institutions on behalf of individuals, companies or organizations whose rights have been infringed.

The CJEU consists of two main courts: the 'Court of Justice' and the 'General Court'. The Court of Justice, formerly known as the European Court of Justice (ECJ), is the supreme court of the European Union. It consists of one judge from each EU member country, as well as 11 Advocates General. Each Advocate General is a non-voting member of the Court who delivers an impartial opinion to the other judges on the legal issues raised in the case. It rules on applications from national courts for 'preliminary rulings', and certain actions for annulment and appeal. The General Court, composed of 47 judges (to be increased to 56 in 2019), hears applications for annulment from individuals, companies and, less commonly, national governments, focusing on competition law, state aid, trade, agriculture and trademarks.

Over the years, the CJEU has decided a few cases that touch upon issues relevant for a discussion of human genome modification. The most relevant ones are *Brüstle* v. *Greenpeace* and *International Stem Cell Corporation* v. *Comptroller General of Patents, Designs and Trade Marks*. In 2012, in *Brüstle* v. *Greenpeace*, the CJEU was called to define the term 'human embryo' for the purpose of the interpretation and application of the Biotech Directive.[219] The case stemmed from a dispute between Greenpeace and Dr Oliver Brüstle, a German scientist known for his research on stem cells. Greenpeace sought the annulment of a German patent for a biotechnological invention made by Dr Brüstle concerning neural precursor cells (i.e. immature body cells capable of forming mature cells in the nervous system, such as neurons) and, specifically, the processes for their production from embryonic stem cells and their use for therapeutic purposes. According to Greenpeace, the patent violated the ban on using human embryos for industrial or commercial purposes contained in the Biotech Directive. The German Federal Court of Justice (*Bundesgerichtshof*), before which the case was pending, raised with the CJEU the issue of what the Biotech Directive means by 'human embryo' and by 'use for industrial or commercial purposes', especially where the embryo is used for the purposes of scientific research.

---

[219]   Case C-34/10. *Oliver Brüstle* v. *Greenpeace eV*. (Reference for a preliminary ruling from the Bundesgerichtshof), Judgment of the Court of Justice (Grand Chamber) of 18 October 2011. E.C.R. 2011 I-09821.

The CJEU started by emphasizing that the letter of Article 6.2 of the Biotech Directive and its object and aim lead to the observation that the concept of 'human embryo' 'constitutes an autonomous concept of Union law which must be interpreted in a uniform manner throughout its territory'.[220] In fact, the lack of a uniform definition would induce the authors of certain biotechnological inventions to seek registration in the jurisdiction with the least restrictive definition of an embryo, which would adversely affect the smooth functioning of the internal market. Then it pointed out that the Biotech Directive seeks to promote investment in the field of biotechnology, while specifying that the use of biological material originating from humans must be consistent with regard for fundamental rights and, in particular, the dignity of the person.[221] In this light, it proceeded to define the 'human embryo' as 'any human ovum ..., as soon as fertilised, ... since ... fertilisation is such as to commence the process of development of a human being'.[222] Moreover, the same applies to a non-fertilized human ovum into which the cell nucleus from a mature human cell has been transplanted and a non-fertilized human ovum whose division and further development have been stimulated by parthenogenesis, because, even if those organisms have not, strictly speaking, been the object of fertilization, due to the effect of the technique used to obtain them, they are 'capable of commencing the process of development into a human being just as an embryo created by fertilization can do so'.[223]

Ultimately, the CJEU left to national courts to determine whether the specific patent application that gave rise to the case could be granted in light of the principles laid out by the Court.[224] In addition, the Court ruled that the exclusion of patentability 'concerning the use of human embryo for industrial or commercial purposes [in the Biotech] Directive also covers use for purposes of scientific research'.[225] However, use of human embryos that is both for therapeutic or diagnostic purposes and useful is patentable.[226] Lastly, the Court specified that the Biotech Directive excludes the patentability of inventions whose production necessitates the prior destruction of human embryos or their use as a base material.[227]

---

[220]   Ibid., para. 26.
[221]   Ibid., para. 32.
[222]   Ibid., para. 35.
[223]   Ibid., para. 37.
[224]   Ibid., para. 38.
[225]   Ibid., para. 46.
[226]   Ibid.
[227]   Ibid., para. 52.

International Stem Cell Corporation v. *Comptroller General of Patents, Designs and Trade Marks* offered the Court a chance to partially backtrack the *Brüstle* decision, which many saw as a threat to stem cell research in Europe.[228] This case was about a lawsuit brought against the UK Patent Office by International Stem Cell Corporation, a California-based 'publicly traded clinical stage biotechnology company',[229] for refusing to register two national patent applications related to the use of chemical or electrical techniques to activate unfertilized human eggs. The legal issue stemmed from the fact that this process, called 'parthenogenetic activation', stimulates ova in a way which is similar, at least initially, to the process by which an embryo forms from a fertilized egg.

Implementing the Biotech Directive, and citing the CJEU decision in *Brüstle*, the UK Patent Office declined registration of the patents on the ground that they violated UK law, and specifically the rule on non-patentability of the commercial and industrial exploitation of human embryos. The Chancery Division (Patents Court) of the High Court of Justice England and Wales referred the matter to the CJEU, requesting it to clarify whether 'unfertilised human ova whose division and further development have been stimulated by parthenogenesis, and which, in contrast to fertilised ova, contain only pluripotent cells and are incapable of developing into human beings' are covered by the Biotech Directive's ban of patentability.[230]

The Court noted that the written observations filed before it in *Brüstle* indicated that parthenotes (activated ova) did have the capacity to develop into a human being.[231] However, in the present case, the parties agreed that, according to current scientific knowledge, parthenotes are not capable of commencing the process of development that leads to a human being.[232] Because of these considerations, the CJEU concluded that parthenotes would not, in and of themselves, constitute human embryos, provided that they are not inherently capable of developing into human beings. It also introduced a caveat when it held that the ruling did not concern parthenotes

---

[228] Case-364/13, *International Stem Cell Corporation* v. *Comptroller General of Patents, Designs and Trade Marks*. Request for a preliminary ruling under Article 267 TFEU from the High Court of Justice (England & Wales), Chancery Division (Patents Court) (United Kingdom), made by decision of 17 April 2013, received at the Court on 28 June 2013. Judgment of the Court of Justice on 18 December 2014, para. 38.

[229] See International Stemcell Corporation, 'Cells for Research and Therapy' http://internatio nalstemcell.com/ accessed 12 October 2018.

[230] Case-364/13, para. 20.

[231] Ibid., para. 32.

[232] Ibid., para. 33.

subjected to additional genetic manipulation.[233] The CJEU said that the question of whether a parthenote is inherently capable of developing into a human being was one which the referring court should determine 'in light of current scientific knowledge'.[234]

This decision altered once more the European patenting regime for human embryonic stem cell (hESC) applications, by stating that moral restrictions against hESC patents are only applicable to such cells derived from embryos that had the potential to develop into a human being. Consequently, human parthenogenic stem cells (hpSCs)-based inventions became patentable in Europe.

Overall, the *International Stem Cell Corporation* revealed a Court that struggles with basic science and that is ready to backtrack previous decisions if it is presented new, undisputed, scientific evidence. The decision was welcomed as a step forward towards striking a balance between protecting human dignity and integrity while granting patent incentives for biomedical research, although some took exception with it because the ruling leaves considerable discretion to national courts. Furthermore, the ruling is limited to very specific human embryonic stem cells and does not clarify if it extends to other non-totipotent human embryonic stem cells, such as stem cells created through somatic cell nuclear transfer.[235]

d Oversight and Supervisory Bodies

Besides the CJEU, several EU bodies, committees and agencies monitor and supervise the Union's legal regime regulating research on human embryos. Even if the main responsibility for enforcement rests with the individual member states, the European Commission, including its Expert Group on Ethics in Science and Technology, has important functions with respect to monitoring, supervision and advice, as does the European Research Council, which is responsible for EU funding, and the Court of Justice of the European Union, as we just saw.

I THE EUROPEAN COMMISSION
EU states are responsible for giving effect to the laws, regulations and administrative provisions necessary to comply with the Biotech Directive, and for

---

[233] Ibid., para. 35.
[234] Ibid., para. 36.
[235] A Nordberg and T Minssen, 'A "Ray of Hope" for European Stem Cell Patents or "Out of the Smog into the Fog"? An Analysis of Recent European Case Law and How it Compares to the US' (2016) 47:2 *International Review of Intellectual Property and Competition Law* 138, 138–177.

informing the Commission about the actions taken.[236] The Commission, for its part, has several reporting obligations to the European Parliament and the Council related to implementation. For a start, every five years, it must report any problems encountered in the relationship between the Biotech Directive and international human rights treaties ratified by the EU states.[237] Moreover, within two years of entry into force of this directive, it must issue a report assessing the implications for basic genetic research of failure to publish, or late publication of, papers on subjects which could be patentable.[238] Also, it must report annually to the same institutions on the development and implications of patent law in the field of biotechnology and genetic engineering.[239] In 2012, the Commission set up an Expert Group to prepare the annual reports.[240] However, this group does not address any ethical issues. These are the remit of the European Group on Ethics in Science and New Technologies, which we will discuss later.

EU states are also responsible for the enforcement of the Clinical Trials Regulation once it enters into force in 2019 by adopting the administrative provisions and penalties necessary to comply with it. Moreover, each EU state must set up an Ethics Committee consisting of healthcare professionals and non-medical members responsible for protecting the rights, safety and well-being of human subjects involved in trials.[241] The Ethics Committees issue opinions on the ethical aspects of clinical trials before they begin, at the request of the competent authority of the member state. A clinical trial cannot go ahead unless the national Ethics Committee has approved it. Written (as opposed to tacit) authorization is a must for clinical trials of medicinal products for gene therapy, somatic cell therapy, including xenogenic cell therapy, and the like.[242]

The Clinical Trials Regulation seeks to improve existing administrative procedures by establishing a streamlined application procedure for patent authorization via the creation of a single entry point – an EU portal and database – for all clinical trials conducted in the European Union. When the Regulation enters into force in 2019, registration via the portal will be a requisite for the assessment of any application, although, again, the final

---

[236] Biotech Directive, art. 15.
[237] Ibid., art. 16.a.
[238] Ibid., art. 16.b.
[239] Ibid., art. 16.c.
[240] Commission Decision of 7.11.2012 on setting up a Commission Expert Group on Development and Implications of Patent Law in the Field of Biotechnology and Genetic Engineering (C(2012) 7686 final).
[241] Clinical Trials Directive, art. 6.
[242] Ibid., art. 9.2.

authorization and oversight of clinical trials will remain the responsibility of member states. The European Commission will supervise this process through the collection and distribution of national reports on trial results. It will also grant market authorizations of new medicinal products following a scientific evaluation made by the European Medicines Agency (EMA), which will manage the EU single portal and database. It should also be mentioned that the EMA and the Commission produce guidelines to harmonize practice. For instance, in 2006, EMA issued a guideline on inadvertent germline transmissions, recommending the conduct of non-clinical studies to prevent involuntary transgressions of the ban on human germline modifications.[243]

II THE EUROPEAN GROUP ON ETHICS IN SCIENCE AND NEW TECHNOLOGIES

In 1991, the Commission set up an expert group called the European Group on Ethics in Science and New Technologies (EGE). The EGE is an independent body of the President of the Commission that gives advice on all aspects of Commission legislation and policies where ethical, societal and fundamental rights dimensions intersect with the development of science and new technologies.[244] It is composed of 15 members – independent experts from diverse academic fields – who are appointed by and report to the Commission President, although it provides advice to the Commission College as a whole.

The EGE has a Secretariat, supported by the Commission, whose mandate is to integrate ethics at the international and inter-institutional levels, including within the Commission itself. The Secretariat also constitutes a platform for the Commission's International Dialogue on Bioethics, a forum that brings together the national ethics councils of 97 states (the 28 EU states, plus the states of the G20 forum and others). Additionally, the Secretariat represents the European Union in liaising with international organizations relevant for the ethical implications of science and new technologies (e.g. the UN and its agencies, the OECD and the Council of Europe). Moreover, it chairs and convenes the inter-service group on Ethics and EU policies, which

---

[243] European Medicines Agency, Guideline on Non-Clinical Testing for Inadvertent Germline Transmission of Gene Transfer Vectors, 16 November 2006, Doc. Ref. EMEA/273974/2005.

[244] See European Commission, 'The European Group on Ethics in Science and New Technologies (EGE) http://ec.europa.eu/research/ege/index.cfm accessed 12 October 2018. For the EGE mandate, see Commission Decision SEC (97) 2404 of 16 December 1997; Amendment to the Remit: Commission Decision C(2001) 691 of 26 March 2001; Renewal of the Remit: Commission Decision 2005/383/EC of 11 May 2005; Renewal of Mandate: Commission Decision 2010/1/EU of 23 December 2009; Renewal of the Mandate: Commission Decision (EU) 2016/835 of 25 May 2016.

coordinates Commission activities in the field of bioethics and ethics of science and new technologies of growing importance for the European Union.

So far, the EGE has published 29 opinions and 3 statements on a range of critical issues, including on animal welfare, genetically modified organisms, biodiversity, nanotechnology, and stem cell research.[245] Of these, some have particular relevance for the purpose of this book, such as the two opinions produced during the process of drafting the Biotech Directive,[246] and its comment on the use of human embryos in EU-funded research.[247] More recently, it issued two statements, one on research integrity and another one on gene editing.[248] Some of these opinions/statements address the ethical defensibility of human germline engineering.

The Commission's drafting of what came to be the Biotech Directive prompted the EGE to issue two opinions. In the first, which it drafted on its own initiative, it promoted an open-minded view on the idea of patenting inventions related to gene editing but also held that certain types of genetic manipulations should be prohibited and that this should be 'mainly a matter to be dealt with under the competent branches of public law dealing with the use and commercialization of research results in respect to public safety, health, environment and animal welfare'.[249] While acknowledging 'the need to reaffirm the ban on genetic engineering for non-therapeutic purposes, contrary to the dignity of man', the EGE also felt that the Biotech Directive was 'not the right place to deal with the very complex issue of the legitimacy of germinal therapy'.[250] Instead, these concerns should be considered mainly in its recitals.

The following year, the EGE issued a second opinion on a new draft of the Biotech Directive. This opinion endorses the patentability of somatic gene therapy for its potential to cure serious diseases and because the use of new therapeutic products in this field could be of great interest for the development of EU biotechnological industry.[251] It then noted that 'because of the

[245] European Commission, 'EGE Reports, Opinions, and Statements' https://ec.europa.eu/res earch/ege/index.cfm?pg=reports accessed 11 October 2018.
[246] EGE Opinion No 3 on Ethical Questions Arising From the Commission Proposal for a Council Directive for Legal Protection of Biotechnological Inventions (1993); and EGE Opinion No 4 on Ethical Implications of Gene Therapy (1994).
[247] EGE Opinion No 12 on Ethical Aspects of Research Involving the Use of Human Embryo in the Context of the 5th Framework Program (1998).
[248] Statement on the Formulation of a Code of Conduct for Research Integrity for Projects Funded by the European Commission (2015); and Statement on Gene Editing (2016).
[249] EGE Opinion No 3, para. 2.2.3.
[250] Ibid., para. 2.2.3.
[251] EGE Opinion No 4, para. 1.6.

important controversial and unprecedented questions raised by germ line therapy, and considering the actual state of the art, germ line gene therapy on humans is not at the present time ethically acceptable'.[252]

Thereafter, it took more than 20 years for EGE to return to this question, and when it did it was because of the rapid development in gene technologies, such as CRISPR-Cas9, which had moved human germline genome modification 'out of the realm of the theoretical', meaning that clinical applications are becoming feasible.[253] This development led the group to issue a statement on gene editing to promote a new debate on how to respond to the challenges posed to the international regulatory environment. The statement stresses that a pressing question is whether germline editing technology research should be suspended, which requires careful consideration given the profound potential consequences of this research for humanity.[254] It also points to the challenge posed by human germline engineering when it blurs the lines between basic and translational research, on the one hand, as well as between therapeutic and enhancement goals in clinical applications, on the other.[255] In response to disagreement among the members of the group whether continued research should be allowed, the EGE called for a debate on the acceptability and desirability of gene editing. This debate should go beyond expert committees and engage civil society and touch upon safety issues and potential health risks or benefits of gene editing technologies as well as human dignity, justice, equity, proportionality and autonomy.[256] So far, this proposal has not generated any EU-sponsored initiatives to promote such a debate.

III THE EUROPEAN RESEARCH COUNCIL

The European Research Council (ERC), which was set up by the European Commission in 2007, is responsible for managing the Horizon 2020 programme. To ensure respect for EU-imposed ethical rules and restrictions, including those under international law,[257] the ERC has set up an Ethical Appraisal Procedure to

---

[252]   Ibid., para. 2.7.
[253]   EGE Statement on Gene Editing (2016) 1.
[254]   Ibid., p. 2.
[255]   Ibid.
[256]   Ibid.
[257]   Horizon 2020 Regulation, art. 19: 'All the research and innovation activities carried out under Horizon 2020 must respect ethical principles and relevant national, EU and international legislation, including the EU Charter and the ECHR and its Additional Protocols'. Furthermore, 'particular attention shall be paid to the principle of proportionality, the right to privacy, the right to the protection of personal data, the right to the physical and mental integrity of a person, the right to non-discrimination and the need to ensure high levels of human health protection.'

review applications for ERC grants.[258] Under this procedure independent experts and/or qualified staff examine the ethical aspects of all proposals considered for funding through the Horizon 2020 programme.[259] It starts with an 'ethics screening' followed by an 'ethics assessment', if appropriate.[260] The review can lead to the inclusion of ethics requirements in the grant agreement. The ERC policy is that Horizon grant applicants and holders have the primary responsibility for the detection of scientific misconduct and for the investigation and adjudication of any breaches of research integrity. However, the ERC attends to all concerns about potential scientific misconduct or suspected breaches of research integrity that may arise during the execution of an ERC project. In case of substantial breach of ethical principles, research integrity or relevant legislation, an 'ethics audit' can be carried out, leading to an amendment of the agreement and, in severe cases, to a grant reduction or to its termination in line with the agreement rules.[261]

3 The Regulation from Bench to Bedside of Human Germline Genome Modification in the European Union

Let us now summarize how EU legislation regulates the so-called translational pipeline, the process of the creation of new medicines, from the bench to bedside, in the case of human germline genome modification.

It must be kept in mind that, since the EU competences over scientific research and its applications are still limited, the EU legal framework is necessarily incomplete. A full account of the substantive provisions that must be respected in any given EU state must take into consideration the national laws of that state. The following analysis centres on EU legislation alone. The national legal frameworks of some selected EU members will follow in the subsequent chapters.

a Basic Research

Although the European Union does not fund research involving the use of techniques that can lead to alterations of the germline identity of human subjects, it does not seem to rule out funding research projects that involve

---

[258] The legal basis for the ethics review is Horizon 2020 Regulation, art. 14.
[259] ERC Rules for Submission and Evaluation. Commission Decision C(2017)4750, Version 3.0, 14 July 2017.
[260] Ibid., ERC Rules for Submission and Evaluation, Annex A.
[261] Multi-Beneficiary General Model Grant Agreement (H2020 General MGA – Multi) Version 5.0 18 October 2017, art. 43 (reduction) and art. 48.1 (suspension).

using genome-modifying technologies on gametes, embryos or embryonic stem cells, as long as this use is consistent with the law of the member state in which the research is carried out.[262]

The European Union has taken a permissive approach to research on adult and embryonic human stem cells. Since 2002, it has funded research on human embryonic stem cells provided this research is carried out in compliance with the European ethical and legal framework for research on stem cells and the laws of the state in which they are doing their research. However, EU funds may not be used for derivation of new stem cell lines, or for research that destroys the embryos (blastocysts), including for the procurement of stem cells.

The procurement of gametes or embryos for research purposes is regulated at the national level. The Human Tissues and Cells Directive, which sets standards of quality and safety for the donation of human tissues and cells, does not extend to in vitro research.[263] Be that as it may, it is understood that Article 3 of the EU Charter, on the right to the integrity of the person, requires the 'free and informed consent of the person concerned' in the fields of biology and medicine. According to the CJEU case law, this right encompasses respect for the rights of donors.[264]

Finally, even where national laws permit creating human embryos only for research purposes, the European Union does not finance this research. As was said, according to Article 19.3(c) of the Horizon 2020 Regulation, the European Union does not fund 'research activities intended to create human embryos solely for the purpose of research or for the purpose of stem cell procurement, including by means of somatic cell nuclear transfer'.[265] Also, the use of EU funds for the derivation of new stem cells, or for research that destroys embryos, including for the procurement of stem cells. Even so, the EU approach was challenged by a pro-life group, One of Us (*Uno di Noi*). In 2012, the group had

---

[262]   Horizon 2020 Regulation, art. 19.4 states: 'Research on human stem cells, both adult and embryonic, may be financed, depending both on the contents of the scientific proposal and the legal framework of the Member States involved. No funding shall be granted for research activities that are prohibited in all the Member States. No activity shall be funded in a Member State where such activity is forbidden.'

[263]   Directive 2004/23/EC of the European Parliament and of the Council of 31 March 2004 on Settings Standards of Quality and Safety for the Donation, Procurement, Testing, Processing, Preservation, Storage and Distribution of Human Tissues and Cells, Recital 11.

[264]   This understanding has been confirmed in a Court of Justice judgment of 9 October 2001, in Case C-377/98 *Netherlands* v. *European Parliament and Council* [2001] ECR-I 7079, at grounds 70, 78 to 80.

[265]   Commission Communication COM (2014) 355 final, of 28 May 2014, on the European Citizens' Initiative 'Uno di Noi', 7. Available at: https://ec.europa.eu/transparency/regdoc/r ep/1/2014/EN/1-2014-355-EN-F1-1.Pdf accessed on 21 December 2018.

brought a petition to the Commission demanding the ban of stem cell research because it causes the destruction of human embryos.[266] In 2014, the Commission replied that it would not take any action in response to the petition, prompting the group to take legal action for annulment of the communication before the General Court.[267] The General Court dismissed the action. Specifically, it held that the claimants could not rely on the *Brüstle* judgment since it was limited to the question of whether a biotechnological invention involving the use of embryos is patentable and does not extend to the question of whether scientific research involving the use (and destruction) of human embryos may be financed by EU funds.[268] Moreover, according to the Court, the Commission's ethical approach to stem cell research cannot be said to involve a manifest error of assessment as required for actions of annulment. In this regard, it highlighted that the Commission had taken 'into account the right to life and human dignity of human embryos, but, at the same time, [considered] the needs of [human embryonic stem cell] research, which may result in treatments for currently-incurable or life-threatening diseases, such as Parkinson's, diabetes, stroke, heart disease and blindness'.[269]

b Preclinical research

Animal welfare is an important value of the Union. This is manifested by the fact that it has had legislation to protect animals used for scientific purposes in place since 1986. The European Union is also a party to the Council of Europe's Convention for the Protection of Vertebrate Animals Used for Experimental and Other Scientific Purposes, which it ratified in 1998.[270]

The most relevant piece of EU legislation in this field is Directive 2010/63/EU of the European Parliament and of the Council of 22 September 2010 on the Protection of Animals Used for Scientific Purposes (Animal Protection Directive), in force since 1 January 2013.[271] This directive applies to any 'use, invasive or non-invasive, of an animal for experimental or other scientific purposes, with known or unknown outcome, or educational purposes, which

---

[266] Ibid.
[267] Case T-561-144 *One of Us and Others* v. *Commission* ECLI:EU:T:2018:210 (Judgment of the General Court of the European Union on April 23, 2018).
[268] Ibid., paras. 174–175.
[269] Ibid., para. 176.
[270] European Convention for the Protection of Vertebrate Animals used for Experimental and Other Scientific Purposes, 18 March 1986, ETS No 123.
[271] Directive 2010/63/EU of the European Parliament and of the Council of 22 September 2010 on the Protection of Animals Used for Scientific Purposes, OJ L 276, 20.10.2010, 33–79.

may cause the animal a level of pain, suffering, distress or lasting harm equivalent to, or higher than, that caused by the introduction of a needle in accordance with good veterinary practice'.[272] The protection revolves around the principle of reduction, replacement and refinement (referred to as the 'Three Rs principle'). In short, it requires that EU states ensure that (1) wherever possible, a scientific method or testing strategy that does not involve the use of live animals is used; (2) the number of animals used in projects is reduced to a minimum without compromising the project's objective; and (3) the refinement of breeding, accommodation and care, and of methods used in procedures, eliminates or reduces to the minimum any possible pain, suffering, distress or lasting harm to the animals.[273]

In line with the directive, procedures on animals may only be carried out for the purposes of basic, translational or applied research if it aims to avoid, prevent, diagnose or treat disease, ill-health or other abnormality or their effects in human beings, animals or plants.[274] As a general rule, endangered species and non-human primates must not be used, unless the purpose of the procedure meets this stated aim, and if there is a scientific justification to the effect that the purpose of the procedure cannot be met by the use of other species or animals.[275] EU member states must set up a national committee for the protection of animals used for scientific purposes, which will advise the competent authorities and animal-welfare bodies on matters dealing with the acquisition, breeding, accommodation, care and use of animals in procedures and ensure sharing of best practices.[276]

The Animal Protection Directive covers any course of action intended or likely to result in the creation and maintenance of genetically modified animals.[277] Besides concerns about the health and welfare of these genetically modified animals, the European Union is concerned about possible adverse effects of introducing such animals into the food chain and the human diet. A basic question at the moment is whether food derived from these animals may be placed on the EU market, which is currently not the case. To anticipate these developments, the EU Food Safety Authority has issued guidelines on how to assess the risks should applications for food and feed

---

[272] Ibid., art. 3.1.
[273] Ibid., art. 4.
[274] Ibid., arts. 5.a and b.i.
[275] Ibid., art. 7 (endangered species) and art. 8 (non-human primates).
[276] Ibid., art. 49.
[277] Ibid., art. 3.1.

derived from these animals be submitted for market authorization in the European Union.[278]

Since the directive only contains minimum standards, EU states may opt for a more extensive protection of animals as long as they are compatible with the TFEU.[279] Also, because the Animal Protection Directive is just a directive, it is not self-executing but must be transposed into national law. In practice, this means that EU researchers must consult relevant national law of the member state in which they are located to find out what the exact rules are.

c Clinical Research/Applications

Clinical research involves a living person whose germline tissue is modified in vivo or who receives germline tissue that was modified ex vivo (by transferring a modified embryo in the uterus of a research participant) to test the safety and efficacy of germline engineering. Should this procedure be permitted at some point in the future, EU legislation concerning safety standards for the process of implanting tissues or cells into a human body will become applicable.[280] The Human Tissues and Cells Directive covers the entire chain of activities: from donation to procurement, testing, processing, preservation, storage and distribution to the site of medical use or to the site where manufactured products are made from human tissues and cells.[281] The concept of 'cells' is defined as 'individual human cells or a collection of human cells when not bound by any form of connective tissue';[282] and 'tissue' means 'all constituent parts of the human body formed by cells', which include fetal tissue, reproductive cells (semen, sperm, and ova) as well as stem cells.[283] The term 'donor' includes 'every human source, whether living or deceased, of human cells or tissues' and 'donation' means 'donating human tissues or cells intended for human applications'.[284] Finally, 'human application'

---

[278] See Guidance on the Risk Assessment of Food and Feed from Genetically Modified Animals and on Animal Health and Welfare Aspects (2012) 10:1 EFSA Journal 2501; and Guidance on the Environmental Risk Assessment of Genetically Modified Animals (2013) 11:5 EFSA Journal 3200.

[279] Animal Protection Directive, art. 2.

[280] Human Tissues and Cells Directive, recital 2.

[281] Directive 2004/23/EC of the European Parliament and of the Council of 31 March 2004 on Settings Standards of Quality and Safety for the Donation, Procurement, Testing, Processing, Preservation, Storage and Distribution of Human Tissues and Cells.

[282] Ibid., art. 3.a.

[283] Ibid., art. 3.b.

[284] Ibid., art. 3.c and d.

means 'the use of tissues or cells on or in a human recipient and extracorporal applications'.[285]

The Human Tissues and Cells Directive contains minimum standards, in the sense that it does not prevent a member state from imposing more stringent measures. For example, member states may well prohibit donations of human tissues and cells to be implanted in a human body.[286] For those states that do not prohibit it, they must designate the competent authority or authorities responsible for implementing the requirements included in the directive, which will supervise compliance, including through inspections and control measures.[287] The directive also imposes obligations on member states regarding the import/export of human tissues and cells, and to set up a system of notification and investigation of adverse reactions.

We have already mentioned EU legislation relating to clinical trials for medicinal products. The Clinical Trials Regulation (replacing the Clinical Trials Directive) ensures that the rights, safety and well-being of human subjects are protected. The current ban on gene therapy trials that can result in modifications to the subject's germline rules out the possibility of testing genetic modification technologies, such as CRISPR-Cas9, on embryos and gametes that are then implanted in a woman to initiate a pregnancy.[288] If this ban is lifted in the future, human germline engineering would likely be regarded as an 'advanced therapy medicinal product' in line with the Advanced Therapy Medicinal Product Regulation.[289] According to this regulation, gene therapy, somatic cell therapy and tissue engineering products qualify as medicinal products whose safety and quality must be tested in clinical trials. In this context, it sets out a definition of 'engineered' cell or tissue. For this, at least one of the following conditions must be fulfilled: (a) the cell or tissue must have been subject to substantial manipulation, so that biological characteristics, physiological functions or structural properties relevant for the intended regeneration, repair or replacement are achieved; or (b) the cell or tissue is not intended to be used for the same essential function or functions in the recipient as in the donor.[290] The Regulation also provides

---

[285]   Ibid., art. 3.1.

[286]   Ibid., art. 4.2.

[287]   Ibid., arts. 6 and 7.

[288]   Clinical Trials Regulation, art. 90.

[289]   Regulation 1394/2007 of the European Parliament and of the Council of 13 November 2007 on Advanced Therapy Medicinal Products, amending Directive 2001/83/EC and Regulation 726/2004, OJ L 324, 10.12.2007, pp. 121–137 ('Advanced Therapy Medicinal Products Regulation'), arts. 1.a and 4.

[290]   Ibid., art. 2.c.

212 Jessica Almqvist and Cesare P.R. Romano

a clarification of the term 'manipulation', which means, among other things, cutting, grinding, shaping, cryopreservation and vitrification.[291]

The general principle is that a clinical trial can be conducted only if the rights, safety, dignity and well-being of subjects are protected and prevail over all other interests, and if it is designed to generate reliable and robust data.[292] Therefore, a clinical trial must be subjected to scientific and ethical review for authorization.[293] It can only be done if 'the anticipated benefits to the subjects or to public health justify the foreseeable risks and inconveniences and compliance with this condition is constantly monitored'. Also, the subjects must give their informed consent in writing, and if not able to give informed consent, then their legal representatives.[294] Moreover, a clinical trial must have been designed to involve as little pain, discomfort, fear and any other foreseeable risk as possible for the subjects, a condition that must be constantly monitored. Specific attention is paid to vulnerable populations, including minors, incapacitated, and breastfeeding or pregnant women.[295] Finally, the sponsors of a clinical trial and the investigator must ensure respect for the protocol and the principles of good clinical practice.[296]

IV CONCLUSIONS

All in all, the current EU/CoE regulatory framework for genome editing of human germline cells is still rather patchy and lacks coherence, hardly facilitating the work of scientists who seek to develop inroads in cures for inheritable genetic diseases. The EU regulatory environment is quite complex, but also limited in scope and probably inadequate to advance scientific progress on germline genome editing. The Biotech Directive does not provide a comprehensive framework regulating biotechnological research. It addresses only the question of the patentability of biotechnological inventions. The Clinical Trials Directive, to be replaced in 2019 by the Clinical Trials Regulation, does not unambiguously rule out research involving genetic modification of human germline cells. It is even unclear whether clinical research on somatic cell therapy can be carried out as long as there is a chance it could cause unintentional modification of the germline genome.

---

[291] Ibid., Annex I.
[292] Ibid., art. 3.
[293] Ibid., art. 6.
[294] Ibid., art. 28.
[295] Ibid., arts. 31, 32 and 33.
[296] Ibid., art. 47.

Some could argue that EU primary law, and specifically the Charter of Fundamental Rights of the European Union, does provide a framework for informing a discussion on European governance of human germline genome modification. After all, it proclaims that human dignity is inviolable, and that it must be respected and protected.[297] It also states that 'everyone has the right to respect for his or her physical and mental integrity'.[298] However, these very generic provisions create more questions than answers. For instance, the Charter does not specify who 'everyone', the beneficiary of the rights, is: does it include human life before birth? Does it include future generations? Also, it is far from evident that germline therapy inherently violates human dignity. Indeed, if it does provide a cure for genetic diseases that condemn scores to a short and painful life, it could be argued that not only does it ensure respect for this right, but also helps protecting it. Indeed, let us not forget that in the European Union 'everyone has the right of access to preventive health care and the right to benefit from medical treatment and that a high level of human health protection shall be ensured in the definition and implementation of all Union policies and activities'.[299]

Of the 28 members of the European Union, 11 have not ratified the Oviedo Convention and are unlikely to do so as long as it remains worded as it is now. This group comprises states like Germany, Austria, Italy, Ireland and Poland who think the Convention does not include sufficient guarantees for the human embryo, and states like the United Kingdom, Belgium, the Netherlands and Sweden, and, outside the European Union, Russia, who believe the Convention excessively restricts research. Seemingly, what prevents the adoption of a comprehensive, clear and consistent 'European' regulatory framework is a fundamental difference of views within the European Union and the CoE on some fundamental issues. As it stands, the 'European' regulatory framework neither blocks research nor facilitates it.

For those states that would like to prevent research on human germline genome modification, the Oviedo Convention is not enough. First, the Convention's prohibitory provisions are vaguely worded and open to interpretation. Moreover, states can attach reservations to their instrument of ratification of the Convention.[300] Second, the Convention leaves it to each

---

[297]   Charter of Fundamental Rights, art. 1.
[298]   Ibid., art. 3.1.
[299]   Ibid., art. 35.
[300]   'Exceptionally and under the protective conditions prescribed by law, the removal of regenerative tissue from a person who does not have the capacity to consent may be authorized provided the following conditions are met ... .ii the recipient is a brother or sister of the donor.' Oviedo Convention, Article 20.2.ii. Croatia, for instance, ratified with

state to adopt the necessary domestic legal instruments to give it effect and establish sanctions for its violations.[301] Third, it is up to national courts to enforce it domestically.[302] Finally, the European Court of Human Rights does not have jurisdiction over the Oviedo Convention. Its jurisdiction is limited to the European Convention on Human Rights only. All it can do is issue advisory opinions at the request of one the parties or the Committee on Bioethics. So far, it has received none.[303] Violations of the Oviedo Convention are instead addressed politically within the Council of Ministers.

Although the Convention allows states to adopt more stringent standards of protection if they wish to do so, it is not prohibitory enough to earn the ratification of those who elevate the human embryo to the status of human being. However, it is too restrictive, especially in Articles 13 and 18, to keep out those states who believe research should not be unduly hampered.

As to the European Union, it does not prevent research on human germline genome modification per se, but it makes it difficult to fund it. The funds in the current Framework Programme (FP8, also known as Horizon 2020) cannot be tapped for research intending 'to modify the genetic heritage of human beings which could make such changes heritable',[304] and 'research activities intended to create human embryos solely for the purpose of research or for the purpose of stem cell procurement, including by means of somatic cell nuclear transfer'.[305] However, although researchers cannot tap EU Framework Programme Funds, they can still be funded nationally or privately.

The Biotech Directive discourages private investment in research because it excludes the patentability of processes that modify the germline genetic identity of human beings and the use of human embryos for industrial or commercial purposes. It remains to be seen what will happen to the prohibition of patentability the day researchers somewhere else in the world develop

---

a reservation to Article 20.2.ii, which allows the removal of regenerative tissue from a person who is not able to consent when no compatible donor with the ability to consent is available, and the recipient is a brother or sister of the donor. Croatia entered a reservation because this provision is not compatible with its law on the Removal and Transplantation of Human Body Parts, which allows the transplantation of regenerative tissue from a minor also for the benefit of his/her parents. Narodne Novine, Official Gazette of the Republic of Croatia, No 53/91.

[301]   Oviedo Convention, art. 25.

[302]   Ibid., art. 23.

[303]   However, the European Court of Human Rights has relied on the text of the Oviedo Convention when adjudicating on violations of the European Court of Human Rights. See, in general, F Seatzu, 'The Experience of the European Court of Human Rights with the European Convention on Human Rights and Biomedicine' (2015) 31:81 *Utrecht Journal of International and European Law* 5, 5–16.

[304]   Horizon 2020 Regulation, art. 19.2.b.

[305]   Ibid., art. 19.2.c.

therapies that modify human germline genome. If foreign pharmaceutical companies are not able to secure patents in Europe, they will simply not commercialize their therapies in the European Union, probably causing the EU public to put pressure on the EU Commission to reconsider the ban. However, for the time being the Biotech Directive stands.

Then, there is the obstacle of the Clinical Trials Regulation. The Regulation, which will come into force in 2019, is about the 'implementation of good clinical practices in the conduct of clinical trials *on medicinal products for human use*'.[306] It bans gene therapy trials 'which can result in modifications to the subject's germ line genetic identity'.[307] Although some argued that because in human germline genome modification no medicinal product is created, and it is rather a process or technique, and, therefore, the Clinical Trials Regulation should not apply,[308] the language of the Regulation and Directive is clear enough to block research moving toward clinical trials.

Future litigation before the Court of Justice of the European Union or intervention by the EU Commission might clarify more the framework, but it is anyone's guess in what direction. Given the cleavage between EU member states on the broader principles, the EU Commission is unlikely to intervene further and risk picking a fight with major member states over a matter that the overwhelming majority of the public does not understand. One should not count on the CJEU either, as, on genetic engineering, it has shown little capacity to assess scientific arguments, and a marked inclination to follow the *vox populi* and above all a strict interpretation of the so-called precautionary principle.[309]

---

[306] Regulation 536/2014 of the European Parliament and of the Council of 16 April 2004 on Clinical Trials on Medicinal Products for Human Use. [Italics of the authors].

[307] Clinical Trials Directive, art. 9.6, last line.

[308] I De Miguel Beriain, 'Legal Issues Regarding Gene Editing at the Beginning of Life: an EU Perspective' (2018) 12 *Regenerative Medicine* 671. Directive 2001/20/EC, on the other hand, refers to 'clinical trials involving medicinal products for gene therapy, somatic cell therapy including xenogeneic cell therapy and all medicinal products containing genetically modified organisms'. Nevertheless, gene editing in embryos does not create a product and still less a 'medicinal product', because it does not create any 'substance' (something separate from the human being in question that is used by or administered to such a human being with a view to restoring, correcting or modifying physiological functions) (citing JL Davies, *The Regulation of Gene Editing in the UK* (2016). On the contrary, it involves the application of a process or technique. Therefore, it is unclear if and how the directive or the regulation that will repeal it will apply to embryonic gene editing, because it may be that some of those modifications will not be considered clinical trials as such (citing J Kipling, *The European Landscape for Human Genome Editing: A Review of the Current State of the Regulations and Ongoing Debates in the EU* (2016)).

[309] On 25 July 2018, the CJEU ruled on the case *Confédération Paysanne and Others* v. *Premier Ministre and Ministre de l'Agriculture, de l'Agroalimentaire et de la Forêt*, C-528/16. Although the case did not regard human cells, it did discuss CRISPR-Cas9 and its effects of the genome of

Europe, both in its EU and CoE meaning, prides itself on its transparent legislative process, one where the public has ample opportunities to weigh in. Recently, both the Parliamentary Assembly of the Council of Europe and the EU European Group on Ethics in Science and New Technologies called for a 'broad and informed public debate',[310] one that goes 'beyond expert committees and engage civil society and touch upon safety issues and potential health risks or benefits of gene editing technologies as well as human dignity, justice, equity, proportionality and autonomy'.[311] So far, these debates have not started. However, if the current regulatory frameworks befuddle legal, scientific and ethical experts, it is hard to see how the wider European public could be able to debate their merits without falling back on entrenched cultural divides that have little to do with science. Given the current political climate in Europe, characterized by rising populism and general distrust for experts of any kind, it might be better to let this dog sleep.

plants in the particular case. Specifically, the case raised the question of whether organisms obtained by mutagenesis, including gene editing techniques such as CRISPR, are subject to the same regulations for genetically modified organisms as transgenic organisms.

The European Union has one of the most stringent regulatory framework for Genetically Modified Organisms, Directive 2001/18/EC on Genetically Modified Organisms (GMO Directive) (OJ 2001 L 106, 1) being the most important instrument. Overall, the European Union relies on a strict interpretation of the 'precautionary principle' to determine what GMOs can be cultivated in Europe. It demands a pre-market authorization for any 'new food' (GMOs and irradiated food) to enter the market and a post-market environmental monitoring, carried out by both the European Food Safety Authority and the member states. Currently, the European Union allows the cultivation of only 62 GMO varieties of six plants (cotton, maize, oilseed rape, soybean, sugar beet and swede rape). It does not allow the cultivation of many GMOs that are commonly cultivated and consumed around the world, which caused other states where these GMOs are cultivated to accuse the European Union of protectionism and challenging, successfully, the restrictive practices before the World Trade Organization.

In the *Confédération Paysanne* case, the Court, ignoring the opinion of its own Advocate General, and to the surprise of many scientists, took the view that organisms obtained by mutagenesis are GMOs within the meaning of the GMO Directive. It did not consider the scientific evidence and instead relied on a narrow interpretation of the 'precautionary principle'. The Court considered that the risks linked to the use of new mutagenesis techniques might prove to be similar to those that result from the production and release of a GMO through transgenesis, since the direct modification of the genetic material of an organism through mutagenesis makes it possible to obtain the same effects as the introduction of a foreign gene into the organism (transgenesis), and those new techniques make it possible to produce genetically modified varieties at a rate greater than those resulting from the application of conventional methods. Thus, the Court concluded that, considering these potential risks, excluding organisms obtained by new mutagenesis techniques from the scope of the GMO Directive would compromise the objective pursued by the directive, and would fail to respect the precautionary principle which the directive seeks to implement.

[310] EGE Statement on Gene Editing (2016), 2.
[311] Ibid.

7

The Regulation of Human Germline Genome Modification in the United Kingdom

James Lawford Davies

I INTRODUCTION

The United Kingdom has a long history of developing novel regulations governing all areas of society. Healthcare in particular has kept lawmakers occupied for many decades, and the United Kingdom has been bold in regulating novel (and sometimes controversial) therapies and technologies. It was one of the first countries in the world to develop a comprehensive framework for the regulation of in vitro fertilisation (IVF) and embryo research, and UK law addresses human cloning, preimplantation genetics, stem cell research and gene therapy – among many other areas.

In this context of broad and enthusiastic regulation, it is perhaps unsurprising that gene-editing technologies fit relatively comfortably within the existing UK regulatory landscape for research and therapy (whether because of national or EU law). It is not, however, without challenges, and this chapter outlines both the regulatory environment in the United Kingdom and some of the controversies that have arisen in relation to germline genome modification.

II A SHORT OVERVIEW OF THE UK LEGAL SYSTEM

The United Kingdom of Great Britain and Northern Ireland, commonly known as the United Kingdom, is a constitutional monarchy with a parliamentary democracy, composed of four distinct but united countries – England, Wales, Scotland and Northern Ireland. Apart from England, each country enjoys a varying degree of autonomy through devolved administrations.

The United Kingdom does not have a single legal system: rather, there are distinct systems of law that apply to specific geographical areas. English law applies in England and Wales; Northern Ireland law applies in Northern Ireland, and Scots law applies in Scotland. There is also a growing body of Welsh law since Welsh devolution, which, of course, applies only in Wales. While these systems diverge in certain areas, there are also substantive fields of law that are common across the United Kingdom. They also have a common supreme court at the apex of their judiciaries, the Supreme Court of the United Kingdom. It is the highest court in the land for all criminal and civil cases in England and Wales and Northern Ireland, and for all civil cases in Scots law.[1]

'English law' is the mother of the Common Law legal tradition and has existed as a distinct legal tradition, separate from the Civil Law legal tradition followed by almost all continental European states, since 1189. As it is well known, in the Common Law judges play a major role in the development of the law, applying statute, precedent and case-by-case reasoning to give explanatory judgments of the relevant legal principles. These judgments are binding in future similar cases (stare decisis).

Unlike most modern states, the United Kingdom does not have a codified single constitution but an unwritten one formed of several Acts of Parliament, court judgments and conventions that have evolved over a long period. The key landmark is the Bill of Rights of 1689, which established the supremacy of Parliament over the Crown.[2]

A further distinguishing feature of the UK legal system is that, unlike many continental European countries, which are 'monist', the United Kingdom is a 'dualist' state.[3] In 'dualist states' a treaty ratified by the government does not become part of the national legal system unless and until it has been incorporated into national law by specific legislation (in the case of the United Kingdom, an Act of Parliament). By virtue of incorporation, the provisions of treaties become 'legislation', which is superior to all other sources of law and may not be challenged in courts. Until then, courts in the United Kingdom have no power to enforce treaty rights and obligations on behalf of the government or a private individual.

Finally, the United Kingdom became a member of the European Community on 1 January 1973 (subsequently the European Union (EU)).

---

[1]   The Supreme Court of the United Kingdom came into being in October 2009, replacing the Appellate Committee of the House of Lords.
[2]   The Bill of Rights 1689 (1 William & Mary, 2nd Session) cl 2.
[3]   On the distinction between dualist and monist legal systems, see Pierre-Marie Dupuy, 'International Law and Domestic (Municipal) Law', *Max Planck Encyclopedia of Public International Law* (updated April 2011). Available online at: http://opil.ouplaw.com/view/10 .1093/law:epil/9780199231690/law-9780199231690-e1056 accessed 19 October 2018.

By virtue of this membership, the United Kingdom is also subject to EU law. In order that Community/Union law was automatically incorporated into the national legal system and to avoid the cumbersome process required by its dualist nature, in 1972 the UK Parliament passed the European Communities Act 1972.[4] The Act provides for the incorporation into UK law of the whole 'acquis communautaire': the Community/Union set of treaties, regulations and directives, together with judgments of the European Court of Justice/ Court of Justice of the European Union.

On 13 July 2017, the government introduced what became the European Union (Withdrawal) Act to Parliament, which, as enacted, makes provision for repealing the 1972 Act on 29 March 2019.[5] Given the numerous uncertainties and ambiguities surrounding the practical consequences of 'Brexit' – including the continuity, harmonisation and post-Brexit interpretation of EU law – this chapter provides an account of the status quo with regard to applicable EU legislation. However, the UK government has already indicated it will seek alignment with EU legislation in relation to medicinal products, medical devices and wider life sciences.

III THE REGULATORY ENVIRONMENT

1 Regulatory Framework

a The Human Fertilisation and Embryology Act 1990 and the Human Fertilisation and Embryology Authority

The United Kingdom has a mature and comprehensive regulatory framework governing the use of human embryos and gametes in treatment and research. Although these regulations do not explicitly address human germline genome modification, the mode of operation of the legislation ensures that, insofar as it is permitted, the modification of the genome of human germline cells may only take place pursuant to a licence granted by a statutory regulator.

The UK regulatory landscape was designed in the 1980s by a 'Committee of Inquiry into Human Fertilisation and Embryology' chaired by the late Dame (subsequently Baroness) Mary Warnock. The Committee was established in July 1982 and reported in June 1984.[6] It is a notable testament to the clarity,

---

[4] European Communities Act 1972, cl. 68.
[5] European Union (Withdrawal) Act 2018, cl.16
[6] Department of Health & Social Security, *Report of the Committee of Inquiry Into Human Fertilisation and Embryology* (HMSO, Cmnd 9314, 1984).

thoroughness and foresight of the 'Warnock Report' that the regulatory framework it proposed is still largely intact today. Most fundamentally, the Report recommended that the use of embryos and gametes in treatment and the use of embryos in research be subject to statutory regulation and licensing, and that a new statutory regulator be established to perform these functions.

The UK government was very slow to take steps to draft legislation to implement the recommendations of the Warnock Report, due in part to the strength and diversity of opinion around certain matters, particularly embryo research. Parliament first debated draft legislation in 1989, and eventually passed it in 1990 with a significant majority, albeit not without obstacles. The Human Fertilisation and Embryology Act 1990 (the 1990 Act) came into force in 1991,[7] and the new statutory regulator, the Human Fertilisation and Embryology Authority (HFEA), opened its doors on 1 August 1991.[8] The 1990 Act has undergone one major revision in 2008, primarily to update it to take into account developments in science and society. The resulting Human Fertilisation and Embryology Act 2008 is largely an amending Act but it also contains some new, freestanding provisions of its own.[9]

The 1990 Act has also been revised and expanded through secondary legislation. Of particular relevance to this chapter are the Human Fertilisation and Embryology (Research Purposes) Regulations passed in 2001.[10] These regulations were introduced following progress in stem cell research and cell nuclear replacement: they extended the list of research purposes which could be licensed by the HFEA to include increasing knowledge about the development of embryos, increasing knowledge about serious disease and enabling any such knowledge to be applied in developing treatments for serious disease.[11]

The substance of the statutory provisions is discussed in detail below. In general terms, the 1990 Act prohibits a range of activities unless they are done 'in pursuance of a licence' granted by the HFEA.[12] It is a criminal offence in the United Kingdom to keep or use gametes or embryos in treatment, or to conduct research on human embryos without, or contrary to the terms of, an HFEA licence.[13] A person found guilty of such an offence is liable to

---

[7]   Human Fertilisation and Embryology Act 1990, c 37 (1990 Act).
[8]   Human Fertilisation and Embryology Authority www.hfea.gov.uk/, accessed 16 October 2018.
[9]   Human Fertilisation and Embryology Act 2008, c 22 (2008 Act).
[10] The Human Fertilisation and Embryology (Research Purposes) Regulations 2001, No. 188.
[11] Ibid. s 2, incorporated into the 1990 Act by the 2008 Act, sch 2, para. 6.
[12] 1990 Act, ss 3-4A.
[13] Ibid. s 3.1A.

a prison sentence of up to two years, a fine or both.[14] It is also an offence to keep or use an embryo for research after 14 days from its creation or from the appearance of the primitive streak (if that appearance is earlier than the 14-day period).[15] This is treated as a more serious offence in the legislation, punishable by a prison sentence of up to 10 years, a fine or both.[16]

There are no specific statutory restrictions on the use of human germline genome modification techniques in research projects, but the legislation does not permit the use of gametes or embryos in treatment if their nuclear DNA has been intentionally altered. The effect of this restriction is to prohibit the use of gene-edited embryos and gametes in treatment. The lawful use of such techniques in treatment would therefore require a change to the legislation, discussed further below.

Related to this, the HFEA grants licences subject to certain standard conditions. Every embryo research licence provides that '[n]o embryo appropriated for the purposes of any project of research shall be kept or used otherwise than for the purposes of such a project'.[17] It follows that no embryo that has been genetically altered in the course of a licensed research project could be lawfully used in treatment – even if the clinical application of human germline genome modification is authorised in the future.

Also, consent is one of the cornerstones of the 1990 Act, and there are detailed provisions in Schedule 3 of the Act setting out what constitutes valid consent for the purposes of both treatment and research. The Act requires that all donors of embryos (or gametes or human cells used to create embryos in vitro) provide their written, signed consent to the use of such material in a licensed research project. The terms of consent may be varied or consent may be withdrawn at any time until the embryos or gametes have been used for the purposes of the research project.

b HFEA Code of Practice and Guidance

In addition to implementing the statutory framework established in the 1990 Act, the HFEA also publishes a Code of Practice that sets out detailed guidance to licensed treatment and research centres on a wide range of

---

[14] Ibid. s 41.4.

[15] Ibid. s 3.3A. There have been recent calls for this 14-day limit to be extended in the United Kingdom. See, for example, I Sample, 'Researchers break record for keeping lab-grown human embryos alive' (*The Guardian*, 5 May 2016) www.theguardian.com/science/2016/may/04/scientists-break-record-for-keeping-lab-grown-human-embryos-alive accessed 16 October 2018.

[16] 1990 Act, s 41.1.b.

[17] Ibid. s 15.4.

issues.[18] Currently in its eighth edition, the Code does not have statutory force, but failure to comply with the Code is taken into account by HFEA licence committees when determining the future of a centre's licence to operate and may be a catalyst for regulatory action.

The majority of the HFEA Code of Practice is concerned with the governance and operation of fertility clinics, but Part 22 concerns research and training. With limited exceptions, these provisions are not technique-specific. They instead set out the HFEA's requirements regarding, inter alia, the information to be provided to donors of gametes and embryos for the use of their material in research before they consent, and also the nature of that consent, supplementing the statutory provisions described above.

So far, the HFEA has not issued a formal policy position on human germline genome modification, though it would certainly do so if it were ever permitted to license the use of human germline genome modification in treatment. However, in 2016, the HFEA did renew an existing research licence to add additional research activities and objectives to that licence, including the use of CRISPR/Cas9 in human embryos – the first such regulatory approval of its kind in the world.[19] This particular project is discussed in more detail below.

2 Definitions

Section 1 of the 1990 Act provides definitions of both 'embryos' and 'gametes'. These do not mirror the usual clinical and scientific definitions but are intended to be broad enough to capture a variety of sources of material, created through a number of different procedures.

a Embryos

In its original terms, the 1990 Act defined an 'embryo' as follows:

1(1) Except where otherwise stated –
   a) embryo means a live human embryo where fertilisation is complete, and
   b) references to an embryo include an egg in the process of fertilisation,

[18] The full Code of Practice is available on the HFEA website: www.hfea.gov.uk/code-of-practice / accessed 16 October 2018.

[19] The full HFEA Licence Committee minutes and papers are available on the HFEA website: www.hfea.gov.uk/media/2444/licence-committee-minutes-14-january-2016.pdf accessed 16 October 2018.

and, for this purpose, fertilisation is not complete until the appearance of a two cell zygote.[20]

This definition was the subject of litigation in 2001–2003 when the ProLife Alliance sought to argue that embryos created by cell nuclear replacement were not 'embryos' for the purposes of the 1990 Act since their creation did not involve fertilisation.[21] Although initially successful, the claimants ultimately failed. As Lord Steyn concluded in the House of Lords judgment rejecting the claim, Section 1 should be interpreted purposively:

> The long title of the 1990 Act makes clear, and it is in any event self-evident, that Parliament intended the protective regulatory system in connection with human embryos to be comprehensive. This protective purpose was plainly not intended to be tied to the particular way in which an embryo might be created. The overriding ethical case for protection was general.[22]

When the 1990 Act was revised and updated in 2008, the UK Parliament broadened the definition in Section 1 to take into account both the possibility of embryos being created through means other than fertilisation and the advent of embryos created using human and animal material ('admixed' embryos). The definition now reads as follows:

1(1) In this Act . . . -
(a) embryo means a live human embryo and does not include a human admixed embryo . . ., and
(b) references to an embryo include an egg that is in the process of fertilisation *or is undergoing any other process capable of resulting in an embryo.*[23]

b Gametes

The 1990 Act defines 'gametes' as follows:

(4) In this Act . . . -
(a) references to eggs are to live human eggs, including cells of the female germ line at any stage of maturity, but (except in subsection (1)(b)) not including eggs that are in the process of fertilisation or are undergoing any other process capable of resulting in an embryo,

---

[20] 1990 Act (unamended), s 1.
[21] *R (on the application of Quintavalle (on behalf of the ProLife Alliance)) v Secretary of State for Health* [2003] UKHL 13.
[22] Ibid. para. 26.
[23] 1990 Act (as amended), s 1.1 (emphasis added).

(b) references to sperm are to live human sperm, including cells of the male germ line at any stage of maturity, and

(c) references to gametes are to be read accordingly.[24]

The 2008 revision expanded the definition to encompass expressly immature gametogenic cells, such as primary oocytes, and spermatocytes, as well as mature eggs and sperm.[25]

c Other Definitions

As mentioned above, the 1990 Act does not generally address specific research techniques or technologies. Instead, the regulatory scheme operates by addressing the purposes for which human embryos may be used in research. Thus, there is currently no reference to 'genome editing' or related technologies in the legislation. However, this is sure to change should Parliament approve legislation that allows use of human germline genome modification in treatment. The amendment of the 1990 Act to take into account the availability of preimplantation genetic diagnosis (PGD) and mitochondrial donation as licensable treatments, discussed further below, is evidence of this. Indeed, this capacity of the regulatory framework to accommodate novel, complex technologies suggests how human germline genome modification may be regulated as licensable treatment in the future.

A final observation in relation to statutory definitions is that the government sought to 'future proof' the legislation by including a regulation-making power in the revised 1990 Act enabling the Secretary of State to expand the definitions of 'embryo' and 'gamete' when this is considered necessary or desirable in light of developments in medicine or science.[26] To date, this power has not been utilised.

3 Oversight Bodies

The UK regulatory framework places all formal responsibility for the oversight of reproductive technologies and embryo research with the HFEA. The same would doubtless follow if the clinical use of human germline genome modification were to become lawful. Still, the HFEA research licencing process does require applicants to obtain ethics approval for their proposed research

[24] Ibid. s 1.4.
[25] Ibid.
[26] Ibid. s 1.6. Because of Section 1.7, this power does not extend to anything containing nuclear or mitochondrial DNA that is not human to be treated as an embryo or gametes.

from a research ethics committee, thus introducing an important element of external input and oversight.[27]

The United Kingdom does not have a national ethics committee as such, but the Nuffield Council on Bioethics (the Council), an independent advisory body supported by charitable funding, is highly influential and policy makers often cite its reports.[28] In 2016, the Council started an inquiry into the ethical issues arising from genome editing and human reproduction. It was published on 17 July 2018, concluding that heritable genome editing 'could be ethically acceptable in some circumstances'.[29] The two principles the Council determined must be satisfied for such applications to be ethically acceptable are

(i) that such interventions are intended to secure, and are consistent with, the welfare of a person who may be born as a consequence; and
(ii) that any such interventions would uphold principles of social justice and solidarity – that they should not produce or exacerbate social division, or marginalise or disadvantage groups in society.[30]

The report further recommends inter alia that there be further public debate about the use and implications of gene-editing technologies, that clinical safety be established before their use and that the adverse consequences for all levels of society be assessed and mitigated. Notably, the report does not seek to distinguish 'therapeutic' interventions from 'enhancement' since germline gene-editing technologies are not straightforwardly therapeutic in the sense of the traditional medical model where a patient is treated to alleviate the symptoms of disease.[31] This distinction is considered further below.

4 Funding

There is no restriction in the United Kingdom on the use of public funds to support research involving human embryo research. Indeed, the United

---

[27] The 1990 Act does not define 'research ethics committee'. The HFEA website indicates that NHS researchers can apply to a NHS research ethics committee, and the Human Fertilisation and Embryology (Disclosure of Information for Research Purposes) Regulations 2010 define it as a 'research ethics committee recognised or established by or on behalf of the Health Research Authority under the Care Act 2014'. There is, however, no regulatory obstacle to a private research centre establishing its own research ethics committee.
[28] Nuffield Council on Bioethics (2018) http://nuffieldbioethics.org accessed 16 October 2018.
[29] Nuffield Council on Bioethics, *Genome Editing and Human Reproduction: Social and Ethical Issues* (Nuffield council on Bioethics 2018) vii.
[30] Ibid.
[31] Ibid., e.g., paras. 3.33 and 3.34.

Kingdom has a long history of funding and facilitating innovative research in reproductive and genetic technologies, supported by successive governments. UK Research and Innovation, a consortium of the seven UK research councils and two other organisations, with a combined budget of £6 billion (about 7.9 billion USD) oversees public funding for research. Of the seven councils, the Medical Research Council and the Biotechnology and Biological Sciences Research Council are most obviously predisposed to fund research involving human germline genome modification. In 2015, these, together with three other UK research organisations, issued a position statement in support of the use of CRISPR/Cas9 and other gene-editing techniques in preclinical research.[32]

As mentioned above, in 2016 the HFEA renewed an existing research licence to add additional research activities, including the use of CRISPR/Cas9 in human embryos. This research is led by Dr Kathy Niakan, at the Francis Crick Institute in London, and their website indicates that her research is primarily funded by the Medical Research Council, the Wellcome Trust and Cancer Research UK – three stalwarts of biomedical research in the United Kingdom, including research in reproductive genetics.[33]

IV SUBSTANTIVE PROVISIONS

1 Regulation of Basic Research

a Embryos

As it was said, with some very limited exceptions, the 1990 Act does not regulate the use of specific techniques in research. There are no statutory provisions directly governing the use of germline genome modification in human embryos in basic research, nor is there any prohibition of research involving such techniques. Rather, the legislation focuses on the *purpose* of research projects and certain other criteria.

The 1990 Act permits the HFEA to issue licences authorising the creation of embryos in vitro and the storage and use of embryos 'for the purposes of

---

[32]   The Academy of Medical Sciences and others, 'Genome editing in human cells – initial joint statement' https://wellcome.ac.uk/sites/default/files/wtp059707.pdf accessed 16 October 2018.

[33]   The Francis Crick Institute, 'Genome editing reveals role of gene important for human embryo development' (20 September 2017) www.crick.ac.uk/news/science-news/2017/09/20/genome-editing-reveals-role-of-gene-important-for-human-embryo-development/ accessed 16 October 2018.

a project of research'.[34] In order to be granted a research licence, applicants must satisfy the HFEA that the proposed project is 'necessary or desirable' for one or more of a list of statutory purposes.[35] These are:

(a) increasing knowledge about serious disease or other serious medical conditions,

(b) developing treatments for serious disease or other serious medical conditions,

(c) increasing knowledge about the causes of any congenital disease or congenital medical condition that does not fall within paragraph (a),

(d) promoting advances in the treatment of infertility,

(e) increasing knowledge about the causes of miscarriage,

(f) developing more effective techniques of contraception,

(g) developing methods for detecting the presence of gene, chromosome or mitochondrion abnormalities in embryos before implantation, or

(h) increasing knowledge about the development of embryos.[36]

Furthermore, a licence application must explain why it is *necessary*, not merely desirable, to use human embryos, as opposed to another source of material.[37]

Besides these statutory provisions, the HFEA has also developed its own additional internal requirements and a process governing how research licence applications will be administered. This includes the requirement that applicants obtain ethics approval from a recognised, independent research ethics committee before they apply for a HFEA research licence. Applicants must then submit an application form setting out the details of their proposed project to the HFEA, together with a fee.[38] The HFEA then sends the proposal for peer review and organises an inspection of the premises and facilities. The report of that inspection, the peer reviews, the application form and the ethics approval are then submitted to an HFEA Licence Committee, which determines whether to grant a licence – or to grant it subject to additional conditions. The HFEA aims to process applications

---

[34] 1990 Act, sch 2, para. 3.1.a and b.

[35] Ibid. sch 2, para. 3A.1.a.

[36] Ibid. sch 2, para. 3A.2. This list was lengthened in 2001 following a government-commissioned report that recommended that the purposes for which human embryos could be used in research should be extended, enshrined in the Human Fertilisation and Embryology (Research Purposes) Regulations 2001.

[37] Ibid. sch 2, para. 3.5.

[38] Currently, between £500 and £750, depending on the project.

within four months, though the process can take longer, particularly where the application is novel or controversial.

As it was said, it is a condition of all HFEA research licences that embryos used or created under the licence cannot be used in treatment.[39] It follows that, if a gene-editing technology were to be applied to an embryo pursuant to an HFEA research licence, that embryo could not then be implanted in a woman; indeed, doing so would constitute a criminal offence.[40] However, it is permissible to use germline genome modification technologies on human embryos for research purposes.

b Gametes

With one very limited exception, the 1990 Act does not address research involving human gametes. The exception is known as the 'hamster test' to assess the viability of sperm. The 1990 Act permits the HFEA to grant a licence that authorises:

> mixing [human] sperm with the egg of a hamster, or other animal specified in directions, for the purpose of developing more effective techniques for determining the fertility or normality of sperm, but only where anything which forms is destroyed when the research is complete and, in any event, no later than the two cell stage.[41]

Otherwise, all the research activities that may be licensed by the HFEA relate to human embryos.

Thus, an HFEA research licence is not required to conduct research on human gametes. A licence would ordinarily be required to store gametes for research purposes, but regulations provide an exception to this licensing requirement for the storage of gametes where such storage is for the purpose of research, developing or testing pharmaceutical or contraceptive products, or teaching.[42] Yet, they cannot be used for treatment services, or in any mixing of eggs and sperm, and must be stored in accordance with certain labelling and access requirements.[43] Storage for other purposes or use in creating embryos could only lawfully take place under a licence. Still, it would be

---

[39] Supra note 17.

[40] Supra note 13.

[41] Ibid. sch 2, paragraph 3.2.

[42] The Human Fertilisation and Embryology (Special Exemptions) Regulations 1991, No. 1588, para. 3.1.

[43] Ibid. paras. 3.2 and 3.4.

permissible to use germline genome modification technologies on human gametes for research purposes, subject to these restrictions.

c Regulation of Preclinical Research

Preclinical research involving animals is permitted in the United Kingdom, but it is subject to stringent regulations. Animal research is governed by the Animals (Scientific Procedures) Act 1986 (ASPA) and an EU Directive which was transposed into UK law in 2012.[44] Researchers must obtain three types of licence from the Home Office: a personal licence to authorise them to conduct regulated procedures; a project-specific licence governing a particular programme of research; and an establishment licence for the research facility. Following the grant of a licence, Home Office inspectors visit research facilities to ensure that the conditions of the establishment's licence are being met.[45]

Although a comprehensive review of the ASPA framework is beyond the scope of this chapter, in broad terms ASPA governs the application of 'regulated procedures' to animals, the breeding and supply of certain species of animals for use in regulated procedures and the methods used to kill animals.[46] 'Regulated procedures' are defined as procedures carried out on animals for a scientific or educational purpose and may cause that animal a level of pain, suffering, distress or lasting harm equivalent to, or higher than, that caused by inserting a hypodermic needle according to good veterinary practice (known as the 'lower threshold').[47]

The Animals (Scientific Procedures) Act 1986 specifically addresses the use of genetic modification where either the modification of an animal's genes

---

[44] Directive 2010/63/EU of the European Parliament and of the Council of 22 September 2010 on the protection of animals used for scientific purposes (Directive 2010/63/EU). Directive 2010/63/EU incorporated into UK law via the Animals (Scientific Procedures) Act 1986 Amendment Regulations 2012, No.3039.

[45] The Home Office takes a risk-based approach to the frequency of inspections: ASPA requires that at least a third of establishments (and all establishments keeping non-human primates) are inspected every year. In practice, the Home Office aims to inspect all establishments at least once a year, and the majority are inspected more frequently (Home Office, '*Animals in Science Regulation Unit Compliance Policy*', December 2017) assets.publishing.service.gov.uk /government/uploads/system/uploads/attachment_data/file/670174/ASRU_Compliance_Poli cy_December_Final.pdf accessed 31 December 2018.

[46] Animals (Scientific Procedures) Act 1986 (ASPA), ss 1.1 and 2 define which animals are 'protected animals' for the purposes of the Act, though animals may also become protected if they are permitted to live until after it attains the stage of its development when it becomes a protected animal.

[47] Ibid. ss 2.1 and Sec. 2.

would cause a level of pain, suffering, distress or lasting harm equivalent to, or higher than the lower threshold, or where the animal is allowed to live until the modification may have the same effect.[48] It follows that both the breeding of genetically modified animals and their use in research is permissible in certain circumstances, subject to strict regulation and oversight.

Following the amendment of ASPA to incorporate the provisions of the EU Directive in 2012, research project evaluation must take into account ethical considerations regarding the use of animals, enshrined in the so-called 3 Rs: replacement, reduction and refinement.[49] These welfare and ethical considerations apply as equally to animals altered through gene-editing techniques as those created through conventional breeding. Related to this, it is likely that the greater efficiency and specificity of gene-editing technologies and methodologies should reduce the number of animals required to generate new lines.

d Regulation of Clinical Research

There is no mechanism under the 1990 Act to permit the clinical use of a new technology or treatment for research purposes. As it was said, embryos that are created or used in licensed research projects cannot be used in treatment. Subject to the restrictions discussed below, embryos and gametes may only be used pursuant to a treatment licence granted by the HFEA, and such licences do not prima facie distinguish between novel and established procedures.[50] Yet, in practice this lacuna has limited practical significance, as illustrated by the recent licensing of a novel – and controversial – therapy.

In 2017, the HFEA granted the first treatment licence of its kind for the clinical use of mitochondrial donation using pronuclear transfer.[51] This licence was granted to Newcastle Fertility Centre following many years of HFEA-licensed research into the development of mitochondrial donation

---

[48] Ibid. ss 2.3 A-C.

[49] Ibid. s 2A.

[50] During the Parliamentary debates on the revision of the 1990 Act in 2008, it was proposed that a new, additional category of HFEA licence be introduced to permit the human use of novel therapies, but this was deemed unnecessary by the government. See, for example, HC Deb (19 May 2008) Vol. 476, cols 85–88.

[51] The minutes of the HFEA Licence Committee meeting at which this licence was granted are available on the HFEA website: HFEA, 'Centre 0017 (Newcastle Fertility at Life) Variation of Licensed Activities to include Mitochondria Pronuclear Transfer' (HFEA, 9 March 2017) https://ifqlive.blob.core.windows.net/umbraco-website/1331/2017-03-09-licence-committee-minutes-variation-of-licensed-activities-to-include-mitochondria-pronuclear-transfer-pnt-centre-0017.pdf accessed 16 October 2018.

techniques using human embryos, including PNT and maternal spindle transfer. The chronology of this project is discussed in more detail below. Suffice to say here that the treatment licence granted to the clinic is subject to specific conditions, and the clinic must make a separate application to the HFEA for permission to use pronuclear transfer for each individual patient. The HFEA's Statutory Approvals Committee considers such applications case by case. Additional monitoring and oversight, accompanied by information and consent requirements, and the clinic's undertaking to conduct follow-up on patients ensure that the early use of this novel technology is careful and cautious. Were germline genome modification ever to be permitted for use in clinical applications, it is very likely to be subject to similar restrictions.

e Regulation of Clinical Applications

So far, clinical application of human germline genome modification is not permitted in the United Kingdom. When the 1990 Act was revised and updated in 2008, the UK Parliament introduced a new approach to the regulation of IVF and other licensed treatments that relies upon the concept of 'permitted' gametes and embryos. The intention was to ensure that embryos created using genetically modified gametes or gametes derived through in vitro gametogenesis, and likewise genetically modified embryos and cloned embryos, could not be used in treatment.[52]

The 1990 revised Act provides that a 'permitted egg' is one:
(a) which has been produced by or extracted from the ovaries of a woman, and
(b) whose nuclear or mitochondrial DNA has not been altered.[53]

Similarly, 'permitted sperm' are those:

(a) which have been produced by or extracted from the testes of a man, and
(b) whose nuclear or mitochondrial DNA has not been altered.[54]

Lastly, an embryo is a 'permitted embryo' if:

(a) it has been created by the fertilisation of a permitted egg by permitted sperm,

---

[52] See Explanatory Notes to the Human Fertilisation and Embryology Act 2008, c 22.
[53] 1990 Act, as revised in 2008, s 3ZA.2.
[54] Ibid. s 3ZA.3.

(b) no nuclear or mitochondrial DNA of any cell of the embryo has been altered, and

(c) no cell has been added to it other than by division of the embryo's own cells.[55]

Under Section 3 of the 1990 amended Act, only 'permitted gametes' and 'permitted embryos' may be transferred to a woman.[56] Thus, it follows that, if the DNA of gametes or embryos have been altered through the application of a gene-editing technology, they cannot be lawfully used in treatment.

Additionally, there is a further restriction in Schedule 2 of the 1990 Act, which states that an HFEA treatment licence 'cannot authorize altering the nuclear or mitochondrial DNA of a cell while it forms part of an embryo'.[57] Through this somewhat circuitous route, the 1990 Act in effect prohibits the use of germline gene modification as a reproductive technology.

There is a noteworthy postscript to this analysis. When Parliament adopted the 2008 amendments of the 1990 Act, taking into account the progress that had been made in research into the use of mitochondrial donation technologies to prevent the transmission of mitochondrial disease, it decided to enable the Secretary of State to designate as a 'permitted embryo' an embryo that had been subject to a process 'designed to prevent the transmission of serious mitochondrial disease'.[58] That power was utilised in 2015, permitting the HFEA to grant licences for the use of mitochondrial donation in treatment. No such power was included to allow for the use of human germline genome modification techniques, nor for any process that would alter the nuclear DNA of an embryo. Thus, the lawful use of gene modification in treatment would require a further change to the legislation. The prospects of the outcome of a Parliamentary debate on this issue are difficult to foresee at the present time.

V CURRENT PERSPECTIVES AND FUTURE POSSIBILITIES

The UK regulatory framework and wider policy context provide a solid foundation for research involving germline genome modification. Further, as we have recently seen with the regulation of mitochondrial donation, the United Kingdom has a well-established process for assessing the safety and efficacy of novel treatments, which may in due course enable the clinical application of germline genome modification as a reproductive technology.

---

[55] Ibid. s 3ZA.4.
[56] Ibid. ss 3.2.a and b.
[57] Ibid. sch 2, para. 1.4.
[58] Ibid. s 3ZA.5.

By way of conclusion, I will briefly consider how gene editing may become a licensable treatment, using the experience of mitochondrial donation as a model, followed by some of the possible challenges to such a development.

1 *From Research to Therapy: The Experience of Mitochondrial Donation*

In May 2004, Newcastle Fertility Centre applied for an HFEA research licence to use pronuclear transfer to avoid the transmission of mitochondrial disease. The HFEA Research Licence Committee rejected the application partly because the proposed research was deemed to breach the restriction in the 1990 Act that '[a] licence … cannot authorize altering the genetic structure of any cell while it forms part of an embryo'.[59] This decision was overturned on appeal to the HFEA Appeal Committee in September 2005 on a number of grounds, including the finding that 'genetic structure' should be interpreted narrowly and centred on the expression of nuclear genes resulting in heritable characteristics.[60]

Licence obtained, the research proceeded and, following promising results, the UK government was invited to include an amendment to the 1990 Act permitting the HFEA to license mitochondrial donation as a treatment as part of the 2008 revision of the Act. Instead, the government introduced a new regulation-making power through the 2008 Act, enabling the Secretary of State to pass regulations allowing the use of mitochondrial donation to avoid the transmission of serious mitochondrial disease.[61] In light of further research data, the Secretary of State was invited to exercise this power in November 2010, and, in February 2011, the HFEA was asked to scope expert views on the effectiveness and safety of mitochondrial transfer. The HFEA duly established an expert panel, which produced its first report in April 2011 concluding that '[t]he evidence currently available does not suggest that the techniques are unsafe' but also recommending further safety studies and experiments.[62]

In June 2012, the Nuffield Council on Bioethics published its supportive ethical review of the technology,[63] and, in September 2012, the HFEA

---

[59] Ibid. sch 2, para. 3.4.

[60] The Appeal Committee's decision is no longer publicly available, but the decision was discussed widely in the media at the time. See, e.g., Cordis, '*HFEA awards licence for pioneering mitochondrial research*' (13 September 2005) https://cordis.europa.eu/news/rcn/24404_en.html accessed 16 October 2018.

[61] 1990 Act, as revised in 2008, s 3ZA.5.

[62] HFEA, *Scientific Review of the Safety and Efficacy of Methods to Avoid Mitochondrial Disease through Assisted Conception, Report provided to the HFEA* (HFEA, 2011).

[63] Nuffield Council on Bioethics, 'Novel techniques for the prevention of mitochondrial DNA disorders: an ethical review' (Nuffield Council on Bioethics, June 2012) http://nuffieldbioethics.org/project/mitochondrial-dna-disorders accessed 16 October 2018.

launched a public consultation, leading to an advice to government, which was published in March 2013. This advice reported that the HFEA's consultation had found general support for permitting mitochondrial donation, providing it was deemed to be safe and carried out within a regulatory framework.[64] Although the advice acknowledged that respondents had expressed strong ethical concerns about the technique, it was held that these were outweighed by the arguments in favour of permitting it as a treatment. The HFEA's expert group also published a second report in March 2013, recommending further safety and efficacy studies.[65]

In February 2014, the government published draft regulations for public consultation, the report of which was published in July 2014.[66] Of the 1,857 responses, 1,152 opposed the implementation of the regulations (though on general grounds, rather than raising any specific response to the consultation questions) and 700 were broadly supportive.[67] The government therefore confirmed the regulations would be presented to Parliament for final approval. In the meantime, the HFEA expert group published its third scientific review in June 2104, again recommending further experiments.[68] In February 2015, both Houses of Parliament voted overwhelmingly in favour of the draft regulations and they were given ministerial approval on 4 March 2015.[69]

On 30 November 2016, the HFEA expert group published its fourth report, concluding that the science was 'now at an acceptable stage for cautious

---

[64] HFEA, *Mitochondria Replacement Consultation: Advice to Government* (HFEA, March 2013).

[65] HFEA, *Annex VIII: Scientific Review of the Safety and Efficacy of Methods to Avoid Mitochondrial Disease through Assisted Conception: Update, Report provided to the HFEA* (HFEA, March 2013).

[66] HFEA, *Government Response to the Consultation on Draft Regulations to Permit the Use of New Treatment Techniques to Prevent the Transmission of a Serious Mitochondrial Disease from Mother to Child* (HFEA, July 2014).

[67] The report states as follows in this regard: 'Although the consultation was about the detail of the draft regulations, not the policy as to whether to allow mitochondrial donation at all, the vast majority of these responses were one-liners either in support or opposing the introduction of the regulations, in principle. Where additional information was provided by the responder in these short responses, it was clear that opposition was informed by faith-based views. Where support was expressed this was generally informed by personal/family experience of mitochondrial disease. There was evidence of a co-ordinated campaign approach. From the total of 1,857 responses, around 700 expressed general support for the regulations and 1,152 opposed the introduction of the regulations with the remainder not expressing a view either way.' Ibid. at 11.

[68] HFEA, *Third Scientific Review of the Safety and Efficacy of Methods to Avoid Mitochondrial Disease through Assisted Conception: 2014 Update, Report provided to the HFEA* (June 2014).

[69] The Human Fertilisation and Embryology (Mitochondrial Donation) Regulations 2015, no. 572.

clinical use'.[70] Shortly afterwards, on 15 December 2016, the HFEA adopted a policy position permitting the 'cautious use' of mitochondrial donation in treatment, and they issued the first treatment licence to Newcastle Fertility Centre in March 2017.[71]

Should clinical application of human genome germline modification become technically feasible, it is highly likely that a similar process of review and consultation will unfold. The pace of research in gene editing is faster than that seen with mitochondrial donation, and the range of potential applications is also greater. While data obtained from international research are of course instructive, the UK government and regulators will still seek to satisfy themselves of the safety and efficacy of any human therapeutic application through expert advice and consultation (whether national or international), and will also no doubt consult the British public on their views about the matter. This process of consideration and debate takes time, as the licensing of mitochondrial replacement testifies. The research licence was first granted to Newcastle Fertility Centre to explore mitochondrial donation in 2005, and it was not until 2017 that they were awarded a licence allowing the technique to be used in treatment. Each step of this process was important, ensuring public, media and Parliamentary support for a novel, controversial therapy. That said, if the United Kingdom is to retain its role as a leader in reproductive and genetic technologies, 12 years might be too long for some researchers to wait to take their innovation from bench to bedside.

2 Gene Editing, Eugenics and Health

The United Kingdom's recent experience in regulating mitochondrial donation is also noteworthy for the legal challenges raised since these are also likely to arise again as and when the clinical application of human germline genome modification comes to be debated.

As it was explained, the UK regulatory framework is based on legislation that is generally drafted in objective and unemotional terms. Of course, while there are underlying rationales and priorities, including a recognition of the special status of the human embryo, these are seldom spelled out in the legislation itself. By contrast, some other European countries (and, of more direct relevance to the United Kingdom, the European Union) have adopted

---

[70] HFEA, *Scientific Review of the Safety and Efficacy of Methods to Avoid Mitochondrial Disease through Assisted Conception: 2016 Update, Report to the HFEA* (November 2016).

[71] HFEA, 'HFEA statement on mitochondrial donation' (15 March 2017) www.hfea.gov.uk/abo ut-us/news-and-press-releases/2017-news-and-press-releases/hfea-statement-on-mitochondrial-donation/ accessed 16 October 2018.

legislation that seeks to limit or prohibit germline interventions and what are frequently referred to as 'eugenic' practices.[72] Yet, the term is open to differing interpretations, which in turn may lead to uncertainty as to the application of such restrictions to human germline genome modification.

This became an issue during the parliamentary debates around the mitochondrial donation regulations. Two particular legal arguments were raised, both of which asserted that EU legislation prevents genetic modification, including mitochondrial donation. The first argument relied on the EU Clinical Trials Directive, which prohibits trials that 'result in modifications to the subject's germ line genetic identity'.[73] During the debates, it was suggested that this prohibition renders mitochondrial donation unlawful, and similar questions have been raised in relation to gene editing. However, the Directive concerns clinical trials 'on medicinal products for human use', whereas gene editing in an embryo involves the application of a *process* or *technique*. Furthermore, editing the DNA of an embryo does not create a 'product', still less a 'medicinal product', because it does not create any 'substance or combination of substances' (i.e. something separate to the human being in question that is either 'presented as having properties for treating or preventing disease in human beings', or used in or administered to such a human being 'either with a view to restoring, correcting or modifying physiological functions by exerting a pharmacological, immunological or metabolic action, or to making a medical diagnosis').[74]

The second argument relied upon a negative rights-based interpretation of the Charter of Fundamental Rights of the European Union.[75] The Charter includes the right to the 'integrity of the person', which incorporates a prohibition of 'eugenic practices, in particular those aiming at the selection

---

[72] For example, Belgian law includes a prohibition on 'research or treatments of eugenic nature, that is to say, focused on the selection or amplification of non-pathological genetic characteristics of the human species' (Act on Research on Embryos In Vitro (2003)) and French law prohibits 'carrying out a eugenic practice aimed at organizing the selection of persons' (French Bioethics Law of 2004). On the Belgian and French regulatory frameworks, see, in this book, Chapters 9 and 14.

[73] Article 9.6 of Directive 2001/20/EC of the European Parliament and of the Council of 4 April 2001 on the approximation of the laws, regulations and administrative provisions of the Member States relating to the implementation of good clinical practice in the conduct of clinical trials on medicinal products for human use, [2001] OJ L 121/34, pp. 34–44.

[74] Article 1 of Directive 2001/83/EC of the European Parliament and of the Council of 6 November 2001 on the Community code relating to medicinal products for human use (as amended) [2001] OJ L 311/67.

[75] Charter of Fundamental Rights of the European Union (proclaimed 7 December 2000, entered into force 7 December 2009) OJ C 326/391, 391–407. See, in this book, Chapter 6.

of persons'.[76] This in turn was said to implement Article 13 of the Council of Europe's Convention on Human Rights and Biomedicine (known as the Oviedo Convention).[77] During the passage of the mitochondrial donation regulations through the UK Parliament, it was argued that this prohibition of eugenic practices precluded the adoption of regulations permitting mitochondrial donation using pronuclear transfer and maternal spindle transfer.

By extension, the same arguments could be made in relation to the clinical use of human germline genome modification. Yet, that would be an erroneous argument. The prohibition of eugenic practices was included in the Charter on the recommendation of the European Group on Ethics in Science and New Technologies to address two concerns:

- Practices which involve for instance forced sterilization, forced pregnancies or abortions, ethnically enforced marriages, etc., all acts which are expressly regarded as international crimes by the Statute signed in Rome on July, 18, 2000 to create the permanent International Criminal Court.
 - Eugenics ... may also involve genetic manipulations on human beings such as the modification of the germ line in view of enhancement, *without any therapeutic aim.*[78]

Article 13 of the Oviedo Convention states as follows:

Interventions on the human genome

An intervention seeking to modify the human genome may only be undertaken for preventive, diagnostic or therapeutic purposes and only if its aim is not to introduce any modification in the genome of any descendants.

The Oviedo Convention has not been ratified by the United Kingdom (and so can have no effect in domestic law), nor by a considerable number of other EU Member States (albeit for different reasons).[79] In any event, to the extent that the Charter and the Oviedo Convention are aligned, it is clear that they are intended to prohibit genetic selection as an end in itself, rather than for therapeutic purposes.

By contrast, though it was not raised during the discussions around mitochondrial donation, the United Kingdom has ratified the United Nations

---

[76]   Ibid., art. 3.2.
[77]   Convention for the Protection of Human Rights and Dignity of the Human Being with Regard to the Application of Biology and Medicine: Convention on Human Rights and Biomedicine (opened for signatures on 4 April 1997, entered into force 12 January 1999) ETS No. 164 (Oviedo Convention).
[78]   European Parliament, Draft Charter of Fundamental Rights of the European Union (2000) CHARTE 4370/00, 22 (emphasis added).
[79]   On this, see, in this book, Chapter 6.

International Covenant on Economic, Social and Cultural Rights, which includes a number of more positive provisions of relevance to the regulation of germline gene editing.[80] In particular, Article 12 of the Covenant enshrines the right to the enjoyment of the highest attainable standard of physical and mental health, including provision for the healthy development of the child, which could be relied upon in support of making safe and effective genome editing available for those affected by heritable disease.[81] Further, Article 15 recognises the right to enjoy the benefits of scientific progress and its applications, including respect for the freedom 'indispensable for scientific research and creative activity'.[82] By virtue of the dualist nature of the UK legal system described above, the Covenant has a limited impact on UK domestic law: as an international treaty it is binding on the United Kingdom in international law only. Further, the scope of Article 15 right in the United Kingdom has never been tested, but its implications are that research into germline genome editing should be permitted in accordance with other human rights and (conversely) that the prohibition of such research would be non-compliant with the Convention – a position reflected in UK law as it stands today.[83]

The above legal arguments about the draft mitochondrial regulations highlight the importance of the distinction between the therapeutic and non-therapeutic application of technologies. UK policy makers and parliamentarians have twice drawn bright lines in law to permit the use of novel genetic technologies, but only in a therapeutic context. The first example is the regulation of PGD. Embryo testing is permissible when there is a *particular* risk that an embryo may have a gene, chromosome or mitochondrial abnormality (usually identified because of the familial genetic history).[84] PGD is done to establish whether that abnormality is present, but only where there is a *significant* risk that a person with the abnormality will have or develop a serious physical or mental disability, a serious illness or any other serious medical condition.[85]

---

[80] International Covenant on Economic, Social and Cultural Rights (adopted 16 December 1966, entered into force 3 January 1976) 993 UNTS 3 (ICESCR) art. 15.1.b. ICESCR was ratified by the United Kingdom on 20 May 1976.

[81] Ibid., arts. 12.1 and 2.b.

[82] Ibid., arts. 15.1.b and 15.3.

[83] For a detailed analysis of the implications of international law on the regulation of genome editing, see R Yotova 'The Regulation of Genome Editing and Human Reproduction Under International Law, EU Law and Comparative Law' (Nuffield Council on Bioethics, June 2017) http://nuffieldbioethics.org/wp-content/uploads/Report-regulation-GEHR-for-web.pdf accessed 18 October 2018.

[84] 1990 Act, sch 2, para. 1ZA, as amended by the 2008 Act.

[85] Ibid.

Similarly, the regulations governing the clinical use of mitochondrial donation provide that the HFEA may permit the use of the technology only when it is satisfied that there is a *particular* risk that an egg may have mitochondrial abnormalities, and there is a *significant* risk that a person with those abnormalities will have or develop a serious mitochondrial disease.[86]

Should embryonic DNA editing become lawful in the United Kingdom in the future, the regulations employed to govern PGD and mitochondrial donation provide compelling models for how the use of new reproductive and genetic technologies can be limited to addressing the risk of serious disease, disability or other serious conditions. This in turn should alleviate concerns about non-therapeutic manipulation and selection, clearly distinguishable from therapeutic aims. The late Dame Anne McLaren, a member of the Warnock Committee, summarised this distinction in the following terms:

> Incidentally, my recollection of the Warnock Committee, which met at a time when transgenic animals were being widely discussed and homologous recombination made gene therapy a plausible long-term goal, is that we all knew what was meant by the 'creation of human beings with specific characteristics', and it would never have occurred to us that 'health' would be considered a specific characteristic.[87]

3 Conclusions

Although the United Kingdom has a reputation of being a liberal jurisdiction for embryo research, it is in fact very tightly regulated and only *potentially* permissive. UK law reflects a compromise that was adopted by Parliament in the 1980s: to permit the creation and use of human embryos in research, but to subject such research to strict scrutiny, licensing and oversight, and to criminalise unlicensed research. That being said, the legislation is drafted in such a way as to facilitate a broad variety of research, as described above, and few procedures are prohibited. Overall, this framework helps allay public and political concern about what is often controversial research, it provides a degree of protection for researchers operating under a licence, and it permits innovation in reproductive and genetic research and technology.

---

[86] Supra note 69, paragraphs 5 and 8.
[87] Personal communication from Dame Anne McLaren, a member of the Warnock Committee, to the author (23 August 2005).

Such a robust framework is particularly valuable when it comes to considering how best to address the clinical application of germline genome modification. In circumstances where UK law is comprehensive and clear in its application to gene editing, there is no merit or purpose in a moratorium on the use of this technology in human embryos, as has been advocated in certain quarters. The successful history of the regulation of preimplantation genetics and mitochondrial transfer in the United Kingdom is testament to the strength of this framework. This success, however, was built upon a vital foundation of open and accessible dialogue between researchers, parliamentarians, policy makers and the public, and it is to be hoped that a similar transparency will be maintained around germline genome modification as it continues to advance.

8

The Regulation of Human Germline Genome Modification in Germany

Timo Faltus

I INTRODUCTION

In the Federal Republic of Germany, intentional alteration of the genome of germline cells and embryos is not just prohibited: it is a crime.[1] The ban is the result of legislation that was adopted more than 25 years ago, and it has not been significantly altered since. However, the rapid emergence of genome manipulation techniques, coupled with recent technological developments, is increasingly exposing the senescence of the regulatory framework.

With the advent of genome editing, there has been a substantive shift in the national debate on the use of genetic engineering. Until recently, the discussion had been more about the application of genetic engineering to crop plants and farm animals (so-called green genetic engineering) than to humans (so-called red genetic engineering), with the application of germline therapy to humans probably being the least discussed aspect. After all, green genetic engineering has been present in the fields for years, and sometimes even on the plate, while most red genetic engineering efforts have been unspectacular cell biology basic research, slowed down by both technical and legal hurdles. However, the emergence of more precise, safer and more predictable genome-editing methods has now brought the issue into German public discourse and, as evidenced by this volume, around the world.

---

[1] However, at the same time, it should be noted that in Germany modification of the genome of somatic cells is legal. There is no ban against it. The legal framework for somatic gene therapy essentially results from pharmaceutical legislation. For an overview of the legal framework of somatic gene therapy: M Carvalho, B Sepodes, AP Martins, 'Regulatory and Scientific Advancements in Gene Therapy: State-of-the-Art of Clinical Applications and of the Supporting European Regulatory Framework' (2017) 4 *Front Med* 182.

II THE REGULATORY ENVIRONMENT

1 Regulatory Framework

a The Federal Constitution and the 1985 Report
of the Benda Commission

Germany is a federal republic, comprising 16 federal states (collectively referred to as *Bundesländer*), each with its own constitution. The Basic Law (*Grundgesetz*) is the Constitution of the Federal Republic of Germany (*Bundesrepublik Deutschland*).[2] Unsurprisingly, the Basic Law makes no mention of genome modification. When it was adopted in 1949, modern genetic technologies were unknown and even unforeseen. Although the Basic Law has been amended several times over the years, explicit references to genome modification have not been added. That being said, the German Constitutional Court (*Bundesverfassungsgericht* – BVerfG) has issued rulings, interpreting and applying the Basic Law, that touch upon the early stages of human life and embryos and deserve to be mentioned.

Of particular interest is a 1975 opinion dealing with the legality of abortion and the rights of life in utero.[3] Specifically, the Court was asked to rule on the constitutionality of the decriminalization of abortion occurring during the first 12 weeks after conception (fertilization). Interpreting Article 2.2 of the Basic Law ('every person shall have the right to life and physical integrity'), in conjunction with Article 1.1 ('human dignity shall be inviolable. To respect and protect it shall be the duty of all state authority'), the Court concluded that the Right to Life enshrined in the Basic Law 'protects, in comprehensive fashion, unborn life as well'.[4] The Court ruled that the state has a duty to protect unborn life by preventing interferences with its development as well as acting proactively to safeguard it. In the words of the Court: 'Where human life exists, human dignity comes with it; it does not matter if the bearer is aware of this dignity and knows how to safeguard it.'[5] Furthermore,

---

[2] Grundgesetz für die Bundesrepublik Deutschland. Ausfertigungsdatum: 23.05.1949, zuletzt geändert am 28. März 2019 = Basic Law for the Federal Republic of Germany. Date of issue: 23.05.1949, last amended March 28, 2019.

[3] Federal Constitutional Court [*Bundesverfassungsgericht*], 39, 1, on Abortion [*Schwangerschaftsabbruch I*] (BVerfGE 39, 1).

[4] BVerfGE 39, 1, s C.I.

[5] '*Wo menschliches Leben existiert, kommt ihm Menschenwürde zu; es ist nicht entscheidend, ob der Träger sich dieser Würde bewußt ist und sie selbst zu wahren weiß.*' BVerfGE 39, 1, s C.I.

to protect unborn life, the state must adopt appropriate criminal provisions.[6] Finally, the Court affirmed that an independent legal entity that must be protected under the Basic Law exists at least 14 days after conception.[7] Although this decision of the Constitutional Court addresses only the legal status of human life 14 days after conception and of embryos in utero, it is a prevailing view among legal scholars that the reasoning of the Court applies also to embryos in vitro.[8] They must be protected to the same degree as embryos in utero are.

As assisted reproductive medicine progressed, the debate about the status of embryos in vitro gained saliency. In the early 1980s, the former Federal Minister of Justice and the Federal Minister for Research and Technology set up an interdisciplinary commission on ethical and legal questions related to the new methods of in vitro fertilization and genome analysis, as well as gene therapy, known as the 'Benda Commission', named after its chairman, Ernst Benda, former judge of the Federal Constitutional Court.[9] The Commission's Working Group issued a report in 1985 concluding that since targeted gene transfer in germ cells was not technically possible, modification of germline genome of humans was not a real issue. The only way to modify genetically germ cells was a random modification followed by a selection of the few successfully transformed embryos. However, since this approach would entail the destruction of human life, it was ultimately deemed to be unacceptable.

Clearly, the Working Group considered future developments unlikely. Specifically, it was sceptical that experiments on animals would ever provide a sufficiently secure basis for targeted gene modification of human germline cells. But, even if that was possible, the Working Group had serious concerns about the blurring of lines between therapy and eugenics, since there was no guarantee that only the most serious genetic defects would be eliminated. In addition, the drafters of the report held that imperfection belongs to the

[6] BVerfGE 39, 1, s C.II.
[7] BVerfGE 39, 1, s C.I.
[8] E.g. Kluth, 'Der *rechtliche* Status des *Menschen* am Beginn seines Seins', (2004) *Zeitschrift für Lebensrecht* (ZfL), 100; Schmid-Jortzig, 'Systematische Bedingungen der Garantie unbedingten Schutzes der Menschenwürde in Art. 1 GG (unter besonderer Berücksichtigung der Probleme am Anfang des menschlichen Lebens' (2001) *Die öffentliche Verwaltung* (DÖV) 925.
[9] Joint Working Group of the Federal Minister for Research and Technology and the Federal Minister of Justice, *Report on In Vitro Fertilization, Genome Analysis and Gene Therapy*, (1985) [*In-vitro-Fertilisation, Genomanalyse und Gentherapie – Bericht der gemeinsamen Arbeitsgruppe des Bundesministers für Forschung und Technologie und des Bundesministers der Justiz, J. Schweitzer Verlag, München, 1985*].

essence of human identity and that the distinction between 'normal' and 'not-normal' human beings offends human dignity.

Because of these considerations, the Commission recommended that genome modification of the germline should be prohibited.[10] That being said, it did not close the door to genetic modification altogether. For the Commission, germline therapy was not to be prohibited per se. At least in the case of the most serious monogenic hereditary diseases, which cause a high degree of suffering, one that it would be inhuman not to prevent, the prohibition of genetic therapy would not be unethical.[11]

b Laws

The Embryo Protection Act (*Embryonenschutzgesetz* – ESchG) is the key law regulating research on human embryos in Germany.[12] It is federal law, and, as such, it applies to all of Germany. The reasoning of the Constitutional Court in the 1975 abortion decision, the work of the Commission and the subsequent social and parliamentary discussion are the background to its adoption in 1990. Animated by a desire to prevent the abuse of artificial reproductive technologies, with the ESchG Germany's Federal Parliament (*Deutscher Bundestag*) created a framework for research on human embryos and germline cells.

The ESchG is discussed in detail below.[13] It suffices to say here that it criminalizes, inter alia, the fertilization of a human egg for a purpose other than inducing a pregnancy of the woman from whom the egg cell was biopsied[14] and any influence on human embryos for purposes other than the preservation of the embryo itself.[15] It also criminalizes the deliberate change of the genetic characteristics of human germline cells.[16] These offences are punishable with imprisonment up to five years or a fine.[17]

---

[10] Ibid. 47.

[11] Ibid. s 46.

[12] Gesetz zum Schutz von Embryonen (Embryonenschutzgesetz – ESchG). Ausfertigungsdatum: 13.12.1990, zuletzt geändert am 21. November 2011 = Embryo Protection Act (ESchG). Date of issue: 13.12.1990, last amended November 21, 2011.

[13] See, in this chapter, Section III 'Substantive Provisions'.

[14] ESchG, § 1.1, no 2.

[15] ESchG, § 2.1.

[16] ESchG, § 5.

[17] The fine is measured in Germany in daily rates. The individual daily rates are then multiplied by a sum of money that takes into account the solvency of the particular person. The background of this regulation is the attempt to adjust the penalties to the different economic circumstances of the accused person.

As we will see, the German Criminal Code (*Strafgesetzbuch*) must also be taken into consideration when discussing modifications of the human germline genome.[18] The relevant provisions will be discussed below. Here it suffices to say that scientists can be prosecuted not only for violations of the ESchG and other relevant statutes carrying criminal penalties, including the Criminal Code, not only for offences that took place in Germany, but also for violations committed abroad that they might aid and abate. Indeed, under Section 9.2.2 of the German Criminal Code, criminal law applies to 'secondary participation' if the 'secondary participant', acting within the territory of the Federal Republic of Germany, contributed to a violation of Germany's criminal laws in a foreign country even if the act is not a criminal offence according to the law of that foreign country. In other words, German scientists, or foreign nationals employed in Germany, participating in cooperative projects with researchers in other states that entail actions that are criminally sanctioned in Germany could be prosecuted in Germany.

c The Covenant on Economic, Social and Cultural Rights

Germany ratified the International Covenant on Economic, Social and Cultural Rights (ICESCR) on 17 December 1973.[19] The Covenant contains provisions such as Articles 15.1.b and 15.2 through 4 (Right to Science and the Rights of Science), Article 12 (Right to Health), and Article 10 (Right to a Family), which may be relevant to the discussion of human germline genome modification. By virtue of its ratification Germany is bound to observe it, as a matter of international law. However, it should be noted that Germany has so far not ratified the Optional Protocol to the Covenant giving individuals access to the communications procedure to raise the question of the alleged violation of their rights before the Committee on Economic, Social and Cultural Rights.[20]

As far as the German legal system is concerned, the Covenant has the rank of federal law.[21] Germany must take into account the requirements of the Covenant, and its provisions might be applied by courts. However, the rights

---

[18] Strafgesetzbuch (StGB). Ausfertigungsdatum: 15.05.1871, in der Fassung der Bekanntmachung vom 13. November 1998 zuletzt geändert 19. Juni 2019 = German Criminal Code (StGB). Date of issue: 15.05.1871, in the version of the announcement of November 13, 1998, last amended June 19, 2019.

[19] See International Covenant on Economic, Social and Cultural Rights (adopted 16 December 1966, entered into force 3 January 1976) 993 UNTS 3 (ICESCR).

[20] Optional Protocol to the International Covenant on Economic, Social and Cultural Rights (adopted 10 December 2008, entered into force 5 May 2013), A/RES/63/117.

[21] Basic Law [Grundgesetz], Art. 59.2.

contained in the Covenant are neither absolute nor supersede the laws of a contracting state per se. For example, Article 4 of the Covenant describes general limitations to the rights included in the Covenant:

> The States Parties to the present Covenant recognize that, in the enjoyment of those rights provided by the State in conformity with the present Covenant, the State may subject such rights only to such limitations as are determined by law only in so far as this may be compatible with the nature of these rights and solely for the purpose of promoting the general welfare in a democratic society.

From this wide possibility of limitations, it follows that the individual member state – here Germany – for example, for purposes of health protection can limit the freedom of science and the freedom of research. Health protection must be seen as part of the welfare in a democratic society, because it serves to promote the common good and general welfare of a society. Applied to the research on germline genome alteration, this means that restrictions of the freedom of research can be justified with the as-yet unforeseeable consequences of interventions in the germline for the individual, but in particular with the consequences that cannot be foreseen for subsequent generations. The questions of a possible uncontrolled spread of artificially introduced, germline-relevant alterations, which is yet to be resolved, may justify a limitation of research on germline genome alteration at least in embryos and living persons. This concerns both interventions with therapeutic intention and interventions in the area of the so-called enhancement.

Research in the field of basic research is permitted within a legal framework. Based on the results of basic research, a different assessment on germline genome alteration may be possible in the future. Thus, in the light of the current state of knowledge, the limitations described below on research on targeted germline alteration are justified in the light of the Covenant.

Furthermore, it is also necessary to emphasize that Article 15.3 protects only the indispensable freedom of scientific research, not any scientific activity. This wording can therefore be interpreted in such a way that the freedom of research of the Covenant in the individual member states can also be limited by reference to public security and order.

Whether Germany's regulatory framework of human germline genome modification is consistent with the obligations it has under the Covenant remains to be seen. To date, the Committee on Economic, Social and Cultural Rights has not discussed the matter in the context of the review of Germany's periodic reports, nor have German courts ever considered the question.

d Governmental Regulations and Professional Guidelines

Obviously, since the ESchG bans germline genome editing, there are no governmental regulations nor professional guidelines regulating it.

e Opinions by National Ethics Committees, Professional Organizations, Scientific Bodies

In recent years, federal lawmakers have become increasingly aware of the importance of genome-editing procedures, including the possibility of targeted germline genome alteration. In 2017, the German Parliament (Bundestag) commissioned the Office of Technology Assessment at the German Parliament (*Büro für Technikfolgen-Abschätzung beim Deutschen Bundestag* – TAB), an independent scientific body that advises the Bundestag and its committees on matters relating to research and technology, to analyse the technical possibilities and the ethical, legal and social consequences of human genome modification.[22] In 2017, TAB invited expert opinions on the application of genome-editing methods to human germline, and it is currently discussing the matter. The TAB report will be published after the German Parliament has assessed the report. Furthermore, the Federal Ministry of Education and Research (*Bundesministeriums für Bildung und Forschung* – BMBF) is currently funding projects on the ethical, legal and social consequences of genome editing, including germline modification, taking into account the evolution of genome-editing methods.[23]

Non-governmental bodies have also contributed to the debate. In 2017, members of the German National Academy of Sciences Leopoldina (*Deutsche Akademie der Naturforscher Leopoldina*) published the discussion paper 'Ethical and Legal Assessment of Genome Editing in Research on Human Cells'.[24] The paper concluded:

> Given the current state of research, no targeted germline modifications that have a direct impact on a subsequently born human being should be undertaken. Before any intervention in the germline can even be considered, the

---

[22] Information about the TAB can be found in English on the TAB's homepage: Office of Technology Assessment at the German Bundestag (TAB) [*Büro für Technikfolgen-Abschätzung beim Deutschen Bundestag*] (2018) www.tab-beim-bundestag.de/en/index.html accessed 02 August 2019.

[23] The author of this contribution is the coordinator of one of these ELSI projects named: GenomELECTION – Genome Editing – Ethical, Legal, and Science Communication Aspects in Molecular Medicine and Crop Plant Engineering, www.tab-beim-bundestag.de/en/index.html accessed 02 August 2019.

[24] Jörg Hacker (ed.), *Ethical and legal assessment of genome editing in research on human cells* (Deutsche Akademie der Naturforscher Leopoldina e.V. 2017).

techniques must be refined until such an intervention represents an accept-
ably low risk in comparison to the hereditary disease it seeks to prevent. The
empirical bases for such a risk assessment and the subsequent normative
evaluation of the risks and opportunities of a germline gene therapy can only
be provided through research. Research on somatic gene therapy and, building on this, research on germline cells and embryos are particularly relevant
here.[25]

Also in 2017, the German Ethics Council (*Deutscher Ethikrat*), an independent council of experts, published a report with recommendations on the
issue of the modification of the genome of human germline cells.[26] The report
calls for a global political debate and international regulation and urges the
German Parliament and the Federal Government to take the initiative, as
soon as possible, to put the question on the agenda of the United Nations, with
the goal of the adoption of binding global legal rules. In May 2019, the
German Ethics Council published its opinion "Intervening in the Human
Germline". For the German Ethics Council, ethical analysis does not result in
a categorical inviolability of the human germ line. In the bottom line of its
opinion, the German Ethics Council concludes that a serious ethical evaluation of germ line interventions – if one considers them to be ethically justifiable – can only be carried out in individual cases. Whether acceptable
minimum standards for such applications can ever be met, however, can
hardly be estimated with the current state of the technology development is
doubtful in many respects for the Council. If this should nevertheless become
reality, the Council has developed a decision tree that considers an opportu-
nity/risk balance as well as ethical standards of orientation.[27]

The members of the German Society of Human Genetics (*Deutsche
Gesellschaft für Humangenetik* – GfH) have stated their opposition to induce
pregnancy with genetically modified embryos on the assumptions that the
health risks of genome editing cannot be sufficiently assessed, that there is still
insufficient evidence of clinical benefits and that ethical and social issues have
not been adequately explored and discussed. However, considering the recent

---

[25] Ibid., s 21.

[26] Deutscher Ethikrat, *Germline intervention in the human embryo: German Ethics Council calls
for global political debate and international regulation: Ad hoc recommendation* (Deutscher
Ethikrat, Berlin 2017).

[27] Deutscher Ethikrat, Eingriffe in die menschliche Keimbahn – Stellungnahme, Berlin 2019.
The German Ethics Council has published a summary of this opinion in English: German
Ethics Council, Intervening in the Human Germline – Opinion: Executive Summary &
Recommendation, Berlin 2019, https://www.ethikrat.org/fileadmin/Publikationen/Stellungn
ahmen/englisch/opinion-intervening-in-the-human-germline-summary.pdf   accessed   02.
August 2019.

new technological possibilities of germline interventions, the GfH auspicates the beginning of a broad social discussion on research on surplus embryos produced during fertility treatments, leading to the adoption of a clear legal framework regulating research on embryos in vitro.[28]

Finally, the Department of Biotechnology of the Association of German Engineers (*Verein Deutscher Ingenieure* – VDI) holds a similar position. It advocates for an interdisciplinary and public dialogue, involving stakeholders from business, science, society, politics and the public sector. The dialogue should address the consequences of the ongoing prohibition of human germline genome editing, possible advantages and disadvantages for the environment and society of lifting the ban, and how the regulation of genome-editing methods can be future-oriented and for the benefit of all.[29]

2 Oversight Bodies

Since germline genome modification is criminally prohibited, detection and punishment of violations of the ban is in the hands of the state, first and foremost. Anyone who identifies, or believes to have identified, an alleged violation can report it to the police or prosecutors, who will then be responsible for investigating the matter. However, unlike everyday criminal offences, such as theft or bodily harm, recognition of these offences requires special knowledge and access to laboratories. Moreover, unlike in the case of garden-variety criminal offences, in the case of germline genome modification, at least initially, there is no injured person who, in turn, could report the alleged crime. This explains why there has not been any criminal prosecution of intentional human germline genome modification so far. In addition, since germline genome modification is criminally prohibited, scientists probably do not engage in such research to begin with.

No specific authorization to carry out permissible research on germ cells and germline cells is required before the start of research. As long as the research activity is genetic engineering work within the meaning of genetic engineering legislation, at most, a general permit for the operation of the laboratory is all that is required.

According to Section 9 of the ESchG, only a licensed physician may carry out artificial fertilization, preimplantation genetic diagnosis (PGD) or transfer of a human embryo into a woman. Moreover, only a licensed physician may

---

[28] German Society for Human Genetics [*Deutsche Gesellschaft für Humangenetik*] (GfH), Opinion on Interventions in the Germline Genome Editing [*Stellungnahme zu Eingriffen in das Keimbahn-Genom*] (13 December 2017).

[29] VDI Society Technologies of Life Sciences (VDI-TLS) (ed), *CRISPR/Cas & Co – New Biotech Tools* [*CRISPR/Cas & Co – Neue Biotech-Werkzeuge*] (VDI, November 2017).

preserve a human embryo or a human egg cell that has been penetrated by a human sperm cell, naturally or artificially. Licensed physicians must also pay attention to additional requirements. According to the Model Professional Code for Physicians in Germany (*Musterberufsordnung für die in Deutschland tätigen Ärztinnen und Ärzte* – MBO), physicians who participate in research on viable human gametes and living embryonic tissue must first obtain the advisory opinion of an ethics committee on the questions of professional ethics and conduct raised by the research.[30] This can be done by a committee established by one of the local Chambers of Physicians (*Ärztekammer*) or another independent, interdisciplinary ethics committee, set up according to state (*Bundesland*) law. However, since the Model Professional Code does not have binding force, physicians must observe the professional code of the state in which they are actually practising. The latter codes entail binding regulations. Violations of the provisions of state professional codes can result in sanctions, including complaints, reprimands, fines or withdrawal of licence, depending on the state in question.

Finally, regarding PGD, which can also be considered to be a special case of germline intervention, since it prevents the creation of germline or promotes a particular germline, it should be pointed out that interdisciplinary ethics committees established at the state levels have to approve the application in the specific case.

Other regulatory authorities have no significance in the context of permissible research on germline interventions. Because clinical studies on human germline alteration are prohibited,[31] such studies cannot be authorized by the regulators. However, for permissible clinical studies in the field of cellular therapy, the Paul-Ehrlich-Institute (PEI) is the competent authority.[32]

In stem cell research, there is a regulatory authority only in relation to human embryonic stem cell research. Obtaining human embryonic stem cells from embryos is prohibited according to Section 2.1 ESchG. However, under certain conditions established in the Stem Cell Act (*Stammzellgesetz* – StZG),[33] human embryonic stem cells may be imported into Germany for research purposes from abroad. Such import authorizations are therefore only

---

[30] Model Professional Code for Physicians in Germany (2018), § 15.
[31] See, in this chapter, Section III.4 'Regulation of Clinical Research'.
[32] The Paul-Ehrlich-Institut (2018) www.pei.de/EN/home/node.html accessed 02 August 2019.
[33] Gesetz zur Sicherstellung des Embryonenschutzes im Zusammenhang mit Einfuhr und Verwendung menschlicher embryonaler Stammzellen (Stammzellgesetz – StZG). Ausfertigungsdatum: 28.06.2002, zuletzt geändert 29.3.2017 = Act to ensure the protection of embryos in connection with the import and use of human embryonic stem cells (Stem Cell Act – StZG). Date of issue: June 28, 2002, last amended March 29, 2017.

relevant in relation to research on germline genome alteration if human germline cells are to be differentiated from human embryonic stem cells. The Robert-Koch-Institute (RKI) is responsible for these import permits.[34]

3 *Funding*

As long as a research project takes place within the legally permissible limits discussed below, public and private funding is available for basic research. However, because basic research projects are not required to obtain prior approval, there is no central directory for these research activities and it is difficult to gauge to what extent they have been actually funded so far.

Since human germline genome modification is generally prohibited by law, there is no funding for it in Germany, neither public nor private. The ban has chilling effects on the funding of borderline research (i.e. research on asexually produced embryos, tripronuclear embryos), too. There seems to be no funding for this research either.

III SUBSTANTIVE PROVISIONS

1 *Key Definitions*

a 'Embryo' and 'Germline Cells'

The ESchG defines the terms 'embryo' and 'germline cells'.[35] Because these definitions are contained in a federal statute, they are paramount for a discussion of germline genome modifications.[36] All prohibitions contained in the ESchG, such as those concerning the destruction of embryos and germline genome modifications, rely on these definitions. Moreover, since the ESchG contains criminally sanctioned prohibitions, these definitions are crucial for the purpose of the application of criminal sanctions, too.

Yet, their importance notwithstanding, their wording is rather complicated and convoluted, giving rise to a series of issues. First, the legal definition of embryo in the ESchG does not correspond to the scientific definition of embryo. The ESchG defines the embryo as a 'human egg cell, fertilized and

---

[34] Robert Koch Institut (2018) www.rki.de/EN/Home/homepage_node.html accessed 02 August 2019.

[35] ESchG, § 8.

[36] The Stem Cell Act (StZG) also defines the 'embryo'. Although it includes asexually produced embryos, it is of no consequence for the legal questions regarding germline genome modifications because the StZG only regulates the import of human embryonic stem cells from abroad.

capable of developing, from the time of fusion of the nuclei, as well as any totipotent cell removed from an embryo that is capable of dividing and developing into an individual under appropriate conditions'.[37] Clearly, the ESchG does not distinguish between embryos and earlier stages of embryo development, such as zygotes, morula and blastocysts.[38] Under the ESchG, an 'embryo' is also 'any totipotent cell removed from an embryo that is capable of dividing and developing into an individual under appropriate conditions'.[39] However, not every fertilized egg (i.e. after the time of fusion of the nuclei) is an embryo falling under the scope of the ESchG. It must be 'capable of development . . . from the moment of the one cell stage'.[40] To determine whether the particular embryo at its one-cell stage has the capacity to develop further, a legal fiction is employed. According to Section 8.2 ESchG, in the first 24 hours after the fusion of nuclei, the fertilized human egg cell is held to be capable of development except if it is established before expiration of this time period that it will not be capable of developing beyond the one-cell stage. This can, e.g., be the case if the cell nucleus apparently decays already in this period of time.

Then, the ESchG defines 'germline cells' (*Keimbahnzellen*) as 'any cells that lead directly from the fertilized egg cell to the egg and sperm cells of the resultant human being and also egg cells from insertion or penetration of the sperm cell until the completion of fertilization by fusion of the nuclei'.[41] This definition can cause problems for germ cells that are artificially produced from stem cells. For such artificially produced germ cells – as in the discussion about the somatic cell nuclear transfer (SCNT) embryos – it is disputed whether they are covered by the definition or not.

Finally, because the ESchG is a criminal statute of sorts, its provisions must be interpreted literally and cannot be applied by analogy. Analogous application, even when comparable interests are to be protected, is forbidden by the Constitution as well as the Criminal Code.[42]

b 'Genome Editing'

German law does not define 'genome editing' or 'genome modification' per se. However, procedures that modify the human genome are captured by the general

---

[37] ESchG, § 8.1.
[38] It should be noted that a zygote may be embryo within the meaning of Section 8.1 and a germline cell within the meaning of Section 8.3 of the ESchG.
[39] ESchG, § 8.1.
[40] Ibid.
[41] ESchG, § 8.3.
[42] Basic Law, Art. 103.2 [*Grundgesetz*]; StGB, § 1.

notion of '[a]nyone who artificially alters the genetic information of a human germ line' contained in Section 5.1 of the ESchG, as well as in the notion of 'genetic engineering work' (*gentechnische Arbeiten*) contained in the Genetic Engineering Act (*Gentechnikgesetz* – GenTG).[43] The overall purpose of the GenTG is to protect humans and the environment from the potential dangers of genetic engineering processes and products, as well as to create a legal framework for research, development, use and promotion of the scientific and technical possibilities of genetic engineering.[44] It applies to genetic engineering, managing genetic engineering facilities, releasing genetically modified organisms and placing products containing or consisting of genetically modified organisms on the market. Although the application of genetically modified organisms to humans is outside the scope of the GenTG,[45] the Act regulates germline manipulation of human cells in vitro, for instance the modification of a primordial sperm cell. That being said, since the ESchG prohibits manipulation of germ cells, the provisions of the GenTG are only relevant for basic research as well as occupational and environmental safety of laboratories during research.

Finally, although the German Criminal Code does not expressly define genome editing, it does define 'bodily harm' (*Körperverletzung*),[46] a concept that, as we will see, impacts research on human germline cells and embryos.

2 Regulation of Basic Research

a Basic Research Using Genome Modification on Gametes

In Germany, basic research that modifies the genome of gametes is permitted as long as the modified gametes are not used for fertilization.[47]

b Basic Research Using Genome Modification on Germline Cells and Embryos

Section 5.1 of the ESchG prohibits research using genome modification on germline cells in vitro. Section 8.3 of the ESchG defines 'germline cell' as 'any

---

[43] § 3, No 2 Gesetz zur Regelung der Gentechnik (Gentechnikgesetz – GenTG). Ausfertigungsdatum: 20.06.1990, in der Fassung der Bekanntmachung vom 16. Dezember 1993, zuletzt geändert 17. Juli 2017 = Act on the regulation of genetic engineering (Genetic Engineering Act – GenTG). Date of issue: 20.06.1990, in the version of the announcement of December of 16, 1993, last amended July 17, 2017.

[44] GenTG, § 1.

[45] GenTG, § 2.3.

[46] StGB, § 223.

[47] ESchG, § 5.4, no 1.

cells that lead directly from the fertilized egg cell to the egg and sperm cells of the resultant human being and also egg cells from insertion or penetration of the sperm cell until the completion of fertilization by fusion of the nuclei'.[48] Section 5.1 of the ESchG relies on this definition, and the artificial alteration of the genetic information of a human germline cell in vitro is punishable with detention up to five years or a fine. This prohibition has therefore a higher punitive threat than the destruction of embryos (up to three years) according to Section 2.1 of the ESchG.

However, there are two exceptions to this rule. First, modifying the genome of a germline cell taken from a dead fetus, a living being or a deceased person, is not a crime as long as the modified cell is not transferred to an embryo, fetus or human being, or not used to create a germ cell. Second, the modification must be intentional. The ESchG does not criminalize the unintentional modification of the genome. Vaccinations, radiological, chemotherapeutic or other treatments that are not intended to alter the genetic information of germline cells are not punishable.[49]

That being said, the legal framework of basic research of intended genome modification in embryos in utero (feasibility issues unconsidered) is more complicated. It is already questionable which legal regulations should cover these actions. It is disputed whether the prohibition of interventions on embryos contained in the ESchG is applicable to embryos in utero. There are valid arguments either way. On the one hand, the ESchG was enacted to regulate in vitro and reproductive medicine technologies, but not in vivo interventions, which are supposed to be covered by the Criminal Code. The object and purpose of the law must be taken into consideration, especially when interpreting provisions of a statute that contains criminal sanctions. On the other hand, Section 8.3 ESchG speaks not only of embryos but also of the human beings who develop from a fertilized egg.

Additionally, Section 2.1 of the ESchG provides that 'anyone who disposes of, or hands over or acquires or uses for a purpose not serving its preservation, a human embryo produced outside the body, or removed from a woman before the completion of implantation in the uterus, shall be punished with imprisonment of up to three years or a fine'. The provision includes both embryos produced in vitro as well as in utero, before implantation in the

---

[48] In very early embryos, which consist of more than one totipotent cell, each cell can also contribute to the germline. These cells are therefore also covered by Section 5.1 of ESchG. In later embryos, which consist of more than one non-totipotent cell, at least one cell can always contribute to the germline. These cells are then again covered by Section 5.1 of ESchG.

[49] ESchG, § 5.4, no 3.

uterus.[50] Since research on embryos invariably results in the destruction of the embryo, unless it is under mere observation, this provision bans all research on human embryos, except for when it is done for the benefit of the specific embryo in question.[51]

It should be noted that the above-mentioned provisions of the Criminal Code regarding bodily harm are not applicable in the case of genome modification of embryos in utero. These provisions apply only to human beings from the beginning of the 'first stage of labor' pains.[52]

However, interventions on the genome of germline cells of embryos in utero, if they occur after implantation and lead to an abortion, might be a crime under Section 218.1 of the Criminal Code. 'Whoever terminates a pregnancy shall be liable to imprisonment not exceeding three years or a fine.' However, the second sentence of the same article continues: 'Acts where the effects of which occur before the conclusion of the nidation shall not be deemed to be an abortion within the meaning of this law.' Thus, feasibility of the procedure aside, germline genome modification occurring before implantation of the fertilized egg into the uterus would not be an abortion within the meaning of the Criminal Code, if also the abortion takes place before implantation. Should the embryo survive the genome modification in utero and result in a birth, the question of criminal liability for the genetic modification would depend on whether, as just discussed, the prohibitions of the ESchG also apply to embryos in utero.

The legal framework of basic research using intended genome modification in born persons is even more complicated. Genome modifications that result in bodily harm of a born person can be construed as criminal acts. The Criminal Code punishes whoever physically assaults or damages the health of another person, or attempts to do so, with imprisonment of up to five years

---

[50] Embryos in vivo are covered because auf Sec. 1.1 No 6 ESchG: 'Anyone who removes an embryo from a woman before completion of implantation in the uterus, in order to transfer it to another woman or to use it for another purpose not serving its preservation shall be punished with imprisonment of up to three years or a fine.'

[51] It is legally unclear whether mere somatic intervention (including non-hereditary genetic interventions) aimed to preserve the embryo in question are prohibited, too. Section 2.1 of the ESchG prohibits only those interventions that do not serve the preservation of the embryo. Therefore, from this wording – considering the prohibition from Section 5.1 – it should be concluded that at least those somatic interventions that preserve the embryo in vitro are permissible.

[52] There are several stages of pain during labour. The 'first stage of labour' is temporally before the 'bearing down pains' (labour pains, coupled with the urge to push) and extends the upper sections of the birth path. The Federal Court of Justice (*Bundesgerichtshof*, BGH) rules that the first stage of labour initiates separation from the maternal organism (cf BGHSt 32, 194, 196).

or a fine.[53] 'Grievous bodily harm' is defined, inter alia, as harm that results in losing the ability to procreate and becoming paralysed, mentally ill or disabled, and is punished with imprisonment of at least one year and up to ten years.[54] Accidental genome modification could be construed as 'negligent assault', which is – in born persons unlike in embryos in vitro – also a crime punishable with imprisonment up to three years and a fine.[55] Consent would not be a valid defence. Under the Criminal Code consent can be given only as long as it does not violate 'good morals', and genome modification might be understood to be contrary to that.[56]

Another question is whether a germline genome modification in a born person is also prohibited and punishable by the ESchG. Since the ESchG was primarily enacted to regulate assisted reproduction rather than medical interventions in born humans, which are regulated by the Criminal Code, on a first view there is no need for additional liability for germline genome alterations by the ESchG. However, as said before, legal scholars also argue that the ESchG should capture germline genome alterations in born humans. For this purpose, the argument that the ESchG refers not only to prenatal embryos but also to (born) humans, who develop from fertilized egg cells, is made.[57]

c Basic Research: Some Special Situations

Since the ESchG was primarily enacted to regulate assisted reproduction rather than genetic engineering, legal loopholes – some less obvious than others – abound. Whether they have any actual practical consequence is hard to tell, since scientists have not sought to exploit them, not just for technical reasons but also probably for fear of incurring criminal prosecution.

I ARRESTED EMBRYOS First, basic research using 'arrested embryos' is, presumably, legally permissible. An 'arrested embryo' is an embryo that has stopped developing, for natural causes, usually because of chromosome damage. Because Section 8.1 requires that the 'fertilized human egg cell is ... capable of development', arrested embryos fall outside the scope of the ESchG.

Leaving aside the question of whether the possibility of conducting research on arrested embryos has any practical meaning, this research carries considerable risk of criminal liability. First, the scientists must be

[53] StGB, § 223.
[54] StGB, § 226.
[55] StGB, § 229.
[56] StGB, § 228.
[57] ESchG, § 2.1, § 8.3.

sure that the embryo they intend to work on is indeed an arrested embryo. Should it subsequently become evident that it was still viable, the scientists might incur in a criminal offence according to Section 2.1 of the ESchG (improper use of embryo by destroying it) and/or Section 5.1 (prohibited germline genome alteration, if germline cells of embryos had been altered). Second, the artificial production of arrested embryos by genetically altering germ cells, including alterations of zygotes, is prohibited by virtue of Section 5.1 of the ESchG, which, again, prohibits the artificial alteration of the genome of germline cells. Finally, Section 2.1 of the ESchG prohibits the genetic alteration of the zygote to transform it into an arrested embryo because this would be an improper use of a human embryo not aimed at its preservation.

II IMPORTING HUMAN EMBRYOS AND GERMLINE CELLS FOR RESEARCH ON GERMLINE GENOME EDITING Importing into Germany embryos created by the fusion of an egg cell and a sperm cell (i.e. 'sexually derived embryos'), and using them for research purposes, is forbidden. The prohibition to change the genetic identity of human embryos contained in Section 5.1 of the ESchG applies not only throughout the federal territory but also to embryos created abroad when they are imported into Germany. The same applies to imported germline cells. However, germ cells and germline cells can be imported into Germany from abroad for work in the field of basic research to change the genetic characteristics of these cells, as long as these germ cells are not used for fertilization,[58] and as long as these altered germline cells are not transferred to another embryo or human being.[59] In all such cases, no prior approval to import is needed.

Since the ESchG – as advocated here – does not prohibit research on, and use of, asexually generated entities, they can be created in Germany as well as be imported. Their import is also not regulated by the Stem Cell Act either, because the Stem Cell Act only regulates the import of pluripotent stem cells, derived from embryos, not the whole embryo.[60]

III SOMATIC CELL NUCLEAR TRANSFER New technical developments for the artificial production of embryos using asexual methods and the non-natural production of germ cells from stem cells create a further legal complication. Somatic cell nuclear transfer (SCNT) is a laboratory strategy for creating a viable embryo from a body cell and an egg cell asexually. The technique

[58] ESchG, § 5.4, no 1.
[59] ESchG, § 5.4, no 2.
[60] StZG, §§ 1 and 2.

258 *Timo Faltus*

consists of taking an enucleated oocyte (egg cell) and implanting a donor nucleus from a somatic (body) cell.

Now – as advocated here – Section 8.1 of the ESchG does not cover asexually produced embryos. Section 8.1 explicitly refers to the 'process of fertilization or the time of pronuclear fusion', neither of which occurs in asexually generated entities with embryo-comparable development.[61] Moreover, the prohibition of the destruction of embryos contained in Section 2.1 of the ESchG covers only embryos within the meaning of the ESchG. Since the ESchG provides for criminal sanctions, and since the Basic Law prohibits the analogous interpretation of criminal law,[62] one must stay within the four corners of that definition and, therefore, conclude that research on asexually generated SCNT entities with embryo-comparable development is not prohibited.

Second, as it was said, Section 5.1 of the ESchG prohibits the artificial alteration of the genetic information of a human germline cell, but Section 8.3 of the ESchG defines 'germline cell' as 'any cells that lead directly from the fertilized egg cell to the egg and sperm cells of the resultant human being and also egg cells from insertion or penetration of the sperm cell until the completion of fertilization by fusion of the nuclei'.[63] Are oocytes used for cell nuclear transfer germline cells in the sense of the ESchG? Before the cell nucleus is removed from the egg cell, egg cells are undoubtedly germ(line) cells within the meaning of the ESchG. But what effect has the removal of the nucleus on this legal assessment? Does the exchange of the cell nucleus of the egg cell by a cell nucleus of a somatic cell cause an inadmissible alteration of the genetic properties of the original egg cell? On the one hand, you can argue that when a germline cell is denucleated, it loses its germline cell properties and, therefore, its 'germline cell' legal status. The manipulation of the remains of a germline cell is not prohibited by Sections 5.1, 5.2 of the ESchG. On the other hand, it has to be seen that the genetic identity of an egg cell does not only consist of the DNA of the cell nucleus but also of the DNA of the mitochondria. An egg cell cannot exist functionally without mitochondria or nucleus. Therefore, it can be argued that removing the nucleus from an egg cell does not cause the cell to lose its legal status as a 'germline cell'. The manipulation of germline cells is prohibited by Section 5 of the ESchG.[64]

---

[61] This only means that there is no embryo in the sense of the law. In the scientific sense, it can be an embryo. The legislator does not have to follow scientific definitions.

[62] See, in this chapter, Section II.1.1 'The Federal Constitution and the 1985 Report of the Benda Commission', and Section II.1.2 'Laws'.

[63] ESchG, § 8.3.

[64] For a detailed analysis of the different opinions, see: S Deuring, 'Die 'Mitochondrienspende' im deutschen Recht' (2017) 35 MedR 215.

However, even if one follows the latter view, SCNT would not be covered by the prohibition of Section 5.1 ESchG as there is an exception to be considered. In any event, the prohibition contained in Sections 5.1, 5.2 of the ESchG does not apply as long as the modified cell is not used for fertilization.[65] Since no fertilization takes place within the SCNT (the transfer of the somatic nucleus into the previously enucleated oocyte cannot be understood as fertilization), the exception of Section 5.4. No. 1 of the ESchG should be considered. The legislator could have envisaged this possibility when the ESchG was drafted,[66] but did not include it in the statute. However, whether SCNT is a crime would have to be considered today as an interpretation by analogy of the ESchG. However, such method of statutory interpretation of criminal law is prohibited by the Constitution.

IV MITOCHONDRIAL REPLACEMENT THERAPY In mitochondrial replacement therapy (MRT), for example, the pronuclei of a fertilized oocyte in the development stage after penetration by the sperm but before fusion of the pronuclei is transferred to a previously enucleated oocyte at the same state of development.[67] Is it lawful in Germany? No, it is not. The exchange of the pronuclei is a prohibited alteration of the genetic identity of a germline cell according to Section 5.1 in conjunction with Section 8.3 of the ESchG. If, on the other hand, one exchanged the nucleus of one zygote for the nucleus of another zygote, this would be – equal to the SCNT – an improper use of an embryo, punishable under Section 2.1 of the ESchG.

V CYTOPLASMIC TRANSFER Another possibility of performing mitochondrial therapy is cytoplasmic transfer. In cytoplasmic transfer, the plasma (with mitochondria) of an oocyte is injected in the ovum of another woman, or, for example, mitochondria from somatic cells of the potential father. This is done to repair an oocyte with at least partially defective mitochondria. Does the addition of cytoplasm from a donor alter the genetic information of the germline cell sufficiently to trigger a violation of Section 5.1 of the ESchG? It depends on the extent to which mitochondrial genome determines the total

---

[65] ESchG, § 5.4, no 1.

[66] Even before the consultations for the ESchG started in 1989, the SCNT technique was also investigated in the sheep model. These studies are thus more than 10 years older than the publications on the SCNT sheep Dolly. However, in the process of adopting the ESchG, the SCNT has obviously not been properly reflected. SM Willadsen, 'Nuclear Transplantation in Sheep Embryos' (1986) 320 *Nature* 63.

[67] The further details and variants of these methods cannot be discussed here due to limited space. For an overview of different methods, see: DP Wolf, N Mitalipov, S Mitalipov, 'Mitochondrial Replacement Therapy in Reproductive Medicine' (2015) 21:2 *Trends in Molecular Medicine* 68.

genome of a person, because Section 5.1 protects 'the genetic information of a human germline cell', not just that of its mitochondria or its nucleus. Since there is no human without mitochondria, it is difficult to argue that altering the mitochondrial genome, or adding mitochondrial genome, does not change the genetic information of a human germline cell. Additionally, cytoplasmic transfer could also be a violation of Section 7.1 of the ESchG, which prohibits uniting 'embryos containing different genetic material to a cell cluster using at least one human embryo', or 'to join a human embryo with a cell that contains genetic information different from the embryo's cells and that will be able to differentiate further'.

VI TRIPRONUCLEAR EMBRYOS Tripronuclear embryos are embryos with three pronuclei. Tripronuclear embryos can occur naturally or be created artificially. They can occur naturally when two sperm cells, not just one, penetrate into one oocyte or when the polar bodies of the egg cell separate incorrectly. Or they can be created artificially by fertilizing an egg cell with two sperm cells, resulting in the so-called triploidy of diandric type.[68] Scientists in countries other than Germany have already started researching on genome modification of tripronuclear embryos.[69] Can this research be carried out lawfully in Germany or by German scientists to study germline genome alteration? Again, the answer lies in the definition of embryo and germline cell in Section 8 of the ESchG, the prohibition of artificial change of the genetic information in Section 5.1 and the prohibition of improper use of embryos in Section 2.1.

The answer is straightforward regarding the consecutive fertilization of an already fertilized egg cell (zygote). Since this endangers the further development of the embryo, it is already prohibited by Section 2.1 of the ESchG. Although tripronuclear zygotes can start dividing as normal embryos do, their abnormal collection of genes invariably generally causes them to stop developing when they only consist of a few cells. Thus, as soon as a tripronuclear zygote stops developing on its own it can be used for research purposes since, as we saw, 'arrested embryos' fall outside the scope of the ESchG.[70] Previous

---

[68] M. Grossmann and others, 'Origin of Tripronucleate Zygotes after Intracytoplasmic Sperm Injection' (1997) 12:12 *Human Reproduction* 2762; C Staessen and AC Van Steirteghem, 'The Chromosomal Constitution of Embryos Developing from Abnormally Fertilized Oocytes after Intracytoplasmic Sperm Injection and Conventional In-vitro Fertilization' (1997) 12:12 *Human Reproduction* 321.

[69] Puping Liang and others, 'CRISPR/Cas9-mediated gene editing in human tripronuclear zygotes' (2015) 6:5 *Protein & Cell* 363.

[70] See, in this chapter, Section III.2.c.i. 'Arrested Embryos'.

destruction would again be a prohibited improper use according to Section 2.1 of the ESchG.

The answer is also straightforward regarding the genetic alteration of germ cells and their use. The genetic modification of germ cells to produce an embryo (regardless of its set of chromosomes, since they are embryos in legal meanings) is prohibited by Sections 5.1 and 5.2 of the ESchG. As long as cell divisions are observed within the entities resulting from the fusion of the oocyte and sperm cell(s), these entities are embryos within the meaning of the ESchG.

VII STEM CELLS It is controversial whether the prohibition of the genetic modification of germline cells applies also to cells that do not originate from germline cells, such as induced pluripotent stem (iPS) cells or human embryonic stem cells.[71] In Germany, iPS cells may be produced and used without special permits. Human embryonic stem cells, which are obtained from blastocysts, cannot be produced, since this would be an embryo destruction prohibited by Section 2.1 of the ESchG. However, under certain conditions, which are mentioned in the Stem Cell Act, human embryonic stem cells may be imported from abroad to Germany and then be used in Germany for research purposes (but not for therapeutic purposes).[72]

A literal reading of Section 8.3 of the ESchG – the only reading not violating the prohibition of analogous interpretation of criminal law – leads to the conclusion that iPS cells, unless they have been derived from germline cells, as well as germ cells derived from iPS cells or from human embryonic stem cells, are not germ cells, legally speaking. Thus, they (as well as germ cells derived from stem cells) are not subject to the prohibition of genome modification contained in Sections 5.1 and 5.2 of the ESchG since these sections only cover germ cells which emerge from natural germline cells. Therefore, it can be argued that they can be genetically modified and used for fertilization without incurring the

---

[71] For an overview of the different views: T Faltus, *Stem Cell Reprogramming – The legal status, the legal handling, and the right-systematic significance of reprogrammed stem cells* [*Stammzellenreprogrammierung – Der rechtliche Status und die rechtliche Handhabung sowie die rechtssystematische Bedeutung reprogrammierter Stammzellen*] (Nomos 2016).

[72] See, T Faltus, *German Legislation Pertaining to International and Transnational Stem Cell Research: Guide to the Design of International and Transnational Research Projects* (Universitätsverlag Halle-Wittenberg 2013); T Faltus, 'No Patent – No Therapy: A Matter of Moral and Legal Consistency Within the European Union Regarding the Use of Human Embryonic Stem Cells' (2014) *Stem Cells Dev.* 23 Suppl 1, 56; U Storz and T Faltus, 'Patent Eligibility of Stem Cells in Europe: Where Do We Stand after 8 Years of Case Law?' (2017) 12:1 *Regen Med.* 37.

criminal sanctions of the ESchG. That being said, scholars have also argued the opposite.[73]

3 Regulation of Preclinical Research

Preclinical research – as far as animal testing is concerned – of germline modification technologies in non-humans is regulated by the Animal Protection Act (*Tierschutzgesetz* – TierSchG).[74] Section 7.2 of the TierSchG defines 'animal testing' as 'interventions or treatments for experimental purposes on animals, if they may be associated with pain, suffering or harm to these animals, [or] on animals that can lead to the birth or hatching of animals suffering pain, suffering or damage, or . . . on the genetic material of animals, if they may be associated with pain, suffering or damage to the mutagenic animals or their carrier animals'.

In general, animal experiments are subject to authorization by the local authorities in the respective state (*Bundesland*).[75] Authorization and notification are regulated by the Animal Protection Act and in the Animal Protection – Experimental Animal Ordinance (*Tierschutz-Versuchstierverordnung* – TierSchVersV).[76] Applicants must provide a scientifically substantiated explanation of why the trials are necessary in the planned form and why other methods are not sufficient. When deciding on the approval of animal experiments, the competent state authority is assisted and advised by a commission of experts, the so-called Ethics Committee, consisting of representatives of animal protection organizations and science.[77] If animal testing is required by law, such as for the development of medicinal products, a simple

---

[73] Deutscher Ethikrat (German Ethics Council), *Stem cell research – new challenges for the ban on cloning and treatment of artificially created germ cells? [Stammzellforschung – neue Herausforderungen für das Klonverbot und den Umgang mit artifiziell erzeugten Keimzellen?]* (Deutscher Ethikrat, 2014), 7.

[74] Animal Protection Act [Tierschutzgesetz] (TierSchG) [amended by Notice of 18 May 2006 and later amended by Article 141 of the Act of 29 March 2017] [Tierschutzgesetz in der Fassung der Bekanntmachung vom 18. Mai 2006 (BGBl. I S. 1206, 1313), das zuletzt durch Artikel 141 des Gesetzes vom 29. März 2017 (BGBl. I S. 626) geändert worden ist.].

[75] TierSchG, § 8.1.

[76] Animal Protection – Experimental Animal Ordinance [Tierschutz-Versuchstierverordnung] (TierSchVersV) [amended by Article 394 of the Ordinance of 31 August 2015] [Tierschutz-Versuchstierverordnung vom 1. August 2013 (BGBl. I S. 3125, 3126), die zuletzt durch Artikel 394 der Verordnung vom 31. August 2015 (BGBl. I S. 1474) geändert worden ist.]. The TierSchVersV serves to implement the EU Directive 2010/63/EU on the protection of animals used for scientific purposes into German law.

[77] TierSchG, § 15.

notification prior to the test suffices or none at all.[78] However, if the test is particularly serious for animals or involves primates, authorization is required.[79]

Since research on human germline genome modification is limited in Germany, these regulations are not really relevant for the subject matter discussed. However, they might become relevant in the development of other therapies, possibly in the context of research into unwanted side effects of germline changes.

4 Regulation of Clinical Research

Clinical studies of germline genome modification are prohibited by virtue of Sections 5.1 and 5.2 of the ESchG,[80] by the prohibition regarding bodily harm according to the Criminal Code,[81] and are excluded by Article 90 of the EU Regulation No 536/2014.[82] The prohibition of clinical research includes also artificially produced germline cells with a genetic identity differing from naturally occurring germ cells.[83]

5 Regulation of Clinical Applications

As it was said, the ESchG was drafted in the late 1980s with the aim to regulate artificial reproductive technology. It applies mainly to in vitro processes. Assuming it applies to in vivo processes too, then clinical applications of human germline genome modification is, again, prohibited by Section 5.1 of the ESchG. Also, genome modification that causes bodily harm is definitely prosecutable under the Criminal Code.[84] Unintentional modifications of the

---

[78] TierSchG, § 8a.1.

[79] TierSchG, § 8a.2.

[80] See, in this chapter, Section III.2.a 'Basic Research using Genome Modification on Gametes', and III.2.b 'Basic Research Using Genome Modification on Germline Cells and Embryos'.

[81] Ibid.

[82] EU Regulation No 536/2014 of the European Parliament and of the Council of 16 April 2014 on clinical trials on medicinal products for human use, and repealing Directive 2001/20/EC. Although the regulation is already in force, it is expected that it will not be applied until the 2020 due to the technical requirements for its application. In addition, clinical studies on the modification of the germline were already prohibited by Article 9.6 of the previous directive 2001/20/EC on clinical trials.

[83] However, since germ cells with different genetic identities already exist naturally, the difference in genetic identity mentioned here in artificial germline cells compared to naturally occurring germ cells can only refer to the fact that such a genetic trait, which cannot occur in naturally developing germline cells, was artificially inserted.

[84] See, in this chapter, Section III.1.b 'Genome Editing'.

genome, as in the case of side effects of a treatment, are in general not prosecutable; if it is a medically acceptable treatment,[85] although, depending on circumstances, it could be construed as 'negligent assault', which is also a crime.[86] Manufacturing and placing on the market a somatic gene therapy medicinal product without the regulatory authorizations required by pharmaceutical legislation is criminally punishable, too.[87]

In clinical practice, PGD as well as sex selection can be construed as an intervention on the germline genome, since they result in eliminating an unwanted genetic line. In principle, Section 3a.1 of the ESchG forbids PGD and sanctions it with a year of imprisonment or a fine. However, there are exceptions. It is permissible if there is a high risk of a serious hereditary disease for the offspring due to the genetic disposition of the woman from which the egg originates and/or to the man from whom the sperm cell originates. In addition, it is permissible if it is used to detect a serious damage to the embryo that is likely to result in stillbirth or miscarriage.[88]

Sex selection via preselected sperm cells is a crime, too.[89] Artificially fertilizing an egg with a sperm cell selected according to the sex chromosome contained therein is prohibited.[90] However, this prohibition does not apply if the selection of the sperm cell serves to protect the potential child from Duchenne muscular dystrophy or a similarly serious sexually transmitted hereditary disease.[91]

IV CURRENT PERSPECTIVES AND FUTURE POSSIBILITIES

Despite its age, the current regulatory framework does not show any significant loopholes through which methods of reproductive medicine treatments and

---

[85]　ESchG, § 5.4, no 3.

[86]　StGB, § 229.

[87]　§ 4b, no 21 Gesetz über den Verkehr mit Arzneimitteln (Arzneimittelgesetz – AMG), Ausfertigungsdatum: 24.08.1976, in der Fassung der Bekanntmachung vom 12. Dezember 2005, zuletzt geändert am 6. Mai 2019 = Act on the Marketing of of Medicinal Products (Medicinal Products Act – AMG). Date of issue 24.08.1976, in the version of the announcement of December 12, 2005, last amended May 6, 2019.

[88]　ESchG, § 3a(2).

[89]　ESchG, § 3.1. Sex selection of in vitro embryos is always prohibited. It would either be a prohibited improper use of in vitro embryos or be a prohibited destruction of in vitro embryos according to § 2.1 ESchG.

[90]　ESchG, § 3.1.

[91]　ESchG, § 3.2. Prior ethical counselling by an ethics committee is not necessary for sex selection via pre-sorted sperm cells. However, the disease threatening the child has to be recognized as being of appropriate severity by the authority responsible according to state (*Bundesland*) legislation.

gene therapy in use today might slip. However, science never stops, loopholes as described above might appear and society changes. German society seems content with the current prohibitions and the associated penalties. So far, there have been only very few prosecutions for violations of prohibitions of the ESchG, probably because the current legal framework, which relies on criminal law, deters attempts to test it. Until now, there has been only little observable lobbying activity or other concerted political push mostly by scientists and physicians asking for the liberalization or at least a modernization of the existing regulatory framework. It is rather that, for example, some civil and church societies advocate for the maintenance of the current regulation or even for further limitations.

However, the rapidly changing biotechnological landscape is begging the question of whether it is time to re-evaluate the prohibition of genome modification of germline cells and embryos, in particular considering the possibility to create germline cells artificially, and to alter their genome. It is not surprising that in view of the evolving technical possibilities scholars who study the scientific, legal, social and ethical implications of biotechnological developments seem to feel a need to re-evaluate the current regulatory framework.[92]

Technology advances and legal and ethical views change. New insights into biology, medicine and technical safety require new assessments of their benefits and their dangers. Insisting on the intangibility of today's moral standards and the connected legislation without further discussion is an unjustified moral imperialism, one that potentially denies potential technical improvements to life due to outdated technological and social evaluations. Of course, after appropriate revaluation, we might conclude that the current regulations might not need to be modified, but 'the slippery slope argument', according to which the first step inevitably leads to further steps and finally to ruin and which too often stifles debate, assumes a priori that society and social discourse lack the ability to differentiate between right and wrong.

Currently, the German Federal Ministry of Education and Research (BMBF) is funding several interdisciplinary Ethical, Legal and Social Implications (ELSI) research projects on genome editing.[93] Some of these are about genome therapy. The results of these research projects will provide stimuli and arguments for this discussion to reassess the legal framework of genome modification of germline cells.

---

[92] For an overview on proposal for changing the current legislation, see: Faltus (n 70) 492–501.

[93] For an overview of this area of research funding by the BMBF, see: Federal Ministry of Education and Research, 'Design Research: Bioethics' (Bundesministerium für Bildung und Forschung, 2018) www.gesundheitsforschung-bmbf.de/de/bioethik.php accessed 02 August 2019.

9

The Regulation of Human Germline Genome Modification in Belgium

Guido Pennings

I INTRODUCTION

Compared to most European states, Belgium stands out for its rather liberal attitude toward research on human embryos. Together with Sweden and the United Kingdom, Belgium is among the few European states that, under certain conditions, allow the creation of embryos for research purposes. It is also one of few European states that has neither signed nor ratified the Oviedo Convention precisely because it deems it excessively restrictive on the subject of embryo research.[1]

Any discussion of how embryo research and its translation into clinical applications are regulated in Belgium must take into consideration how the overall legal and political system is structured. Belgium is a federal state, composed of Communities and Regions.[2] The three Communities are the Flemish, French, and German communities. The Regions are Flanders (Flemish speaking), Wallonia (French speaking), and the capital city of Brussels. A particular complexity of the Belgian federal model is that the territorial areas of the Communities and Regions to a certain extent overlap and are not identical.

Communities and Regions, as political entities, are responsible for different subject matters. The Federal State has competence over justice, defense, federal police, social security, public debt and other aspects of public finances, nuclear energy, and state-owned companies. It is also responsible for the obligations of Belgium and its federalized institutions toward international

---

[1] Convention for the Protection of Human Rights and Dignity of the Human Being with regard to the Application of Biology and Medicine (opened for signature April 4, 1997, entered into force December 1, 1999) ETS No. 164 (the Oviedo Convention). On the Oviedo Convention, see, in this volume, Part 3, Section III, Chapter 6.

[2] Constitution of Belgium, 1994, art. 1.

organizations. It controls substantial parts of public health, home affairs, and foreign affairs. Research on human embryos, including germline modification, falls under the competence of the Federal State.

Communities exercise competences only within linguistically determined geographical boundaries: culture, education, and the use of the relevant language. They have also jurisdiction over matters less directly linked to culture and language, such as health policy (curative and preventive medicine) and social security.

Finally, Regions have authority in fields connected with their territory in the widest meaning of the term, such as the economy, employment, agriculture, water policy, housing, public works, energy, transport, the environment, and town and country planning.

II THE REGULATORY FRAMEWORK

1 *Key Laws*

Currently, Belgium regulates research on human embryos, including germline modification, mainly through two laws: the 2003 Law regarding Research on Embryos In Vitro (Embryo Research Law)[3] and the 2007 Law regarding Medically Assisted Reproduction and the Disposition of Embryos and Gametes (MAR Law).[4] Clinical research is regulated by the 2004 Law regarding Experiments on the Human Person[5] and by the 2017 Law on Clinical Trials.[6]

As far as international and European law is concerned, Belgium is a member of the European Union (EU), and therefore it is subject to EU directives and regulations.[7] In this regard, it should be recalled here that the

---

[3] Law regarding Research on Embryos In Vitro (Loi relative à la recherche sur les embryons in vitro – Wet betreffende het onderzoek op embryo's in vitro) 2003 (Embryo Research Law). The law was amended twice on May 30, 2005, and on January 6, 2014. The amendments regarded only Article 9 concerning the Federal Commission for Medical and Scientific Research on Embryos In Vitro.

[4] Law regarding Medically Assisted Reproduction and the Disposition of Embryos (Loi relative à la procréation médicalement assistée et à la destination des embryons surnuméraires et des gametes – Wet betreffende de medisch begeleide voortplanting en de bestemming van de overtallige embryo's en de gameten) 2007.

[5] Law regarding Experiments on the Human Person (Loi relative aux expérimentations sur la personne humaine – Wet inzake experimenten op de menselijke persoon) 2004.

[6] Law regarding Clinical Trials with Drugs for Human Consumption (Loi relative aux essais cliniques de médicaments à usage humain – Wet betreffende klinische proeven met geneesmiddelen voor menselijk gebruik) 2017.

[7] On the relevant EU norms, see, in this volume, Part 3, Section III, Chapter 6.

storage of tissues and cells is regulated by Directive 2004/23 of the European Parliament and of the Council of March 31, 2004, on setting standards of quality and safety for the donation, procurement, testing, processing, preservation, storage, and distribution of human tissues and cells. The use of animals in experiments is governed by Directive 2010/63/EU of the European Parliament and of the Council of September 22, 2010, on the protection of animals used for scientific purposes; and clinical trials are regulated at the European level by Regulation (EU) No. 536/2014 of the European Parliament and of the Council of April 16, 2014, on clinical trials on medicinal products for human use.[8] Article 90 of that regulation reads: "No gene therapy trials may be carried out which results in modifications to the subject's germline genetic identity."[9] It is still unclear whether this article will have much impact on the development of germline modification technology in Belgium in particular and the European Union in general, since it can be argued that no medicinal product is created; it is rather a process or technique being applied.[10] Future litigation before the European Court of Justice or intervention by the EU Commission might clarify the matter. Remarkably, neither the 2004 Law regarding Experiments on the Human Person nor the 2017 Law on Clinical Trials reproduces Article 90 of the EU Directive. The 2004 Law regarding Experiments on the Human Person contains the rather cryptic Article 17 that states: "No clinical trials for gene therapy are allowed that lead to the modification of the genetic identity of the participant; the experiment should in other words not be directed at the selection or enhancement of non-pathologic genetic characteristics of the human species."[11]

Belgium is also a member of the Council of Europe but, so far, has decided not to ratify the Convention for the Protection of Human Rights and Dignity of the Human Being with regard to the Application of Biology and Medicine (the Oviedo Convention).[12] I will explain the reasons for Belgium not to do so later

---

[8]   European Parliament and of the Council Regulation (EU) No. 536/2014 of the European Parliament and of the Council of April 16, 2014, on clinical trials on medicinal products for human use, and repealing Directive 2001/20/EC [2014] OJ L158/1-L158.76.

[9]   Council Regulation (EU) No. 536/2014, art. 90.

[10] I De Miguel Beriain, "Legal Issues Regarding Gene Editing at the Beginning of Life: an EU Perspective" (2018) 12 *Regenerative Medicine* https://doi.org/10.2217/rme-2017–0033 accessed October 13, 2018.

[11] Law regarding Experiments on the Human Person (Loi relative aux expérimentations sur la personne humaine – Wet inzake experimenten op de menselijke persoon) 2004.

[12] Convention for the Protection of Human Rights and Dignity of the Human Being with regard to the Application of Biology and Medicine: Convention on Human Rights and Biomedicine (opened for signature April 4, 1997, entered into force December 1, 1999), ETS No. 164 (the Oviedo Convention).

on. For now, it suffices to say that, unless the Convention is fundamentally amended, it is unlikely that Belgium will ratify it. Finally, Belgium has ratified the Covenant on Economic, Social and Cultural Rights,[13] on April 21, 1983,[14] as well as the Optional Protocol,[15] on May 20, 2014, giving individuals the chance to bring complaints for violations of the Covenant before the Committee on Economic, Social and Cultural Rights.[16] Within the Belgian legal system, it is widely accepted that ratified treaties have precedence over laws, federal and below.[17] It is more controversial whether they also prevail over the Constitution, but since, so far, no provision in the Belgian Constitution touches upon research on embryos or, more generally, on biomedical research, I will not address the question.

2 Key Bodies and Institutions

When it comes to research involving human embryos, in Belgium there are three main bodies that need to be discussed: local ethics committees; the Federal Commission for Medical and Scientific Research on Embryos In Vitro (Federal Commission); and the Advisory Committee on Bioethics. Scientific research is mostly funded by the National Fund for Scientific Research.

i Local Ethics Committees

The main task of local university ethics committees is to review proposals for clinical trials. They are expert independent bodies, linked to a faculty

---

[13] International Covenant on Economic, Social and Cultural Rights (adopted December 16, 1966, entered into force January 3, 1976) 993 UNTS 3 (ICESCR).

[14] Status of Treaties, "International Covenant on Economic, Social and Cultural Rights" (United Nations Treaty Collection, August 24, 2018) https://treaties.un.org/pages/ViewDetails .aspx?src=IND&mtdsg_no=IV-3&chapter=4&clang=_en accessed August 24, 2018.

[15] Optional Protocol to the International Covenant on Economic, Social and Cultural Rights (adopted December 10, 2008, opened for signature September 24, 2009) A/RES/63/117.

[16] Status of Treaties, "Optional Protocol to the International Covenant on Economic, Social and Cultural Rights" (United Nations Treaty Collection, August 24, 2018) https://treaties.un.org/pa ges/viewdetails.aspx?src=ind&mtdsg_no=iv-3-a&chapter=4&lang=en accessed August 24, 2018.

[17] Cour de Cassation, May 27, 1971, Pasicrise (Pas.) 1971, 1, 886; Arresten van het Hof van Cassatie at 1971, p. 959. For doctrinal opinions in this regard, see, D Heirbaut, M E. Storme, "The Belgian Legal Tradition: From a Long Quest for Legal Independence to a Longing for Dependence" (2006) 14 *Eur. Rev. Private L.* 645, 665–666; B Peeters, "Interpretation of Double Tax Conventions" (1993) Kluwer Vol. LXXVIII.a 221, 223–223; F Swennen, Y-H Leleu, "Belgium" (2011) 19 *Am. U. J. Gender Soc. Pol'y & L.* 57, 59, para. 4; R Torfs, "The Permissible Scope of Legal Limitations on the Freedom of Religion or Belief in Belgium" (2005) 19 *Emory Int'l L. Rev.* 637, 638; S Snacken, "Penal Policy and Practice in Belgium" (2007) 36 *Crime & Just.* 127, 206.

of medicine or to a "scientific association of general practitioners" and licensed by the Ministry of Public Health, Safety of the Food Chain and Environment.[18] Each faculty of medicine or scientific association of general practitioners can have one, and one only.[19] Currently, there are 162 ethics committees in Belgium.[20] Embryo research can be done only in a university program that is recognized as a center for reproductive medicine or a center for medical genetics. There are 17 so-called B centers in Belgium, that is, centers for reproductive medicine that are licensed according to the standards set by the Crown Order of February 15, 1999, on the so-called Care Programs for Reproductive Medicine.[21] There are also eight licensed centers of medical genetics that are situated in hospitals with B centers. Only a few of them currently do embryo research.[22]

ii Federal Commission for Medical and Scientific Research on Embryos In Vitro

The Federal Commission (*Commission fédérale pour la recherche médicale et scientifique sur les embryons in vitro – Federale Commissie voor medisch en wetenschappelijk onderzoek op embryo's in vitro*) consists of 14 experts on the medical, legal, ethical, societal, and scientific aspects of research on human embryos. Its mission is to guarantee the application of the Embryo Research Law and to formulate recommendations for law-making initiatives as well as for the application of the law by the local ethics committees.

Under the Embryo Research Law, all research that involves embryos in vitro can be performed only if approved by the ethics committee of the university where the research is done.[23] Research projects approved by local ethics committees are then subject to review by the Federal Commission.[24] It is noteworthy that, once approval has been obtained locally, approval by the Federal Commission is presumed. The project can be blocked only by vote of

---

[18] Royal Decree of August 12, 1994, Belgian Official Gazette (Belgisch Staatsblad), September 27, 1994. This Decree contains a legal obligation to establish an ethics committee in every Belgian general and psychiatric hospital. Royal Decree of August 12, 1994, B.S. September 27, 1994.

[19] Herman Nys, "Part II. The Physician: Patient Relationship" (Suppl. 71, Kluwer Law International, the Netherlands 2012) 99–189.

[20] Ibid., 174.

[21] March 25, 1999, Zorgprogramma's "reproductieve geneeskunde," B.S. March 25, 1999, art. 2.

[22] Vrije Universiteit Brussel, Université libre de Bruxelles, Ghent University, Antwerp University and Catholic University of Leuven.

[23] Embryo Research Law, art. 7.

[24] Ibid., art. 9 and 10.

two-thirds of the members of the Federal Commission within two months after the protocol has been presented to them.[25]

iii The Belgian Advisory Committee on Bioethics

The Belgian Advisory Committee on Bioethics was established on January 15, 1993, by an agreement between the Federal Government, the French-speaking Community, the Flemish Community, the German-speaking Community, and the Joint Commission for Community Matters.[26] The Committee is composed of 43 members with an equal representation of men and women, Dutch and French speaking, and members from scientific and medical domains and from philosophical, legal, and human sciences. The Belgian Advisory Committee's mission is twofold: to provide advisory opinions on the problems raised by research and research applications in the fields of biology, medicine, and healthcare, and to inform the public and the government authorities about them.[27]

Its advisory opinions are nonbinding. However, they carry a certain weight in shaping policy discussions. More often than not, the Advisory Committee adopts its opinions by majority vote, not by consensus. Since the Committee is designed to represent the various ideological and philosophical movements in the Belgian society in a balanced way, consensus is rarely reached. Rather, the Committee's statements aim at reflecting the various positions present among its members.

To date, the Committee has given two opinions that are relevant for the issue of human germline modification. The first was issued in 1997, at the time the Convention of Oviedo was adopted,[28] and reflects the diversity of views among its members on Article 13, on germline gene therapy, and Article 18, on research on embryos, of the Oviedo Convention. Regarding the former, some members argued that the current prohibition on germline modification contained in the Oviedo Convention should not be an obstacle to Belgium's ratification because the Convention has an amendment procedure and can be reconsidered five years after its entry into force.[29] Others

[25] Ibid.
[26] Belgisch Staatsblad (Belgium's Official Gazette), 12.05.1993, pp. 10824–10828.
[27] Ibid.
[28] Belgian Advisory Committee 1997, "Opinion No. 2 of 7 July 1997 concerning the Convention Human Rights and Biomedicine of the Council of Europe" (Belgian Advisory Committee on Bioethics, July 7, 1997) www.health.belgium.be/sites/default/files/uploads/fields/fpshealth_theme_file/opinion_2_web.pdf accessed August 24, 2018.
[29] Oviedo Convention, art. 32.

believed, correctly, that agreeing on amendments would turn out to be difficult, if not impossible, and that Belgium should not bind itself to a treaty that prohibits germline modification. The discussion regarding Article 18 focused mainly on the prohibition on creating embryos for research. For some members, this prohibition expressed respect for human life and a rejection of a utilitarian framework, while others considered the possibility of embryo creation as necessary for new developments in medically assisted reproduction.

The second opinion relevant to human germline modification was issued in 2005 and regards the topic of somatic and germinal line gene modification.[30] It discussed at length the distinction between therapeutic and nontherapeutic applications and the concept of eugenics. The two topics are connected in the Embryo Research Law because research and acts with a "eugenic purpose" are forbidden. The law defines "eugenic purpose" as "selection or improvement of non-pathological characteristics of the human species."[31] The only exception is selection of preimplantation embryos on their HLA (human leukocyte antigen) type. This selection is known as the "savior sibling," where embryos are selected to be compatible with an existing sibling who needs a transplantation of hematopoietic stem cells.[32] Since this selection has no medical benefit for the child that will be born from this embryo, this exception was added to the MAR Law.[33] Unsurprisingly, no consensus was reached within the Advisory Committee on the germline genome modification and the concept of eugenics. The three main positions were: those who believe that germline modification should be allowed with decisions made on a case-by-case basis, assuming that the risks are minimal; those who are generally against germline modifications based on risks related to eugenics; and those who are not in principle against germline modification but do not see this as a priority or a real political issue. All positions broadened the arguments to societal issues such as society's views on disability, the future of medicine, and

---

[30] Belgian Advisory Committee 2005, "Opinion No. 33 of 7 November 2005 concerning somatic and germinal line gene modification with a therapeutic and/or enhancement purpose" (Belgian Advisory Committee on Bioethics, November 7, 2005) www.health.belgium.be/site s/default/files/uploads/fields/fpshealth_theme_file/opinion_33_web.pdf accessed August 24, 2018.

[31] Embryo Research Law, art. 5.4.

[32] G Pennings, R Schots, I Liebaers, "Ethical Considerations on Preimplantation Genetic Diagnosis for HLA Typing to Match a Future Child as a Donor of Haematopoietic Stem Cells to a Sibling" (2002) 17 *Hum Reprod* 534.

[33] Law regarding medically assisted reproduction and the destination of surplus embryos and gametes (Loi relatif à la procréation médicalement assistée et à la destination des embryons surnuméraires et des gametes) 2007 (MAR Law), art. 68.

problems of regulation. All in all, it looks as if no member of the Committee wanted to be cast as being firmly against germline modification.

iv The National Fund for Scientific Research

Funding for embryo research is mostly provided by the National Fund for Scientific Research. The National Fund has no a priori limitations on the kind of research it supports. Obviously, all research on human embryos must comply with the laws and be approved by the committees mentioned above. Universities or hospitals can also decide to fund embryo research with their own money, but they evidently prefer to obtain external money.

It is very difficult to estimate how much funding is available and has been disbursed for this kind of research, since some research is funded by general research money. Moreover, laboratory equipment and such go through different channels of funding. Be that as it may, I am not aware of anyone who, in Belgium, is currently working on human embryos for germline modification.

III SUBSTANTIVE PROVISIONS

1 *The Embryo Research Law and the Law Regarding Medically Assisted Reproduction and the Disposition of Embryos and Gametes*

The Embryo Research Law was drafted at the beginning of the 2000s and adopted in 2003.[34] The MAR Law was drafted in 2005 and adopted in 2007.[35] In Belgian politics, especially regarding bioethical issues such as medically assisted reproduction and euthanasia, an important distinction exists between Catholics and Humanists (the latter are split between Socialist and Liberal parties). These topics were largely avoided by every government that included the centrist Christian Democracy Party and the Socialists. These parties were present in every after-war government until 1999. However, in 1999, a new coalition was constituted without the Christian Democrats.[36] Between 1999 and 2008, a number of laws regarding ethically sensitive issues such as experiments on embryos in vitro, medically assisted reproduction, and euthanasia were introduced by a coalition of Socialists, Liberals, and Greens.[37]

---

[34] Embryo Research Law, art. 4.5.
[35] MAR Law, art. 68.
[36] Verhofstadt I Government.
[37] I Nippert, *Präimplantationsdiagnostik – ein Ländervergleich'*, Berlin: *Friedrich Ebert Stiftung* (Stabsabteilung der Friedrich-Ebert-Stiftung 2006).

Article 3 of the Embryo Research Law stipulates that "research on embryos" is allowed as long as the following conditions are met: research must (1) have a therapeutic objective or the aim to advance knowledge about fertility, infertility, organ or tissue transplants, the prevention or treatment of diseases; (2) be based on the most recent scientific knowledge and meet the requirements of a correct methodology of scientific research; (3) be carried out in an approved laboratory connected to an academic program of reproductive healthcare or human genetics and in the appropriate material and technical circumstances (research in reproductive healthcare programs that are not based in a university can be performed only in collaboration with an academic institution); (4) be carried out under the supervision of a medical specialist or a doctor of science and by persons having the required qualifications; (5) be performed on an embryo during the first 14 days of development, excluding the freezing period; and (6) have no alternative of comparable effectiveness.[38]

Belgium allows the creation of embryos for research, but only as a last resort. Article 4 prohibits the creation of embryos for research except when the research goal cannot be achieved by other means, including resorting to supernumerary embryos.[39] This is a significant deviation from Article 18 of the Oviedo Convention, which prohibits the creation of human embryos for research purposes, with no exceptions. The departure from the standard set by the Oviedo Convention is deliberate. The Belgian legislators rejected the fundamental moral distinction, implicitly made by the drafters of the Oviedo Convention, between embryos created for research and supernumerary embryos.[40] As later discussed, under Belgian law, both research embryos and supernumerary embryos can be used in research, albeit under different conditions. Because of the rejection by Belgian lawmakers of the moral distinction between the two kinds of embryos, the State Council (Conseil d'État) stated, in the aftermath of the enactment of the Belgian Embryo Research Law, that, should Belgium ever ratify the Oviedo Convention, it could do so only with a reservation to Articles 13 and 18.[41]

Further constraints can be found in Article 5. It prohibits (1) the transfer of embryos used in research in humans – unless the research is for the benefit of the embryo itself, with a therapeutic objective, or mere observation that does not harm the embryo's integrity;[42] and (2) certain activities (i.e., reproductive

[38] Embryo Research Law, art. 3.
[39] Embryo Research Law, art. 4.
[40] Belgian Senate, *Legislative Proposal Concerning the Research on Embryos in Vitro* (Explanation 2–695/1, 2001) 4–6.
[41] State Council, *Advice of the State Council* (33.641/3, 2002), Belgian Senate (2–695/16, 2002).
[42] Embryo Research Law, art. 5.2.

cloning, eugenics, sex selection for nonmedical reasons, implantation of human embryos in animals, and creation of chimaeras or hybrids).[43] It is commonly accepted that this list is an exhaustive one and not merely illustrative. This interpretation stems from the basic premise that everything that is not prohibited by the Embryo Research Law is allowed. The legislator explicitly prohibited germline therapy for eugenic purposes and thus allowed corrective germline modification. This conclusion finds support in the parliamentary debates preceding the law.[44] The legislative history of the Embryo Research Law shows that the only distinction the drafters considered necessary was between "medical" and "nonmedical" research. It is also clear that the distinction is not always clear-cut and the members of the Parliament were aware of this problem.

In sum, based on the analysis of Articles 3, 4, and 5 of the Embryo Research Law, it is reasonable to conclude that in Belgium research using germline modification in human embryos is permitted as long as the conditions set forth in Article 3 are met and no genetically modified embryo is implanted. Contrary to most other countries in Europe, Belgium allows the creation of embryos for research when the research goal cannot be reached by research on supernumerary embryos and when the research meets the criteria stipulated by the Embryo Research Law.[45]

Just to be clear, the Embryo Research Law draws no categorical difference between research embryos and supernumerary embryos. Yet, the lawmakers implicitly drew a distinction when they adopted the "subsidiarity principle" (i.e., research must be carried out on spare embryos, if possible, and only on research embryos, if necessary). This suggests that they believed that there is a gradual difference between using existing embryos for research and creating embryos for research.[46] When applying for permission to do research on human embryos, the researchers are asked to specify in their project proposals whether embryos will be created,[47] and the Federal Commission verifies whether sufficient reasons have been given to justify the creation of embryos. The fact that the research cannot be performed by using other methods or material, such as animal embryos or human embryos that are too far developed, would be a sufficient reason.

---

[43] Ibid., art. 5.1, 5.4 and 5.5.

[44] Senate, *Parliamentary Documents* (2001–01, no 2–695/1, 2000), 5–6. See also Special Commission, *Preliminary Report* (discussion of amendment no 127) 154.

[45] Embryo Research Law, art. 4.1.

[46] G Pennings, A Van Steirteghem. "The Subsidiarity Principle in the Context of Embryonic Stem Cell Research" (2004) 19 *Hum Reprod* 1060.

[47] Embryo Research Law, art. 7.1.

When drafting the Embryo Research Law, the lawmakers were fully aware that, because of technical developments such as in vitro maturation and freezing techniques, it would be necessary to create some embryos.[48] They turned out to be correct. From 2007 to 2015, about 16,000 embryos have been used in 36 research projects.[49] About 66% were fresh supernumerary embryos (i.e., embryos that have been labeled as unfit for transfer or for freezing, for example due to chromosomal abnormalities, abnormal fertilization, or, predominantly, inferior quality); 26% were frozen supernumerary embryos (i.e., embryos that have been cryopreserved with the aim of future transfer but where the parental project is either accomplished or abandoned or the legally determined period of cryopreservation has expired). Only 8% (1,236) were created ad hoc for research. It is a small percentage and it stayed small because of a practical rather than legal constraint: a shortage of oocytes. Oocytes become available mainly after retrieval during infertility treatment. Four categories of oocytes can be identified: (1) oocytes that are immature at the moment of retrieval but that mature later on; (2) mature oocytes in the absence of sperm (although this source has dried up since oocytes can be vitrified); (3) oocytes of egg bank donors who are primarily interested in donating them for reproduction but who may agree to donate some for research; and (4) oocytes of participants in studies of in vitro maturation of oocytes.[50] However, all these sources are relatively limited. The donors who originally donated for reproduction and agree to direct some to research receive the same financial compensation (maximum 2,000 euros) as donors donating for reproduction.

In all cases, donors must consent to the use of their oocytes in research. This was a concern at the time the law was drafted. In fact, the lawmakers added a paragraph to the article allowing the creation of embryos for research, saying that oocyte donors must meet certain age requirements, must give written informed consent, and hormonal stimulation must be scientifically justified.[51] These conditions were considered sufficient to prevent exploitation of women in the context of oocyte donation for research.

Embryo procurement is regulated both by the Embryo Research Law and the MAR Law. The Embryo Research Law requires specific consent. This means that couples who intend to donate embryos to research must be

---

[48] Belgian Senate, *Parliamentary Documents* (2001–01, no 2–695/1, 2000) 5–6.
[49] G Pennings, S Segers, S Debrock, B Heindryckx, V Kontozova-Deutsch, U Punjabi, H Van De Velde, A Van Steirteghem, H Mertes, "Human Embryo Research in Belgium: An Overview" (2017) 108 *Fertil Steril* 96, 98.
[50] Ibid. 106.
[51] Embryo Research Law, art. 4.2.

informed, among other things, of the purpose and methodology of the research that is expected to be performed on the donated embryos.[52] The MAR Law regulates the disposition of supernumerary embryos. It offers couples three options for the disposition of their embryos at the end of the storage period: donation to others, destruction, or donation to research.[53] Couples must express their preference as part of the consent process that takes place before in vitro fertilization/intracytoplasmic sperm injection (IVF/ICSI) treatment begins.[54]

The MAR Law requires specific consent, too. However, practical circumstances might make it difficult to meet the specific consent requirement. The standard storage period for supernumerary embryos is five years. However, due to rules by funding agencies, most research projects last four years. This creates a gap between the length of the consent under the MAR Law and the usual length of research protocols. This means that the consent obtained at the outset of IVF/ICSI treatment may not cover research protocols yet to be approved. The result is that patients cannot be informed in advance of research projects that might take place toward the end of the storage period. To address this dilemma, the Federal Commission decided that generic consent is sufficient for cryopreserved embryos.[55] Patients can indicate general areas of research for which they accept their embryos to be used. On the other hand, specific consent is required for "fresh supernumerary embryos," that is, supernumerary embryos that have yet to be frozen. Consent to use these embryos in research must indicate the specific research project for which the embryos will be used.[56]

At the moment, and to our knowledge, no protocol for research on germline genome editing has been submitted for approval. However, to anticipate future developments, the informed consent forms for frozen supernumerary embryos will be amended to include a special entry on genome editing. In the new form, patients will be given the possibility to indicate whether they want their embryos to be used for research on genome editing.

---

[52] Ibid., art. 8.

[53] G Pennings, "Belgian Law on Medically Assisted Reproduction and the Disposition of Supernumerary Embryos and Gametes" (2007) 14 *Eur J Health L* 251.

[54] MAR Law, art. 7–8. To be noted that patients can revisit their decision and change the destination of their embryos at any time as long as both partners agree. Still, this article conflicts with art. 8 of the Embryo Research Law that states that objection by one of the donors (providers of the gametes) is sufficient to prevent use for research.

[55] Federal Commission, *Activity Report 2006–2007*, Belgian Senate and Chamber of Representatives (Session 2008–2009, April 1, 2009).

[56] Ibid. 24.

2 Definitions and Associated Challenges

The Embryo Research Law defines an "embryo" as "a cell or coherent whole of cells with the capacity to grow into a human,"[57] an "embryo in vitro" as "an embryo outside the female body,"[58] and a "supernumerary embryo" as "an embryo that is created in the context of medically assisted reproduction but that is not implanted into the woman."[59] It defines "research" as "scientific tests or experiments on embryos *in vitro*."[60] During the parliamentary debate, it was pointed out that the law did not cover research on embryos in utero.[61]

These definitions present challenges. First, the definition of "embryo" is not sufficiently specific. The fact that they must have "the capacity to grow into a human"[62] raises the question of how embryos that do not have the capacity to grow into a human should be categorized. This is the case of parthenotes (which is an organism produced from an unfertilized ovum that is incapable of developing beyond the early embryonic stages), aneuploid embryos,[63] and abnormally fertilized oocytes.[64] This problem was noted during the parliamentary debate that led to the adoption of the law.[65] The issue was resolved by the Federal Commission, which concluded that parthenotes lack the capacity to grow into a human and therefore must not be treated as embryos.[66] However, since there is some uncertainty about the status of aneuploidy embryos and abnormally fertilized oocytes, the Federal Commission, in an effort to respect the spirit of the law, decided that these entities must be considered to be "embryos."

A second problematic term is "research." The Embryo Research Law defines it as "scientific tests or experiments on embryos in vitro."[67] However, sometimes there can be confusion as to whether something is "research" or just a study, for instance, measuring the impact of the use of a new technique,

---

[57] Embryo Research Law, art. 2.1.
[58] Ibid., art. 2.2.
[59] Ibid., art. 2.3.
[60] Ibid., art. 2.4.
[61] Belgian Senate, *Parliamentary Documents* (2001–01, no 2–695/1, 2000) 5–6.
[62] Ibid., art. 2.1.
[63] An aneuploid embryo is an embryo with an abnormal number of chromosomes (deviating from 46) in its cells. Aneuploid embryos cannot normally result into a life born child (with the exception of trisomies for chromosome 13, 18, and 21).
[64] Abnormally fertilized embryos appear to be either unfertilized (displaying zero pronuclei) or abnormally fertilized (displaying one or three pronuclei) at the time they are checked for fertilization.
[65] Belgian Senate, *Parliamentary Documents* (2001–2002, no. 2–695/12) 119.
[66] Federal Commission, *Report 76*, Plenary meeting of February 26, 2018.
[67] Embryo Research Law, art. 2.4.

which would not require authorization. The Federal Commission has indicated that the concept of "research" is not applicable when devices or methods are used that follow the standard methodology and the guidelines of good clinical practice described in guidelines of recognized scientific societies.[68] Other examples of projects that the Federal Commission has not considered "research" include the case of a center that planned to move from slow freezing to vitrification, at a moment when vitrification was no longer considered experimental. Another case involved a center collecting data on a new incubator to compare it to an older type with no intention to acquire new scientific knowledge and without destroying embryos. Research should also be distinguished from training. Medical staff need to learn how to perform certain interventions such as intracytoplasmic sperm injection and embryo biopsy. Such embryos are destroyed after the intervention, but the initial act was not part of a "research program." These activities do not need to obtain approval from the Federal Commission.

3 Clinical Applications

Clinical applications would be understood by the Federal Commission as modification of embryos that are to be transferred in order to grow into new persons. The embryo research law suggests that some clinical applications are permitted. Article 5 makes it lawful to transfer in uterus embryos used in research if the research is "for the benefit of the embryo itself" or is limited to "a mere observation that does not harm the embryo's integrity." However, this intermediate category (neither fully research, nor therapy) would certainly need approval by the local ethics committee since it would be considered as an experiment on a human being. In these cases, the authority to review protocols belongs to the local ethics committee rather than the Federal Commission.

The Belgian legislation on research on embryos in vitro is based on two important values: the defense of freedom of research and the acceptance of the ethical pluralism that characterizes the Belgian society. The words of one of the proponents of the Embryo Research Law, Mr. Philippe Mahoux, expressing his conviction in scientific progress just before the vote on the law in the Belgian Senate are particularly telling. Mahoux said: "It is impossible, useless and dangerous to try to block the progress of knowledge ... It is difficult to

---

[68] Federal Commission, *Report 20*, Plenary meeting of December 1, 2008. These societies include the European Society for Human Reproduction and Embryology (ESHRE), National Institute for Health and Care Excellence (NICE), Royal College of Obstetricians and Gynaecologists (RCOG), or American Society for Reproductive Medicine (ASRM).

abandon such responsibility when it is not inevitable, unless one adheres to a mystic approach that glorifies nature and has more trust in chance than in reason."[69]

Human embryo research is firmly established in Belgian society. Neither politicians nor the public show any intention to change the present law. Moreover, compared to other countries, a large percentage of IVF patients donate their embryos to science.[70] More than half of these patients chose donation for research.[71] As a consequence, Belgium has a surplus of embryos for research at the moment, unlike many other countries in Europe. This is unlikely to change in the future as long as research pursues therapeutic purposes, that is to say, it focuses on preventing or curing diseases, including genetically inherited diseases, rather than nontherapeutic purposes (enhancement).

However, things may change in the future. Due to the growing trend of extended culture of blastocyst for transfer and cryopreservation, early cleavage-stage embryos (day 1–3) are no longer available for research. These modifications imply that embryos are grown for longer before a biopsy for genetic analysis is performed. In addition, with the development of trophectoderm biopsy (instead of blastomere biopsy), and subsequent cryopreservation until the genetic diagnosis is known, fresh good-quality, genetically abnormal embryos are no longer available. The declining number of supernumerary embryos in their early stages of development that are available for research may lead to an increase in the creation of embryos for research by scientists who want to study early embryo development as it will be easier for scientists to justify their creation as required by Article 3 of the Embryo Research Law.

The possibility to create embryos for research may prove to be crucial for germline applications. For instance, mosaicism, which occurs when not all cells of the embryo carry the same genetic information, is a significant problem for germline modification. The intended germline modification might not affect all cells of the embryo, frustrating the intended result. As Ma et al. showed, a possible solution would be to apply CRISPR simultaneously with ICSI.[72] But in order to find out which method works, one needs to create embryos.

---

[69] Belgian Senate, *Plenary Sessions* (2–249, 2002) 36.

[70] V Provoost, G Pennings, P De Sutter, A Van de Velde, M Dhont, "Trends in Embryo Disposition Decisions: Patients' Responses to a 15-year Mailing Program" (2012) 27 *Hum Reprod* 506.

[71] V Provoost, G Pennings, P De Sutter, M Dhont, "A Private Matter: How Patients Decide What to Do with Cryopreserved Embryos after Infertility Treatment" (2012) 15 *Hum Fertil* 210. Provoost et al. say 50.8 percent.

[72] H Ma et al., "Correction of a Pathogenic Gene Mutation in Human Embryos" (2017) 548 *Nature* 413.

10

The Regulation of Human Germline Genome Modification in Sweden

Santa Slokenberga and Heidi Carmen Howard

I INTRODUCTION

The Kingdom of Sweden is a democratic state, based on the rule of law. It has a tradition of encouraging and nurturing research and innovation and, despite the long-lasting historical links between the state and the church,[1] at least in modern times, religion has not played a decisive role in steering scientific research in any particular direction.[2] Sweden is also a country that prides itself on its good standing in the international community and its desire to abide by the international legal obligations that it has voluntarily accepted.[3] Thus, although the overall framework for biotechnology research in the country is rather liberal (for instance, human embryos can be created for research

---

[1] Instrument of Government 1634 (*Regeringsformen* 1634) ('The unity of religion and the proper worship service is the most powerful foundation for a lawful, coherent and lasting regiment') para. 1. The Swedish Church Act (*Kyrkolagen* 1686) governed the relationship between Swedish state and church until 1992 (adopted 1991, applicable from 1993), when the new Church Law (Kyrkolag 1992:300) came into force (in effect until 2000). Only in the year of 2000, the Swedish state was separated from the church. S Ekström, 'Svenska kyrkan i utveckling historia, identitet, verksamhet och organisation' (SvK, 30 March 2011) www.sorenekstrom.se/Documents/SvKBildFarg.pdf accessed 9 September 2018. Essential in this regard is Lag (1998:1591) om Svenska kyrkan. See Swedish Government, 'Regeringens proposition 1997/98:116' (*Regeringen*, 12 March 1998) www .regeringen.se/49bb8e/contentassets/092607777fo6a4fbcab6f550a8bb31dbe/staten-och-trossamfun den–bestammelser-om-svenska-kyrkan-och-andra-trossamfund accessed 15 May 2018 (Prop. 1997/ 98:116); Swedish Government, 'Regeringens proposition 1995/96:80' (Regeringen 5 October 1995) www.riksdagen.se/sv/dokument-lagar/dokument/proposition/andrade-relationer-mellan-staten- och-svenska_GJo38o/html accessed 15 May 2018.

[2] Prop. 1997/98:116 (n 1); Lag (1991:115) Law on measures for research or treatment purposes with human eggs (*om åtgärder i forsknings- eller behandlingssyfte med ägg från människa*) SFS nr: 1991:115.

[3] However, of course, tensions internally exist. For example, historically in regard to the ECHR. See H Wenander, 'Sweden: European Court of Human Rights Endorsement with Some Reservations' in Patricia Popelier, Sarah Lambrecht and Koen Lemmens' (eds.), *Criticism of the European Court of Human Rights* (Intersentia 2016).

purposes), by virtue of its external commitments, as well as national regulatory autonomy, research on human germline modification in Sweden cannot advance to clinical trials and beyond. Indeed, currently, both the national law on genetic integrity and Sweden's external commitments prohibit modification of the human germline in clinical trials as well as any use of germline-editing technology in humans. Furthermore, based on the expressed concerns about uncertainties it involves for future generations in the preparatory work of the national law,[4] it is not obvious that should European laws become more permissive, so would Swedish national law. The most recent legal developments, which underline a broader scope of the national prohibition than that stemming from the European Union (EU) law, could be argued to support this claim.[5]

This chapter discusses the overall regulatory framework specific to genetic research (in human embryos) in Sweden (Section II), including laws and regulatory bodies, and, specifically, the regulatory framework for human germline modification (Section III). Section IV introduces emerging policy responses to germline modifications and puts forward some tentative conclusions on the future of research on human germline modification in Sweden.

II THE OVERALL REGULATORY ENVIRONMENT

1 *The Legal Framework*

Sweden's regulatory environment pertaining to human germline modification is affected by the international and European regional legal orders and organizations Sweden is a member of. The obligations Sweden has depend on the international organization or legal order in question, its competences and authority, as well as how it has been used in the area of human germline modification.

a Sweden's International Legal Obligations

Sweden is a member of several international organizations that have discussed and/or taken a stance on human germline genome modification (regulation), such as the United Nations (UN) and its agency the United Nations

---

4 Regeringens proposition 2005/06:64 Genetisk integritet m.m., 138.
5 Regeringens proposition 2017/18:196 Anpassningar av svensk rätt till EU- Prop. förordningen om kliniska läkemedelsprövningar, Regeringen (21 mars 2018).

Educational, Scientific and Cultural Organization (UNESCO), and the Organisation for Economic Co-operation and Development (OECD).[6]

I UNITED NATIONS In 1946, Sweden became a member of the UN and committed to further the purposes of the UN in the area of human rights.[7] Among the UN human rights instruments that are relevant to regulating modification of the human germline is the International Covenant on Economic Social and Cultural Rights (ICESCR or the Covenant),[8] which Sweden ratified in 1972.[9] The Covenant contains several provisions that could be relevant to a discussion on human germline genome modification: for example, Articles 15.1.b and 15.2 through 4 (collectively known as the Right to Enjoy the Benefits of Scientific Progress), Article 12 (Right to the Highest Attainable Standard of Health), for example, if human germline genome modification were done for health-related purposes, and Article 10 (Right to the Protection and Assistance to the Family).[10] Although under Article 2 of the same Covenant, Sweden is required to 'take steps' 'with a view to achieving progressively the full realization of the rights recognized in the present Covenant',[11] the content of these provisions and their applicability to human germline modification remain contestable and ambiguous.[12] As of

---

[6]   Kungörelse (1974:152) om beslutad ny regeringsform (Publication on a new form of government), SFS nr: 1974:152. Chapter 1, Section 10 states: 'Sweden is a member of the European Union. Sweden also participates in the framework of the United Nations and the Council of Europe and in other contexts of international cooperation.'

[7]   Charter of the United Nations (adopted 17 December 1963, entered into force 31 August 1965) 1 UNTS XVI (UN Charter), s 1.3; United Nations Treaty Collection, 'Chapter 1: Charter of the United Nations and Statute of the International Court of Justice, Declarations of Acceptance of the Obligations contained in the Charter of the United Nations – Admission of States to Membership in the United Nations in accordance with Article 4 of the Charter' (United Nations) https://treaties.un.org/pages/ViewDetails.aspx?src=TREATY&mtdsg_no=I-2&chapter=1&clang=_en accessed 1 October 2018.

[8]   International Covenant on Economic, Social and Cultural Rights (adopted 16 December 1966, entered into force 3 January 1976) 993 UNTS 3 (ICESCR).

[9]   United Nations Treaty Collection, 'Status International Covenant on Economic, Social and Cultural Rights, 'United Nations Treaty Collection' (United Nations) https://treaties.un.org/Pages/ViewDetails.aspx?src=IND&mtdsg_no=IV-3&chapter=4&clang=_en accessed 31 August 2018.

[10]   Right to Family Life is protected within the realm of civil and political rights, under Article 17 of the Covenant on Civil and Political Rights which Sweden ratified also on 6 December 1971.
       International Covenant on Civil and Political Rights (adopted 16 December 1966, entered into force 23 March 1976) 999 UNTS 171 (ICCPR).

[11]   See Article 2 ICESCR (n 8); see ICESCR 'General Comment 3: The Nature of States Parties' Obligations" (14 December 1990) UN Doc E/1991/23.

[12]   For an in-depth analysis of this right, see AR Chapman, 'Towards an Understanding of the Right to Enjoy the Benefits of Scientific Progress and Its Applications' (2009) 8 *Journal of Human Rights* 1.

now, the International Committee on Economic Social and Cultural Rights, which is competent to carry out the monitoring functions of ICESCR, has not interpreted the respective provisions in the context of modification of the human germline to place any concrete obligations on the parties of the Covenant,[13] nor has Sweden interpreted the Covenant as placing any obligations to it in relation to the modification of the human germline.

II UNESCO Sweden has been a member of UNESCO since 1950.[14] Although UNESCO has not addressed the question of the modification of the human germline in hard law instruments, which would place obligations on Sweden, it has adopted declarations that are relevant to germline editing.[15] These declarations shape the legal environment in which Sweden is operating, and Sweden shall account for them by virtue of its membership in UNESCO. The declarations do not *expressis verbis* prohibit human germline editing; yet, reproductive cloning is prohibited,[16] and they also indicate that human germline interventions could be contrary to human dignity.[17] In regard to the latter, very recently, the UNESCO International Bioethics Committee has taken a rather firm stand and has recommended 'a moratorium on genome editing of the human germline', emphasizing that currently 'the concerns about the safety of the procedure and its ethical implications

---

[13] The Committee was established to carry out the monitoring functions assigned to the United Nations Economic and Social Council in Part IV of the Covenant. See Economic and Social Council, 'Review of the composition, organization and administrative arrangements of the Sessional Working Group of Governmental Experts on the Implementation of the International Covenant on Economic, Social and Cultural Rights' (28 May 1985) Resolution 1985/17. Work on a general comment has begun, which might bring some clarity about the scope of the right. See ICESCR, 'General discussion on a draft general comment on article 15 of the International Covenant on Economic, Social and Cultural Rights: on the right to enjoy the benefits of scientific progress and its applications and other provisions of article 15 on the relationship between science and economic, social and cultural rights' (OHCHR, 2018) www.ohchr.org/EN/HRBodies/ICESCR/Pages/Discussio n2018.aspx accessed 1 October 2018.

[14] UNESCO, 'List of the 195 Members (and the 11 Associate Members) of UNESCO and the date of which they became Members (or Associate Members) of the Organization' (UNESCO, 30 August 2018) www.unesco.org/eri/cp/ListeMS_Indicators.asp accessed on 30 August 2018.

[15] UNESCO, Universal Declaration on the Human Genome and Human Rights (adopted 11 November 1997, endorsed by UN General Assembly in Resolution AIRES/53/152 on 9 December 1998) 53/152; UNESCO, International Declaration on Human Genetic Data (adopted 16 October 2003) C/RES/23; UNESCO, Universal Declaration on Bioethics and Human Rights (adopted 10 January 2005) 23 C/RES/24.

[16] Universal Declaration on Human Genome and Human Rights (n 15) art. 11.

[17] Universal Declaration on the Human Genome and Human Rights (n 15) art. 24.

are so far prevailing'.[18] Further to the questions on human genome, genetic data and bioethics, UNESCO contributions in other areas that relate to human germline genome modification should also be taken into consideration.[19]

III OECD In recent years, the OECD Working Party on Biotechnology, Nanotechnology and Converging Technologies·has focused on policy issues in emerging technology fields including gene editing.[20] In 2016, it noted that the development of synthetic biology faces several obstacles, including biohazard concerns and ethical issues, and noted the existing call of representatives from several countries on 'scientists around the world to abstain from germline gene editing research until the risks are better assessed and a broad societal consensus about the appropriateness of these techniques is reached'.[21] So far the OECD has not yet adopted a policy on the issue.[22] Moreover, as in the case of UNESCO, actions of the OECD in other areas that relate to human germline genome modification should also be taken into consideration (i.e. privacy and animal welfare).[23]

---

[18] International Bioethics Committee, 'Report of the IBC on Updating Its Reflection on the Human Genome and Human Rights' (UNESCO, 2015) SHS/YES/IBC-22/15/2 REV2, para. 118. Sweden joined UNESCO on 23 January 1950. UNESCO, 'Countries: UNESCO Member States List' (UNESCO, 19 July 2018) https://en.unesco.org/countries/s accessed 19 July 2018.

[19] For example, UNESCO Universal Declaration of Animal Rights (adopted 15 October 1978). However, status of this declaration has been subject to discussions. LW Roeder and A Simard, *Diplomacy and Negotiation for Humanitarian NGOs* (Springer 2013) 149.

[20] Sweden acceded to the OECD on 28 September 1961. OECD, 'List of OECD Member countries – Ratification of the Convention on the OECD' (OECD, 19 July 2018) www .oecd.org/about/membersandpartners/list-oecd-member-countries.htm accessed 19 July 2018.

[21] OECD, *OECD Science, Technology and Innovation Outlook* (OECD Publishing 2016) 106.

[22] The OECD has launched a project on gene editing with a view to produce a forum conducive to evidence-based discussion across countries on the many issues of shared concern. See OECD BNCT, 'Project on Gene Editing' (Innovation Policy Platform) www .innovationpolicyplatform.org/project-gene-editing-oecd-bnct accessed 19 July 2018.

[23] 'The welfare of laboratory animals is important; it will continue to be an important factor influencing the work in the OECD Chemicals Programme. The progress in OECD on the harmonization of chemicals control, in particular the agreement on Mutual Acceptance of Data, by reducing duplicative testing, will do much to reduce the number of animals used in testing. Such testing cannot be eliminated at present, but every effort should be made to discover, develop and validate alternative testing systems.' OECD, 'Animal Welfare' (OECD) www.oecd.org/chemicalsafety/testing/animal-welfare.htm accessed 15 May 2018. OECD, 'OECD Guidelines on the Protection of Privacy and Transborder Flows of Personal Data' (OECD, 2013) www.oecd.org/sti/ieconomy/oecdguidelinesontheprotectionofprivacyandtrans borderflowsofpersonaldata.htm accessed on 30 August 2018.

b Sweden's Regional Legal Obligations

Key regional actors that affect Sweden's regulatory autonomy in the area of human germline modification are the Council of Europe (CoE) and the European Union.

I COUNCIL OF EUROPE Within the CoE, when it comes to germline modification of the human genome, the most relevant legal instruments are the European Convention on Human Rights (ECHR), from which positive obligations on the part of the state stem to account for risks to the rights protected under the ECHR;[24] the Convention for the Protection of Human Rights and Dignity of the Human Being with regard to the Application of Biology and Medicine; the Convention on Human Rights and Biomedicine (Oviedo Convention),[25] in which Article 13 *expressis verbis* addresses the question of interventions on the human genome;[26] and some of its additional protocols,[27] including the Additional Protocol to the Convention on Human Rights and Biomedicine, concerning Biomedical

---

[24] Sweden ratified the Convention in 1952. CoE, 'Chart of Signatures and Ratifications of Treaty 005. Convention for the Protection of Human Rights and Fundamental Freedoms' (Treaty Office, 19 July 2018) www.coe.int/en/web/conventions/full-list accessed 19 July 2018. Overall, Sweden has had a very complex relationship with the ECHR. Although the convention was ratified in 1952, it was not until 1995 when in connection with Sweden's accession to the EU, the ECHR was incorporated into Swedish law. See Swedish Government, 'Regeringens proposition 1993/94:117' (Regeringen, 9 December 1993) http://data.riksdagen.se/dokument/GH03117 accessed 15 May 2018. See Konstitutionsutskottet, 'Inkorporering av Europakonventionen och andra fri- och rättighetsfrågor, betänkande 1993/94:KU24' (Konstitutionsutskottet, 14 April 1994) www.riksdagen.se/sv/dokument-lagar/arende/betankande/inkorporering-av-europakonventio nen-och-andra_GH01KU24/html accessed 15 May 2018. Law on the European Convention for the Protection of Human Rights and Fundamental Freedoms, SFS 1994:1219 (*Lag* (1994:1219) *om den europeiska konventionen angående skydd för de mänskliga rättigheterna och de grundläggande friheterna*). The Instrument of Government (n 1) c 2, art 19. Currently, the ECHR is not only Sweden's external commitment, it is also part of a national law and a constitutional obligation. This combination created the ECHR's unclear status on the hierarchy of norms. See Wenander (n 3) 244.

[25] Convention for the Protection of Human Rights and Dignity of the Human Being with Regard to the Application of Biology and Medicine: Convention on Human Rights and Biomedicine (opened for signatures on 4 April 1997, entered into force 12 January 1999) ETS No. 164 (Oviedo Convention); CoE, 'Chart of Signatures and Ratifications of Treaty 164. Convention for the Protection of Human Rights and Dignity of the Human Being with Regard to the Application of Biology and Medicine: Convention on Human Rights and Biomedicine' (Treaty Office, 19 July 2018) www.coe.int/en/web/conventions/full-list accessed 19 July 2018.

[26] Oviedo Convention (n 25).

[27] Additional Protocol to the Convention on Human Rights and Biomedicine concerning Biomedical Research (opened for signatures 25 January 2005, entered into force 9 January 2007) CETS No. 195 (Additional Protocol).

This Additional Protocol is a technology-neutral instrument and seeks to 'protect the dignity and identity of all human beings and guarantee everyone, without discrimination, respect for their

Research (Additional Protocol on Biomedical Research). The Additional Protocol on Biomedical Research is a technology-neutral instrument and seeks to 'protect the dignity and identity of all human beings and guarantee everyone, without discrimination, respect for their integrity and other rights and fundamental freedoms with regard to any research involving interventions on human beings in the field of biomedicine'.[28] Other legal instruments that are relevant to the modification of the human germline include the data protection framework, in particular the Convention for the Protection of Individuals with regard to Automatic Processing of Personal Data,[29] and the animal protection framework, in particular the European Convention for the Protection of Vertebrate Animals used for Experimental and other Scientific Purposes,[30] both of which Sweden has signed and ratified.

> integrity and other rights and fundamental freedoms with regard to any research involving interventions on human beings in the field of biomedicine'. Additional Protocol to the Convention on Human Rights and Biomedicine Concerning Genetic Testing for Health Purposes (opened for signatures 27 November 2008, entered into force 7 January 2018) CETS No. 203; Additional Protocol to the Convention on Human Rights and Biomedicine Concerning Transplantation of Organs and Tissues of Human Origin, CETS No. 186; Additional Protocol to the Convention for the Protection of Human Rights and Dignity of the Human Being with regard to the Application of Biology and Medicine on the Prohibition of Cloning Human Beings (opened for signatures 12 January 1998, entered force 3 January 2001) CETS No. 168. None of these have been ratified by Sweden. However, Sweden signed the protocol on biomedical research in 2005, see CoE, '*Chart of Signatures and Ratifications of Treaty 195. Additional Protocol to the Convention on Human Rights and Biomedicine, Concerning Biomedical Research*' (Treaty Office, 19 July 2018) www.coe.int/en/web/conventions/full-list/-/conventions/tr eaty/195/signatures?desktop=true accessed 19 July 2018. Sweden signed the protocol on cloning in 1998, see CoE, 'Chart of Signatures and Ratifications of Treaty 168. Additional Protocol to the Convention for the Protection of Human Rights and Dignity of the Human Being with Regard to the Application of Biology and Medicine, on the Prohibition of Cloning Human Beings' (Treaty Office, 19 July 2018) www.coe.int/en/web/conventions/full-list/-/conventions/treaty/173/signatures? desktop=true accessed 19 June 2018.

[28] Additional Protocol to the Convention on Human Rights and Biomedicine Concerning Biomedical Research (n 27) art. 1.

[29] Convention for the Protection of Individuals with regard to Automatic Processing of Personal Data (opened for signatures 28 January 1981, entered into force 10 January 1985) CETS No. 108. Sweden ratified it in 1982, see CoE, 'Chart of Signatures and Ratifications of Treaty 108. Convention for the Protection of Individuals with Regard to Automatic Processing of Personal Data' (Treaty Office, 19 July 2018) www.coe.int/en/web/conventions/full-list accessed 19 July 2018. Sweden has signed the Protocol amending the Convention for the Protection of Individuals with regard to Automatic Processing of Personal Data, see CoE, 'Chart of Signatures and Ratifications of Treaty 223. Protocol amending the Convention for the Protection of Individuals with regard to Automatic Processing of Personal Data' (Treaty Office, 25 November 2018) www.coe.int/en/web/conventions/full-list/-/conventions/treaty/223 /signatures accessed 25 November 2018.

[30] European Convention for the Protection of Vertebrate Animals used for Experimental and other Scientific Purposes, CETS No. 123. Sweden ratified it in 1988, see CoE, 'Chart of Signatures and Ratifications of Treaty 123. European Convention for the Protection of

One could say that, because Sweden has not ratified the Oviedo Convention, this legal instrument has no relevance to a discussion of the Swedish legal framework. However, it would be a mistake to dismiss it summarily. First, Sweden has signed it, and Article 18 of the Vienna Convention on the Law of Treaties obliges states to refrain from acts that would defeat the object and purpose of a treaty that has been signed pending its eventual ratification.[31] Therefore, one could argue that Sweden cannot legislate or interpret its existing legislation in such a way to reach a result that defeats the object and purpose of the Oviedo Convention. As Article 13 of the Oviedo Convention is intrinsically related to protecting humanity, it is directly anchored in the values enshrined in Article 1 of the Convention: dignity and integrity.[32]

Second, one could argue that the ECHR, which has been ratified by Sweden, must be read considering the Oviedo Convention, particularly for those issues over which the ECHR is silent, and where the states enjoy a narrow margin of appreciation.[33] Although Sweden has not ratified the Oviedo Convention, in so far as human germline modification could be anchored in human integrity as protected under the ECHR, Sweden's positive obligations in mitigating human integrity risks should be shaped by similar solutions as those under the Oviedo Convention.[34]

Third, the CoE includes all the Member States of the EU. Although the EU has not been primarily concerned with human rights, it has long been

Vertebrate Animals Used for Experimental and Other Scientific Purposes' (Treaty Office, 19 July 2018) www.coe.int/en/web/conventions/full-list accessed 19 July 2018.

[31] Vienna Convention on the Law of Treaties (adopted on 23 May 1969, entered into force on 27 January 1980) 1155 UNTS 331. Sweden signed the Vienna Convention on the Law of Treaties in 1970 and ratified it in 1975. Vienna Convention on the Law of Treaties 'United Nations Treaty Collection' (UN, 1 October 2018) https://treaties.un.org/pages/ViewDetailsIII .aspx?src=TREATY&mtdsg_no=XXIII-1&chapter=23&Temp=mtdsg3&clang=_en accessed 1 October 2018.

[32] Oviedo Convention (n 25) art. 1.

[33] On margin of appreciation, see, e.g., G Letsas, 'Two Concepts of the Margin of Appreciation' (2006) 26:4 Oxford Journal of Legal Studies 705. On how margin of appreciation relates to the national regulatory autonomy, see S Slokenberga, European Legal Perspectives on Health-Related Direct-to-Consumer Genetic Testing (Jure 2016) c 5, 9.

[34] Human germline modification could trigger integrity as protected under Article 3 and (or) 8 ECHR. Article 1 ECHR places obligation on Sweden to ensure protection of rights set forth in the ECHR in its jurisdiction. On the one hand, the Council of Europe is an international legal order following the principles of public international law. On the other hand, the ECtHR through its key doctrines has engaged in extensive interpretation of the provisions of ECHR, often blurring the line between the State's obligations under the ECHR and State's discretion not to commit itself to the approach Council of Europe has taken in a particular area or in a particular issue. Slokenberga (n 33) c 2 and 9.

established that the protection of fundamental rights is an integral dimension to the exercise of the EU competences.[35] Thus, given the overlap between the competences of the EU and CoE in the area of germline editing and, accordingly, overlapping commitments placed on Sweden, it is essential that the EU and CoE norms are not conflicting. Until now, the European Court of Human Rights (ECtHR) has appeared willing to accommodate EU human rights protection under the ECHR regime.[36] For the modification of the human germline in so far as human rights are concerned, it could mean that the EU regulatory responses could be accommodated in the CoE system, and, in so far as the EU has legislated on the matter, the fact that Sweden has not ratified the Oviedo Convention becomes irrelevant.

II EUROPEAN UNION Upon accession to the EU in 1995, Sweden transferred competence to the EU to legislate in several areas that touch on aspects of human germline genome modification.[37] This is the case, for example, with public health and internal market, within which EU legislation on clinical trials and advanced medicinal products has been adopted,[38] as well as data protection,[39] animal welfare,[40] and tissues and

---

[35] See Ibid. 177–194. For an in-depth analysis on fundamental rights in the EU legal order, see C Amalfitano, *General Principles of EU Law and the Protection of Fundamental Rights* (Edward Elgar Publishing 2018).

[36] This is commonly known as the Bosphorus presumption or test, whereby the ECtHR is willing to acknowledge equivalent human rights protection, albeit to do so on a case-by-case basis. See *Bosphorus v. Ireland* [2005] GC ECHR 440. This presumption has further been refined in the subsequent case law.

[37] For the principle of conferral and EU competences, see Article 5 Treaty on the European Union (TEU) and Articles 2–6 Treaty on the Functioning of the European Union (TFEU). Consolidated versions of the Treaty on European Union and the Treaty on the Functioning of the European Union [2016] OJ C202/1.

[38] An internal market legislation: European Parliament and of the Council Directive 2001/20/EC on the approximation of the laws, regulations and administrative provisions of the Member States relating to the implementation of good clinical practice in the conduct of clinical trials on medicinal products for human use [2001] OJ L 121/34. An internal market and a public health legislation: European Parliament and Council Regulation (EU) No 536/2014 on clinical trials on medicinal products for human use, and repealing Directive 2001/20/EC Text with EEA relevance [2014] OJ L 158/1. An internal market legislation: European Parliament and Council Regulation (EC) No 1394/2007 of the on advanced therapy medicinal products and amending Directive 2001/83/EC and Regulation (EC) No 726/2004 (Advanced Medicinal Products Regulation) [2007] OJ L 324/121.

[39] A data protection legislation: European Parliament and Council Regulation (EU) 2016/679 on the protection of natural persons with regard to the processing of personal data and on the free movement of such data, and repealing Directive 95/46/EC (General Data Protection Regulation) [2016] OJ L 119/1.

[40] An internal market legislation: European Parliament and Council Directive 2010/63/EU on the protection of animals used for scientific purposes [2010] OJ L 276/33.

cells.[41] As is further discussed in detail, particularly the regulatory frame-work of clinical trials is of direct relevance to human germline modifica-tion, which expressly precludes gene therapy trials which result in modifications to the subject's germline genetic identity. How legislation in these areas relates to Sweden and scientists depends on such considera-tions as the type of harmonization that is used, whether further action on the part of an EU Member State (Sweden) is permissible and the type of instrument that is used. While a regulation is directly applicable, a directive requires a national transposition on the part of Sweden, and hence national law implementing EU law is generally applied.[42] In so far as the EU law is being applied, the rights protected in the Charter of Fundamental Rights of the European Union shall also be applied, and in particular Article 1, which protects human dignity, Article 2, which protects life, Article 3, which protects integrity, and Article 4, which prohibits torture, inhuman or degrading treatment or punishment.[43]

The EU clinical trials regulatory framework comprises the old framework, Directive 2001/20/EC (Clinical Trials Directive), which has been transposed

[41] Although Directive 2004/23/EC (also known as the Human Tissue and Cells Directive) applies to quality and safety for human tissues and cells, EU Member States are free to decide whether any specific type of human cells can or cannot be used in research. Whenever national law permits use of a particular type of cells, the Human Tissue and Cells Directive applies and is transposed into national law (in this case, Swedish law) accordingly. This Directive should not interfere with decisions made by Member States concerning the use or non-use of any specific type of human cells, including germ cells and embryonic stem cells. If, however, any particular use of such cells is authorized in a Member State, this Directive will require the application of all provisions necessary to protect public health, given the specific risks of these cells based on the scientific knowledge and their particular nature, and guarantee respect for fundamental rights. Moreover, this Directive should not interfere with provisions of Member States defining the legal term 'person' or 'individual'. European Parliament and Directive 2004/23/EC on setting standards of quality and safety for the donation, procurement, testing, processing, preservation, storage and distribution of human tissues and cells [2004] OJ L 102/48, Rec. 12.

[42] There are rare exceptions when an individual could rely directly on a directive. See S Robin-Olivier, 'The Evolution of Direct Effect in the EU: Stocktaking, Problems, Projections' (2014) 12 International Journal of Constitutional Law 165.

[43] Charter of Fundamental Rights of the European Union [2012] OJ C 326/391. Despite the CFREU humble wording, which suggests its applicability to the Member States when 'implementing' EU law, it has been well-established that it is generally applicable whenever EU law applies. This finding in the case of Åkerberg Fransson has generated controversy and follow-up case law that shapes the scope of application of the CFREU. See C-617/10 Åkerberg Fransson ECLI:EU:C:2013:280. Post Åkerberg Fransson see, for example, Case C-390/12, Pfleger, EU:C:2014:281; Case C-206/13, Siragusa, EU:C:2014:126; Case C-56/13, Érsekcsanádi Mezőgazdasági, EU:C:2014:352. For analysis of the scope of EU law see M Dougan, 'Judicial Review of Member State Action under the General Principles and the Charter: Defining the Scope of Union Law' (2015) 52:5 Common Market Law Review 1201.

in national laws in the EU Member States,[44] and the recently adopted Regulation 536/2014 (Clinical Trials Regulation).[45] The Clinical Trials Regulation has been adopted and has entered into force but is not being applied yet. As specified in Article 99.2 of the Clinical Trials Regulation, its application depends on the functionality of the EU portal (a single entry point for the submission of data and information relating to clinical trials in accordance with the regulation)[46] and the EU database (an EU-level database for the purposes of containing data and information that are submitted in accordance with the regulation).[47] Once the EU portal and the EU database have achieved full functionality, the European Commission shall publish a notice in the *Official Journal of the European Union*;[48] six months after this notice, the Regulation will be applicable.[49] As reported by the European Medicines Agency, it is expected to happen sometime in 2019.[50] Once applicable, the Regulation will apply directly throughout the EU. At that point, the Clinical Trials Directive will be repealed;[51] however, during a transition period, rules of the Clinical Trials Directive and the Clinical Trials Regulation will apply.[52] Both, the Clinical Trials Directive and the Clinical Trials Regulation expressly preclude clinical trials that result in modifications to the subject's germline genetic identity.[53] This ban is also upheld in regard to

[44] Clinical Trials Directive (n 38).
[45] Clinical Trials Regulation (n 38).
[46] Ibid., art. 80.
[47] Ibid., art. 81.
[48] Ibid., art. 82.3.
[49] Ibid., art. 99.
[50] European Medicines Agency, 'Clinical Trial Regulation' www.ema.europa.eu/human-regulatory /research-development/clinical-trials/clinical-trial-regulation accessed 1 October 2018.
[51] Ibid., art. 96.1.
[52] Ibid., art. 98.
[53] Clinical Trials Directive (n 38) art. 9.6. Clinical Trials Regulation (n 38) rec 75 and art. 90. Article 2 (a) of the Clinical Trials Directive states that 'clinical trial' is 'any investigation in human subjects intended to discover or verify the clinical, pharmacological and/or other pharmacodynamic effects of one or more investigational medicinal product(s), and/or to identify any adverse reactions to one or more investigational medicinal product(s) and/or to study absorption, distribution, metabolism and excretion of one or more investigational medicinal product(s) with the object of ascertaining its (their) safety and/or efficacy'.
  Article 2.2 of the Clinical Trials Regulation states that 'Clinical trial' means a clinical study which fulfils any of the following conditions:
'(a) the assignment of the subject to a particular therapeutic strategy is decided in advance and does not fall within normal clinical practice of the Member State concerned;
(b) the decision to prescribe the investigational medicinal products is taken together with the decision to include the subject in the clinical study; or
(c) diagnostic or monitoring procedures in addition to normal clinical practice are applied to the subjects.'

advanced medicinal products.[54] This means that the Swedish law that implements the Clinical Trials Directive must preclude these trials; once the Clinical Trials Regulation is applicable, this prohibition stemming from the EU law will be directly applicable in Sweden. The practical significance for clinical trials and human germline modification is rather limited, as the Regulation continues to uphold the ban that has been enshrined in the directive and implemented in Swedish law.[55]

In sum, each of these international and European legal orders and organizations of which Sweden is a member has addressed germline gene editing in humans in some fashion. The CoE has adopted treaties that address human germline genome modification in an overarching manner. UNESCO has exercised its soft competence and the OECD has accounted for the issue while stating the need for further research and debate. The EU has legislated in a number of areas that relate to human germline genome modification, including harmonizing clinical trials, thus making it difficult, if not impossible by pre-emption, for Sweden to adopt national legislation on those issues. Nonetheless, in so far as the EU has not pre-empted Sweden from legislating, for example, in the areas that fall beyond the scope of clinical trials regulatory framework, Sweden retains competence to legislate nationally with due regard to its external commitments.

c Sweden's National Legal Framework

The question of heritable genetic modification in humans has been considered, albeit using different terms, in Sweden's legislative history. It is currently regulated under the Genetic Integrity Act (*Lag om genetisk integritet* – GIA),[56] which was promulgated on 18 May 2006. However, it

---

European Parliament and Council Directive 2001/83/EC on the Community code relating to medicinal products for human use [2001] OJ L 311/ 67. European Commission, 'Directive 2001/83/EC of the European Parliament and of the Council of 6 November 2001 on the Community Code Relating to Medicinal Products for Human Use' (European Commission, 2012) https://ec.europa.eu/health//sites/health/files/files/eudralex/vol-1/dir_2001_83_con sol_2012/dir_2001_83_cons_2012_en.pdf accessed 15 May 2018, see M9.

54   Advanced Medicinal Products Regulation (n 38) art. 4.1.

55   The ban can be found in the Genetic Integrity Act (GIA) (*Lag (2006:351) om genetisk integritet m.m*) SFS 2006:351. However, this law is not listed on EUR-Lex (eur-lex.europa.eu accessed 15 May 2018) as national legislation transposing the Clinical Trials Directive.

56   GIA (n 55).

was also part of the previous legislative framework, which was in effect between 1991 and 2006.[57] As some of the questions raised by human germline editing relate to human dignity and the right to private life, it can be argued that they anchor in Sweden's constitutional values and constitutional objectives.[58]

Apart from the Genetic Integrity Act, the question of human germline editing is not addressed specifically in other Swedish legislation. Nonetheless, other legal instruments can be applicable in different contexts and for different actions relating to the modification of the human genome. For example, requirements for ethical approval of research involving humans, their cells or data are regulated by the Act Concerning the Ethical Review of Research Involving Humans (*Lag om etikprövning av forskning som avser människor* – ERRIH).[59] Personal data are protected by the EU GDPR, which is complemented by the Law with Additional Provisions to the EU Data Protection Regulation (*Lag med kompletterande bestämmelser till EU:s dataskyddsförordning*)[60] and the Public Access to Information and Secrecy Act (*Offentlighets- och sekretesslag*).[61] Collection, storage and use of human biological material is regulated, inter alia, by the Biobanks in Medical Care Act (*Lagen om*

---

[57] Lag (1991:115) om åtgärder i forsknings- eller behandlingssyfte med ägg från människa), (Law on measures for research or treatment purposes with human eggs), SFS 1991:115.

[58] Instrument of Government (n 1) c 1, art. 2 of the Instrument of Government defines basic principles of the form of government; it requires exercising public power (including legislative power) with respect for dignity of the individual. Chapter 2 of the Instrument of Government addresses fundamental rights, including privacy (c 2, art. 6). Likewise, it is a constitutional requirement that no law may violate ECHR (c 2, art. 19). In that regard, Article 8 ECHR, which protects privacy and the convention, builds on dignity. To what extent gene editing could be seen as part of private life under the ECHR is a margin of appreciation question. It could be argued that this is a question that leaves narrow margin of appreciation, which does not allow for manoeuvre at the national level.

[59] Lag (2003:460) om etikprövning av forskning som avser människor (the Act Concerning the Ethical Review of Research Involving Humans), SFS 2003:460. There has been a proposal to revise the current law. See Swedish Government, 'Prop. 2017/18:45 En ny organisation för etikprövning av forskning som avser människor' 116' (Regeringen, 23 November 2017) www .regeringen.se/4adfba/contentassets/94fda6517a0b4b94b3cddgaf39e10de7/en-ny-organisation-for-etikprovning-av-forskning.pdf accessed 15 May 2018; Law with supplementary provisions to the EU Data Protection Regulation (Lag (2018:218) med kompletterande bestämmelser till EU:s dataskyddsförordning') SFS 2018:218.

[60] Law with supplementary provisions to the EU Data Protection Regulation (Lag (2018:218) med kompletterande bestämmelser till EU:s dataskyddsförordning') SFS 2018:218.

[61] Offentlighets- och sekretesslag (2009:400) (Publicity and Privacy Act), SFS 2009:400.

biobanker i hälso- och sjukvården m.m.).[62] Finally, the 1988 Animal Welfare Act (*Djurskyddslag*) protects animals used in research.[63]

2 Definitions and Related Challenges

In Swedish law, key terms, such as 'embryos' or 'gametes', are not defined, nor are the terms 'basic research' or 'preclinical research' used in legislation. The Genetic Integrity Act, however, mentions 'sperm', 'egg' and 'fertilized eggs'. The law also mentions 'eggs used for somatic cell nuclear transfer',[64] and 'somatic cell nucleus transfer' is defined as 'replacing the cell nucleus of an egg with the nucleus from a body cell',[65] which is intended to refer to 'cloning' practices.

In the Genetic Integrity Act, 'gene therapy' is the term used to address modifying the genome. It is defined as 'a treatment that involves introducing, with the use of a carrier (vector), a healthy gene into the cells of an individual with a genetic disease'.[66] Chapter 2, Sections 3 and 4, of the Act sets forth the rules governing gene therapy. Section 3 states that 'experiments for the purposes of research or treatment that entail genetic changes that can be inherited in humans may not be carried out'. Section 4 states that 'treatment methods that are intended to bring about genetic changes that can be inherited in humans may not be used'. Although the Genetic Integrity Act narrowly defines 'gene therapy' as those approaches specifically using a vector (Chapter 1, Section 5) to deliver DNA, the detailed rules in Section 3 and 4 of Chapter 2 seem to suggest that these prohibitions can also be extended to methods of genetic modification other than those involving traditional approaches with vectors inserting a "healthy" gene (e.g. CRISPR-Cas9 system can be delivered as a ribonucleoprotein; and the insertion is not of a healthy gene per se, but of nucleotides). In other words, the substantive provisions of the Genetic Integrity Act regulates consequences (i.e. whether the genetic changes that are introduced/created can or cannot be inherited in humans), rather than means (i.e. how the genetic change, and consequent possibility of inheritance is achieved).

---

[62]   Lag (2002:297) om biobanker i hälso- och sjukvården m.m. (Law on biobanks in health care, etc.), SFS 2002:297. The biobank regulatory framework is currently being reviewed, see Statens Offentliga Utredningar, *Framtidens biobanker* (SOU 2018:4). Several changes in the act come into effect on 1 January 2019.

[63]   Djurskyddslag (1988:534) (The Animal Welfare Act), SFS 1988:534. A new act on animal protection has been adopted. See Djurskyddslag (2018:1192) (The Animal Welfare Act), SFS 2018:1192. It will enter into force on 1 April 2019, when the current act loses its effect.

[64]   GIA (n 55). c 5, s 3.

[65]   Ibid., c 1, s 5.

[66]   Ibid.

This ambiguity, however, is not the only one that renders the scope of the Genetic Integrity Act to the modification of the human germline unclear. The Genetic Integrity Act defines 'gene therapy' as the introduction of 'a healthy gene into the cells of an individual with a genetic disease'.[67] Over the course of the last decades, knowledge of the human genome has advanced, allowing for the identification of genetic variants that predispose individuals to genetic disorders; moreover, genes have also been associated with various non-disease traits. If the term 'genetic disease' is interpreted *stricto sensu* referring to a gene-related condition that has manifested in an individual, the prohibition enshrined in the Genetic Integrity Act has a limited scope. Chapter 2, Sections 3 and 4, of the Act does not offer clarifications. Section 3 addresses 'research or treatment', while Section 4 addresses 'treatment methods'. Depending on how 'treatment methods' are defined, one could argue that certain human germline genome-editing interventions without a therapeutic purpose (i.e. for pure enhancement of traits) fall outside the scope of the Genetic Integrity Act.[68] In either case, the legislature has emphasized that the scope of Chapter 2, Section 3 GIA, is broader than that of the Clinical Trials Regulation, without exhaustively marking the legislature's intended scope of application.[69]

On the surface, these ambiguities may seem to be a matter of semantics only. However, they could have an important impact on the interpretation of the scope of the Genetic Integrity Act for human germline modification and consequently on the matters that are (un)regulated under the Act. The matters that could be argued to not to fall under the Act could be regulated through other provisions. They include those regulating biobanks, data protection and ethical approvals and EU Clinical Trials framework. Moreover, they could also be addressed through obligations under the Vienna Convention on the Law of Treaties not to defeat the aim and object of those treaties Sweden has acceded to, even when Sweden has not yet proceeded with their ratification.

---

[67] GIA (n 55) c 1, s 5.

[68] The notion of pure enhancement is infrequent, albeit existing to connote use of an enhancer for non-medical purposes. See, e.g., R Merkel and others, *Intervening in the Brain: Changing Psyche and Society* (Springer 2007) 398.

[69] Swedish Government, 'Regeringens proposition 2017/18:196 Anpassningar av svensk rätt till EU- Prop. förordningen om kliniska läkemedelsprövningar' (Regeringen, 21 mars 2018) 156 www.regeringen.se/496172/contentassets/2171dade90524a8a84f8296f391fa88c/anpassningar-av -svensk-ratt-till-eu-forordningen-om-kliniska-lakemedelsprovningar-prop.-2017_18_196.pdf accessed 1 October 2018.

3 Oversight Bodies

As there are different laws affecting research on human germline genome editing in Sweden, so there is an array of public actors (oversight bodies) that may oversee or address this issue. No body has been established specifically to oversee or enforce the Genetic Integrity Act. General questions pertaining to genetic technologies in the context of health and research fall within the competence of the National Board of Health and Welfare (*Socialstyrelsen*).[70] It is the main administrative authority with regulatory and oversight powers in the area of health and social care.[71] The Medical Products Agency (*Läkemedelsverket*) is the competent authority for control and supervision of medicinal products and devices.[72] The Swedish Data Protection Authority (*Datainspektionen*) has the task to protect the privacy of individuals in the information society.[73] Finally, when animals are used in research, the responsible authority is the Swedish Board of Agriculture (*Jordbruksverket*).[74]

The Swedish Gene Technology Advisory Board (*Gentekniknämnden*) is entrusted to promote, through advisory activities, an ethically justifiable and safe use of genetic engineering 'to protect human and animal health and the environment'.[75] The Swedish National Council on Medical Ethics (*Statens medicinsk-etiska råd*) is an advisory board to the Swedish government and parliament on ethical issues raised by scientific and technological advances in biomedicine.[76]

When research involves humans, human tissues or animals, ethics committees are also involved. Research involving humans is vetted by the Ethical Review Agency (*Etikprövningsmyndigheten*).[77] Until 31 December 2018, the Central Ethical Review Board (*Centrala etikprövningsnämnden*) was the

[70] Socialstyrelsen (National Board of Health and Welfare) www.socialstyrelsen.se/ accessed 15 May 2018.

[71] Förordning (2015:284) med instruktion för Socialstyrelsen (Ordinance (2015: 284) with instruction to the National Board of Health), SFS 2015:284, s 1.

[72] Läkemedelsverket (Medical Products Agency) www.lakemedelsverket.se accessed 15 May 2018. See Förordning (2007:1205) med instruktion för Läkemedelsverket (n 72) s 1.

[73] The Swedish Data Protection Authority (*Datainspektionen*) www.datainspektionen.se/other-languages/ accessed 15 May 2018.

[74] The Swedish Board of Agriculture (*Jordbruksverket*) www.jordbruksverket.se/ accessed 15 May 2018.

[75] The Swedish Gene Technology Advisory Board (*Gentekniknämnden*) www.genteknik.se/ accessed 15 May 2018.

[76] The Swedish National Council on Medical Ethics (*Statens medicinsk-etiska råd*) www .smer.se/about-us/ accessed 15 May 2018.

[77] ERRIH (n 59) s 24.

enforcing body of the Ethical Review of Research Involving Humans Act.[78] For research involving animals, regional animal ethics boards are relevant, as well as the Central Animal Ethics Board (*Centrala djurförsöksetiska nämnden*), which inter alia hears appeals of decisions taken by the regional animal ethics boards.[79]

Finally, similarly to the Genetic Integrity Act,[80] the Ethical Review of Research Involving Humans Act[81] and the 1988 Animal Welfare Act[82] envisage sanctions for violating prohibitions relating to gene therapy, as well as relating to research. These sanctions are subject to criminal enforcement, which brings into the picture the authorities involved in handling criminal procedures.

4 Funding

In Sweden, there are considerable possibilities to seek funding for research, including research on human germline genome modification. In principle, funding can be sought through both the State and through private foundations. These include, for example:

- The Swedish Research Council (*Vetenskapsrådet*) provides funding for research in all disciplinary domains.[83]
- Vinnova, the Swedish government agency that administers state funding for research and innovation, provides funding for various research domains.[84]
- The Wallenberg Foundations, the set of public and private foundations funded by individual members of the Wallenberg family, provide funding for a variety of scientific domains.[85]

---

[78]   Regional ethics review boards were created under Section 24 of ERRIH as of 31 December 2018. There were five regional boards. Ethics Committee (*Etikprövingsnämderna*) www.epn.se/start/ accessed 22 July 2018.
[79]   The 1988 Animal Welfare Act (n 23) s 21.b.
[80]   GIA (n 55) c 8, ss 2–7.
[81]   ERRIH (n 59) s 38.
[82]   The 1988 Animal Welfare Act (n 23) ss 36–37.
[83]   Swedish Research Council, 'Research for a Wiser World' www.vr.se/english.html accessed 19 July 2018.
[84]   Vinnova, 'Vinnova' www.vinnova.se/sok-finansiering/hitta-finansiering accessed 19 July 2018.
[85]   Wallenberg, 'The Wallenberg Foundations' (The Wallenberg Foundations) www.wallenberg.org/en accessed 19 July 2018.

- The Ragnar Söderbergs Foundation (*Ragnar Söderbergs stiftelse*) provides funding for research on medicine, law and economics.[86]
- Forte, a research council and a government agency under the Swedish Ministry of Health and Social Affairs, finances scientific research relating to human health, work life and welfare.[87]

Generally, funders request compliance with the legal requirements. Some hold the principal investigator responsible to ensure that research conducted with funding from a specific source complies with the terms and conditions specified in Swedish law.[88] In so far as different sources of funding have been reviewed, we could not identify a funder that explicitly excludes human germline gene modification from the list of projects that can be funded; however, this does not exclude that such rules exist.

Furthermore, since Sweden is a Member of the EU, researchers in Sweden can apply for EU research funding (e.g. for this current period, the Horizon 2020 Framework Program).[89] As part of the application, the applicant needs to reflect on the ethical issues raised by the project.[90] Article 19 of the EU Horizon 2020 Regulation, laying down the rules for participation and knowledge dissemination, in Horizon 2020, places a general obligation on the funding seekers to comply inter alia with the ethical principles or any applicable legislation. Recital 29 of the Regulation emphasizes that these are 'fundamental ethical principles'. It is not, however, further delineated what these principles are or how to identify what would be regarded as an ethical principle or a fundamental ethical principle under this regulation. The question on the notion of applicable legislation is not less ambiguous. Any applicable legislation under Article 19 means 'relevant national, Union and international legislation, including the Charter of Fundamental Rights of the European Union and the European Convention on Human Rights and its Supplementary Protocols'. This means that the EU law needs not to be

---

[86]   Ragnar Söderbergs Stiftelse, 'Ragnar Söderbergs Stiftelse' www.ragnarsoderbergsstiftelse.se accessed 19 July 2018.

[87]   Swedish Research Council, 'Swedish Research Council for Health, Working life and Welfare' https://forte.se accessed 15 May 2018.

[88]   Eg Ethical guidelines, see Swedish Research Council, 'Vetenskapsrådet' www.vr.se/english .html accessed 20 July 2018.

[89]   European Parliament and Council Regulation (EU) No 1290/2013 laying down the rules for participation and dissemination in 'Horizon 2020 – the Framework Programme for Research and Innovation (2014–2020)' and repealing Regulation (EC) No 1906/2006 (Horizon 2020 Regulation) [2013] OJ L 347/ 81.

[90]   Ibid., art. 13. For considerations on challenges in that regard, see J Reichel, 'Alternative Rule-Making within European Bioethics–Necessary and Therefore Legitimate?' (2016) 21 *Tilburg Law Review* 169.

applicable to the subject matter; it could also be other laws attributable to the issue in some way. In so far as the EU law regulates human germline modification, the rules become clearer and more straightforward. For example, the EU clinical trials legislation prohibits gene therapy trials that result in modifications to the subject's germline genetic identity.[91] This prohibition is also echoed in the Horizon 2020 ethics self-assessment document, which states that 'research activity intended to modify the genetic make-up of human beings that could make such changes heritable (apart from research relating to cancer treatment of the gonads, which may be financed)' is not eligible for funding.[92] A proposal that fails to meet the applicable requirements 'may be excluded from the evaluation, selection and award procedures at any time'.[93]

III THE SPECIFIC REGULATORY ENVIRONMENT

1 *Basic Research*

The Genetic Integrity Act does not define the terms 'basic research' or 'pre-clinical research'. The exact legal requirements that are applicable to research generally depend not only on the aim, nature and content of experiments but also on a number of practicalities, for example, how the necessary biological materials are obtained. However, as far as the Genetic Integrity Act is concerned, Chapter 2, Section 3, is of particular importance. It states that 'experiments for the purposes of research or treatment that entail genetic changes that can be inherited in humans may not be carried out'.[94] While the wording does not *expressis verbis* address basic research, the preparatory works for the GIA indicate that the Swedish legislator adopted the Act to make the previous legal framework on research more liberal and enable some research in the area.[95] If one takes into consideration the historical evolution of the Act and the intention of Swedish legislature, basic research on human germline genome modification is not prohibited by Chapter 2, Section 3, of the Genetic Integrity Act and is therefore legal in Sweden with fertilized eggs up to 14 days.

Beyond consideration of the Genetic Integrity Act, depending on the circumstances surrounding a particular basic research project involving

---

[91] See Clinical Trials Regulation (n 38) art. 90. Clinical Trials Directive (n 38) art. 9.4.
[92] European Commission, Directorate-General for Research & Innovation, Horizon 2020 Programme Guidance How to complete your ethics self-assessment, Version 5.3 of 21 February 2018.
[93] Horizon 2020 Regulation (n 90) art. 13.4.
[94] GIA (n 55) c 2, s 3.
[95] Prop. 2005/06:64 (n 4) 134.

human biological material, ethical approval may be necessary.[96] Ethical approval is regulated by a separate law. ERRIH.[97] It applies inter alia to the research involving biological material that has been taken from a living human being and can be traced back to that person.[98] Also, it applies to the research involving biological material that has been taken for a medical purpose from a deceased person and can be traced back to that person.[99] However, if the material is taken from a living or deceased person but cannot be traced back to that person, and tissues from a biobank are not involved in research, the ERRIH is not triggered and approval is not necessary under this law.

The ERRIH sets forth key principles for ethical review. Five of them are of relevance to basic research. First, only research that respects human dignity can be approved.[100] Second, human rights and fundamental freedoms shall always be taken into consideration in ethics review, while also taking into account the interest that new knowledge can be developed through research. As a rule, though, people's welfare must be given precedence over the needs of society and science.[101] Third, the risk to the researchers and possible scientific value of the research must be assessed.[102] Fourth, ethical review must consider whether there are alternative ways to achieve the intended outcome as well as data protection issues.[103] Fifth, the research must be carried out by a researcher with appropriate qualifications.[104] The decision on ethics review is subject to appeal.[105]

2 *Preclinical Research*

There is no widely held consensus on the definition of preclinical research by different stakeholders, including scientists, legal scholars or policy makers.

---

[96]   ERRIH (n 59) ss 4.3 and 4.5. Moreover, if materials from a biobank are being used, the Law on Biobanks in Health Care, etc. (*Lag om biobanker i hälso- och sjukvården m.m.*) applies. Lag (2002:297) om biobanker i hälso- och sjukvården m.m. (Law on biobanks in health care, etc.), SFS 2002:297, c 1, s 3.

[97]   ERRIH (n 59).

[98]   Ibid., s 3.

[99]   Ibid., s 4.

[100] Ibid., s 7.

[101] Ibid., s 8.

[102] Ibid., s 9.

[103] Ibid., s 10.

[104] Ibid., s 11.

[105] Until December 31, 2018: Regional Ethics Review Boards and the appeal body is Central Ethics Review Board. From January 1, 2019: Ethical Review Agency and the appeal body is Board of Appeal. Ibid., s 31.

Indeed, many experiments considered by some authors as preclinical could easily be considered as basic research by others. While the editors of this book have defined preclinical as studies involving the modification of animal germline genome that either takes place in vivo or ex vivo with subsequent transfer in the animal (i.e. a modified embryo is implanted in a female animal), neither the term nor this definition is used in the Swedish legislation presented below.[106]

That being said, preclinical research, in the sense of research performed to gather information needed to go on to conduct clinical trials, often involves animals. The animal rights legal framework is currently undergoing changes. Currently, the key legal act to address animal welfare is the 1988 Animal Welfare Act, which regulates the use of animals in scientific research.[107] However, on 1 April 2019 the new Animal Welfare Act (*Djurskyddslagen*) of 2018 (the 2018 Animal Welfare Act) comes into effect. Generally, the 1988 Animal Welfare Act and the 2018 Animal Welfare Act set forth principles and foundation rules for ensuring animal welfare; however, further requirements can be adopted in the delegated acts. Among these acts, the Animal Welfare Ordinance (*Djurskyddsförordning*) and the Swedish Board of Agriculture's regulations and general guidelines on laboratory animals (*Statens jordbruks-verks föreskrifter och allmänna råd om försöksdjur*) are of particular importance. The former sets forth general requirements for ethical approvals regarding animal use in research and provides authority to the Swedish Board of Agriculture (*Jordbruksverket*) to further delineate the rules;[108] the latter lays down conditions for the protection of laboratory animals, as well as conditions for the protection of other animals kept in laboratory animal facilities.[109] Although under the 2018 Animal Welfare Act similar room for delegated acts is left, as of now they have not been adopted; consequently, it is too early to discuss what, if any, changes the new law will bring for subjecting animals to germline gene-editing research with a view to gaining knowledge

---

[106]   Nonetheless, this concept has been accounted for in the legislative initiatives and government reports. It has been indicated that '[i]n medical research, pre-clinical research is often referred to as developmental work in, for example, laboratory or animal studies, and clinical research, ie research involving clinical studies on humans (ethical assessments in the border between health and research, SMER Report 2016: 1 p. 21)'. Statens Offentliga Utredningar, *Etikprövning – en översyn av reglerna om forskning och hälso- och sjukvård* (SOU 2017: 104), 339.

[107]   The 1988 Animal Welfare Act (n 23).

[108]   Djurskyddsförordning (1988:539) (Animal Welfare Ordinance), SFS 1988:539.

[109]   Statens jordbruksverks föreskrifter och allmänna råd om försöksdjur (SJVFS 2017:40) (Swedish Board of Agriculture's regulations and general guidelines on laboratory animals), c 1, s 1.

about possible clinical applications, should the bans that are in effect in Sweden eventually be lifted.

Under the 1988 Animal Welfare Act, animals used for research fall in the category of 'laboratory animals' or animals that are used or are intended to be used for animal experiments,[110] where research is one of that type.[111] The Swedish Board of Agriculture's regulations further delineate use of animals for research and set forth a ban with several exceptions on the use of primates and other endangered animals for research.[112] To use animals in research, researchers must obtain permission from the competent authority: the Swedish Board of Agriculture.[113] Furthermore, before animals are used for the experiments, a regional animal research ethics board (*Regional djurförsöksetisk nämnd*) must give its approval, too.[114] Failure to comply with the requirements concerning permission or failure to obtain an ethical approval could be sanctioned with a fine or an imprisonment for up to two years.[115]

Research involving germline gene editing on animals is not prohibited in Sweden. The 1988 Animal Welfare Act describes research animals as 'maps and models' for human health and needs,[116] and addresses gene therapy in that regard.[117] However, not all animals may be used for research. For example, it is prohibited to use apes (*Pongidae*) and gibbons (*Hylobatidae*), except for public display and experiments that are in compliance with the provisions set forth in Chapter 10, Section 8.[118] Threatened species and apes, except human primate monkeys (*Pongidae*) and gibbons (*Hylobatidae*), may be used

---

[110]   The 1988 Animal Welfare Act (n 23) ss 1 and 1 b.

[111]   Ibid., s 1 c.

[112]   Swedish Board of Agriculture's regulations and general guidelines on laboratory animals (n 110) c 10, ss 8 and 9. The exceptions set forth in Chapter 10, Section 9 include permitting research on primates, including endangered primates except for human apes (Pongidae) and gibbons (Hylobatidae), as well as other threatened species inter alia in order to avoid, prevent, diagnose or treat disabling or potentially life-threatening clinical conditions in humans. 'Disabling clinical condition' means a reduction in an individual's physical or psychological functional capacity.

[113]   The 1988 Animal Welfare Act (n 23) s 19.a; Animal Welfare Ordinance (n 23) s 40.a.

[114]   The 1988 Animal Welfare Act (n 23) s 21.

[115]   Ibid., s 36.

[116]   Statens Offentliga Utredningar, *Etisk prövning av djurförsök – genteknik och bioteknik på djur* (SOU 2003:107) 66.

[117]   Ibid. 68.

[118]   Swedish Board of Agriculture's regulations and general guidelines on laboratory animals (n 110) c 10, s 8. These requirements are, first, that 'the animals are not exposed to anything that can cause suffering to the animal'; or, second, it is observation studies that may be carried out without a business permit or ethical approval of animal experiments under Chapter 2, Section 21 of the Swedish Board of Agriculture's regulations and general guidelines on laboratory animals; or, third, a regional animal experimental committee has given permission to do so.

in animal experiments provided that (1) there are scientific reasons explaining why the purpose of animal testing cannot be achieved by using species other than threatened species or primates; (2) the animals are only used in animal experiments for the purposes as stated in the Swedish Board of Agriculture's regulations and general guidelines on laboratory animals; as well as (3) a regional animal experimental committee has given permission to do so.[119]

The *traveaux preparatoires* discusses in detail the question of genetic engineering technologies (*genteknik*),[120] and Section 19 of the 1988 Animal Welfare Act applies to gene editing as part of animal testing.[121] However, whether gene editing of an animal's germline cells to study possible clinical applications for humans is permissible, is unclear. Generally, in order for animals to be used for research, an ethical approval is necessary. Research on animals can be permitted inter alia in (a) basic research;[122] (b) for the purposes of studying the effects of diseases, ill health or other deviating conditions on human beings;[123] as well as (c) research involving evaluation, detection, regulation or modification of physiological conditions in humans.[124] As far as (b) and (c) are concerned, animals may be used for the development, manufacture or testing of quality, efficacy and safety of medicines, foods, feed and other substances or products.[125]

Although germline gene editing on animals is not prohibited in Sweden, in so far as research relates to gaining knowledge about inheritable changes in humans, it is unclear whether the study could pass the ethical analysis and be approved. The ethical analysis involves assessment of the animal's suffering, which is weighed against the expected benefit of the animal experiment for humans, animals or the environment,[126] and is done on a case-by-case basis by a regional animal ethics board.[127] Should any of the boards permit germline genome modification to gain knowledge about health applications for humans at a time when clinical trials and clinical studies are prohibited, one could

---

[119] Ibid., c 10, s 9. This provision sets forth detailed further rules for using primates, threated primates and other threated species for different research categories.

[120] SOU 2003:107 address in detail address the question of 'genteknik'. See SOU 2003:107 (n 117) c 2.1, 33, as well as c 5, 63–86.

[121] The 1988 Animal Welfare Act (n 23) sec. 1 c. Detailed requirements regarding genetically modified animals and ethical approval application are in Swedish Board of Agriculture's regulations and general guidelines on laboratory animals (n 110) c 14, ss 7–9.

[122] Swedish Board of Agriculture's regulations and general guidelines on laboratory animals (n 110) c 4, s 1.

[123] Ibid., c 4, s 2.

[124] Ibid., c 4, s 3.

[125] Ibid., c 4, s 5.

[126] Ibid., c 7, s 13.

[127] On practicalities relating to the applications, see Ibid., c 2, s 14.

question the consistency of the Swedish regulatory framework as well as the necessity to subject animals to experiments when the knowledge cannot be applied in practice.

3 *Clinical Research*

Swedish national law is clear on banning experiments for the purposes of research or treatment that entail genetic changes that can be inherited in humans.[128] Clinical trials of therapies involving human germline genome modification cannot be done in Sweden because of the combined effect of EU legislation and the Genetic Integrity Act. In the EU, and therefore in Sweden, clinical research for medicinal products falls within the domain of shared competence between the EU and Member States as part of internal market and common safety concerns in public health matters.[129] As mentioned above, in the EU, clinical trials are currently regulated by the Clinical Trials Directive[130] and the Clinical Trials Regulation.[131] Although the Clinical Trials Regulation has been adopted and is legally in force, it has not been applied yet for the reasons discussed previously in Part 1. Once it is legally in force, which is estimated to be sometime in 2019,[132] it will be directly applicable throughout the EU.

The Clinical Trials Directive states that 'no gene therapy trials may be carried out which result in modifications to the subject's germ line genetic identity'.[133] This is echoed by the Clinical Trials Regulation.[134] These prohibitions in EU law are then transposed into Swedish law by the Genetic Integrity Act, Chapter 2, Section 3, which states that 'experiments for the purposes of research or treatment that entail genetic changes that can be inherited in humans may not be carried out'.[135] Moreover, once the Clinical Trials Regulation is applicable, a special provision in the Genetic Integrity Act shall also come into effect that makes a clear reference to Article 90 of the Clinical Trials Regulation.[136] In that way the legislature has intended that not

---

[128]   GIA (n 55) c 2, s 3.
[129]   Consolidated version of the Treaty on the Functioning of the European Union (n 37) art 4.2.k.
[130]   Clinical Trials Directive (n 38).
[131]   Clinical Trials Regulation (n 38).
[132]   See Section I of this chapter.
[133]   Clinical Trials Directive (n 38) art. 9.6.
[134]   Clinical Trials Regulation (n 38) rec 75 and art. 90.
[135]   Prop. 2005/06:64 (n 4), 128.
[136]   See GIA (n 55) c 2, s 3. The provision enters into effect when the Government decides so. It states '[p]rohibition of conducting gene therapy clinical trials, which results in the genetic identity of the genital identity of the subject, is contained in Article 90 of Regulation (EU) No 536/2014 of the European Parliament and of the Council of 16 April 2014 on clinical trials for

only those clinical trials prohibited under the EU law are banned in Sweden but also other clinical trials.[137] Disregarding this intention, one can, however, question the scope of this ban in light of the terminological ambiguities presented in Section II of this chapter. Failure to comply with the prohibition is sanctioned with a fine or imprisonment for up to six months, unless the actions in question are punishable by a more severe penalty under the Swedish Penal Code.[138]

Furthermore, the Genetic Integrity Act prohibits introducing in a woman's body a fertilized egg that has been used for experiments. The prohibition also 'applies if the egg, before fertilization, or the sperm used for fertilization have been used for such an experiment or if the egg has been subject to somatic nuclear transfer'.[139] A violation of a prohibition shall be sentenced to a fine or imprisonment for up to one year.[140]

4 Clinical Application

As we already said, Chapter 2, Section 4, of the Act clearly states that 'treatment methods that are intended to bring about genetic changes that can be inherited in humans may not be used'.[141] Moreover, the prohibition on introducing a fertilized egg that has been subject to scientific experiments, and the use of an egg and sperm if the egg, before fertilization, or the sperm used for fertilization has been used for an experiment or if the egg has been subject to somatic nuclear transfer that was discussed above, applies and thus bars foreseeable clinical applications.[142] This prohibition applies to medicinal products as covered by the EU clinical trials regulatory framework; moreover, disregarding the ambiguities with the scope of the GIA, *mutatis mutandis* to the prohibition of clinical trials, the prohibition of clinical application should also apply to situations not covered by the EU law. Medicinal products that are developed and applied contrary to this prohibition must be deemed being contrary to EU law. Furthermore, any violation of the prohibition set forth in Chapter 2, Section 4, of the Act can be expected to be sanctioned under

medicinal products for human use and repealing Directive 2001/20/EC'. See also Prop. 2017/18:196 (n 59).
[137] Prop. 2017/18:196 (n 59) 156.
[138] GIA (n 55) c 8, s 2.
[139] Ibid., c 5, s 5.
[140] Ibid.; ibid., c 8, s 3.
[141] Ibid., c 2, s 4.
[142] Ibid., c 5, ss 3 and 5.

Chapter 8, Section 2, of the Act with up to six months of imprisonment, unless the offence is subject to stricter penalties under the Swedish Penal Code.[143]

IV CURRENT PERSPECTIVES AND
FUTURE POSSIBILITIES

Currently, the Swedish national regulatory framework applicable to human germline genome modification revolves around the Genetic Integrity Act, which is complemented by other legislation, such as the Ethical Review of Research Involving Humans Act and the 1988 Animal Welfare Act. Moreover, EU law and treaties that have been signed (e.g. the Oviedo Convention) or ratified by Sweden (e.g. the ECHR), as well as positions of international actors such as UNESCO and the OECD, contribute to shaping the overall Swedish legal landscape on human germline genome editing.

The Genetic Integrity Act, as currently worded, contains many ambiguities. It is also outdated with respect to some technological developments. For example, as regards to human germline modification the Act does not allow to easily identify which matters fall outside the scope of it. For instance, is it applicable beyond 'gene therapy' to research for purposes other than therapy (e.g. for enhancement)? A narrow reading of Chapter 2, Sections 3 and 4, of the Genetic Integrity Act suggests that gene-editing technologies that do not use vectors (in the traditional sense) to introduce a healthy gene and do not target a 'genetic disease' might fall outside the scope of the Act. We caution that such an interpretation may defeat the purpose of key provisions of the Act.[144] Moreover, limiting the definition of gene therapy to a certain technology and excluding non-traditional approaches (i. e. not exactly as those described in the Act) could also result in a situation where Sweden has transposed the Clinical Trials Directive more narrowly than intended by the EU legislature.[145] This, in itself, is problematic from the perspective of EU law. In these situations, solutions need to be sought. As the jurisprudence of the Court of Justice of the EU suggests, in such cases extensive reading needs to be given to the national provision in question to meet the aim of the respective EU law[146] and consequently to ensure that no clinical trials as defined in the Clinical Trials Directive that involve human germline modification are carried out.

---

[143] Ibid., c 8, s 2.
[144] See Section III of this chapter.
[145] The Clinical Trials Directive does not make any distinction in relation of technology. It prohibits clinical trials. Under EU law, when transposing a directive, Sweden is required to give full effect of the provisions therein. See Clinical Trials Directive (n 38) art. 9.4.
[146] Case C-106/89, *Marleasing SA v La Comercial Internacional de Alimentacion SA*, ECLI:EU: C:1990:395, paras. 8–9.

Swedish law permits research on fertilized eggs, including the creation of embryos specifically for research purposes, although this is prohibited by the Oviedo Convention.[147] Moreover, we argue that basic research using germline genome modification on human sperm and eggs, as well as fertilized eggs and eggs that have been subject to somatic nuclear transfer, is legal in Sweden as long as the embryos are destroyed by day 14. However, the overall legal framework prohibits research beyond that stage.

At the time that the Genetic Integrity Act was drafted, Swedish legislature expressed doubts about reviewing the bans under Chapter 2, Sections 3 and 4.[148] To the extent to which it transferred competences to the EU on clinical trials, including advanced medicinal products, Sweden is pre-empted from legislating on the issue nationally. As long as the current EU clinical trials regulatory framework (i.e. Clinical Trials Directive and Clinical Trials Regulation) remains as it is, clinical trials cannot be funded, authorized or performed in Sweden. Furthermore, in so far as the ban in the Oviedo Convention extends to a reading of the provisions of the European Convention on Human Rights, Sweden's international and constitutional obligation to comply with the ECHR could be argued to also restrict this possibility.

At the time of this writing, several Swedish national regulatory agencies are actively debating the issue of human germline modification, technological advances and whether and how the national legal framework can accommodate them, as well as whether the national law still sets an adequate balance between various rights, interests and values. For instance, in 2015, the Swedish National Council on Medical Ethics commented on the technology used in gene editing.[149] In December 2017, it held a conference to reflect on scientific advances in the field, and the challenges they pose to national and international law, as well as the ethical issues.[150] Furthermore, the Swedish Gene Technology Advisory Board, which has the mandate to promote, through advisory activities, ethically justifiable and safe use of genetic engineering to protect human and animal health and the environment,[151] has also addressed

---

[147] Oviedo Convention (n 25) art. 18. This prohibition, however, is not among those listed in Article 26.2, which would preclude nationally adopted restrictions.

[148] Prop. 2005/06:64 (n 4) 138–142.

[149] SMER, 'Smer kommenterar 2015:1, Tekniken CRISPR/Cas9 och möjligheten att redigera det mänskliga genomet' www.smer.se/smer-kommenterar/smer-kommenterar-20151-tekniken -crisprcas9-och-mojligheten-att-redigera-det-manskliga-genomet/ accessed 8 April 2018.

[150] SMER, 'Smers etikdag 2017: Se presentationerna och diskussionerna' www.smer.se/nyheter/ smers-etikdag-2017-se-presentationerna-och-diskussionerna/ accessed on 31 August 2018.

[151] Förordning (2007:1075) med instruktion för Gentekniknämnden (Regulation with instruction for the Gentekniknämnden), SFS 2007:1075.

the issue in its 2016 report.[152] The Swedish National Council on Medical Ethics has suggested that the Ministry of Health and Social Affairs (*Socialdepartementet*) revise the existing regulatory framework, and in particular the Genetic Integrity Act.[153] However, whether the review will take place depends of the National Board of Health and Welfare.

[152]   Gentekniknämnden, 'Genteknikens utveckling 2016' www.genteknik.se/wp-content/uploads/2017/05/001_2017-Genteknikens-utveckling-2016.pdf accessed 15 May 2018.
[153]   See SMER, 'Tillsätt en parlamentarisk utredning för att se över lagstiftningen på genteknikområdet' www.smer.se/skrivelser/tillsatt-en-parlamentarisk-utredning-for-att-se-over-lagstiftningen-pa-genteknikomradet/ accessed on 31 August 2018.

11

The Regulation of Human Germline Genome Modification in the Netherlands

Britta van Beers, Charlotte de Kluiver, and Rick Maas

I INTRODUCTION

When it comes to medical law and medical ethics, the Netherlands has a reputation of being pragmatic and progressive, emphasizing values such as tolerance, pluralism and personal autonomy.[1] The Dutch legal framework for euthanasia, which made the Netherlands the first country in the world to legalize it, is generally regarded as a prime example of this legal and ethical tradition.

This approach is also reflected on an academic level. Until recently, Dutch scholars regarded self-determination as the central principle of medical law and medical ethics. In this vein, Henk Leenen's classic *Handbook on Health Law* (*Handboek gezondheidsrecht*) traditionally described self-determination not only as a central legal principle but even as a patient's individual right.[2] This focus on personal autonomy originated in the 1970s, when medical law and medical ethics were emerging academic disciplines in the Netherlands. In that era, a paternalistic attitude from medical professionals towards their patients was not uncommon. To counterbalance the power of the medical profession, Dutch legal scholars and ethicists called for the recognition of patients' right to self-determination. This idea proved to be influential: it was crucial for the recognition of patients' rights in Dutch medical contract law.

However, during the past decade, the traditional focus on self-determination has been reconsidered. Dutch medical law and ethics scholars have started to also pay attention to values and principles other than the one of self-determination, such as the principles of protection, equality and

---

[1] Rendtorff and Kemp, *Basic Ethical Principles in European Bioethics and Biolaw* (Centre for Ethics and Law 2000) 209–216.

[2] E.g. HJJ Leenen and JKM Gevers, *Handboek gezondheidsrecht* (4th edn, Boom Juridisch 2000) 33.

human dignity.[3] Similarly, in more recent editions of Leenen's *Handbook*, self-determination is no longer described as a right or as the cornerstone of health law, but only as one of several guiding principles of medical law and ethics.[4]

How can this shift in focus be explained? One of the most important reasons is the emergence of biomedical technologies and their accompanying regulatory frameworks. Indeed, consistent with emerging international regulation, such as the Council of Europe's Oviedo Convention (1997),[5] the Dutch legislature adopted a restrictive and prohibitive approach to biomedical technologies, as exemplified by the Embryo Act (2002)[6] and other biomedical laws. As we will discuss at length below, according to current Dutch law, human embryos and gametes cannot be sold;[7] embryos cannot be created for scientific purposes;[8] and preimplantation genetic diagnosis is only allowed under strict conditions for very serious diseases.[9] More to the point, the Embryo Act prohibits human germline genome editing of cells with which impregnation might take place.[10]

As the *travaux préparatoires* and explanatory memoranda to these laws make clear, the legal restrictions and prohibitions are underpinned by values such as respect for human life, non-commercialization and human dignity.[11] The Dutch legislature takes these principles to be of such importance that they can outweigh other important values in these contexts, such as self-determination and scientific progress. However, the tensions between these principles remain. Whenever the Dutch legal order is challenged by new biomedical developments, the tensions between these values and principles resurface.

---

[3] E.g. AC Hendriks, *In beginsel. De gezondheidsrechtelijke beginselen uitgediept* (inaugural Leiden) (NJCM boekerij 2005); AC Hendriks, BJM Frederiks and MA Verkerk, 'Het recht op autonomie in samenhang met goede zorg bezien' (Tijdschrift voor Gezondheidsrecht 2008), 2–18; BC van Beers, *Persoon en lichaam in het recht. Menselijke waardigheid en zelfbeschikking in het tijdperk van de medische biotechnologie* (dissertation VU) (Boom Juridische uitgevers 2009).

[4] HJJ Leenen and others, *Handboek gezondheidsrecht* (6th edn, BJu 2014).

[5] Convention for the Protection of Human Rights and Dignity of the Human Being with Regard to the Application of Biology and Medicine: Convention on Human Rights and Biomedicine (opened for signatures on 4 April 1997, entered into force 12 January 1999) ETS No. 164 (Oviedo Convention).

[6] Act of 20 June 2002 Relating to the Use of Gametes and Embryos (*Wet houdende regels inzake handelingen met geslachtscellen en embryo's*) (Embryo Act).

[7] E.g. Act of 24 May 1996 relating to providing Organs (*Wet houdende regelen omtrent het ter beschikking stellen van organen*) (Organ Donation Act), art. 2; Embryo Act, art. 27.

[8] Embryo Act, art. 24.a.

[9] Ibid., arts. 26.1 and 26.2.

[10] Ibid., art. 24.g.

[11] See parliamentary document *Kamerstukken II* 2000/01, 27 423, no 3, 3, 16, 41–49, 64.

The rise of gene-editing tool CRISPR/Cas9 offers a striking illustration of that dynamic. Since CRISPR/Cas9 took the life sciences by storm, human germline genome editing has become a topic of heated debates in Dutch politics and academia. These debates make clear that the Dutch prohibitive approach to germline editing is under increasing pressure and that personal autonomy and self-determination are returning to the discussion. For example, over the past few years, several organizations, politicians and academics have proposed to lift the ban on germline editing as soon as the technology is safe for introduction in the clinic, with the purpose of enabling prospective parents to use the technology for therapeutic purposes.[12] In these discussions, personal autonomy is often invoked as an important argument.[13]

In this chapter we first discuss the general regulatory and institutional framework (Section II). We then focus on the most important legal provisions within the regulation of human germline editing (Section III). Finally, we offer an overview and analysis of current public and political debates on this technology in the light of the tensions between self-determination and reproductive autonomy on the one hand and human dignity and respect for human life on the other (Section IV).

II THE REGULATORY FRAMEWORK

This section outlines the Dutch legal framework and the institutional environment regulating research involving human gametes and embryos. First, we describe the broad legal framework within which research on human germline genome modification takes place in the Netherlands. Subsequently, we turn to the specific national legal framework, of which the 2002 Act Relating to the Use of Gametes and Embryos (the Embryo Act) is the most important one. While discussing the relevant legislation, we will also describe the relevant regulatory authorities and advisory bodies.

1 The General Legal Framework

The broad legal framework within which research on human germline genome modification takes place in the Netherlands comprises the Constitution and international law as incorporated by the Constitution, including EU law, some provisions of the Civil Code and the Criminal Code and the legal doctrine of progressive legal protection.

[12] See, in this chapter, Section IV 'Current perspectives and future possibilities'.
[13] *Id.*

a The Constitution, International Law and EU Law

The Constitution of the Kingdom of Netherlands contains several provisions regarding the place of international law (treaties, customary international law and acts of international organizations) in the Dutch legal system.[14] Under the Constitution, the Dutch government has a general obligation to 'promote the development of the international legal order'.[15] The States General (*Staten-Generaal*), the Kingdom's bicameral Parliament, approves the ratification and denunciation of treaties.[16] Under Article 91.3, treaties that conflict with the Constitution or lead to conflicts with it must be approved by the Houses of the States General by two-thirds majority. Article 92 provides: 'Legislative, executive and judicial powers may be conferred on international institutions by or pursuant to a treaty, subject, where necessary, to the provisions of Article 91.3.' Lastly, ratified treaties and 'resolutions by international institutions which may be binding on all persons by virtue of their contents' have binding legal effect within the Dutch legal system[17] and displace statutory regulations in force within the Kingdom.[18]

In general, the Netherlands can be regarded as a law-abiding member of the international community, hosting on its territory several major international courts and tribunals. It is a member of all major international organizations, both global and regional. It is a member of the United Nations and its specialized agencies, including the United Nations Economic, Social and Cultural Organization (UNESCO). Within Europe, it is also a member of the European Union and the Council of Europe. By virtue of Article 92 of its Constitution, binding legal instruments adopted by those organizations are part of its national legal system.

The Netherlands ratified the International Covenant on Economic, Social and Cultural Rights on 11 December 1978, which is, therefore, part of its legal system.[19] So far, it has just signed but not ratified the Optional Protocol giving individuals access to the Committee on Economic, Social and Cultural Rights to claim violations of their rights under the Covenant.[20]

---

[14] Constitution of the Kingdom of the Netherlands of 24 August 1815 (*Grondwet voor het Koninkrijk der Nederlanden van 24 augustus 1815*) (Constitution), arts. 93 and 94.
[15] Ibid., art. 90.
[16] Ibid., art. 91.
[17] Ibid., art. 93.
[18] Ibid., art. 94.
[19] International Covenant on Economic, Social and Cultural Rights (adopted 16 December 1966, entered into force 3 January 1976) 993 UNTS 3 (ICESCR).
[20] Optional Protocol to the International Covenant on Economic, Social and Cultural Rights (adopted 10 December 2008, entered into force 5 May 2013), A/RES/63/117.

Within the Council of Europe, as every member, the Netherlands is party to the European Convention on Human Rights and subject to the jurisdiction of the European Court of Human Rights.[21] Among the relevant provisions of the European Convention on Human Rights, Article 8 (the right to respect for private and family life) deserves special mention in the context of reproductive medicine. For over a decade, the European Court of Human Rights has ruled that the right to private life also protects certain reproductive rights and interests, including 'the right to respect for both the decisions to become and not to become a parent',[22] 'the right of a couple to conceive a child and to make use of medically assisted procreation for that purpose'[23] and 'the desire to conceive a child unaffected by [a] genetic disease [...] and to use assisted reproductive technologies and PGD to this end'.[24]

However, the Netherlands is not party to the Convention for the Protection of Human Rights and Dignity of the Human Being with Regard to the Application of Biology and Medicine (Oviedo Convention), even though the Netherlands signed it already in 1997.[25] In 2016, the Dutch government decided not to ratify it, mainly because of the prohibitions contained in Article 13, which bans human germline editing, and Article 18.2, which bans the creation of human embryos for research purposes.[26] Relying heavily on two legislative evaluation reports,[27] the government concluded that these prohibitions could hold back further advances in reproductive medicine, more specifically research into early embryonic development and preclinical research into safety of reproductive techniques.[28] The decision not to ratify the Oviedo Convention was heavily criticized in Dutch legal literature.[29] Some authors argued that the Convention is a comprehensive document, with respect to patients' rights, privacy, human dignity, equitable access to healthcare and non-discrimination, and that by failing to ratify it, the Netherlands had put itself outside of the international

---

[21] Convention for the Protection of Human Rights and Fundamental Freedoms, ETS No. 5 (European Convention on Human Rights, as amended) (ECHR).

[22] *Evans v. United Kingdom* app no 6339/05 (ECHR [Grand Chamber], 10 April 2007) para 71.

[23] *S.H. v. Austria* app no 57813/00 (ECHR [Grand Chamber], 3 November 2011) para 82.

[24] *Costa & Pavan v. Italy* app no 54270/10 (ECRM 28 August 2012) para 57.

[25] Parliamentary document on the Embryo Act 14, 33508, 9, p. 31; 2012/13, 33508, 3, 13; 2012/13, 33400 VII, 83, 11.

[26] Ibid.

[27] HB Winter and others, *Evaluatie Embryowet en Wet donorgegevens kunstmatige bevruchting* (ZonMW 2012); E.T.M. Olsthoorn-Heim, *Evaluatie Embryowet* (ZonMW 2006).

[28] Parliamentary document 2014/15, 34000 XVI, 106.

[29] M Buijsen, 'Ratificatie van het Biogeneeskundeverdrag: kwestie van menselijke waardigheid' (2015) S&D 4.

legal order on biomedicine, healthcare and ethics.[30] Nevertheless, as we discuss below, the current Dutch legislation on research on human gametes and embryos still corresponds, in large lines, with the provisions of the Oviedo Convention, including the bans on human germline editing and on the creation of embryos for research purposes.

Finally, the Netherlands is a member of the European Union. As such, and by virtue of Article 93 of its Constitution, it is bound by all EU legislation, as well as case law of the Court of Justice of the European Union.[31] Among relevant EU legal instruments, one must mention Regulation 536/2014,[32] on the implementation of good clinical practice in the conduct of clinical trials on medicinal products for human use, and Regulation 1394/2007, on advanced therapy medicinal products.[33]

b The Civil Code and Criminal Code

The Civil Code (*Burgerlijk Wetboek*)[34] and the Criminal Code (*Wetboek van Strafrecht*)[35] contain provisions that must be taken into account when discussing the general legal framework for the regulation of acts affecting embryos and fetuses in the Netherlands. Article 1.2 of the Civil Code states: 'A child of which a woman is pregnant, is regarded to have been born already as often as its interests require so. If it is born lifeless, it is deemed to have never existed.' This legal fiction, known as the '*nasciturus* fiction', dates back to Roman law and many European legal orders of the Romano-Germanic tradition have incorporated it in their civil codes. It does not imply legal personhood of unborn life, nor does it aim to determine the legal status of various types of unborn human life. Instead, once the child is born, one may, for the purposes of the law, act as if the child was born at an earlier moment than it actually was,

---

[30] JCJ Dute, 'Buiten de (mensenrechten)orde? Over het niet ratificeren van het Biogeneeskundeverdrag door Nederland' (2015) *Magazine for Health law* (TVG) 39, 394, at 401; COGEM/Health Council of the Netherlands, *Editing Human DNA: Moral and Social Implications of Germline Genetic Modification* (Bilthoven 2017), p. 42.

[31] Case C-26/62 *Van Gend en Loos* [1963] ECR; Case C-6/64 *Costa/ENEL* [1964] ECR.

[32] On EU laws regulating human germline genome modification, see, in this volume, Part 2, Section II, Chapter 6. EU Regulation No 536/2014 of the European Parliament and of the Council of 16 April 2014 on clinical trials on medicinal products for human use, and repealing Directive 2001/20/EC Text with EEA relevance, OJ L 158, 27.5.2014,1–76.

[33] EC Regulation No 1394/2007 of the European Parliament and of the Council of 13 November 2007 on advanced therapy medicinal products and amending Directive 2001/ 83/EC and Regulation (EC) No 726/2004 (Text with EEA relevance), OJ L 324, 10.12.2007, 121–137.

[34] 1992 New Dutch Civil Code (*Nieuw Burgerlijk Wetboek van 1992*) (Civil Code).

[35] Criminal Code of 3 March 1881 (*Wetboek van Strafrecht van 3 maart 1881*) (Criminal Code).

if 'its interests require so'. In other words, the *nasciturus* fiction is used to protect several legal interests of an already born child. For instance, in accordance with its Roman law roots, Article 1.2 of the Civil Code is most commonly invoked for the purposes of inheritance law, to enable a child to be recognized as the beneficiary of an inheritance, even if the child was not born yet at the time of the deceased's death. More to the point, recently, the legal fiction has been used to justify prenatal child protective measures, giving rise to much discussion among legal scholars, as this new interpretation does seem to imply prenatal protection.[36] Another example is the discussion about the duty of care of healthcare providers towards unborn human life. In 2005, the Dutch Supreme Court was confronted with the question whether an embryo can be considered a 'patient' under Dutch law. The Court ruled that this could be the case if the embryo is in vivo and the pregnant woman has explicitly entered into a treatment contract on behalf of her unborn child on the basis of the Civil Code.[37]

The Criminal Code is relevant mainly because of the criminal law provisions on abortion. According to Article 82.a of the Criminal Code: 'taking the life of a person or of an infant at birth or shortly afterwards shall include: the killing of a fetus which might reasonably be expected to have the potential to survive outside the mother'.[38] In other words, once a fetus has reached the stage when it could survive outside the womb (the so-called viability limit), abortion is a crime. This criminal law provision has led to the enactment of the 1981 Termination of Pregnancy Act (*Wet Afbreking Zwangerschap*). According to the Act, abortion can be performed by a certified clinic or hospital at any point between conception and viability, which was originally set to be at 24 weeks of the pregnancy.[39] Although medical technology has advanced in the meantime, making it possible to keep children alive who are born already after 22 weeks of pregnancy,[40] in 2011, the Minister of Health decided that the

---

[36] P Vlaardingerbroek and others, *Het hedendaagse personen- en familierecht* (Deventer 2011), pp. 26–27; See, for an interesting analysis of both this judicial interpretation and the ensuing academic discussions, LT Haaf, 'Unborn and Future Children as New Legal Subjects: An Evaluation of Two Subject-Oriented Approaches – The Subject of Rights and the Subject of Interests', 18 *German Law Journal* 5, 1091–1119.

[37] *Baby Kelly* [2005] Dutch Supreme Court, ECLI: NL: HR: 2005: AR5213.

[38] Translation provided by Legislationline, 'Criminal Codes: Netherlands' www .legislationline.org/documents/section/criminal-codes/country/12/Netherlands/show accessed 26 October 2018.

[39] Act of 1 May 1981 concerning regulations on the termination of pregnancy [*Wet van 1 mei 1981, houdende regelen met betrekking tot het afbreken van zwangerschap*].

[40] A Reerink, 'Wel/niet levensvatbaar' (NRC Handelsblad, 3 februari 2011) www.nrc.nl/nieuws/2011/02/03/welniet-levensvatbaar-11993573-a427619 accessed 26 October 2018.

traditional limit of 24 weeks would be maintained.[41] During the first 24 weeks, the mother can request abortion, but only within the limits of the Termination of Pregnancy Act.[42] Abortion outside the limits established by the Termination of Pregnancy Act and after viability is a crime punished under Article 82.a of the Criminal Code.

In sum, within Dutch law, and the Dutch Civil Code and Criminal Code in particular, embryos and fetuses are not regarded as legal subjects with independent legal rights. However, unborn human life, even in its earliest stages, is not treated as just a legal object either.

c The Legal Doctrine of Progressive Legal Protection

Dutch health law scholars have developed the legal doctrine of 'progressive legal protection' as a theoretical framework to explain the legal status of unborn life.[43] It posits that as unborn life develops over time, it becomes increasingly worthy of legal protection. There are several embryonic development phases that are deemed legally relevant.

The first phase, between fertilization (either in utero or in vitro) and nidation (i.e. the organic process whereby a fertilized egg becomes implanted in the lining of the uterus of placental mammals), is called the *status potentialis*. At this early stage, the guarantees and requirements of the Termination of Pregnancy Act do not apply yet.[44] Moreover, the Embryo Act makes it clear that during this early phase, the interests and values of others (e.g. patients who might benefit from research, or the parents) can outweigh those of the embryos in vitro.[45]

The second phase is the *status nascendi*, which starts once nidation is completed.[46] Within the *status nascendi* phase, because of the aforementioned criminal law provisions relating to abortion, one can distinguish two separate sub-

---

[41] 'Abortusgrens blijft staan op 24 weken' (Volkskrant, 19 April 2011) www.volkskrant.nl/mensen/ abortusgrens-blijft-staan-op-24-weken~bbbf9c18/ accessed 26 October 2018.

[42] TPA, art. 2–5.

[43] HJJ Leenen and others, *Handboek gezondheidsrecht* (6th edn, BJu 2014) 139; AM te Braake, 'De juridische status van het embryo: een stevig aangemeerde leer' (1995) 19: 2 *Tijdschrift voor Gezondheidsrecht* 32.

[44] Termination of Pregnancy Act, art. 1.2.

[45] Parliamentary document *Kamerstukken II* 2000/01, 27 423, no 3, 5 (Explanatory memorandum).

[46] According to Leenen's *Handbook on Health Law*, nidation constitutes the beginning of pregnancy, although this remains contested among health law scholars. HJJ Leenen and others, *Handboek gezondheidsrecht* (6th edn, BJu 2014), s 4.3.2; W van der Burg, 'De juridische "status" van het embryo: een op drift geraakte fictie' (1994) 7 *Tijdschrift voor Gezondheidsrecht* 386, 386–401.

stages: before and after viability.[47] During this phase, the embryo still does not have the status of a legal subject. This remains the case even when the embryo is beyond the viability limit and abortion is no longer allowed.

The doctrine of progressive legal protection explains some features of the Dutch regulatory framework with regard to unborn life, such as the difference in legal protection before and after the viability limit. However, recent technological developments have created dilemmas that cannot be adequately resolved by resorting to the idea of progressive legal protection. The Embryo Act created various categories of embryos and fetuses with various legal regimes of protection that cannot be explained in terms of different stages in biological development. For example, the Embryo Act created different regimes of protection for embryos in vitro that are intended to be implanted for pregnancy and for those that are not.[48] It also distinguishes between embryos that are left over from IVF treatments and embryos that were deliberately created for research purposes, an act that is prohibited. As the Minister of Health stated during the parliamentary discussions that led up to the Embryo Act: 'To us the intention with which embryos are created are decisive for the degree to which acts with embryos are to be permitted.'[49] Accordingly, several legal scholars argue that the doctrine of progressive legal protection has severe limitations[50] and fails to offer a satisfying theoretical framework to explain the status of embryos in an era of biomedical technologies.[51] In sum, the Embryo Act demonstrates that not only the stage of biological development matters for the Dutch legal framework surrounding acts with embryos and fetuses but also the intentions with which these entities were created and the circumstances in which they were placed.[52]

2 The Specific Regulatory Framework

a The Embryo Act

The Act Relating to the Use of Gametes and Embryos (*Wet houdende regels inzake handelingen met geslachtscellen en embryo's*), better known as 'the

---

[47] HJJ Leenen and others, *Handboek gezondheidsrecht* (6th edn, BJu 2014) 134.
[48] Embryo Act, arts. 16 and 10 respectively.
[49] Parliamentary document *Handelingen II* 2001/02, 336.
[50] W van der Burg, 'De juridische "status" van het embryo: een op drift geraakte fictie' (1994) 7 *Tijdschrift voor Gezondheidsrecht* 386, 386–401.
[51] BC van Beers, 'De mysterieuze status van het embryo' (2005) 13 *Nederlands Juristenblad* 678, 678–685; BC van Beers, *Persoon en lichaam in het recht. Menselijke waardigheid en zelf-beschikking in het tijdperk van de medische biotechnologie* (dissertation VU) (Boom Juridische uitgevers 2009) 244–250.
[52] Ibid.

Embryo Act', is the most important national law with respect to research with human embryos. As its official name makes clear, it regulates the use of both human embryos and gametes. It was adopted on 20 June 2002, after a long deliberative and legislative process. The bill was drafted under a government consisting of political parties with different political outlooks: a conservative liberal political party (VVD), a liberal-democratic party (D66) and the labour party (PvdA). Policy formation under this government was a balancing act, and it shows in the Embryo Act.[53]

The Embryo Act strikes a difficult balance between, on the one hand, limiting the use of embryos for research or medical purposes in accordance with the principles of human dignity and respect for human life, and, on the other hand, supporting scientific research to promote the health of those who are ill and the welfare of infertile couples.[54] Moreover, the Explanatory Memorandum mentions the welfare of the future child as an important perspective that was taken into consideration by the legislator. Thus, the overarching idea underlying the Embryo Act is that respect for human life and dignity calls for caution and restraint when human embryos are involved.[55] Even if instrumental use of embryos is allowed, embryos still have a certain special legal standing that distinguishes them from mere objects. Therefore, embryos can only be used for certain purposes and under strict conditions. As the Explanatory Memorandum to the Embryo Act explains: 'Exactly because we give much weight to the principle of respect for human life, we subject the use of gametes and embryos to certain conditions and restrictions, and restrict the purposes for which gametes and embryos may be used.'[56]

The Act defines an 'embryo' as a 'cell or coherent whole of cells with the capacity to grow into a human being'.[57] A fertilized egg certainly qualifies as such. However, how this definition relates to other embryo-like entities remains unclear. For instance, as the Act is worded, the creation of human–animal hybrids and synthetic embryos, as long as they do not have 'the capacity to grow into a human being', does not fall under the scope of this law. Although some have bemoaned the vagueness of the definition of embryo,[58] it is not unlike, nor certainly less precise than, the one provided for by the

[53] BC van Beers, 'De mysterieuze status van het embryo' (2005) 13 *Nederlands Juristenblad* 678, 683.
[54] Parliamentary document *Kamerstukken II* 2000/01, 27 423, no 3, 5–6 (Explanatory memorandum).
[55] Ibid. 5.
[56] Ibid. 6.
[57] Embryo Act, art. 1.c.
[58] HB Winter and others, *Evaluatie Embryowet en Wet donorgegevens kunstmatige bevruchting* (ZonMW 2012); E.T.M. Olsthoorn-Heim, *Evaluatie Embryowet* (ZonMW 2006).

legislation of several other states around the world. The definition of 'fetus' contained in the Embryo Act is more peculiar: an 'embryo in the human body'.[59] In this chapter, however, we use the term 'embryo' to refer to all types of unborn life, both in vitro and in vivo.

The Act permits, under strict conditions, the scientific use of embryos that are left over after an IVF treatment (so-called surplus embryos), but, in any event, not beyond 14 days after fertilization.[60] The creation of embryos for research is prohibited.[61] The first condition is consent of the individuals who underwent the fertility treatment.[62] Article 8 limits the range of options that these individuals have when choosing the destiny of the surplus embryos: they can donate them to others who wish to become pregnant (par. 1.a); to scientists for research (par. 1.b and c); or they can opt for destruction (par. 3).[63] A second requirement is that research on embryos must take place in accordance with a protocol that provides a complete description of the planned research.[64] The research protocol must be approved by a national oversight body called the Central Committee on Research Involving Human Subjects before the scientific research involving the surplus embryos can move forward.[65] A similar procedure applies to research involving 'fetuses'[66] and gametes.[67]

We will discuss how the Embryo Act regulates human germline editing in more detail below.[68] For now, it suffices to say that it prohibits germline modification of nuclear DNA for reproductive purposes,[69] reproductive cloning,[70] sex selection[71] and bringing together human and animal gametes for the purpose of creating a hybrid.[72] Finally, it bans selling embryos and gametes.[73] In case of violation of these prohibitions, the Act provides for criminal sanctions varying from a fine to imprisonment.[74]

---

[59] Embryo Act, art. 1.d.
[60] Ibid., art. 25.b.
[61] Ibid., art. 24.a.
[62] Ibid., art. 8.
[63] Ibid., art. 24.c, prohibiting using embryos for purposes other than the ones mentioned in Article 8.
[64] Ibid., art. 3.1.
[65] For more information about this committee, see infra Section II.2.c.
[66] Embryo Act, art. 3.3.
[67] Ibid., art. 3.1.
[68] See, in this chapter, Section III, 'Substantive Provisions'.
[69] Embryo Act, art. 24.g.
[70] Ibid., art. 24.f.
[71] Ibid., art. 26.1.
[72] Ibid., art. 25.a.
[73] Ibid., art. 27.
[74] Ibid., arts. 28 and 29 sanctioned with a prison sentence of one year maximum.

b Regulation Preimplantation Genetic Diagnosis 2009

In the Netherlands, preimplantation genetic diagnosis (PGD) is only allowed for couples that are at a high risk of giving birth to children with a severe, hereditary disorder. The legal framework consists of a general law, the Special Medical Procedures Act (*Wet op de Bijzondere Medische Verrichtingen*), that allows for the creation of administrative decrees to regulate certain special medical procedures, such as PGD.[75] For PGD the following decree is currently in force: Regulation Preimplantation Genetic Diagnosis 2009.[76] Based on this decree, the PGD National Indications Committee (*Landelijke Indicatiecommissie PGD*) was called into existence to draft guidelines on the question as to which genetic disorders are serious enough to justify PGD.[77] Section 2 of the Special Medical Procedures Act gives the power to the Ministry of Public Health to license permits to hospitals to perform PGD. So far, only the Maastricht University Medical Centre has been licensed to do so.

c Medical Research on Human Subjects Act, the Central Committee on Research Involving Human Subjects, and the Minister of Health, Welfare and Sport

The Medical Research on Human Subjects Act (*Wet medisch-wetenschappelijk onderzoek met mensen* – WMO), adopted in 1998, regulates medical research on human subjects.[78] It establishes various regulatory bodies, including the Central Committee on Research Involving Human Subjects (*centrale commissie voor medisch-wetenschappelijk onderzoek*),[79] the governmental body in charge of implementing the Medical Research on Human Subjects Act as well as the Embryo Act.[80] It is composed of up to 15 doctors and persons who are experts in the field of embryology, pharmacology, pharmacy, nursing, behavioural sciences, legal science, the methodology of scientific research and ethics, as well as a person who specifically assesses the scientific research from the perspective of the subject.[81]

---

[75] Preimplantation Genetic Diagnosis.
[76] 2009 Regulation Preimplantation Genetic Diagnosis (*Regeling preïmplantatie genetische diagnostiek*) (Regulation PFD) 2009.
[77] PGD, 'What is PGD?: PGD National Indications Committee' (PGD Nederland) www .pgdnederland.nl/en/pgd-national-indications-committee accessed 26 October 2018.
[78] Act of 26 February 1998, regarding rules on medical scientific research on humans (*Wet van 25 februari 1998, houdende regelen inzake medisch-wetenschappelijk onderzoek met mensen*).
[79] Medical Research Act, art. 14.
[80] https://english.ccmo.nl, accessed 24 November 2018.
[81] Medical Research Act, art. 14.

All research that falls under the scope of the Medical Research on Human Subjects Act or the Embryo Act must be reviewed by the Central Committee. The Central Committee protects subjects taking part in medical research by reviewing the research against the statutory provisions and taking into account the interests of medical progress. The Central Committee reports to the Minister of Health, Welfare and Sport.[82] The Minister can suspend research on human subjects in case of unacceptable risks for the research subjects' health.[83] In case of scientific research on medicine concerning gene therapy, somatic cell therapy, xenogeneic cell therapy or medicine that contains genetically modified organisms (GMOs), research is only allowed after explicit consent of the Minister and/or the Central Committee.[84]

The Central Committee also reports annually to the Minister of Health on the application of the Embryo Act 'with special attention being paid to new developments concerning actions involving germ cells and embryos, insofar as these are apparent from the submitted research protocols'.[85] The Minister of Health sends their annual report to the two chambers of the States General and gives its opinion on the new developments identified by the Central Committee.[86]

d The Environmental Management Act and the Commission on Genetic Modification

The Environmental Management Act (*Wet Milieubeheer*) was adopted in 1979 to govern general subjects of environmental protection.[87] It establishes many bodies, the most relevant for this chapter being the Commission on Genetic Modification (*Commissie genetische modificatie* – COGEM).[88] The Commission is composed of 20 members, appointed by the Minister of Infrastructure and Water Management. It operates under the umbrella of the Ministry of Health. It advises the Minister of Health on notifications and applications for a licence relating to the production of or activities involving GMOs and on safety measures to be taken to protect public health and the

---

[82] Embryo Act, art. 4.
[83] Medical Research Act, art. 3.a.
[84] Ibid., art. 13i.4.
[85] Embryo Act, art. 4.1.
[86] Ibid., art. 4.2.
[87] Act of 13 June 1979 regarding rules on several general topics relating to environmental hygiene (*Wet van 13 juni 1979, houdende regelen met betrekking tot een aantal algemene onderwerpen op het gebied van milieuhygiëne*) (Environmental Management Act) [amended last in 2018], preamble.
[88] Environmental Management Act, art. 2.26.

environment.[89] It advises the administrative authority authorized to approve licences relating to research establishments working on GMOs.[90] It also advises the administrative authority in charge of monitoring the production of or activities involving GMOs on matters related to its monitoring tasks.[91] And, finally, it informs the relevant ministers when the production of or activities involving GMOs have ethical or social implications which the Commission considers to be important.[92]

e The Health Act and the Health Council

The Health Act (*Gezondheidswet*) was adopted in 1956 and is the general statute regulating public health in the Netherlands.[93] It establishes the Health Council (*Gezondheidsraad*), an advisory body composed of about 100 members, appointed by the Crown and operating under the umbrella of the Ministry of Health.[94] Government ministers can use the advice of the Health Council to substantiate policy decisions.[95] The Health Council informs the Minister of Health, Welfare and Sport periodically of the current state of public health and health-related research.[96] The Health Council also has an independent and 'alerting' function: it can give unsolicited advice[97] by issuing advisory reports on health-related scientific developments.[98]

3 Funding

In the Netherlands, research on gene editing and human embryos is publicly funded by two organizations: the Netherlands Organization for Scientific Research (*Nederlandse Organisatie voor Wetenschappelijk Onderzoek –* NWO)[99] and the Organization Health Research Netherlands (*Organisatie*

---

[89] Environmental Management Act, art. 2.27.1.a.
[90] Environmental Management Act, art. 2.27.1.b.
[91] Environmental Management Act, art. 2.27.1.c.
[92] Environmental Management Act, art. 2.27.2.
[93] Act of 18 January 1956, on new legal regulations on the organization of health care (*Wet van 18 januari 1956, houdende nieuwe wettelijke voorschriften met betrekking tot de organisatie van de gezondheidszorg*) (Health Act) [amended last in 2018].
[94] Health Act, art. 21.
[95] Health Act, art. 22.
[96] Ibid.
[97] Website Health Council: www.healthcouncil.nl/task-en-procedure/legal-task, accessed 24 November 2018.
[98] Website Health Council: www.healthcouncil.nl/task-en-procedure/independence, accessed 24 November 2018.
[99] NWO (2018) www.nwo.nl accessed 7 November 2018.

ZorgOnderzoek Nederland – ZON).[100] The NWO was established in 1987 by the Act on the Netherlands Organization for Scientific Research (*Wet op de Nederlandse organisatie voor wetenschappelijk onderzoek*).[101] Its task is furthering the quality of scientific research and stimulating new developments in scientific research in general.[102] The ZON focuses specifically on research and development in the fields of health, prevention and care. It was established in 1998 by the Act on the Organization of Health Research in the Netherlands (*Wet op de organisatie ZorgOnderzoek Nederland*).[103]

Funding for health research comes from the Netherlands Organization for Health Research and Development (*Nederlandse organisatie voor gezondheidsonderzoek en zorginnovatie*),[104] an organization in which NWO and ZON collaborate. ZonMw is tasked to solve problems and challenges in health care and research and promote the actual use of scientific knowledge.[105] Currently, no research into human gene editing is funded in the Netherlands by either NWO or Horizon 2020.[106]

III SUBSTANTIVE PROVISIONS

Having presented the regulatory environment in general terms, it is now possible to explain the Dutch legal rules on germline genome modification. For the regulation of human genome germline editing, two provisions of the Embryo Act are essential: the prohibition on creating human embryos for research (Article 24.a) and the prohibition on genetically modifying human embryos and gametes (Article 24.g). These prohibitions mirror the prohibitions contained in Articles 18.2 and 13 of the Oviedo Convention, even though the Netherlands has not ratified it.

---

[100] ZonMw (2018) www.zonmw.nl accessed 7 November 2018.
[101] Act of 7 July 1987 on the regulation of the Dutch Organization of Scientific Research (*Wet van 7 juli 1987, houdende herziene regeling van de Nederlandse organisatie voor zuiverwetenschappelijk onderzoek*) (*Act on the Netherlands Organization for Scientific Research*) [amended last in 2017].
[102] NWO Act, art. 3.1.
[103] Act of 14 February 1998, on the Act on the Organization of Health Research in the Netherlands (*Wet van 14 februari 1998*) (*Act on the Organization of Health Research in the Netherlands*) [amended last in 2011]; ZON Act, art. 2.1.
[104] ZonMw, 'ZonMw in the Netherlands' (ZonMw 2018) www.zonmw.nl/en/about-zonmw/zonmw-in-the-netherlands/ accessed 24 October 2018.
[105] ZonMw in English (2018) www.zonmw.nl/en/ accessed 7 November 2018.
[106] According to the latest project lists on: European Commission, 'Examples of EU funded projects' (European Commission) http://ec.europa.eu/budget/euprojects/search-projects_en last accessed 7 November 2018 and NWO, 'Research & Results' (NWO 2018) www.nwo.nl/en/research-and-results accessed on 7 November 2018.

1 *The Prohibition to Create Human Embryos for Research Purposes*

Research on embryos can be done with surplus embryos if certain conditions are met, namely: (i) the research must be likely to lead to new insights in the field of medical science;[107] (ii) it cannot be conducted in any another manner but by using surplus embryos;[108] (iii) its methodology must be convincing;[109] (iv) the donor couple must give informed and written consent, after a 'sufficient time of reflection';[110] (v) cells grown from an embryo, as well as reproductive cells and embryos, cannot be used for purposes other than those for which they may be made available;[111] and (vi) the embryo must not be allowed to develop outside the human body for more than 14 days.[112]

However, scientists cannot create embryos for research purposes. Article 24. a of the Embryo Act prohibits 'deliberately creating embryos and using deliberately created embryos for scientific research and other purposes than initiating a pregnancy'. The ban is phrased in such a way that Dutch scientists are also not allowed to import and use embryos that have been created for research purposes abroad, as confirmed by the Explanatory Memorandum of the Embryo Act.[113] As will be discussed below, this prohibition forms the main legal obstacle for research involving human germline editing.

It should be noted that the legislator had in mind at least four considerations when drafting the ban. The first was the need to respect human life, demanded both by the Dutch legal system and by prevailing societal attitudes. According to the government, 'the creation of embryos for research purposes constitutes a graver interference with the principle of respect for human life than in case of using embryos that already existed, such as embryos that are left over from IVF'.[114] The second was the need to ensure that scientific research is not unduly hindered.[115] The third was taking into consideration the views on the matter of the public. The government observed in that context 'that, outside scientifically oriented circles, there hardly is public support within society for the creation of embryos for scientific purposes'.[116] And, finally, the fourth was keeping the

---

[107] Embryo Act, art. 10.a; Medical Research (Human Subjects) Act, arts. 14.1 and 16.
[108] Embryo Act, art. 10.b; Medical Research (Human Subjects) Act, arts. 14.1 and 16.
[109] Embryo Act, art. 10.c.
[110] Embryo Act, art. 8; Medical Research (Human Subjects) Act, arts. 14.1 and 16.
[111] Embryo Act, arts. 24.d and 24.h.
[112] Embryo Act, art. 24.e; Medical Research (Human Subjects) Act, arts. 14.1 and 16.
[113] Parliamentary document *Kamerstukken II* 2000/01, 27 423, no 3, 57–58 (Explanatory memorandum).
[114] Parliamentary document *Kamerstukken II* 2000/01, 27 423, no 3, 24 (Explanatory memorandum).
[115] Ibid. 25–26.
[116] Ibid. 27.

Netherlands in line with prevailing international and European standards.[117] At the time, only the United Kingdom permitted creating embryos for research purposes.

At the same time, the government underlined that scientific and societal developments in this field usually take place at a rapid pace. To enable the Embryo Act to respond to these developments, the ban on the creation of embryos for other purposes than a pregnancy should, according to the government, not be written in stone. Therefore, the ban can be lifted by mere Royal Decree, an act of the government that does not require a parliamentary vote.[118] Should that day arrive, the government has already made it clear that it will follow the so-called not allowed, unless approach,[119] namely, embryos will not be created ad hoc for research unless: (i) the scientific research leads to new and fundamental insights in infertility, artificial reproduction, transplantation medicine, or hereditary or congenital disorders; (ii) the specific research cannot be conducted without creating embryos, for example by using surplus embryos from IVF;[120] and (iii) the same requirements already in place for research on surplus embryos are met.[121]

2 The Ban on Human Germline Genetic Modification

Article 24.g of the Embryo Act prohibits 'deliberately modifying the genetic material of the *nucleus* of human germ cells with which a *pregnancy* will be established'.[122] These words suggest that human genetic modification is prohibited only for *reproductive* purposes, and only where *nuclear* DNA is concerned. In other words, Article 24.g does not prohibit *research* on germline editing. It also does not prohibit modification of *mitochondrial* DNA for either reproductive or research purposes. Therefore, in theory, scientists in the Netherlands can modify the genome of both human embryos and gametes for research purposes. Moreover, the technology of so-called human nuclear genome transfer[123] (also

---

[117] Ibid. 27–28.
[118] Embryo Act, art. 33(2).
[119] Parliamentary document *Kamerstukken II* 2000/01, 27 423, no 3, 30 (Explanatory memorandum).
[120] See first draft of the Embryo Act, art. 11 (Parliamentary document *Kamerstukken II* 2000/01, 27 423, no 1–2).
[121] Embryo Act, arts. 5, 6 and 7.
[122] Emphasis added.
[123] Nuclear genome transfer is a procedure that can be used to prevent passing on mitochondrial diseases to future children in the following way: one removes the nuclear DNA from eggs that originate from a woman with dysfunctional mitochondrial DNA and transfers the resulting nuclear DNA into the enucleated eggs from a third party who donated her eggs for this procedure.

known as 'mitochondrial replacement therapy')[124] is not explicitly prohibited, even for reproductive purposes.

However, even though the Embryo Act does not explicitly prohibit scientific research involving human germline editing or nuclear genome transfer, the prohibition of the creation of human embryos for research contained in Article 24.a makes this kind of research practically impossible. As Dutch scientists have emphasized repeatedly, surplus embryos cannot be used for this kind of research; it takes embryos that have been created ad hoc.[125] As a result, the Embryo Act contains, what could be called, a de facto prohibition on research in the field of human germline editing, since the creation of research embryos is indispensable for the most important types of research in this field.

What the Embryo Act does not prohibit, de jure or de facto, is research involving human germline editing or nuclear genome transfer that does not involve the creation of embryos for research purposes, such as editing the nuclear DNA of gametes for research purposes. Additionally, the Embryo Act, at least in theory, does not impede the introduction of nuclear genome transfer in clinical contexts. In practice, this will not be authorized, as long as research involving nuclear genome transfer cannot take place in the Netherlands. This was confirmed by the Dutch Minister of Health in answer to questions from Members of Parliament about the technology.[126]

Originally, the ban on human germline editing was repealable through a simple Royal Decree, just like the prohibition of the creation of embryos for research.[127] However, in response to the EU Clinical Trials Directive, which states that 'no gene therapy trials may be carried out which result in modifications to the subject's germ line genetic identity',[128] and which came into force during the parliamentary discussion of the Act, the government changed its

---

[124]   We agree with Baylis that the term 'mitochondrial replacement' is misleading, and will therefore, like the Dutch Health Council and COGEM (see COGEM/Health Council of the Netherlands, *Editing Human DNA: Moral and Social Implications of Germline Genetic Modification* (Bilthoven 2017)), use the term 'nuclear genome transfer' instead (see F Baylis, 'Human Nuclear Genome Transfer (So-Called Mitochondrial Replacement): Clearing the Underbrush' (January 2017) 31:1 Bioethics 7, 7–19.

[125]   As evidenced in a report commissioned by the Dutch government: J Eeuwijk and others, *Onderzoek naar speciaal kweken*, (Pallas 2015).

[126]   Parliamentary document *Kamerstukken II* 2016/17, 29 323, no 105, 12–13.

[127]   See first draft of the Embryo Act, art. 32.3 (parliamentary document *Kamerstukken II* 2000/01, 27 423, no 1–2).

[128]   Article 9.6 of Directive 2001/20/EC of the European Parliament and of the Council of 4 April 2001 on the approximation of the laws, regulations and administrative provisions of the Member States relating to the implementation of good clinical practice in the conduct of clinical trials on medicinal products for human use, [2001] OJ L 121, 34–44.

mind and amended the text of the proposed law accordingly.[129] As a result, a parliamentary vote is now needed in order to lift the ban on germline modification.

In its Explanatory Memorandum, the government stressed the need for further ethical reflection on human germline modification. Questions that need to be addressed include 'the question whether human dignity implies the right to inherit a genetic pattern that has not been modified as a result of intentional human interventions, or that germline therapy would, on the contrary, be required by that same principle'.[130] Furthermore, the government mentioned the irreversibility of human germline modification and the possible risk that human diversity would decrease as a result. Nevertheless, the government raised these questions without offering any answers. When it came to, for example, the ban on reproductive cloning, contained in Article 24.f of the Embryo Act, the government was much more outspoken: 'we completely share the fundamental argument that cloning human individuals violates human dignity'.[131]

IV CURRENT PERSPECTIVES AND FUTURE POSSIBILITIES

As discussed in the previous section, two bans dominate the national legislative and policy framework for human germline editing: the ban on intentionally modifying the genetic material of the nucleus of human germline cells for reproductive purposes and the ban on the creation of embryos for research purposes. So far, political debates in Parliament have tended to focus on the latter. The rise of CRISPR/Cas9 gave new impetus to the debate, but the issue has so far remained unresolved. A change of government in September 2017, with a different position on medical-ethical issues than the previous government, has further complicated the discussion.[132] A discussion of the debate will shed light on current perspectives and future possibilities for the evolution of the regulatory framework for embryo research and human germline genome modification in the Netherlands.

---

[129]   Parliamentary document *Kamerstukken II* 2000/01, 27 423, no 5, 99–100.
[130]   Parliamentary document *Kamerstukken II* 2000/01, 27 423, no 3, 45 (Explanatory memorandum).
[131]   Ibid. 42.
[132]   The previous government consisted of a conservative liberal political party (VVD) and the labour party (PvdA). The current government consists of a conservative liberal political party (VVD), a liberal-democratic party (D66), a center-right Christian-democratic party (CDA) and another Christian-democratic party (ChristenUnie).

Following up on the findings and recommendations of two evaluation studies on the Embryo Act that had been published since this legislation came into force,[133] in May 2016, the Minister of Health, Mrs Edith Schippers, sent a letter to Parliament proposing to lift the ban on creating embryos for research.[134] According to Schippers, who is a prominent member of the People's Party for Freedom and Democracy (*Volkspartij voor Vrijheid en Democratie* – VVD), a conservative-liberal political party, important scientific research is hindered by the prohibition. The Minister proposed to lift the ban only for some types of research. She stressed that, although the principle of respect for human life is an important one, other interests should also be taken into account. Through a revision of the Embryo Act, she aimed to offer relief to 'infertile couples and couples who risk passing on hereditary diseases to their offspring'. Describing her approach to research embryos as a 'not allowed, unless' policy,[135] she proposed to replace the existing blanket ban with a provision that allows the creation of embryos for certain types of research under certain conditions:

1. the research must be designed to generate new insights in the field of infertility, assisted reproductive technologies and hereditary or congenital deficiencies and must be directly relevant to clinical practice;
2. the research cannot be performed by using surplus embryos;
3. the research design and research activities must meet the relevant quality standards for scientific research;
4. the medical objective must outweigh the objections to creating embryos specifically for scientific research.[136]

In her letter, she identified three types of research that would fulfil these conditions: in vitro maturation (IVM), in vitro gametogenesis (IVG) and nuclear genome transfer (NGT). The 2016 letter to Parliament made headlines, received much media attention and led to diverging reactions.[137] In

---

[133] HB Winter and others, *Evaluatie Embryowet en Wet donorgegevens kunstmatige bevruchting* (ZonMW 2012); E.T.M. Olsthoorn-Heim, *Evaluatie Embryowet* (ZonMW 2006). Schippers had commissioned an inquiry into the matter. The resulting report had concluded that the legal ban on the creation of embryos for research purposes hampered clinically relevant developments in the field of medical technologies. J Eeuwijk and others, *Onderzoek naar speciaal kweken* (Pallas 2015) 15.

[134] Parliamentary document 2015/16, 29 323, no 101.

[135] Parliamentary document *Kamerstukken II* 2015/16, 29 323, no 101, 4.

[136] Ibid. 5.

[137] Eg G de Wert, 'Embryowet blijft te beperkt', (NRC Handelsblad, 31 May 2016) www.nrc.nl/nieu ws/2016/05/31/embryowet-blijft-te-beperkt-1623987-a557164 accessed 7 November 2018; BC van Beers, 'Debat over embryo's hoort op hoogste politieke niveau' (Volkskrant, 2 June 2016) www.volkskrant.nl/columns-opinie/debat-over-embryowet-hoort-op-hoogste-politieke-niveau~b5

a subsequent 2017 letter to the Parliament,[138] she also added human germline genome editing to the list. The 2017 letter made clear that Schippers considered human germline genome editing viable as soon as it is proven safe for clinical applications. However, to prove safety, clinical research is needed, and, according to her, this research should take place also in the Netherlands. Thus, she proposed to include germline genome editing in the list of types of research for which the ban on creating embryos should be lifted. The Minister stressed the importance of a public debate on the matter, although, in her view, it should focus on the question as to *how* to regulate the use of this technology, and not as to *whether* this technology is desirable in the first place. Indeed, she was 'optimistic that a regulatory framework – comparable to the existing regulatory framework for preimplantation genetic diagnostics (PGD) – could work well and could warrant that only morally and socially acceptable applications of germline modification would take place'.[139] As discussed, in the Netherlands, PGD is permitted only for serious hereditary diseases. Apparently, according to Schippers, human germline modification does not raise additional issues compared to PGD, apart from the safety issues. The Minister's position is that human germline editing should be available for prospective parents as soon as the technology no longer poses any health risks, and as long as it is only used for the elimination of serious hereditary diseases.

Barely a month later, the Dutch Health Council and the Netherlands Commission on Genetic Modification issued a joint advisory report on human germline genetic modification with exactly the same recommendation: lift the ban on creating embryos for research and on human germline genome modification for therapeutic purposes as soon as the technology is safe for introduction in the clinic.[140] In the report's chapter on the ethical dimensions of the issue, the two bodies discussed the ethical concerns about producing embryos for research, instead of using surplus embryos. The Netherlands Commission on Genetic Modification and Health Council argued that the fear of 'instrumentalization' of embryos could not justify a blanket ban on creating embryos for research. They emphasized that germline modification

---

b712fb/ accessed 7 November 2018; W Dondorp, G de Wert and S Repping, 'Instrumenteel gebruik van embryo's is al lang geaccepteerd' (Volkskrant, 9 June 2016) www.volkskrant.nl/col umns-opinie/instrumenteel-gebruik-van-embryo-s-is-allang-geaccepteerd~b796f9bb/ accessed 7 November 2018; NOS, 'Embryo's zijn mensen, die kweek je niet' (NOS, 27 May 2016) nos.nl/artikel/2107661-embryo-s-zijn-mensen-die-kweek-je-niet.html accessed 7 November 2018.

[138] Parliamentary document *Kamerstukken II* 2016/17, 29 323, no 110, 2.
[139] Ibid.
[140] COGEM/Health Council of the Netherlands, *Editing Human DNA: Moral and Social Implications of Germline Genetic Modification* (Bilthoven 2017).

would be able to eliminate serious diseases, also in the few cases in which PGD would no longer be an option.

Additionally, they discussed several arguments that have been raised against human germline editing based on human dignity, designer babies and slippery slopes, and equality and justice, and offered counterarguments for each, finding that using human germline editing to eliminate grave diseases does not conflict with human dignity but rather respects it, as the future child and future generations will be permanently rid of unhealthy genes. Moreover, both organizations argued that if the Dutch legislator does not prohibit the use of NGT, then it should also not prohibit the use of germline editing to prevent serious diseases.

Both Schippers's proposal and the advisory report were discussed in media at length.[141] Both positive[142] and negative[143] reactions were expressed by several ethicists and legal scholars. Moreover, several important organizations and institutions responded. The Dutch Council of State (*Raad van State*), a constitutionally established advisory body to the Dutch government and States General, was very critical. In its advisory report on the Minister's proposal, it wrote that it is not convinced of the necessity of lifting the ban on creating embryos to enable research in the field of reproductive technologies.[144] Moreover, it stressed the importance of the principle of respect for human life and human dignity, and pointed out that Schippers's policy could have certain negative side effects for society.

Similarly, the Rathenau Instituut, a government-sponsored organization that performs research relating to the societal aspects of science, innovation and new technologies, expressed concerns about these developments. In

---

[141] E.g. M Keulemans, 'Gezondheidsraad adviseert: legaliseer het genetisch bewerken van embryo's', (Volkskrant, 28 March 2017) www.volkskrant.nl/wetenschap/gezondheidsraad-adviseert-legaliseer -het-genetisch-bewerken-van-embryo-s~b53e7492/ accessed 24 November 2018.

[142] E.g. A Bredenoord, quoted in: D Waterval, 'Nederland is volwassen genoeg om aanpassing DNA te reguleren', (Trouw, 29 March 2017) www.trouw.nl/home/-nederland-is-volwassen-genoeg-om-aanpassing-dna-s-te-reguleren~a53c82f9/ accessed 24 October 2018; Y Buruma, 'Genetische modificatie. Fundamentele vragen voor het recht' (2017) 1754 NJB 32, 2310; EJ Oldekamp and MC de Vries, 'Nieuwe procreatietechnieken. Achterhaalde juridische kaders?' in *Nieuwe techniek, nieuwe zorg*. Preadviezen (Handelingen Vereniging voor Gezondheidsrecht. Deel 2018–1) (Den Haag: Sdu 2018) 15–88.

[143] D Pessers, 'De aanbidding van het DNA' (De Groene Amsterdammer, 6 September 2017) www.groene.nl/artikel/de-aanbidding-van-het-dna last accessed 24 October 2018; B van Beers, 'We zijn blij met de assemblage van onze iZoon: Designbaby's of de voortplanting van de toekomst' (2018) 6 De Groene Amsterdammer 42; T Vaessen, 'D66 maakt weg vrij voor industriële fabricage van baby's' (Financieele Dagblad, 22 December 2017) fd.nl/weekend/1229975/d66-maakt-weg-vrij-voor-industriele-fabricage-van-baby-s accessed 24 October 2018.

[144] Raad van State, 'Advice W13.16.0202/III' (4 November 2016) www.raadvanstate.nl/adviezen/zoeken-in-adviezen/tekst-advies.html?id=13060 accessed 24 October 2018.

a report entitled 'Rules for the Digital Human Park',[145] the Rathenau Instituut called for a broader discussion, in which more collective values and interests also would be involved, such as the rights and interests of future generations, human dignity and the protection of the human genome as the common heritage of humanity. Clearly, the authors were inspired by German philosopher Peter Sloterdijk's famous essay *Rules for the Human Zoo*.[146] Like Sloterdijk, they argued that new technologies, such as germline editing and persuasive technologies, could evolve into practices of breeding and taming human beings. Hence, according to them, the possible prospect of a self-domestication of the human species offers an important perspective for debates on germline modification.

Both Schippers's letters and the report written by the Health Council and the Netherlands Commission on Genetic Modification remained undiscussed in Parliament for a long time. The reason is that in March 2017 elections took place. Parliamentary discussions on 'controversial' topics like human germline editing were postponed until the new government was sworn in September 2017. Schippers did not return as Minister of Health. She was succeeded by Mr Hugo de Jonge, a member of Christian Democratic Appeal (*Christen-Democratisch Appèl* – CDA). The coalition agreement, which serves as the basis for the current government, immediately made clear that Minister De Jonge will not continue Schippers's line of policy with regard to human gene editing and creating embryos for research.[147] The coalition agreement emphasized the need for further public debate about these issues before further political decisions are made.[148] This was confirmed through several letters that De Jonge sent to Parliament in 2017 and 2018.[149]

Overall, in current debates on germline genome editing two conflicting ethical perspectives seem to dominate in the Netherlands. The report 'Rules

---

[145]   R Van *Est and others*, Regels voor het digitale mensenpark. 'Telen' en 'temmen' van de mens via kiembaanmodificatie en persuasieve technologie (Den Haag: Rathenau Instituut 2017). An English version was published earlier: *Rules for the digital human park. Two paradigmatic cases of breeding and taming human beings: human germline editing and persuasive technology* (11th Global Summit 2016) www.globalsummit-berlin2016.de/programme/GlobalSumm it2016DiscussionPapers.pdf accessed 24 October 2018.

[146]   P Sloterdijk, 'Rules for the Human Zoo: A response to the Letter on Humanism' (2009) 27 *Environment and Planning D: Society and Space* 12.

[147]   The coalition agreement was translated in English, see 'Confidence in the Future', www .kabinetsformatie2017.nl/documenten/verslagen/2017/10/10/coalition-agreement-confidence-in-the-future, accessed 24 October 2018.

[148]   'Confidence in the Future', 22.

[149]   Parliamentary documents *Kamerstukken II* 2017/2018, 34 775 XVI, no 46; Amendment Parliamentary documents 2017/18, 1141, 1; Parliamentary documents *Kamerstukken II* 2017/ 2018, 34 990, no 1.

for the Digital Human Park' calls these two opposing views the 'human rights regime' and the 'medical ethics regime'.[150] The 'human rights regime' goes back to the legal-ethical approach that underpins the founding international conventions and declarations in the field of biolaw. The Council of Europe's Oviedo Convention and UNESCO's Universal Declaration on the Human Genome and Human Rights, each in its own way, embody the thought that biomedical developments touch on the question of what it means to be human and 'that the misuse of biology and medicine may lead to acts endangering human dignity', to quote the Oviedo Convention's recital.

This approach can be distinguished from the 'medical ethics regime', which the Rathenau Instituut's report describes as follows: 'The basic question in this regime is whether a particular intervention in the human body satisfies criteria of safety, informed consent, and, in the context of reproductive medicine, also parental rights and reproductive freedom. In these terms, human germline engineering may be deemed ethically acceptable, especially when a particular intervention may alleviate potential suffering of a (future) human individual.'[151] This line of thinking can, for example, be recognized in the international calls for a moratorium on human gene editing from scientists working in the field. These groups' prime concerns are that 'the precise effects of genetic modification to an embryo may be impossible to know until after birth' and 'potential problems may not surface for years'.[152] Their main recommendations are more research and better education of the public by experts 'about this new era of human biology'.[153]

The Embryo Act shares the same legal-ethical outlook as the Oviedo Convention when it comes to human germline editing and research into human germline editing. However, this prohibitive 'human rights regime' approach is under increasing pressure. Instead, the 'medical ethics regime' approach, which focuses on the prevention of clinical risks and the principle of self-determination, is becoming more dominant in this context. Schippers's letters can serve as a good illustration of the current tendency to focus on safety risks instead of the legal-ethical principles that are at stake. By proposing to lift the ban on research embryos without first investigating whether IVM, IVG, NGT and germline editing are the way forward in the first place, Schippers seems to presuppose that the only possible moral objection against these technologies is that they are not clinically safe yet. Yet, this approach ignores some of the most

[150]   *Rules for the Digital Human Park* (n 145) 16–17.
[151]   Ibid. 16.
[152]   E Lanphier and others, 'Don't edit the human germline' (2015) 519 *Nature* 411.
[153]   D Baltimore and others, 'A prudent path forward for genomic engineering and germline gene modification' (2015) 348:6230 *Science* 38.

important questions raised by the prospect of germline genome editing. Interestingly, the European Group on Ethics in Science and New Technologies (EGE), which is an independent advisory body of the President of the European Commission, warns exactly about this tendency in a 2016 statement on gene editing:

> The EGE cautions against reducing the debate to safety issues and the potential health risks or health benefits of gene editing technologies. Other ethical principles such as human dignity, justice, equity, proportionality and autonomy are clearly at stake and should be part of this necessary reflection towards the international governance of gene editing.[154]

This brings us to a second line of critique on the manner in which the public debate on germline editing currently plays out in the Netherlands. From a human rights perspective, the legal principle of human dignity is of central importance to the debate on germline editing. Yet, in many of the policy documents, little reflection is offered on human dignity. A striking example is the position paper on genome editing that was written by the Royal Netherlands Academy of Arts and Sciences. This paper does not make any mention of human dignity.[155] Instead, the Academy's main concerns about human germline applications are about various types of health risks.

The report of the Health Council and the Netherlands Commission on Genetic Modification does engage with the principle of human dignity, albeit quite briefly. However, its interpretation is strikingly one-dimensional. It is common among scholars of bioethics and biolaw to distinguish between two dimensions of, or perspectives on, human dignity: human dignity as 'empowerment' and human dignity as 'constraint'.[156] The first perspective understands respect for human dignity as respect for personal autonomy and self-determination. The second interprets human dignity as a principle that protects individuals against dehumanization, objectification or commodification (even if these individuals consent to their dehumanizing treatment). What seems to underlie the ban on germline editing in the Oviedo Convention is mainly the second view of human dignity.[157] The Convention's Explanatory Report, in

---

[154] European Group on Ethics in Science and New Technologies, 'Statement on gene editing' (2016) ec.europa.eu/research/ege/pdf/gene_editing_ege_statement.pdf accessed 24 October 2018.

[155] Royal Netherlands Academy of Arts and Sciences, *Position paper on genome editing* (Koninklijke Nederlandse Akademie Van Wetenschappen, November 2016) www.knaw.nl/en/news/publications/genome-editing/@@download/pdf_file/Genome%20Editing%20Positi on%20Paper%20KNAW%20November%202016.pdf accessed 24 October 2018.

[156] Brownsword/Beyleveld, *Human Dignity in Bioethics and Law* (Oxford University Press 2001).

[157] For a further discussion, see BC van Beers, 'Imagining future people in biomedical law: From technological utopias to legal dystopias within the regulation of human genetic modification

its comments on Article 13, identifies as its ultimate fear the use of 'intentional modification of the human genome so as to produce individuals or entire groups endowed with particular characteristics and required qualities'.[158]

The Health Council and the Netherlands Commission on Genetic Modification, on the contrary, only highlight the other dimension of human dignity: empowerment. According to the report, if human germline editing is used to prevent suffering by removing the genetic cause of a disease, it should be considered a form of respect for human dignity.[159] From their perspective, dignity is primarily about alleviating suffering.

Since Schippers presented her ideas on germline editing, the political composition of the government has changed. It remains to be seen how this will affect the political debate on germline editing in the Netherlands. It is clear that this government is more cautious in its approach and prefers to await further public and political debate about the issue. How the current government intends to stimulate this debate is still unknown. It would be our suggestion that, in order to make this debate more inclusive and balanced, both the human rights perspective and the medical-ethical perspective should be properly represented. This means that not only the safety risks of the technology but also the legal and ethical principles that are at stake must be discussed; that not only scientists but also citizens should be heard; that not only individual but also collective interests and values should be addressed in the discussion; and that human dignity should not only be understood as respect for autonomy but also as protection against dehumanization.

technologies', in M Ambrus, R Rayfuse and W Werner (eds.), *Risk and the Regulation of Uncertainty in International Law* (Oxford: Oxford University Press) 117–140.

[158]   Explanatory Report to the Convention for the Protection of Human Rights and Dignity of the Human Being with Regard to the Application of Biology and Medicine: Convention on Human Rights and Biomedicine (Explanatory Report), para 89.

[159]   COGEM/Health Council of the Netherlands, *Editing Human DNA: Moral and Social Implications of Germline Genetic Modification* (Bilthoven 2017) 55.

The Regulation of Human Germline Genome Modification in Italy

Ludovica Poli

I INTRODUCTION

For historical, social, political, and religious reasons, Italy has traditionally approached life sciences, especially those involving humans, with great caution. The main law touching on human germline genome modification, called Law 40/2004 "Rules on Medically Assisted Reproduction,"[1] bans most "experimentation" on human embryos but, at the same time, fails to define what an embryo is and, surprisingly, might have left the door open to clinical application of therapies that modify the human germline genome.

This chapter first discusses the general regulatory environment for scientific work involving human germline genome modification. After having introduced the hierarchy of norms in the Italian legal system, it considers in detail relevant provisions of the Italian Constitution, European Union (EU) law, international treaties ratified by Italy, and Law 40/2004. The second part of the chapter aims at assessing whether and to what extent research using human germline genome modification technologies, as well as clinical applications of such techniques, is permitted under the Italian legal system. Finally, it offers some remarks on future perspectives.

II THE GENERAL LEGAL SYSTEM

In the Italian legal system, sources of law are arranged hierarchically. The Constitution of the Italian Republic (*Costituzione della Repubblica Italiana*)[2]

---

[1] Law 19, No. 40 (Rules on Medically Assisted Procreation) 2004.
[2] Constitution of the Italian Republic (entered into force on January 1, 1948) (Italian Constitution).

is the apex of the system, the fundamental source of law providing the legal basis for other sources, as well as for the activities of the public authority.[3]

International law becomes part of the Italian legal system in different ways, depending on whether it is customary, treaty, or EU law. According to Article 10 of the Constitution, the national legal system "conforms to the generally recognized rules of international law." As confirmed by the Constitutional Court, the Constitution refers to international customary rules in general, with the exclusion of international treaties.[4] International customary law prevails over both ordinary laws and the Constitution itself, with the only exception for the fundamental principles stated therein.[5]

Norms contained in treaties other than those of the EU legal system become part of the domestic legal system through a special incorporation procedure, by an act of Parliament.[6] Although they become part of the Italian legal system by virtue of an ordinary law, they assume a rank superior to statutory law (ordinary laws and sub-legislative acts) but not the Constitution.[7] Treaties that have been incorporated in the Italian legal system must conform to every article of the Constitution, not just to its core values. The Italian Constitutional Court might be called to verify the constitutional legitimacy of norms contained in treaties that have been ratified by Italy, with the aim to strike a "reasonable balance between the duties flowing from international law obligations, as imposed by Article 117.1 of the Constitution, and the safeguarding of the constitutionally protected interests contained in other articles of the Constitution."[8]

As far as EU law is concerned, among the legislative acts listed under Article 288 of the Treaty on the Functioning of the European Union, regulations (as well as decisions addressing persons other than states) are directly applicable in all EU member states: they do not need any implementing legislation by national parliaments. Directives and decisions addressing states, on the contrary, require national parliaments to enact proper legislation for implementation. Be that as it may, both directives and decisions might still have direct

---

[3] In particular, constitutional law and laws of constitutional reform; ordinary law, law decree and legislative decree; regional law; government regulations; habit or custom.

[4] Constitutional Court Judgment Nos. 15/1996, 348/2007 and 349/2007.

[5] Constitutional Court Judgment No. 48/1979, para. 3.

[6] The so-called order of execution (*ordine di esecuzione*) is contained in an ordinary law requiring the implementation of the treaty.

[7] These conclusions are based on the interpretation of Art. 117.1 of the Italian Constitution, providing that "[l]egislative powers shall be vested in the State and the Regions in compliance with the Constitution and with the constraints deriving from EU legislation and international obligations"; Constitutional Court Judgment Nos. 348/2007 and 349/2007.

[8] Constitutional Court Judgment No. 349/2007, para. 4.7.

effects to the extent they impose an unconditional and sufficiently clear and precise obligation on member states, for instance when they create individual rights which national courts must protect.[9]

Crucially, EU law has primacy over Italian law. Not only must Italian law be interpreted consistently with it, but, as confirmed by the Italian Constitutional Court, Italian judges must set aside domestic law, whenever incompatible with EU law.[10] The only exception to the primacy of EU law is a subset of fundamental principles and constitutional rights, the core values of the Italian legal system, which can never be breached by EU law.[11] In case of conflict between EU law and these fundamental principles, the Italian Constitutional Court will exercise its power to verify the constitutional legitimacy of the EU law and might set it aside to give precedence to the Constitution.[12]

III REGULATORY ENVIRONMENT

1 *Constitution of the Italian Republic, EU Law, and Treaties*

Human germline genome modification technologies, which have enormous therapeutic potential, can further certain values enjoying a constitutional status under the Italian legal system, such as the promotion of scientific progress and the protection of health. In fact, according to the Constitution, Italy "promotes the development of culture and of scientific and technical research"[3] and "safeguards health as a fundamental right of the individual and as a collective interest."[14] Both are to be considered core values of the Italian legal system.

The right to science and the right to health are also guaranteed under the International Covenant on Economic, Social and Cultural Rights (ICESCR), which Italy ratified in 1978.[15] Article 12 provides "the right of everyone to the

---

[9]   R Adam, A Tizzano, *Manuale di Diritto dell'Unione Europea* (Giappichelli 2017) 173–183.
[10] Constitutional Court Judgment No. 170/1984, *Granital*. The Court has based its conclusions on the interpretation of Art. 11 of the Italian Constitution, allowing limitations of sovereignty that may be necessary to participate in international organizations aiming at ensuring peace and justice among nations.
[11] Constitutional Court Judgment Nos. 168/1991 and 115/1993.
[12] This position has been confirmed by the Court of Justice of the European Union in the Judgment of December 5, 2017 (Grand Chamber), case C-42/17 M.A.S. and M.B (Taricco II), released upon the request for a preliminary ruling from the Italian Constitutional Court (Order No. 24/2017).
[13] Italian Constitution, art. 9.
[14] Ibid., art. 32.
[15] International Covenant on Economic, Social and Cultural Rights (adopted 16 December 1966, entered into force January 3, 1976) 999 UNTS 3 (ICESCR).

enjoyment of the highest attainable standard of physical and mental health," while Article 15 enshrines the right "to enjoy the benefits of scientific progress and its applications," as well as the "the freedom indispensable for scientific research and creative activity." The right to health is recognized in numerous international instruments and has already been the object of exhaustive elaboration (including in the General Comment No. 14,[16] adopted by the Committee on Economic, Social and Cultural Rights in 2000). On the contrary, the contents of the right to science are still debated and its centrality in the human rights discourse is rarely invoked. However, as stressed by the Special Rapporteur in the Field of Cultural Rights, Ms. Farida Shaheed, "scientific innovations are changing human existence in ways that were inconceivable a few decade ago."[17] Science not only offers solutions to individual, social, economic, and developmental issues, but it also has an autonomous standing among other fundamental rights. The meaning of science, thus, is not just instrumental to the realization or improvement of other rights, it has also an inherent value strictly linked to the idea of dignified human life that goes beyond survival and security needs.[18]

While a General Comment on Article 15 of the ICESCR is being drafted, some indications on the normative content of the right come from the above-mentioned report by the Special Rapporteur, as well as from the *Venice Statement on the Right to Enjoy the Benefits of Scientific Progress and Its Applications*.[19] In particular, the definition of the scope of the right to enjoy the benefits of scientific progress and its applications is of key relevance for the purposes of the present study. In this regard, it is useful to insist on two specific points.

First, the expression "benefits of scientific progress" makes it clear that these benefits "encompass not only scientific results and outcomes but also the scientific process, its methodologies and tools."[20] In other words, even the

---

[16] Committee on Economic, Social and Cultural Rights, "General Comment No. 14 – The Right to the Highest Attainable Standard of Health (Art. 12)" (August 11, 2000) E/C.12/2000/4.

[17] HRC "Report of the Special Rapporteur in the Field of Cultural Rights, Farida Shaheed on the Right to Enjoy the Benefits of Scientific Progress and its Applications" (14 May 2012) A/HRC/20/26, para. 1 (Report of the Special Rapporteur in the Field of Cultural Rights).

[18] L Shaver, "The Right to Science: Ensuring that Everyone Benefits from Scientific and Technological Progress" (2015) 411 *Journal européen des droits de l'homme – European Journal of Human Rights* 416.

[19] The Venice Statement on the Right to Enjoy the Benefits of Scientific Progress and its Applications is the outcome of three expert meetings held between June 2007 and July 2009 under the auspices by UNESCO in collaboration with the Amsterdam Centre for International Law, the Irish Centre for Human Rights, and the European Inter-University Centre for Human Rights and Democratization (Venice Statement) www.aaas.org/sites/defa ult/files/VeniceStatement_July2009.pdf accessed February 27, 2018.

[20] Venice Statement, para. 8.

general enhancement of the conditions for further scientific activity is to be considered a benefit. This is a key issue as it permits to stress that basic research also, and not only applied research, is covered by the right to science.

Second, the term "enjoyment" has to be considered as covering not only the dissemination of the outcomes of the scientific progress[21] but also the participation in its development. This has been confirmed by the Special Rapporteur in the Field of Cultural Rights, who stated that the normative content of the right envisaged under Article 15 of the ICESCR includes, along with the "access to the benefits of science by everyone, without discrimination," the "participation of individuals and communities in decision-making" and also "opportunities for all to contribute to the scientific enterprise and freedom indispensable for scientific research."[22]

Thus, any total ban of research on human germline genome modification techniques would certainly imply a violation of Article 15 of the ICESCR, interfering with scientific progress and impairing the right of many people to enjoy the benefits of science and its applications.

However, as the debate over the human germline genome modification demonstrates, while the benefits of intervening on gametes or on early-stage embryos to treat genetic diseases are evident, there is also fear that germline modifications might leave room for abuses.[23] For example, human germline genome modification techniques could be used to enhance traits with no therapeutic need. This generates "a public discomfort [. . .] whether for fear of exacerbating social inequities or of creating social pressure for people to use technologies they would not otherwise choose."[24] Moreover, along with other genetic technologies, human germline genome modification might be considered as having an implicit and inherent eugenic nature. Eugenics, as such,

---

[21]   The access to scientific knowledge, even in the form of right to science education, is a pivotal aspect of the right to science and it is also a precondition for its full realization, as well as for the concrete possibility to make informed decisions. Report of the Special Rapporteur in the Field of Cultural Rights (n 17) para. 27; Ibid. para. 22.

[22]   Ibid. para. 25.

[23]   J Sugarman, "Ethics and Germline Gene Editing" (2015) EMBO reports http://embor .embopress.org/content/16/8/879 accessed 27 February 2018; T Ishii, "Germline Genome-Editing Research and its Socioethical Implications" [2015] 21:8 *Trends in Molecular Medicine* 473; T Ishii, "Reproductive Medicine Involving Genome Editing: Clinical Uncertainties and Embryological Needs" (2017) 27 *Reproductive BioMedicine* Online www.rbmojournal.com/arti cle/S1472-6483(16)30549-1/fulltext accessed 27 February 2018; MH Porteus, CT Dann, "Genome Editing of the Germline: Broadening the Discussion" (2015) 23:6 *Molecular Therapy* 980; KS Bosley and others "CRISPR Germline Engineering – The Community Speaks" (2015) 33 *Nature Biotechnology* 478.

[24]   National Academies of Science, Engineering and Medicine, *Human Genome Editing. Science, Ethics, and Governance* (National Academies Press 2017) 9.

"is concerning because it could be used to reinforce prejudice and narrow definitions of normalcy in our societies."[25] It "sends a message about the 'fitness' of [certain] traits or conditions, thereby reflecting on the worth and value of people who have that trait in our society."[26]

Eugenic practices, and in particular "those aiming at the selection of persons," are explicitly prohibited under Article 3.2 of the Charter of Fundamental Rights of the European Union,[27] which affirms the right to the integrity of the person in the fields of medicine and biology. In the Italian legal system, the Charter has the same value as EU treaties and enjoys primacy over Italian law. According to the explanations attached to the Charter, Article 3.2 refers to "possible situations in which selection programmes are organized and implemented, involving campaigns for sterilization, forced pregnancy, compulsory ethnic marriage among others, all acts deemed to be international crimes in the Statute of the International Criminal Court adopted in Rome on 17 July 1998."[28] The drafters of the EU Charter, thus, had the intention to ban projects aimed at the improvement of the human race based on the selection by the State of who can procreate. They meant to ban "traditional" applications of eugenics, one that harks back to the horrors of World War II. However, concerns about a potential new method of selection, realized through genome manipulation, are not completely ill-founded. Future parents are now in a position to know in advance the genetic makeup of their children, and human germline genome modification could therefore open the door to the "production" of "designer babies," with parents making decisions about their children's genes in view of an improvement that goes beyond the treatment of medical disorders. Indeed, to some extent, these technologies allow a new form of eugenics, transforming parents from "victims" of a State's impositions to be "responsible" for choices imposed on their children.[29]

[25] KE Ormond and others, "Human Germline Genome Editing" (2017) 101 *Am J Hum Genet* 167, 171.

[26] Ibid. 172.

[27] Charter of the Fundamental Rights of the European Union (ratified December 7, 2000, entered into force December 1, 2007) (2000) OJ C 364/1.

[28] Explanations Relating to the Charter of Fundamental Rights (2007) OJ C 303. Although they do not have the status of law, the explanations are considered a valuable tool of interpretation intended to clarify the provisions of the Charter.

[29] C Campiglio, "Eugenetica e diritto internazionale" in N Boschiero (ed.), *Ordine internazionale e valori etici: VIII Convegno, Verona, 26–27 giugno 2003* (Editoriale Scientifica 2004) 453, 461. See also DJ Galton, "Eugenics: Some Lessons from the Past" (2005) 10:1 *Reproductive BioMedicine Online* 133 www.rbmojournal.com/article/S1472-6483(10)62222-5/pdf accessed February 27, 2018; JN Missa, "From State Eugenics to Private Eugenics" (1999) 13:4 *Best Practice & Research Clinical Obstetrics & Gynaecology* 533.

However, such a controversial issue has to be contextualized. Human germline genome modification needs to be discussed in a more comprehensive framework, including all the possible techniques potentially determining a selection of individuals, and keeping in mind human rights. This is the approach followed by the European Court of Human Rights (ECtHR) in the case *Costa and Pavan v. Italy*.[30] In that case, the Court was called to decide on the compatibility of Italy's ban of preimplantation genetic diagnosis (PGD) with the European Convention. In its decision, the Court stressed that concerns regarding "eugenic" uses of PGD would not be sufficient to justify banning it. Selectivity is not just a distinctive element of PGD, and therefore it does not represent a sufficient reason to exclude its applicability. This position was subsequently confirmed by the Italian Constitutional Court in Judgment No. 96/2015.[31]

A more explicit limit to human germline genome modification comes from the Convention for the Protection of Human Rights and Dignity of the Human Being with Regard to the Application of Biology and Medicine, adopted within the Council of Europe in 1997 (also known as the Oviedo Convention).[32] Article 13 of the Oviedo Convention entitled "interventions on the human genome" reads as follows: "An intervention seeking to modify the human genome may only be undertaken for preventive, diagnostic or therapeutic purposes and only if its aim is not to introduce any modification in the genome of any descendants."[33] The provision, along with others contained in Chapter IV on the "human genome," is clearly inspired by the precautionary principle and it is strictly connected to the "slippery slope argument"

---

[30] European Court of Human Rights *Costa and Pavan v. Italy* (2012) application no. 54270/ 10, para 63. For a more detailed analysis: L Poli, "La diagnosi genetica pre-impianto al vaglio della Corte Europea dei diritti dell'uomo" (2013) 1 *Rivista di Diritto Internazionale* 119; L Poli, "Pre-implantation Genetic Diagnosis under the European Court of Human Rights' Review: An Opening toward a Wider Acceptance of the Technique in Europe?" (2013) 4 *CYIL* 141.

[31] Constitutional Court Judgment No. 96/2015, declaring the unconstitutionality of Art. 1.1, 1.2, and 4.1 of Law No. 40/2004. According to the Court, the exclusion of fertile people carriers of transmittable genetic diseases from medically assisted reproduction technology was in breach of Art. 3 of the Constitution, guaranteeing the principle of equality. The Court found that this exclusion unreasonably balanced the interests at stake and violated the criterion of reasonableness of the legal order, considering that while Law 40/2004 prohibited women from acquiring information about the embryo through preimplantation genetic diagnosis, any fetus affected by genetic disease could be legally aborted.

[32] Convention for the Protection of Human Rights and Dignity of the Human Being with Regard to the Application of Biology and Medicine (adopted April 4, 1997, entered into force December 1, 1999) ETS 164 (Oviedo Convention).

[33] Ibid., art. 13.

developed in bioethics.[34] As the Explanatory Report of the Oviedo Convention makes clear, the provision aims at addressing the fear that misuse of the progress of science in the field of genome editing "may endanger not only the individual but the species itself," through the "intentional modification of the human genome so as to produce individuals or entire groups endowed with particular characteristics and required qualities."[35] Although from a bioethical perspective germline editing is widely perceived as a technology that might "cross a line many have viewed as ethically inviolable,"[36] it remains unclear to what extent an intervention seeking to impede the transmission of a serious disease would "endanger the species itself" as mentioned in the Explanatory Report. The fear of potential eugenics drifts is traceable in the provision, along with the uncertainty surrounding the balance between risk and benefits connected to the application of human germline genome modification techniques.

Be that as it may, the Oviedo Convention is not currently binding for Italy. Although the Italian Government signed it, and authorization to ratification was given by the Parliament with Law 145/2001, it was never ratified.[37] Granted, Article 18 of the 1969 Vienna Convention on the Law of Treaties contains an obligation not to defeat the object and purpose of a treaty prior to its entry into force.[38] This "pre-conventional obligation" is a key principle of international law, being the expression of "good faith taken in the sense of

[34] IR Pavone, *La Convenzione europea sulla biomedicina* (Giuffrè Editore 2009) 72–73. See also R Andorno "The Oviedo Convention: A European Legal Framework at the Intersection of Human Rights and Health Law" (2005) 2:4 *JIBL* 133.

[35] Explanatory Report to the Convention for the Protection of Human Rights and Dignity of the Human Being with Regard to the Application of Biology and Medicine: Convention on Human Rights and Biomedicine (1997) European Treaty Series No. 164, para. 89.

[36] National Academies of Science, Engineering and Medicine, *Human Genome Editing* (n 24) 7; Parliamentary Assembly of the Council of Europe Recommendation 2115: The Use of New Genetic Technologies in Human Beings (October 12, 2017) (Parliamentary Assembly of the Council of Europe, *The Use of New Genetic Technologies in Human Beings*).

[37] Law No. 145 /2001, "Ratification and implementation of the Council of Europe Convention for the Protection of Human Rights and Dignity of the Human Being with regard to the Application of Biology and Medicine: Convention on Human Rights and Biomedicine, adopted in Oviedo on 4 April 1997, as well as the Additional Protocol of 12 January 1998, n. 168, on the prohibition of cloning of human beings" (*Ratifica ed esecuzione della Convenzione del Consiglio d'Europa per la protezione dei diritti dell'uomo e della dignità dell'essere umano riguardo all'applicazione della biologia e della medicina: Convenzione sui diritti dell'uomo e sulla biomedicina, fatta a Oviedo il 4 aprile 1997, nonché del Protocollo addizionale del 12 gennaio 1998, n. 168, sul divieto di clonazione di esseri umani*).

[38] Vienna Convention on the Law of Treaties (adopted May 23, 1969, entered into force January 27, 1980) 1155 UNTS 331. Art. 18 reads as follow: "A State is obliged to refrain from acts which would defeat the object and purpose of a treaty when: (a) It has signed the treaty or has exchanged instruments constituting the treaty subject to ratification, acceptance or

protection of legitimate expectations of a minimal loyal behavior."[39] One could argue that, even in the absence of a ratification, Italy is called nonetheless to act consistently with the object and purpose of the Oviedo Convention, namely "the protection of the dignity and identity of all human beings and the guarantee for everyone, without discrimination, of the respect for integrity and other rights and fundamental freedoms with regard to the application of biology and medicine."[40] However, the indeterminacy of the notion of "human dignity," which not only is unsatisfactorily explained in the Explanatory Report[41] but is also hardly definable as an objective value, taking the shape of "a right of humanity,"[42] does not allow us to conclude that human germline genome modification applications per se are contrary to the object and purpose of the Oviedo Convention.

In sum, while international law seems to pose concrete limitations to the development and application of human germline genome modification, a deeper analysis demonstrates that the need to avoid misuses of these techniques is not incompatible with an opportune regulation of these methods. This is also the approach recently taken by the Council of Europe. In October 2017, the Parliamentary Assembly of the Council of Europe adopted a resolution recognizing that, while the Oviedo Convention prohibits intervention on germline genome, nothing precludes a possible amendment of the treaty, to be implemented in line with the obligation contained in Article 28 of the Convention, under which States parties must ensure that "the fundamental questions raised by the developments of biology and medicine are the subject of appropriate public discussion in the light, in particular, of relevant medical, social, economic, ethical and legal implications, and that their possible

approval, until it shall have made its intention clear not to become a party to the treaty; or (b) It has expressed its consent to be bound by the treaty, pending the entry into force of the treaty and provided that such entry into force is not unduly delayed". O Dörr, "Art. 18. Obligation not to defeat the object and purpose of a treaty prior to its entry into force," in O Dörr and K Schmalenbach (eds.) *Vienna Convention on the Law of Treaties: A Commentary* (Springer 2018) 243.

39   R Kolb, *Good Faith in International Law* (Hart Publishing 2017) 43.

40   Oviedo Convention, art. 1.1.

41   The Explanatory Report does not provide a definition of "human dignity," rather it focuses on the extension of the protection, in particular stressing, at para 19, that "[t]he Convention (...) uses the expression 'human being' to state the necessity to protect the dignity and identity of all human beings. It was acknowledged that it was a generally accepted principle that human dignity and the identity of the human being had to be respected as soon as life began."

42   P De Sena, "Dignità umana in senso oggettivo e diritto internazionale" (2017) *Diritti umani e diritto internazionale* 573. See also C McCrudden, "Human Dignity and Judicial Interpretation of Human Rights" (2008) 19:4 *EJIL* 655; P Sykora, A Caplan, "Germline Gene Therapy is Compatible with Human Dignity" [2017] 18 *EMBO reports* 2086.

application is made the subject of appropriate consultation."[43] Additionally, the Parliamentary Assembly, while urging "member States which have not yet ratified the Oviedo Convention to do so without further delay, or, as a minimum, to put in place a national ban on establishing a pregnancy with germline cells or human embryos having undergone intentional genome editing," has also explicitly encouraged the development of a common regulatory and legal framework to balance "potential benefits and risks of these technologies aiming to treat serious diseases, while preventing abuse or adverse effects of genetic technology on human beings."[44]

An opportune regulation would be therefore in line with international sources calling for caution toward possible misuses of the techniques and would represent at the same time a full realization of the right to science as provided for both in the ICESCR and in the Italian Constitution.

2 *Law 40/2004 on Rules on Medically Assisted Reproduction*

In Italy there is no law regulating human genome modification per se.[45] However, there is a law ruling on medically assisted reproduction (Law 40/2004), which is relevant for human germline genome editing because it regulates research on "human embryos." The key provision of Law 40/2004 is Article 13, entitled "Experimentations on Human Embryos." It reads:

1. Experimentations on human embryos are prohibited.

2. Clinical and experimental research on human embryos is permitted provided it pursues only therapeutic and diagnostic purposes, aimed at protecting the health and development of the embryo itself, and where alternatives are not available.

3. The following are in any case prohibited: a) the production of human embryos for research or experimentation purposes or in any case for purposes other than the one set by this law; b) any form of eugenic selection of embryos and gametes or interventions that, through selection techniques, manipulation or artificial procedures, aim at altering the genetic heritage of the embryo

---

43 Oviedo Convention, art. 28.
44 Parliamentary Assembly of the Council of Europe, *The Use of New Genetic Technologies in Human Beings* (n 36) para 5.
45 However, Article 81 "Quinquies" of the Industrial Property Code – Legislative Decree No. 30 of 10 February 2005 "Codice della proprietà industriale, a norma dell'articolo 15 della legge 12 dicembre 2002, n. 273" (Code of industrial property, in accordance with Article 15 of Law No. 273 of December 12, 2002) states that procedures for modifying the germline genetic identity of the human being are excluded from patentability, and thus, from commercial exploitation. It implements Directive 98/44/EC of the European Parliament and of the Council of July 6, 1998 on the legal protection of biotechnological inventions [1998] *OJ* L213/13.

or the gamete, or set genetic features, with the exception of interventions for diagnostic and therapeutic purposes, as per paragraph 2 of this article; c) cloning by nucleus transfer or early embryo splitting or ectogenesis whether for procreative or research purposes; d) the fertilization of a human gamete with a gamete of different species and the production of hybrids or chimeras.

4. The violation of the prohibitions referred to in paragraph 1 shall be punished with imprisonment from two to six years and with a fine ranging from 50.000 to 150.000 euros. In the case of violation of one of the prohibitions referred to in paragraph 3, the penalty is increased. Mitigating circumstances cannot be considered equivalent or prevalent on the aggravating circumstances envisaged by paragraph 3.

5. Healthcare professionals who are found guilty of one of the offenses referred to in this article are suspended from exercising their profession for a term between one to three years.

Law 40/2004, approved after an intense debate not only in the Parliament but also within civil society, filled a significant normative gap.[46] It also opened "a Pandora's box of ethical and technical issues,"[47] fueling an already intense debate over challenging and sensitive issues. Rather than providing clear directions for the application of medically assisted reproduction techniques, it just aimed at strictly restraining their use. It posed limits on the number of embryos that could be created at any one time and banned embryo cryopreservation, as well as technologies using a third party and PGD of the embryos. Eventually and unsurprisingly, the Italian Constitutional Court found several of its provisions unconstitutional,[48] while the ECtHR declared the prohibition of PGD

---

[46]   JA Robertson, "Protecting Embryos and Burdening Women: Assisted Reproduction in Italy" (2004) 19:8 *Human Reproduction* 1693; A Boggio, "Italy Enacts New Law on Medically Assisted Reproduction" (2005) 20:5 *Human Reproduction* 1153; I Riezzo and others, "Italian Law on Medically Assisted Reproduction: Do Women's Autonomy and Health Matter?" (2016) 16 *BMC Women's Health* 44.

[47]   V Fineschi and others, "The New Italian Law on Assisted Reproduction Technology (Law 40/ 2004)" (2005) 31 *Journal of Medical Ethics* 536, 539.

[48]   The Constitutional Court, with Judgment No. 151/2009, has declared unconstitutional Art. 14.2 (insofar as it imposed the creation of a limited number of embryos – maximum of three – and the duty to implant them simultaneously *in utero*), and Art. 14.3 (insofar as it did not provide that the transfer of the embryos should be made without prejudice to the health of the woman). With Judgment No.162/2014, the Court has then declared unconstitutional Art. 4.3, Art. 9, sections 1 and 3 (limited to the phrase "in breach of the prohibition laid down by Article 4.3"), and Art. 12.1, insofar as they ruled out recourse to heterologous techniques in the event of medically established sterility or infertility. Finally, Judgments Nos. 96/2015 and 229/2015 (n 31) considered the exclusion of fertile people carriers of transmittable genetic diseases from medically assisted reproduction technology.

in violation of Article 8 of the European Convention on Human Rights in the case *Costa and Pavan v. Italy*.[49]

In Judgment 229/2015, the Constitutional Court declared the unconstitutionality of Articles 13.3.b and 13.4 because they criminalized the selection of embryos even when it is aimed at avoiding the implantation of embryos affected by genetic transmissible diseases that meet the severity criteria set forth in Article 6.1.b of the Law 194/1978 regulating abortion.[50] This decision was consistent with the conclusions reached in its previous Judgment 96/2015 and by the European Court in the case *Costa and Pavan*.

However, on the question of the ban on research on cryopreserved supernumerary embryos, both the Italian Constitutional Court and the ECtHR have been less trenchant. First, in *Parrillo v. Italy*, in the light of a lack of European consensus on the matter, the European Court recognized that States have a wide margin of appreciation on this sensitive issue. Thus, it ruled out that such prohibition violates Article 8 of the European Convention.[51] Then, in Judgment 84/2016 the Constitutional Court rejected as inadmissible a request of a ruling on the constitutionality of the ban, issued by the Tribunal of Florence on December 7, 2012. In the decision, the Constitutional Court affirmed that it is within the powers of the legislator, "acting as the interpreter of the general will, ... to strike a balance through legislation between the fundamental values that are in conflict, taking account of the views and calls for action that it considers to be most deeply rooted at any given moment in time within the social conscience."[52]

After these two decisions, Article 13 of the Law 40/2004 remains an extremely ambiguous provision as far as genome editing is concerned. On the one hand, the Article prohibits experimentations on embryos in general terms (paragraph 1) and explicitly bans interventions that, "through selection techniques, manipulation or artificial procedures, aim at altering the genetic heritage of the embryo or the gamete, or set genetic features" (paragraph 3). On the other hand, it leaves the door open for possible diagnostic or therapeutic application of gene editing (paragraphs 2 and 3).

---

[49] *Costa and Pavan v. Italy* (n 30).
[50] Law No. 194/1978 "Rules for the social protection of motherhood and voluntary termination of pregnancy" (*Norme per la tutela sociale della maternità e sull'interruzione volontaria della gravidanza*).
[51] European Court of Human Rights, *Parrillo v. Italy* (2015) application no. 46470/11, para 174–176.
[52] Constitutional Court Judgment No. 84/2016, para 11.

3 The Opinion on Ethical Considerations Surrounding Gene Editing and CRISPR-CAS9 Technique of the National Committee for Bioethics

The overview of the Italian regulatory framework would not be complete without mentioning the *Opinion on Ethical Considerations Surrounding Gene Editing and CRISPR-CAS9 Technique*, released by the Italian Committee for Bioethics (*Comitato Nazionale per la Bioetica* – ICB) on February 23, 2017 ("2017 Opinion").[53] The ICB was established in 1990 by the Council of Ministers, with the task of advising the Presidency of the Council of Ministers, and even drafting laws, on ethical and legal problems arising out of advances in biological scientific research and technological applications.[54] While ICB documents are not per se binding, they contribute to stimulate and drive political and legal debate over ethically controversial issues.

In its 2017 Opinion, the Committee expresses its favor for animal experimentation (as long as it is performed according to international rules) with the aim to test the safety and effectiveness of gene editing on germline and embryos.[55] While encouraging research on gene editing on human somatic cells,[56] the Committee unanimously rejects experimentation on gametes that are intended to be used in reproduction and on human embryos to be implanted.[57]

With regard to gene editing on gametes that are not to be used in reproduction and/or on embryos that are not to be implanted, the Committee's members are split. One view is that the moratorium on clinical research should not be extended to in vitro basic research on these gametes and embryos because this would mean "precluding research aiming at improving [gene editing]" and because the findings of this research are necessary to assess the scientific feasibility and ethical acceptability of clinical applications in the future.[58] The other view is that basic research is not currently justified, considering that, from a clinical perspective, gamete selection is preferable to gamete editing and that assessing the effectiveness and safety of in vitro gene

---

[53] Italian Committee for Bioethics (ICB), "L'editing genetico e la tecnica CRISPR-CAS9: considerazioni etiche" (February 23, 2017). http://bioetica.governo.it/it/documenti/pareri-e-ris poste/l-editing-genetico-e-la-tecnica-crispr-cas9-considerazioni-etiche accessed July 31, 2018.

[54] National Committee for Bioethics http://bioetica.governo.it/it accessed July 31, 2018.

[55] Ibid. para. 4.1.

[56] Ibid. para. 4.2.

[57] Ibid. para. 5.

[58] Ibid. para. 5.1. This is the position of nine members of the Committee (Professors Battaglia, Canestrari, Casonato, de Curtis, Di Segni, Flamigni, Garattini, Toraldo di Francia, Zuffa).

editing on embryos is not possible, since results of the genetic modification can be assessed only at birth or even later.[59]

4 Oversight Bodies for Research on Human Germline Modification

In Italy, a number of bodies oversee biomedical research. Governmental bodies include the Ministry of Health (*Ministero della Salute*) and the Ministry of University and Scientific Research (*Ministero dell'Università e della Ricerca Scientifica*).

Additionally, a number of independent bodies (*Comitati Etici*) have been established in universities, in public health facilities and in the so-called healthcare research institutes (*Istituti di ricovero e cura a carattere scientifico*), where clinical research activities are carried out. While the organizational set up of Committees varies from region to region, all Committees are called to evaluate and approve clinical trial protocols. In addition, their powers extend to issues concerning the use of medicinal products and medical devices as well as of surgical and clinical procedures.[60] In more detail, according to Article 1 of Ministry of Health Decree of February 8, 2013, the Ethics Committees

> 1. [. . .] have the responsibility to ensure the protection of the rights, safety and well-being of people taking part to experimentations and to provide public guarantees of such protection.
>
> 2. Where not already assigned to specific bodies, ethics committees can also perform advisory functions in relation to ethical issues related to scientific and welfare activities, in order to protect and promote fundamental values. The ethics committees, moreover, can propose training initiatives for health personnel regarding issues related to bioethics.[61]

The general framework of Ethics Committees has been modified in time. The most recent reform took place in 2018 when Law 3/2018[62] reduced the

---

[59] Ibid. para. 5.2. This is the position of nine members of the Committee (Professors Amato, Caltagirone, Dallapiccola, D'Agostino, Gensabella, Morresi, Palazzani, Proietti, Scaraffia).

[60] Law No. 189 /2012 "Conversion into law, with amendments, of the law decree n. 158 of 13 September 2012, containing urgent provisions to promote the development of the country through a higher level of health protection" (*Conversione in legge, con modificazioni, del decreto-legge 13 settembre 2012, n. 158, recante disposizioni urgenti per promuovere lo sviluppo del Paese mediante un più alto livello di tutela della salute*), art 12.10.c.

[61] Ministry of Health Decree of February 8, 2013, "Criteria for the composition and functioning of ethics committees" (*Criteri per la composizione e il funzionamento dei comitati etici*).

[62] Law No. 3/2018, "Delegation to the Government in the matter of clinical trials of medicines, containing provisions for the reorganization of the health professions and for the health leadership of the Ministry of Health" (*Delega al Governo in materia di sperimentazione clinica*

number of the Committees to a maximum of 40 throughout the country and established the "National Coordinating Center for Territorial Ethics Committees" (*Centro di coordinamento nazionale dei comitati etici territoriali*), with the mission to coordinate, guide, and monitor the activities of all Ethics Committees. In exceptional cases, the National Coordination Center might be involved in the evaluation of studies requiring a review for security reasons.

Finally, as already mentioned, the Italian Committee for Bioethics might provide opinions on ethical and legal problems arising out of advances in biological scientific research and technological applications.

Still, despite the existence of numerous entities with oversight competences, there are no governmental or nongovernmental bodies explicitly tasked to monitor research involving human genome editing per se.

5 Funding Opportunities

At this point in history, research on human germline genome modification cannot be funded with public money. It cannot be funded with EU money either. EU Regulation No. 1291/2013,[63] establishing "Horizon 2020," the Framework Programme for Research and Innovation (2014–2020), explicitly excludes from funding research activities intended to modify the genetic heritage of human beings that could make such changes heritable.[64] While the European Union left the door open for a possible review of this exclusion "in the light of scientific advances," to date no exceptions have been made.[65]

Whether research on human germline genome modification is privately funded and carried out in Italy is unknown, but it is unlikely because researchers and funders would be under threat of criminal prosecution under paragraphs 4 and 5 of Article 13 of Law 40/2004.

---

di medicinali nonché disposizioni per il riordino delle professioni sanitarie e per la dirigenza sanitaria del Ministero della salute).

[63] EU Regulation No. 1291/2013 of the European Parliament and of the Council of December 11, 2013 establishing Horizon 2020 – the Framework Programme for Research and Innovation (2014–2020) and repealing Decision No. 1982/2006/EC (2013) OJ L 347/104.

[64] With the sole exception of research relating to cancer treatment of the gonads: EU Regulation No. 1291/2013, art. 19.

[65] Communication from the Commission to the European Parliament, the Council, the European Economic and Social Committee and the Committee of the regions, "Horizon 2020 interim evaluation: maximizing the impact of EU research and innovation", COM (2018) 2 final, (January 11, 2018).

IV SUBSTANTIVE PROVISIONS

Having presented the regulatory environment in general terms, it is now possible to determine if, and to what extent, research using germline modification is permitted in the Italian legal system at each step of the research cycle.

1 Basic Research

a Genome Modification in Human Embryos

As we saw, Article 13 of Law 40/2004 provides:

1. Experimentations on human embryos are prohibited.

2. Clinical and experimental research on human embryos is permitted provided it pursues only therapeutic and diagnostic purposes, aimed at protecting the health and development of the embryo itself, and where alternatives are not available.

Although Article 13 of Law 40/2004 is a very ambiguous norm, at least it is clear it does not allow human germline genome modification basic research on embryos (paragraph 1). By "basic research," we mean research done in laboratory, in vitro, aimed at improving scientific knowledge and "performed without thought of practical ends."[66] Considering that Law 40/2004 does not provide a definition of "embryo," the ban of experimentation contained in Article 13 possibly includes also zygotes (pre-embryos), which, since 2009,[67] can be frozen as part of in vitro fertilization, whenever the physician believes that the transfer in utero is not compatible with the health conditions of the woman.[68]

Article 13.3 explicitly prohibits the production of human embryos for research or experimentation purposes (as well as human cloning and the creation of hybrids or chimeras). Moreover, Article 14 prohibits the removal of embryos.[69] Thus, not only supernumerary healthy

---

[66] V Bush, *Science: The Endless Frontier* (United States Government Printing Office, 1945), c 3. See also N Roll-Hansen, "Why the Distinction between Basic (Theoretical) and Applied (Practical) Research is Important in the Politics of Science" (2009) Centre for the Philosophy of Natural and Social Science Contingency and Dissent in Science Technical Report 04/09, 4 https://pdfs.semanticscholar.org/62fo/dced123c24c7bc89b7d0d72bfcf885634a43.pdf accessed February 27, 2017).

[67] Due to the exception – *de facto* introduced by Constitutional Court Judgment No. 151/2009 – to the prohibition of cryopreservation of embryos (n 48).

[68] Private communication, Dr. Gianluca Gennarelli, Senior Consultant, responsible person for medically assisted fertilization procedures at Ospedale Sant'Anna, Turin.

[69] Law 40/2004, art. 14.

embryos[70] but even embryos affected by a genetic disorder (which, for this reason, will not be transferred in utero) are kept in cryopreservation sine die, since they cannot be destroyed nor used for research. Italian scientists cannot create embryonic stem cell lines because doing so requires the destruction of at least one embryo. However, they are not prohibited from working on imported lines, and they do so.

Such a legal framework not only generates unnecessary hindrances for scientists who operate in Italy, but it might be also considered as a violation of Article 15 of ICESCR. In 2017, a communication was submitted to the Committee on Economic Social and Cultural Rights, raising the question of the compatibility of the Italian legislation with the right to science guaranteed under the Covenant. In particular, as far as Article 15 is concerned, Law 40/2004 has been challenged by a couple of healthy carriers of a serious hereditary disease who could not donate to science their embryos affected by such a disorder. The applicants argued that, by prohibiting research on embryos, even those affected by genetic disorders, Law 40/2004 interferes with scientific progress, slowing down the search for a cure for various diseases; secondly and consequently, in doing so, Law 40/2004 violates their right to enjoy the benefits of scientific progress and its applications and prevents them from participating in scientific research by donating their embryos affected by a genetic disorder. The Commitee, however, has declared inadmissible the communication in relation to this issue, considering that the authors (who have not claimed the intention to perform themselves any scientific research) cannot be considered victims of a violation of their freedom of research.[71]

b Germline Modification in Human Gametes

As far as gametes are concerned, some issues are clearly regulated, while others remain ambiguous. The prohibition contained in Article 13.3 of Law 40/2004 refers only to the production of embryos for research. Thus, the production of "artificial" gametes (i.e., gametes derived from pluripotent cells via in vitro gametogenesis) is permitted in Italy. Basic research on "artificial" and

---

[70] According to the 2015 guidelines on medically assisted procreation, all embryos which are not immediately transferred in utero have to be frozen and cryopreserved at the centers where the techniques were carried out and the related charges are borne by the same centers: Ministry of Health Decree of 1 July 2015, "Guidelines containing indications for medically assisted procreation procedures and techniques" (*Linee guida contenenti le indicazioni delle procedure e delle tecniche di procreazione medicalmente assistita*), 117.

[71] Committee on Economic, Social and Cultural Rights, *CSA and GP* v. *Italy*, Communication No. 22/2017, *Views adopted by the Committee under the Optional Protocol of the Covenant*, 7 March 2019, UN Doc. E/C.12/65/D/22/2017, para 6.19.

"natural" gametes is also permitted, considering that it is not banned in clear words unlike the prohibition of experimentations on embryos.

As far as gametes procurement is concerned, the Ministry of Health is working on a regulation to transpose the relevant EU directives,[72] specifying quality and safety standards for donated human tissues and cells, in the Italian legal system. The guidelines adopted by the Ministry of Health in 2015, which are rather directed to couples accessing ART (assisted reproduction technologies) treatments, are silent as to gametes donors. In case of heterologous fertilization, they prohibit intended parents from choosing particular phenotypic characteristics of the donor to avoid illegitimate eugenic selection.[73]

As it will be recalled, Article 13.3.b of Law 40/2004 prohibits

> any form of eugenic selection of embryos and gametes or interventions that, through selection techniques, manipulation or artificial procedures, aim at altering the genetic heritage of the embryo or the gamete, or set genetic features, with the exception of interventions for diagnostic and therapeutic purposes, as per paragraph 2 of this article.

A key interpretative doubt emerges from the reading of this norm and concerns the definition of "intervention." If "intervention" refers exclusively to clinical activity (i.e., in vivo, on patients), the prohibition of manipulation does not apply to in vitro research on gametes, and their manipulation is allowed. This interpretation would be consistent with the way the word "intervention" is used elsewhere in Law 40/2004, where it is used in the context of medically assisted reproduction procedures and not research as such.[74] It would be also consistent with the other prohibition listed in the provision, namely the "selection for eugenic purposes." It is clear, indeed, that this kind of selection can be realized only in a clinical activity (i.e., in a medically assisted reproduction procedure), since the eugenic purpose per se implies the transfer in utero as it is connected with the potential birth of a new individual.

---

[72]   Directive 2004/23/EC of the European Parliament and of the Council of March 31, 2004 on setting standards of quality and safety for the donation, procurement, testing, processing, preservation, storage and distribution of human tissues and cells (2004) OJ L 102/48; Commission Directive 2006/17/EC of February 8, 2006 implementing Directive 2004/23/EC of the European Parliament and of the Council as regards certain technical requirements for the donation, procurement, and testing of human tissues and cells (2006) OJ L 38/40; Commission Directive 2006/86/EC of October 24, 2006 implementing Directive 2004/23/EC of the European Parliament and of the Council as regards traceability requirements, notification of serious adverse reactions and events, and certain technical requirements for the coding, processing, preservation, storage, and distribution of human tissues and cells (2006) OJ L294/32.

[73]   Ministry of Health Decree of July 1, 2015, 113.

[74]   See arts 2, 4, 10, 15, 16.

However, Article 13 deals with research and not with clinical activity. For this reason the term "intervention" might need to be given a special meaning, different from the one given to it in other sections regulating other stages of research. If that is the case, the legitimacy of gametes manipulation depends on whether or not they are manipulated for therapeutic or diagnostic purposes: Only manipulation pursuing such aims would be allowed.

Another relevant issue is whether and to what extent mitochondrial manipulation techniques (MMT)[75] are permitted under Law 40/2004, considering that "the transfer of female embryos created via MMT can impact future generations through the maternal transmission of mtDNA."[76] Preclinical research using MMT seems to be permitted as long as it concerns oocytes not destined to be implanted once fertilized. In addition, clinical applications should also be considered compatible with Article 13, if they aim at protecting the health and development of the embryo itself. This point is of foremost importance, since in the near future MMT might be used "to prevent the onset of serious mitochondrial diseases in offspring as a reproductive option for prospective parents who wish to have a genetically related child and who have experienced PGD failure due to the high load of mtDNA mutations in the oocytes."[77]

2   Preclinical Research Using Germline Genome Modification Technologies

The Legislative Decree No. 26, of March 4, 2014,[78] implementing Directive 2010/63/EU on the Protection of Animals Used for Scientific Purposes, implicitly allows germline genome modification research on animals.[79] In particular, this conclusion can be derived from the definition of "procedure" detailed under Article 3.1.a of the Legislative Decree, as

> any use, invasive or non-invasive, of an animal for experimental or other scientific purposes, with known or unknown outcome, or educational

---

[75]   "Thus, MMT can alter the mtDNA content of human oocytes or zygotes through CT, karyoplast transfer (which includes carryover mtDNA) or autologous mitochondrial transfer (which might undergo mutagenesis during preparation) to treat intractable infertility or prevent mitochondrial disease in offspring": T Ishii and Y Hibino, "Mitochondrial Manipulation in Fertility Clinics: Regulation and Responsibility" (2018) 5 *Reproductive BioMedicine and Society Online* 93, 94 https://doi.org/10.1016/j.rbms.2018.01.002 accessed July 31, 2018.

[76]   Ibid. 95.

[77]   Ibid. 107.

[78]   Legislative Decree No. 26/ 2014, "Implementation of Directive 2010/63/EU on the protection of animals used for scientific purposes" (*Attuazione della direttiva 2010/63/UE sulla protezione degli animali utilizzati a fini scientifici*).

[79]   Directive 2010/63/EU of the European Parliament and of the Council of September 22, 2010 on the protection of animals used for scientific purposes (2010) OJ L276/33.

purposes, which may cause the animal a level of pain, suffering, distress or lasting harm equivalent to, or higher than, that caused by the introduction of a needle in accordance with good veterinary practice. This includes any action that intends or may lead to a series of genetically modified animals with a phenotype suffering in these conditions.[80]

As in any research on animals, limits and conditions are imposed to germline modification technologies. In particular, animals can be only used for scientific purposes when no other experimental method or strategy not involving the use of living animals and that is scientifically valid and reasonably and practically applicable is available.[81] Additionally, in Italy, raising genetically modified animals is allowed under the authorization of the Ministry of Health, provided the risks and benefits, the extent to which the manipulation is needed, the possible impact on the welfare of the animals, and the potential risk for human health, animal health, and the environment are assessed in advance.[82] Moreover, specific provisions apply to research on endangered species (Article 7), nonhuman primates (Article 8), animals taken from the wild (Article 9), stray and feral animals of domestic species, including dogs and cats (Article 11).

3   *Clinical Research Using Germline Genome Modification Technologies*

Clinical research using human germline genome modification is explicitly prohibited by Legislative Decree No. 211 of June 24, 2003,[83] implementing Directive 2001/20/EC on good clinical practice in the conduct of clinical trials on medicinal products for clinical use.[84] Article 9.6 of the Legislative Decree requires the "sponsor"[85] to obtain written authorization by the Ministry of

---

[80]   Also, Art. 17.1 states that "[a] procedure is considered terminated (. . .) when, in the case of new lines of genetically modified animals, transmission of genetic alteration has not given rise to or is not expected to give rise, with regard to descendants, to a level of pain, suffering, distress or prolonged injury equivalent or superior to that caused by the insertion of a needle."

[81]   Legislative Decree No. 26/ 2014, art 1.2.

[82]   Ibid., art. 10.4.

[83]   Legislative Decree No. 211 /2003, "Implementation of Directive 2001/20/EC on the application of good clinical practice in the execution of clinical trials of medicinal products for clinical use" (*Attuazione della direttiva 2001/20/CE relativa all'applicazione della buona pratica clinica nell'esecuzione delle sperimentazioni cliniche di medicinali per uso clinico*).

[84]   Directive 2001/20/EC of the European Parliament and of the Council of 4 April 2001 on the approximation of the laws, regulations, and administrative provisions of the Member States relating to the implementation of good clinical practice in the conduct of clinical trials on medicinal products for human use (2001) OJ L121/34.

[85]   Namely, "an individual, company, institution or organization which takes responsibility for the initiation, management and/or financing of a clinical trial": Legislative Decree No. 211/ 2003, art 2.3.e.

Health before the beginning of clinical trials involving the use of gene therapy but affirms that "trials of gene therapy that lead to changes in the germline genetic properties of the subject cannot be performed."[86] The provision replicates almost *verbatim* Article 9.6 of the Directive 2001/20/EC, which requires authorization by the competent authority in Member States "before commencing clinical trials involving medicinal products for gene therapy, somatic cell therapy including xenogenic cell therapy and all medicinal products containing genetically modified organisms" and prohibits any gene therapy trials "which result in modifications to the subject's germ line genetic identity."[87]

In 2014, the Directive was repealed by EU Regulation 536/2014.[88] However, implementation of EU Regulation 536/2014 depends on the development of a fully functional EU clinical trials portal and database, which is estimated to occur in 2020. Echoing the Directive, Article 90 of the Regulation, entitled "Specific requirements for special groups of medicinal products," provides that "no gene therapy clinical trials may be carried out which result in modifications to the subject's germ line genetic identity."

4 Clinical Applications of Germline Genome Modification Technologies

Counterintuitively, Article 13 of Law 40/2004 leaves the door open to clinical applications of germline modification technologies. As it has been recalled, Article 13 allows clinical and experimental research on human embryos, including "selection techniques, manipulation or artificial procedures, aim[ing] at altering the genetic heritage of the embryo or the gamete, or set genetic features,"[89] provided research "pursues only therapeutic and diagnostic purposes, aimed at protecting the health and development of the embryo itself, and where alternatives are not available."[90]

Although extremely ambiguous, this part of Article 13 ought to be interpreted as dealing with clinical applications and not with clinical research as such. We intend clinical application as referring to approved therapies that cure embryos from genetic defects and ensure their development. These therapies would be consistent with the "therapeutic purposes," on which the

---

[86] Ibid., art. 9.6.
[87] Directive 2001/20/EC, art. 9.6.
[88] EU Regulation No. 536/2014 of the European Parliament and of the Council of April 16, 2014 on clinical trials on medicinal products for human use, and repealing Directive 2001/20/EC (2014) OJ L 158/1.
[89] Law 40/2004, art. 13.3.b.
[90] Ibid., art. 13.2.

provision insists. On the contrary, clinical research remains prohibited, also considering that, due to the principle of primacy of EU law over domestic sources referred to above, Article 13 of Law 40/2004 must be interpreted consistently with the mentioned EU norms (in particular, EU Regulation 536/2014), prohibiting gene therapy trials resulting in modifications to the subject's germline.

As such, clinical applications of germline modification technologies are not prohibited by Law 40/2004.

V CURRENT PERSPECTIVES AND FUTURE POSSIBILITIES

For historical, social, and religious reasons, Italy has always been very conservative when it comes to life sciences. This exceedingly cautious attitude is reflected in Law 40/2004, at least in its original version before being chipped away by the Constitutional Court and ECtHR. Thus, it seems bizarre that Law 40/2004 prohibits basic research and clinical trials of human germline genome modification but does not prohibit its clinical application. It is probably an accidental paradox, the result of badly and hastily drafted legislation. However, it might be a serendipitous omission, one that is consistent with the overall aim of Law 40/2004: to protect the embryo. Curing it from genetic diseases would further this goal.

However, given the prevailing attitude among Italian lawmakers, particularly in the field of life sciences, there are still many doubts about the concrete possibility that the Ministry of Health might give the green light to clinical applications that lead to a modification of the human germline genome. It is also impossible, at present, to define whether these techniques can be considered as medicines – and, therefore, follow the process that, through the role of Ethical Committees referred to above, terminates with the authorization granted by the Italian Agency for Drugs (*Agenzia Italiana del Farmaco*) – or whether they should be classified differently.

In any event, recent developments and new possibilities in the field of human germline genome modification call for regulations that, while setting limits to contain possible abuses, do not wholly frustrate scientific and technological progress.[91] As it has been correctly stated, we are witnessing "a complex – more implicit than explicit – negotiation process between

---

[91] EM Kane, "Human Genome Editing: An Evolving Regulatory Climate" (2017) 57 *Jurimetrics J.* 301.

science and society taking place regarding which technique should be developed and used and for which reason this should or could be done."[92]

As far as Italy is concerned, considering that the ban on clinical research using germline modification technologies comes from EU law, the only space for a normative change concerns basic research applying human germline genome modification on embryos. This is currently outlawed by Law 40/2004, but it should be permitted, under certain conditions (such as the informed consent of people whose gametes have been used). Only the improvement of basic research on gametes would in fact permit future decisions on whether, and possible identification of which, clinical applications are scientifically feasible and ethically acceptable.[93] Italy's prohibition of research on cryopreserved supernumerary embryos, which, in any case, are not meant to be implanted and are therefore destined to remain stored sine die, unreasonably hampers two fundamental rights: the right to science and the right to health.

The approach toward the regulation of human germline genome editing should also be reconsidered at the international level. So far, concerns for potential misuses and abuses of human germline genome modification have determined an extremely cautious attitude toward these technologies, as illustrated, among other documents, by the Report on Pre-implantation Genetic Diagnosis and Germline Intervention prepared by the International Bioethics Committee of UNESCO.[94] However, it is paramount to develop greater awareness of the potential of these scientific innovations and, while maintaining a high level of attention for possible dangerous drifts, to limit the constraints to research in this field, with the aim to promote a full realization of the right to science.

---

[92] M Braun and P Dabrock, "'I bet you won't': The Science-society Wager on Gene Editing Techniques" (2016) 17:3 *EMBO reports* 279. According to the authors, "[t]he current debate on gene editing could perhaps be understood as a wager between science and society: one side – mostly scientists – is trying to guess and cater for the possible reaction of the other side, while the other side – mostly the public – is trying to discern the underlying intentions and goals of the other."

[93] D Baltimore and others, "A Prudent Path Forward for Genomic Engineering and Germline Gene Modification" (2015) 348:6230 *Science* 36; Italian Committee for Bioethics (ICB), *L'editing genetico* (n 53) 15.

[94] UNESCO – International Bioethics Committee, "Report of the IBC on Pre-implantation Genetic Diagnosis and Germ-line Intervention" (2003) SHS-EST/02/CIB-9/2(Rev.3).

13

The Regulation of Human Germline Genome Modification in Spain

Iñigo de Miguel Beriain and Carlos María Romeo Casabona

I INTRODUCTION

Spain's regulatory framework for human gene editing, whether that be for somatic or germline cells, is particularly complex for several reasons. First, Spain is, at the same time, a member of regional international and supranational organizations (i.e. the Council of Europe and the European Union (EU)) and a unitary state (the Kingdom of Spain), but one where many competencies have been devolved at the regional level to "autonomous communities" (*Comunidades Autónomas*).[1] Thus, a discussion of how Spain regulates human germline editing necessarily spans EU law, treaties adopted by the Council of Europe, and national and regional laws. Second, current laws are inadequate. Mostly, they were passed before the development of gene-editing technologies, and they have not yet been revised to ensure that the full potential of these discoveries is reaped. Moreover, certain aspects of these laws are unclear, especially those relating to the definitions of embryo and pre-embryo. A multiplicity of oversight bodies further confuses the regulatory framework. They include not only national but also regional bodies. Finally, there is also confusion about whether clinical testing of gene editing techniques is allowed and lack of information about funding for this kind of research. All of this makes it considerably difficult for scientific research in Spain to advance beyond basic and preclinical trials of human germline genome modification.

In the following pages, we endeavor to bring as much clarity as possible to the matter and give as accurate a depiction as we can of how gene editing is

---

[1] In Spain, there are seventeen autonomous communities and two autonomous cities. See, Romeo Casabona, C.M. et al., "Spain," *International Encyclopedia of Laws (IEL)*. Kluwer Law Online. 2016. See Carlos M. Romeo-Casabona, et al., "Spain," *International Encyclopaedia for Medical Law* (Kluwer Law International, The Netherlands) (2016) www.kluwerlawonline.com/abstract.php?id=MEDI20160475 accessed August 30, 2018.

actually regulated in Spain at this point in history. Of course, several key issues are still debated and the views presented in this chapter might not necessarily reflect those of the majority of scholars.

II THE REGULATORY ENVIRONMENT

The general regulatory environment for biomedical research in Spain is made of several different sources of law and regulatory agencies (Table 13.1).

1 *The General Legal Framework*

First, Spain, as a member of the EU, is subject to the regulations issued by the EU institutions. Within this framework, three main legal documents refer to gene editing. First, there is the Charter of Fundamental Rights of the European Union, whose Article 3, entitled "right to the integrity of the person," bans eugenic practices in relation to human beings. It recites: "In the fields of medicine and biology, the following must be respected in particular: (...) b) the prohibition of eugenic practices, in particular those aiming at the selection of persons."[2] Of course, this is a generic reference, and only concerns one of the possible purposes of genetic editing. However, it is important to bear this article in mind, if only because this is the only place where the subject is mentioned in the Charter.

Besides this, there are two more sets of key legal instruments on the matter in the EU. The first one is the EU Directive 98/44/EC on Biotechnological Inventions. Article 6.1 provides: "Inventions shall be considered unpatentable where their commercial exploitation would be contrary to ordre public or morality; however, exploitation shall not be deemed to be so contrary merely because it is prohibited by law or regulation."[3] Paragraph 2 of the same article continues: "On the basis of paragraph 1, the following, in particular, shall be considered unpatentable ... b) processes for modifying the germline genetic identity of human beings shall be considered unpatentable ... d) processes for modifying the genetic identity of animals which are likely to cause them suffering without any substantial medical benefit to man or animal, and also animals resulting from such processes."[4]

---

[2] Charter of Fundamental Rights of the European Union. 2012/C 326/02.
[3] European Parliament and Council Directive (EC) 98/44 on the Legal Protection of Biotechnological Inventions, art. 6.1.
[4] Ibid.

TABLE 13.1 *Spain: the legal framework regulating human germline genome editing*

Source	Name of the law	Most relevant clauses
Spain	Spanish Constitution	Articles 1 and 2
EU	Charter of Fundamental Rights of the European Union	Article 3
EU	Directive 98/44/EC on Biotechnological Inventions	Article 6.b, d.
EU	Regulation 536/2014 on Clinical Trials of Medicinal Products for Human Use	Article 90
EU	Directive 2001/20/EC, of April 4, 2001, of the European Parliament and of the Council on the approximation of the laws, regulations and administrative provisions of the Member States relating to implementation of good clinical practice in the conduct of clinical trials on medicinal products for human use	Article 9
Spain	Convention for the Protection of Human Rights and Dignity of the Human Being with regard to the Application of Biology and Medicine: Convention on Human Rights and Biomedicine (Oviedo Convention)	Article 13
Spain	Law 24/2015, of July 24, 2015, on Patents	Article 13
Spain	Criminal Code	Title V and Article 349
Spain	Law 14/2007, of July 3, 2007, on Biomedical Research	Articles 30 and ff and 74
Spain	Law 14/2006, of May 26, 2006, on Assisted Reproductive Technology	Articles 13–15
Spain	Royal Decree 1716/2011 establishing the basic requirements for the authorization and operation of biobanks for biomedical research and the treatment of biological samples of human origin, and regulates the operation and organization of the National Register of Biobanks	
Spain	Royal Decree 53/2013, of February 1, 2013, establishing basic rules for the protection of animals used for experimental and other scientific purposes, including teaching	Article 21, articles 33 and ff.
Spain	Royal Decree 1090/2015, of December 4, 2015, regulating clinical trials of medicines, the Ethics Committees for Research on Medicinal Products and the Spanish Register of Clinical Studies	Article 2

The second one is Directive 2001/20/EC, which reads: "No gene therapy trials may be carried out which result in modifications to the subject's germline genetic identity."[5] This was repeated in the subsequent EU Regulation 536/2014 on Clinical Trials of Medicinal Products for Human Use, which repealed it.[6] The prohibition has also been included in the Spanish law. Article 17.3 of the Royal Decree 1090/2015 Regulating Clinical Trials of Medicines, the Ethics Committees for Research on Medicinal Products, and the Spanish Register of Clinical Studies reads: "Clinical trials with gene therapy medicinal products that cause changes in the gene identity of the person's germline are prohibited."[7]

While EU regulations are directly applicable in Member States, directives oblige Member States to develop their own legislation to give them effect.[8] Therefore, to implement Directive 2001/20/EC, Spain enacted Law 24/2015 on Patents, which is currently in force. Article 5 (exceptions to patentability) includes the following clauses: "(b) Procedures for the modification of the germline genetic identity of the human being; and (d) Procedures for the modification of the genetic identity of animals which involve suffering without substantial medical or veterinary use to humans or animals, and animals resulting from such procedures."[9]

Spain is not only subject to EU laws. It also has obligations under the treaties adopted by the Council of Europe, a regional international organization to which Spain belongs. The Council is responsible for sponsoring the drafting of international treaties aimed at strengthening international compliance with human rights. Of these treaties, one is of particular importance for gene editing. We are talking, obviously, about Article 13 of the Oviedo Convention, which reads: "An intervention seeking to modify the human genome may only be undertaken for preventive, diagnostic or therapeutic purposes and only if its aim is not to introduce any modification in the genome

---

[5] European Parliament and Council Directive (EC) 2001/20 on the approximation of the laws, regulations, and administrative provisions of the Member States relating to implementation of good clinical practice in the conduct of clinical trials on medicinal products for human use, art. 9.6.

[6] European Parliament and Council Regulation (EU) No 536/2014 on clinical trials on medicinal products for human use and repealing Directive 2001/20/EC, art. 90.

[7] Royal Decree 1090/2015 regulating clinical trials with medicines, the Research Ethics Committees on Medicines, and the Spanish Registry of Clinical Studies (*Real Decreto 1090/2015, por el que se regulan los ensayos clínicos con medicamentos, los Comités de Ética de la Investigación con medicamentos y el Registro Español de Estudios Clínicos*), BOE No. 307.

[8] See, Europa, "Regulations, Directives, and other acts" https://europa.eu/european-union/eu-law/legal-acts_en accessed May 30, 2018.

[9] Law 24/2015 *On Patents* (*Ley 24/2015, de Patentes*), BOE No. 177, 62765–62854.

of any descendants."[10] The Oviedo Convention was ratified by Spain and became part of our national law on January 1, 2000.[11] According to Article 96.1 of the Spanish Constitution, in Spain, ratified treaties become part of Spanish law as soon as they have been published in the Official Bulletin of the state. Also, human rights treaties are relevant in the interpretation of national law.

At the national level, one must start by noting that the Spanish Constitution includes several references to values, principles, and rights that might be affected by gene editing, but makes no explicit mention of eugenics, germline modification, or similar ideas.[12]

However, there are several laws that are relevant to this issue. To begin with, the Criminal Code, in Title V, entitled Felonies Related to Genetic Engineering (Delitos relativos a la ingeniería genética), includes several articles referring to gene editing, such as the following:[13]

Article 159

1. Those who manipulate the human genes so as to alter the genome for purposes other than eliminating or decreasing serious flaws or diseases, shall be punished with a sentence of imprisonment from two to six years and special barring from public employment and office, profession or trade, from seven to ten years.

2. Should alteration of the genome be perpetrated due to serious negligence, the punishment shall be a fine from 6 to 15 months and special barring from public employment and office, profession or trade from 1 to 3 years

Article 160

1. The use of genetic engineering to produce biological weapons or those intended to exterminate human beings shall be punished with a sentence of imprisonment of 3 to 7 years and special from public employment office, profession or trade for a term of 7 to 10 years

---

[10] Convention for the Protection of Human Rights and Dignity of the Human Being with regard to the Application of Biology and Medicine: Convention on Human Rights and Biomedicine (opened for signature April 4, 1997, entered into force December 1, 1999), ETS No. 164 (the Oviedo Convention), art. 13.

[11] Instrumento de Ratificación del Convenio para la protección de los derechos humanos y la dignidad del ser humano con respecto a las aplicaciones de la Biología y la Medicina (Convenio relativo a los derechos humanos y la biomedicina) 20638, BOE No. 251, 36825.

[12] Some of its articles mention values, principles, and rights that might be affected by gene editing. For instance, Article 1 states: "Parties to this Convention shall protect the dignity and identity of all human beings and guarantee everyone, without discrimination, respect for their integrity and other rights and fundamental freedoms with regard to the application of biology and medicine." In the same sense, Article 2 reads: "The interests and welfare of the human being shall prevail over the sole interest of society or science."

[13] Carlos M. Rome-Casabona, "Genética Biotecnología y Ciencias Penales" (Pontificia Universidad Javeriana-Bogotá: Grupo Editorial Ibáñez 2009) 116.

Article 162

In the felonies defined in this Title, the judicial authority may impose any one or number of the consequences foreseen in article 129 of this code when the offender belongs to a company, organization or assembly which engages in these activities, even though only transitionally.[14]

In addition, the Criminal Code deals with issues related to genetically modified organisms, even if only tangentially, in some other parts. Article 349 states:

[t]hose who breach the established safety measures in the handling, transport or possession of organisms, specifically endangering the life, physical integrity or health of persons or the environment, shall be punished with imprisonment of six months to two years, a fine from 6 to 12 months, and a special barring from public employment and office, profession or trade for a term of 3 to 6 years.[15]

Biomedical research is also regulated by various legal instruments. Law 14/2007 on Biomedical Research ("Biomedical Research Law" or "Law 14/2007")[16] regulates basic and clinical research and clinical use of gene editing involving human gametes, pre-embryos, and embryos.[17] Royal Decree 1090/2015 regulates clinical trials for medicines and sanitary products.[18] Finally, Law 14/2006 on human-assisted reproduction techniques regulates basic and clinical research involving gametes or pre-embryos and aimed to improve human assisted reproduction techniques.[19]

Article 74 of Law 14/2007 imposes fines for the misuse of germline gene editing: "the carrying out of any intervention aimed at the introduction of a modification in the genome of the descent" is considered as a very serious infraction that is punishable with a fine of between €10,001 and €1,000,000.[20]

---

[14]   Organic Law 10/1995 on the Criminal Code (*Ley Orgánica 10/1995 del Código Penal*).
[15]   Ibid.
[16]   Law 14/2007 on Biomedical Research (*Ley 14/2007 de Investigación biomédica*), BOE No. 159. See, in general, Carlos M. Rome-Casabona, "La Ley de Investigación Biomédica: un nuevo mapa normativo para la investigación científica en el Sistema Nacional de Salud" (2008) 16 *DS no extraordinario*, 63–74.
[17]   Law 14/2007 on Biomedical Research (*Ley 14/2007 sobre la investigación biomédica*), BOE No. 159 art. 30 and ff.
[18]   Royal Decree 1090/2015 regulating clinical trials with medicines, the Research Ethics Committees on Medicines, and the Spanish Registry of Clinical Studies (*Real Decreto 1090/2015, por el que se regulan los ensayos clínicos con medicamentos, los Comités de Ética de la Investigación con medicamentos y el Registro Español de Estudios Clínicos*), BOE No. 307.
[19]   Law 14/2006 on Assisted Reproductive Technologies (*Ley 14/2006 sobre técnicas de reproducción humana asistida*), BOE No. 126.
[20]   Law 14/2007, art. 74.

Other articles of this key law specify what the oversight bodies and supervisory procedures are.

2 Major Regulatory Agencies

There is a multitude of bodies having the competence to regulate aspects of human germline editing in Spain, and many include a number of further sub-bodies (Table 13.2). Therefore, it is extremely difficult to produce an exhaustive catalogue. Moreover, this type of information would be of little use, as most of them would hardly be involved in a genetic editing process, at least at this stage. In any case, we will try to do our best to describe the supervisory bodies that deal with this matter in an appropriate way.

In Spain, competences are divided between the state and the regional authorities (*Comunidades Autónomas*), but when it comes to biological research on humans, all key regulatory powers rest somehow with the state. This is due to a simple fact: the state's main advisory bodies, as opposed to the regional ones, have a decisive influence in determining what type of research or clinical application is approved by the relevant administration.

At the state level, the main administrative authority is the Health Institute Carlos III (*Instituto de Salud Carlos III*),[21] which is part of the Ministry of Health and Consumption (*Ministerio de Sanidad y Consumo*). A number of different committees advise the Institute and the Ministry, depending on whether it is basic and preclinical research or clinical trials, whether it is for research and therapy or for reproductive purposes, and whether it involves animals. Their favorable advice is necessary for the Health Institute and the Ministry to issue authorizations. The main ones are as follows:

- Commission of Guarantee for the Donation and Use of Human Cells and Tissues (*Comisión de Garantías para la Donación y Utilización de Células y Tejidos Humanos*).[22] Under Articles 34 and 35 of the Royal Decree 1527/2010,[23] the Commission has administrative authority over research involving

---

[21]   Ministerio de Ciencia, Innovación y Universidades, "Instituto de Salud Carlos III" www.isciii.es accessed May 14, 2018.

[22]   Ministerio de Ciencia, Innovación y Universidades, "Comusión de Garantías para la Donación y Utilización de Células y Tejidos Humanos" www.isciii.es/ISCIII/es/contenidos/fd-el-instituto/fd-organizacion/fd-estructura-directiva/fd-subdireccion-general-investigacion-terapia-celular-medi cina-regenerativa/fd-comites/comision-de-garantias.shtml accessed May 14, 2018.

[23]   Royal Decree 1527/2010 regulating the Guarantees Committee for the Donation and Use of Human Cells and Tissues and the Registry of Research Projects (*Real Decreto 1527/2010 por el que se regulan la Comisión de Garantías para la Donación y Utilización de Células y Tejidos Humanos y el Registro de Proyectos de Investigación*), BOE No. 294.

TABLE 13.2 *Spain: principal regulatory agencies*

Research stage	Supervising body	Administrative authority	Applicable regulation
Preclinical (animals)	Ethics Committees on Animal Experimentation (*Comités de ética de experimentación animal*)	Regional Authorities One in each Comunidad Autónoma	Royal Decree 53/2013, of February 1, 2013, establishing the basic rules applicable for the protection of animals used in experimentation and other scientific purposes, including teaching
Research involving human gametes, pre-embryos, or embryos for purposes other than human reproduction	Commission of Guarantee for the Donation and Use of Human Cells and Tissues (*Comisión de Garantías para la Donación y Utilización de Células y Tejidos Humanos*) Research Ethics Committee of the Given Research Center	Regional Authority (*Consejería de Salud de la Comunidad Autónoma*) or National Authorities (*Ministry of Health*) Direction of the Research Centre	Royal Decree 1527/2010, of 15 November 2010, regulating the Guarantees Committee for the Donation and Use of Human Cells and Tissues and the Registry of Research Projects Law 14/2007, of July 3, 2007, on Biomedical Research
Research related to the development or application of assisted reproduction technology (ART)	National Commission on Human-Assisted Reproduction (*Comisión Nacional de Reproducción Humana Asistida*)	The corresponding administrative authority, determined by the autonomous community where the research is to be performed Usually, the Council of Health Care of the Autonomous Community (*Consejería de Salud de la Comunidad Autónoma*). Regional Authority	Royal Decree 42/2010, of January 15, which regulates the National Commission on Assisted Human Law 14/2006 on human-assisted reproduction techniques regulates basic and clinical research involving gametes or pre-embryos and aimed to improve human-assisted reproduction techniques

(*continued*)

TABLE 13.2 (continued)

Research stage	Supervising body	Administrative authority	Applicable regulation
Clinical trials	Drugs Research Ethics Committees (*Comités Éticos de Investigación con Medicamentos*)	Spanish Agency for Medicines and Health Products (*Agencia Española de Medicamentos y Productos Sanitarios*)	Royal Decree 1090/2015, of December 4, 2015, regulating clinical trials of medicines, the Ethics Committees for Research on Medicinal Products and the Spanish Register of Clinical Studies

human gametes, pre-embryos, and embryos. Indeed, research on these entities can only be performed if the research project obtains a favorable report by the corresponding Guarantees Commission for the Donation and Use of Human Cells and Tissues. Apart from that, we must highlight that research projects using human gametes, pre-embryos, or embryos must be approved by the management of the research center in which the research takes place and receive a favorable assessment from the Research Ethics Committee corresponding to that center.

- Research related to the development or application of assisted reproduction technology (ART) falls under the jurisdiction of the National Commission on Human-Assisted Reproduction (*Comisión Nacional de Reproducción Humana Asistida*).[24] Under Royal Decree 42/2010,[25] research on the reproductive uses of human gametes and pre-embryos must receive the Commission's positive assessment before it can be submitted to the corresponding administrative authority, determined by the autonomous community where the research is to be performed, for approval.

- Spanish Agency for Medicines and Health Products (*Agencia Española de Medicamentos y Productos Sanitarios*).[26] This body is a national agency, under the Ministry of Health, that gives final approval for clinical trials, according to Article 17 of Royal Decree 1090/2015 of December 4, 2015. Articles 11, 12 and 17 regulate the aims and supervision of the ethics committees supervising clinical trials and the requirements applicable to clinical trials.

- Drugs Research Ethics Committees (*Comités Éticos de Investigación con Medicamentos*).[27] These regional bodies, under the autonomous communities, oversee clinical trials involving drugs and medicinal products.[28] Clinical trials involving gene editing can only be conducted if one of these

---

[24] Ministerio de Sanidad, Consumo y Bienestar Social, "Comisión Nacional de Reproducción Humana Asistida" www.cnrha.mscbs.gob.es accessed May 14, 2018.

[25] Royal Decree 42/2010, which regulates the National Commission on Assisted Human Reproduction (*Real Decreto 42/2010 por el que se regula la Comisión Nacional de Reproducción Humana Asistida*), BOE No. 30.

[26] Ministerio de Sanidad, Consumo y Beinestar Social, "Agencia Española de Medicamentos y Productos Sanitarios – AEMPS" www.aemps.gob.es/home.htm accessed May 14, 2018.

[27] Agencia Española de Medicamentos y Productos Sanitarios, "Listado de Comités de Ética de la Investigación que pueden evaluar estudios clínicos (ensayos clínicos o estudios observacionales) con medicamentos o con productos sanitarios" (September 24, 2018) www.aemps.gob.es /investigacionClinica/medicamentos/docs/listado-comites-investigacion-clinica.pdf accessed September 24, 2018.

[28] Royal Decree 1090/2015 regulating clinical trials with medicines, the Research Ethics Committees on Medicines, and the Spanish Registry of Clinical Studies (*Real Decreto 1090/2015, por el que se regulan los ensayos clínicos con medicamentos, los Comités de Ética*

committees has expressed its favorable opinion. Furthermore, trials also need the approval of the Spanish Agency for Medicines and Health Products (*Agencia Española de Medicamentos y Productos Sanitarios*) to be conducted.[29]

- Also, the ethics committees related to animal welfare should be mentioned. There are different types of committees. The most popular ones are the Ethics Committees on Animal Experimentation (*Comités de ética de experimentación animal*). These committees give advice to the administrative authorities about research involving animals.[30] It is then an administrative authority of the corresponding autonomous community (Regional Authority), for instance the council member in charge of the Health Care Department of the Autonomous Community, which is responsible for the final decision to authorize and oversee the research. Apart from these committees, we have the Spanish Committee for the Protection of Animals used for scientific purposes (*Comité español para la protección de los animales utilizados con fines científicos* – CEPAFIC).[31] This is a national committee, under the Ministry of Agriculture and Fisheries, Food and Environment (*Ministerio de agricultura y pesca, alimentación y medio ambiente*). It is a consultative body, which advises the ministry, the Ethics Committees on Animal Experimentation, the regional governments, and so on.

III THE REGULATORY ENVIRONMENT FOR RESEARCH ON HUMAN GERMLINE MODIFICATION

1 Definitions

Article 3 of Law 14/2007 defines an embryo and a pre-embryo as follows:

l. "Embryo": phase of embryonic development from the moment at [sic] which the fertilized oocyte is found in the uterus of a woman until the

---

de la Investigación con medicamentos y el Registro Español de Estudios Clínicos), BOE No. 307, art. 11, 12, and 17.

[29]   Ibid., art. 11.2 and 17.

[30]   Royal Decree 53/2013 establishing the basic rules applicable for the protection of animals used in experimentation and other scientific purposes, including teaching (*Real Decreto 53/2013 por el que se establecen las normas básicas aplicables para la protección de los animales utilizados en experimentación y otros fines científicos, incluyendo la docencia*), BOE No. 34, art. 35 and ff.

[31]   Ministerio de Agricultura, Pesca y Alimentación, "Comité español para la protección de los animals utilizados con fines científicos" www.mapama.gob.es/es/ganaderia/temas/produccion-y-mercados-ganaderos/bienestanimal/en-la-investigacion/CEPAFIC.aspx accessed May 14, 2018.

beginning of organ genesis and which ends 56 days from the moment of fertilization, with the exception of the computation of those days in which the development could have been stopped.

s. "Pre-embryo": embryo constituted in vitro that is formed by the group of cells that are the result of the progressive division of the oocyte from the time it is fertilized until 14 days after.[32]

These definitions are important, since they allow us to differentiate two key entities: the embryo and the pre-embryo. As we shall see, their legal status is different. However, the language used to describe them is extremely problematic.[33] The embryo is defined as a "phase of embryonic development." However, a pre-embryo is an embryo, too, even if this does not sit well with the definition of embryo. If an embryo is defined as "a phase of embryonic development from the moment in which the fertilized oocyte is found *in the uterus of a woman*" (emphasis in original), it seems clear that a pre-embryo, while still in vitro, cannot fit with this description.

Second, it is good to keep in mind that there is no legal definition of genome editing in Spanish law as such. Spanish law relies instead on the definition of "gene therapy medicinal product" included in Directive 2001/83/EC,[34] and incorporated by Royal Decree 1090/2015:

Gene therapy medicinal product means a biological medicinal product which has the following characteristics:

(a) it contains an active substance which contains or consists of a recombinant nucleic acid used in or administered to human beings with a view to regulating, repairing, replacing, adding or deleting a genetic sequence;

(b) its therapeutic, prophylactic or diagnostic effect relates directly to the recombinant nucleic acid sequence it contains, or to the product of genetic expression of this sequence.

Gene therapy medicinal products shall not include vaccines against infectious diseases.[35]

---

[32] Law 14/2007, art. 3.

[33] Iñigo de Miguel Beriain, *La Clonación, diez años después* (Comares, Granada 2008); Iñigo de Miguel Beriain, *Bioethics and New Biotechnologies in Human Health* (Gijón 2009) (International Prize 2008: "House of Representatives of the Principality of Asturias – International Society of Bioethics (SIBI)").

[34] European Parliament and Council Directive (EC) 2001/83 on the Community Code Relating to Medicinal Products for Human Use, formerly assumed by European Parliament and Council Regulation (EC) 1394/2007 on Advanced Therapy Medicinal Products and amending Directive (EC) 2001/83 and Regulation (EC) 726/2004.

[35] Royal Decree 1090/2015, art. 2.1.n.

As already mentioned, modifications involving human gametes or pre-embryos in the context of assisted reproduction are, by contrast, regulated by Law 14/2006. Article 14 of Law 14/2006 regulates use of gametes for research purposes. Article 13 addresses the use of therapeutic techniques in the pre-embryo. However, if modifications are made for research purposes, they are ruled by Article 15 of this same Law 14/2006. Both articles will be extensively analyzed in Section III.3 of this chapter.

Finally, it is important to mention the Royal Decree 1716/2011, which establishes the basic requirements for the authorization and operation of biobanks for biomedical research and the treatment of biological samples of human origin, and regulates the operation and organization of the National Register of Biobanks (*Registro Nacional de Biobancos*) for biomedical research.[36] The decree applies to the storage and use of embryonic stem cell lines for research or clinical purposes, including gene editing, in a biobank.

Besides the statutory provisions aimed at protecting humans, one must also take into account the legislation on the protection of animals against genetic modification. Often preclinical studies involve animals. This issue is generally regulated by Law 32/2007 for the care of animals on their holding, transport, testing, and slaughter.[37] However, when it comes to genetic editing, the most relevant norm is Article 21 of the Royal Decree 53/2013 of February 1, 2013 establishing basic rules for the protection of animals used for experimental and other scientific purposes, including teaching.[38] This Decree implements EU Directive 2010/63/EU on the Protection of Animals Used for Scientific Purposes.[39]

We will now move on to analyze the specific Spanish legal framework regarding basic or clinical research in human embryos or gametes using germline modification and its clinical application. However, before tackling

---

[36] Royal Decree 1716/2011 establishing the basic requirements for the authorization and operation of biobanks for the purpose of biomedical research and the treatment of biological samples of human origin, and regulating the operation and organization of the National Registry of Biobanks for biomedical research (*Real Decreto 1716/2011 por el que se establecen los requisitos básicos de autorización y funcionamiento de los biobancos con fines de investigación biomédica y del tratamiento de las muestras biológicas de origen humano, y se regula el funcionamiento y organización del Registro Nacional de Biobancos para investigación biomédica*), BOE No. 290.

[37] Law 32/2007 for the care of animals, in their exploitation, transportation, experimentation and sacrifice (*Ley 32/2007 para el cuidado de los animales, en su explotación, transporte, experimentación y sacrificio*) BOE No. 268.

[38] Royal Decree 53/2013, art. 21.

[39] European Parliament and Council Directive (EU) 2010/63 on the Protection of Animals Used for Scientific Purposes (L 276/33).

this complicated issue, we need to explain how research on animals is regulated in Spain.

2 *Regulation of Research on Animals*

Royal Decree 53/2013 establishes the basic rules for the protection of animals used for experimental and other scientific purposes.[40] This becomes relevant, for instance, when preclinical research involves research on animals. First, under Article 21, several species of primates cannot be used, namely the gorilla (*Gorilla*), the chimpanzee (*Pan troglodytes*), the bonobo, dwarf chimpanzee, gracile chimpanzee, pygmy chimpanzee (*Pan paniscus*), and the Bornean orangutan (*Pongo pygmaeus*).[41] Other primates can be used only if two conditions are met:

(a) The procedure is aimed at clinical research carried out for the purpose of preventing, preventing, diagnosing or treating disabling or potentially life-threatening diseases or to ensure the conservation of the species;

(b) It has been scientifically justified that the purpose of the procedure cannot be achieved by using animals other than non-human primates or animals of species not included in Annex A to Council Regulation (EC) No 338/97 of 9 December 1996.[42]

Second, Articles 32 and following of the Royal Decree 53/2013 detail the procedure that research projects involving animals must follow.[43] All research projects involving animals must include detailed information on:

4. Relevance and justification for the following:

(a) Use of animals, including their origin, estimated number, species and life stages.

(b) Procedures.

5. Application of methods to replace, reduce and refine the use of animals in procedures.

6. Use of anesthetics, pain relievers, and other means to relieve pain.

7. Measures to reduce, avoid and alleviate any form of suffering of the animals throughout their life, where appropriate.

8. Use of humanitarian endpoints.

[40]   Royal Decree 53/2013.
[41]   Ibid., art. 21.
[42]   Ibid.
[43]   Ibid.

9. Experimental or observational strategy and statistical model to reduce the minimum number of animals used, the pain, suffering, distress and impact environmental management, where appropriate.

10. Reuse of animals and their cumulative effect on the animal.

11. Proposal for the classification of procedures according to their severity.

12. Measures to avoid unjustified repetition of procedures, if any.

13. Housing, husbandry and animal care conditions.

14. Euthanasia methods.

15. Training of the people involved in the project.[44]

3 Regulation of Basic and Clinical Research and Clinical Application of Germline Modification in Human Gametes, Pre-Embryos, and Embryos

At this point in history, the prevailing view among scholars is that, in Spain, any intervention seeking to modify the human genome that is not for preventive, diagnostic, or therapeutic purposes is prohibited.[45] According to this view, basic and clinical research and clinical application of germline editing of human gametes, pre-embryos, and embryos are prohibited by the combined effect of Article 13 of the Oviedo Convention and Article 74 of Law 14/2007 on Biomedical Research. This view has chilling effect on Spanish researchers, who, currently, are not engaging in research in this direction.[46]

We beg to differ. We do not believe the existing legal framework is as restrictive as it is considered to be. Let's recall Article 13 of the Oviedo Convention: "[a]n intervention seeking to modify the human genome may only be undertaken for preventive, diagnostic or therapeutic purposes and only if its aim is not to introduce any modification in the genome of any descendants."[47] At the same time, Article 74 of Law 14/2007 makes "[t]he carrying on of any intervention aimed at the introduction of a modification in the genome of the descendant" a crime, punished with a fine from €10,001 to €1,000,000. The wording of these articles leaves room for interpretation. Neither do they specify what "human genome" or "descendant" encompasses.

---

[44]   Ibid. Annex X.

[45]   Vicente Bellver Capella, "Los Diez Primeros Años del Convenio Europeo sobre Derechos Humanos y Biomedicina: Reflexiones y Valoración" (2008) XIX no 3 *Cuadernos de Bioética* 401; Vera Lúcia Raposo, "The Convention of Human Rights and Biomedicine Revisited: Critical Assessment" (2016) 20:8 *The International Journal of Human Rights* 1277.

[46]   However, the editing of somatic cells is perfectly acceptable according to Spanish law, and it can be funded and carried out if all guarantees and procedures are met.

[47]   Oviedo Convention, art. 13.

Moreover, it is unclear what "aimed at" in the context of these provisions means.[48]

Let us start with the term "human genome." We are aware that many consider the genetic modification of germline a modification of the human genome.[49] However, although we are only jurists and not biologists, it seems clear to us that the human genome changes only when a new gene is *added* to the vast pool of human genetic diversity.[50] However, human germline editing can be limited to replacing a pathologic gene with its healthy expression. Nothing new is added to the human genome pool.[51] We are just replacing a human gene with another human gene.[52] If "modification in the human genome" means the introduction of a new gene into our common genetic heritage, then interventions to substitute a defective gene with a healthy one are not banned.[53] We are not alone in this interpretation. For instance, as Ana Nordberg et al. noted:

> [t]he explanatory report to the Oviedo Convention, which has interpretative value, stresses that the primary goal and general interpretative guiding principle of the convention is to "protect the dignity and identity of all human

[48] Ana Nordberg, Timo Minssen, Sune Holm, Maja Horst, Kell Mortensen, Birger Lindberg Møller "Cutting Edges and Weaving Threads in the Gene Editing (Я)Evolution: Reconciling Scientific Progress with Legal, Ethical, and Social Concerns" (2018) 5:1 *Journal of Law and the Biosciences* 1, 26.

[49] Vicente Bellver Capella, "Los Diez Primeros Años del Convenio Europeo sobre Derechos Humanos y Biomedicina: Reflexiones y Valoración" (2008) XIX no 3 *Cuadernos de Bioética* 401, 408; Roberto Andorno, "The Oviedo Convention: A European Legal Framework at the Intersection of Human Rights and Health Law" (2005) 2:4 *Journal of International Biotechnology Law* 133, 140; Vera Lúcia Raposo, "The Convention of Human Rights and Biomedicine Revisited: Critical Assessment" (2016) 20:8 *The International Journal of Human Rights* 1277.

[50] Iñigo de Miguel Beriain, "Should human germ line editing be allowed? Some suggestions on the basis of the existing regulatory framework" (2018) *Bioethics* https://onlinelibrary.wiley.com /doi/pdf/10.1111/bioe.12492 accessed May 28, 2018.

[51] Nicolae Morar, "An Empirically Informed Critique of Habermas' Argument from Human Nature" (2015) 21:1 *Science and Engineering Ethics* 95.

[52] This subtle distinction has been extremely well addressed by Japanese bioethicist Tetsuya Ishii, who has written that "the functional correction of a small mutation in the embryo via HDR [homology-directed repair] along with a short DNA template appears to be acceptable because this form of genetic modification can leave a wild-type gene, which is in a natural genetic state, and would fall outside of one of the ethical objections against germline gene modification: transgression of the natural laws. The copying of a naturally occurring variant via HDR along with a short DNA template might be considered to be natural." Tetsuya Ishii, "Germline Genome Editing in Clinics: The Approaches, Objectives and Global Society" (2015) 16:1 *Briefings in Functional Genomics* 46.

[53] Iñigo de Miguel Beriain, "Should human germ line editing be allowed? Some suggestions on the basis of the existing regulatory framework" (2018) *Bioethics* https://onlinelibrary.wiley.com /doi/pdf/10.1111/bioe.12492 accessed May 28, 2018.

beings," understood as protecting the biological and genetic identity of the species and inspired by the principle of the primacy of the human being. Arguably, this principle is honored insofar as the gene-editing objective is to prevent or correct genetic mutations, which by themselves are a threat to the integrity and future of human identity.[54]

Again, if this interpretation is acceptable, we should conclude that a modification of human genome in the gametes, the embryo, or the pre-embryo genome, during basic, preclinical or clinical research, using germline modification technologies and the clinical application of these techniques (at least in pre-embryos and embryos) is illegal in Spain only if it involves the deliberate introduction of novel genetic material. Otherwise, it is permitted, under the conditions we will discuss further below.

One could go one step further and argue that even modifications of human genome by causing a novel gene expression in our common pool might be acceptable, as long as there is no attempt to use them for reproductive purposes. This is not only due to the arguments already mentioned, but also for an additional fundamental reason: the term "descendant" is not defined in the Spanish legal system, a circumstance that opens up room for debate. Indeed, the term "descendant" might be understood in several different ways. Does it refer to the embryo or the pre-embryo as such, or does the prohibition apply to them only insofar as it is meant to avoid the creation of a genetically modified person (the person the embryo might become)? Both interpretations are radically different and involve different conclusions on the acceptability of clinical research and clinical application of the gene edition.

In our opinion, it seems much more reasonable to conclude that the prohibition of modification of the human genome is meant to protect not the embryos as such, but rather the descendants of the embryo or, at least, the human beings that might emerge if embryos are transferred into a uterus and pregnancy continues. This interpretation seems fairly well aligned with the doctrine of the Spanish Constitutional Court, which approaches the protections to be guaranteed to a human *nasciturus*, which hinges on the idea of "gradualism."[55]

---

[54] Nordberg, Timo Minssen, Sune Holm, Maja Horst, Kell Mortensen, Birger Lindberg Møller "Cutting Edges and Weaving Threads in the Gene Editing (Я)Evolution: Reconciling Scientific Progress with Legal, Ethical, and Social Concerns" (2018) 5:1 *Journal of Law and the Biosciences* 1, 26.

[55] Prenatal human life should receive a lesser or a greater degree of legal protection according to its degree of maturity. Thus, an embryo deserves less protection than a thirty-week-old fetus; for example, see Pedro J. Femenía López, "Embrión" *Enciclopedia de Bioderecho y Bioética* http://enciclopedia-bioderecho.com/voces/135 accessed May 28, 2018.

If we are right, then neither the Oviedo Convention nor Spanish Law 14/2007 on Biomedical Research ban basic or clinical research using germline modification technologies. As long as the embryo is not transferred into uterus, the human genome will not change.[56] The prohibition of modification of the human genome affects only clinical application of gene editing, but only to the extent to which it involves the introduction of new gene material in the human genome.

Again, we believe that in Spain basic and clinical research using germline modification technologies and clinical application of these techniques are legal as long as they do not involve the introduction of new genetic material into the human genome, nor intend to change the human genome (even if they cause this final effect). If the pre-embryos' germline is modified on purpose, then we should consider that clinical application of these techniques is banned, while basic and clinical research might yet be considered as allowed as long as the pre-embryos are not transferred into a human uterus (since they will be destroyed and, thus, no modification is caused in our common heritage).

That being said, the Spanish Criminal Code bans any intervention aimed at manipulating human genes for purposes other than "eliminating or decreasing serious flaws or diseases."[57] Even if we were to concede that clinical applications of gene editing might be acceptable according to the Oviedo Convention and Law 14/2007, under the conditions discussed in the previous paragraphs, the Criminal Code presents a significant obstacle to clinical research. When gene editing research moves on to clinical trials, it can do so only for the purpose of eliminating or decreasing serious flaws or diseases, not just any flaw or disease. Yes, the Code does not describe which flaws or diseases should be considered as "serious." Therefore, a certain amount of uncertainty remains after all.

Make no mistake. We are not saying that research involving gene editing can take place freely in Spain, even when limited to curing serious flaws or diseases. There are still significant restrictions. Law 14/2007 addresses clinical research on human embryos in vivo. Article 30, entitled "limitations on research with live embryos and fetuses in the uterus," reads: "Interventions on the live embryo or fetus in the uterus shall be exclusively authorized when their purpose is diagnostic or therapeutic in its own interest, notwithstanding what is legally established on the voluntary interruption of a pregnancy."[58]

---

[56] Bartha Maria Knoppers, et al., "Human Gene Editing: Revisiting Canadian Policy, (2017) 2.1 *NPJ Regenerative Medicine* 3.

[57] Criminal Code, art. 159.1.

[58] Ibid.

Moreover, Law 14/2007 expressly prohibits the creation of human pre-embryos and embryos exclusively for the purpose of experimentation.[59] However, it allows research that does not entail the creation of a pre-embryo or of an embryo exclusively for this purpose, consistently with the "gradualist" approach to the protection of human life set out by our Constitutional Court's jurisprudence.[60] Research can be done on supernumerary embryos from assisted human reproduction techniques.[61]

4 Specific Rules Concerning Human Gametes and Pre-Embryos Used for Research or Clinical Application in the Context of Assisted Reproduction Techniques

According to Article 14 of Law 14/2006 on Assisted Human Reproduction Techniques, gametes may be used for research purposes but cannot be used for originating pre-embryos for reproductive purposes.[62] However, it must be noted that Spanish law does not have a general ban on clinical application of gene editing or similar procedures to human gametes. The recent approval of an autologous germline mitochondrial transfer (AUGMENT)[63] procedure in a clinical trial, discussed in the last part of this section, is evidence of that.[64]

Research with human pre-embryos is regulated by Article 15 of Law 14/2006, entitled "Use of Pre-Embryos for Research Purposes." It reads:

1. Research or experimentation with surplus pre-embryos from the application of assisted reproduction techniques shall only be authorized if it complies with the following requirements:

---

[59] Law 14/2007, Preamble, III.
[60] For example, Spanish Constitutional Court Judgment 53/1985; Spanish Constitutional Court Judgment 212/1996; Spanish Constitutional Court Judgment 116/1999.
[61] Law 14/2006, art. 15.
[62] This regulation is consistent with the Oviedo Convention Explanatory Report: "Medical research aiming to introduce genetic modifications in spermatozoa or ova which are not for procreation is only permissible if carried out in vitro with the approval of the appropriate ethical or regulatory body." EC Treaty No 164, "Explanatory Report to the Convention for the protection of Human Rights and Dignity of the Human Being with regard to the Application of Biology and Medicine: Convention on Human Rights and Biomedicine" (1997) ETS 164, para 91. It is remarkable that the writers of the Report omitted any allusion to the embryo or the pre-embryo.
[63] AUGMENT is a treatment to boost the egg's energy levels for embryo development. It uses the energy-producing mitochondria from the same woman's egg precursor cells, which are immature egg cells found in the protective lining of the ovaries, to supplement the existing mitochondria in the egg. See Augment, "IVF success? It starts with egg health" https://augmenttreatment.com accessed May 14, 2018.
[64] Tetsuya Ishii, Yuri Hibino, "Mitochondrial Manipulation in Fertility Clinics: Regulation and Responsibility" (2018) 5 *Reproductive Biomedicine and Society Online* 93.

c) (. . .) that the research is carried out in authorized centers. In any case, the projects will be carried out by qualified scientific teams under the control and monitoring of the competent health authorities.

d) That they are carried out on the basis of a project duly presented and authorized by the competent health authorities, following a favorable report from the National Commission on Assisted Human Reproduction in the case of research projects related to the development and application of assisted reproduction techniques, or from the competent body in the case of other research projects related to the obtaining, development and use of embryonic stem cell lines (. . .)

2. Once the project has been completed, the authority that granted the authorization must transfer the result of the experiment to the National Commission on Assisted Human Reproduction and, where appropriate, to the competent body that informed it.[65]

Moreover, according to Article 13 of Law 14/2006, interventions for therapeutic purposes on live in vitro pre-embryos may only be aimed at treating a disease or preventing its transmission, with reasonable and proven guarantees.[66] Further, such therapy is permitted only if the following requirements are met:

a) That the couple or, if applicable, the woman alone, has been duly informed about the procedures, diagnostic tests, possibilities and risks of the proposed therapy and has previously accepted them.

b) That they are pathologies with a precise diagnosis, with a serious or very serious prognosis, and that offer reasonable possibilities of improvement or cure.

c) That non-pathological hereditary characteristics are not modified and that the selection of individuals or race is not sought.

d) That it is carried out in authorized health centers and by qualified teams equipped with the necessary resources, as determined by Royal Decree.

3. The performance of these practices in each case will require the authorization of the corresponding health authority, following a favorable report from the National Commission on Assisted Human Reproduction.[67]

We must also add that clinical application of gene modification on human embryos or pre-embryos for therapeutic purposes must be considered a clinical trial if the clinical applications of these techniques involve the intent to create a person, that is, a born human being. If the intervention does not have this purpose, then we should not consider it a clinical application,

---

[65]   Law 14/2006 on Assisted Human Reproduction Techniques, art. 15.
[66]   Ibid., art. 13.
[67]   Ibid.

and, therefore, Law 14/2007 or Law 14/2006 but not Royal Decree 1090/2015 would apply. In other words, if the embryos or pre-embryos are to be destroyed, this would not be considered a clinical application of a medicinal product and, thus, Law 14/2007 or Law 14/2006 would be applicable. Of course, the destruction is much more likely to happen in the case of pre-embryos, which are in vitro and not in a female uterus. Therefore, in legal terms, while it is hard to conceive clinical trials in embryos (since they might change the germline of a born human being), trials in pre-embryos are conceivable, since the pre-embryos are destroyed afterward and no person is created at all (keeping in mind that a pre-embryo can hardly be considered as a person in the Spanish legal framework, as showed in previous sections).

However, if the intervention is done with the aim to create a healthy person, then the Royal Decree 1090/2015 applies. This regulation states, "Clinical trials with gene therapy medicinal products that cause changes in the gene identity of the person's germline are prohibited."[68] It follows that clinical trials of gene modification techniques that do not cause "changes in the gene identity" – whatever this means[69] – are allowed. Of course, the research protocol must still be approved by the Spanish Agency for Medicines and Health Products (*Agencia Española de Medicamentos y Productos Sanitarios*), since gene editing is considered an "advanced medicinal drug."[70]

IV CURRENT PERSPECTIVES AND FUTURE POSSIBILITIES

In conclusion, it is clear that the Spanish legislation on the matter is lacking and requires modifications for various reasons. First, it lacks clarity. Most key clauses are open to different interpretations, some of them contradictory. Second, the definition of the concepts suffers from contradictions and gaps. Thus, for example, we do not know at present what exactly "genetic modification" means in normative terms. We also do not know whether a modification that does not involve the introduction of novel genetic material into the human species is a prohibited practice.

---

[68] Royal Decree 1090/2015, art. 17.3.

[69] As Rosario Isasi et al. have noted, genetic identity "has yet to be defined, and we need to look for an approach to genome editing that can lead toward compromise or consensus." Rosario Isasi, Erika Kleiderman, Bartha Maria, "Editing Policy to Fit the Genome?" (2016) 351:6271 *Science* 337.

[70] Royal Decree 477/2014 regulating the authorization of advanced therapy medicinal products for nonindustrial manufacture (*Real Decreto 477/2014 por el que se regula la autorización de medicamentos de terapia avanzada de fabricación no industrial*) BOE No. 144, 45068–45078.

Third, the regulatory regime is cumbersome, with several bodies at various levels with, sometimes, overlapping competences. Moreover, the information available on the functioning of these bodies and their decisions is wanting. Overall, it is necessary to make it possible to reconcile more efficiently the needs of research with gene-editing techniques with the necessary respect for human rights, including the right to health, the right to form a family, and the right to benefit from scientific progress, all rights codified in treaties that have been ratified by Spain.

Obviously, some of the modifications required are easier to implement than others, but the biggest obstacle of all is Article 13 of the Oviedo Convention. Spain, as any sovereign country, could withdraw from the Oviedo Convention. But this is a dramatic step that is not easy to undertake, both for legal reasons and, fundamentally, for political reasons. Until Article 13 of the Oviedo Convention is amended, far-reaching changes in the Spanish legislation on gene editing of human genome are unlikely.

The Regulation of Human Germline Genome Modification in France

Alessandro Blasimme, Dorothée Caminiti, and Effy Vayena

I INTRODUCTION

The current Fifth French Republic was established on October 4, 1958, by initiative of the then prime minister of the Fourth Republic, General Charles De Gaulle. This entailed a new Constitution and a transition from a parliamentary system in which the members of the government were accountable to the members of the French National Assembly, to a semi-presidential system in which the executive power is more independent and overall stronger than in the previous phase. The president of the Republic – who is elected by the citizens for a five-year term – is the executive head of state and shares executive powers with the prime minister – whom he appoints. In turn, the prime minister nominates the ministers of the government.

Legislative power belongs to the National Assembly – directly elected by the French people – and to the Senate – whose members are elected indirectly by departmental and municipal representatives (the so-called *grands électeurs*). The judiciary power is independent from the other two and follows a civil law system administered in three levels (first instance, appeal and cassation). A Constitutional Council (*Conseil Constitutionnel*) oversees the regularity of popular vote and the conformity of new laws with the French Constitution.

Finally, a Council of State (*Conseil d'État*) acts as the supreme court for administrative justice and also as a legal advisor to the executive. The Council of State is composed of high-level government officials, many of whom are selected from the prestigious National School of Public Administration (*École Nationale d'Administration* – ENA). As we will see in this chapter, the Council of State has played an important role in regulatory matters around biomedicine and biotechnology.

II THE REGULATORY ENVIRONMENT

1 *The Legal Framework*

In France, research involving or potentially leading to human germline genome modification is regulated by the Constitution, international law (including EU law), and the Civil, Public Health, and Criminal Codes.

a The Constitution, Treaties, and EU Law

According to Article 55 of the Constitution of the Fifth Republic, duly ratified or approved treaties or international agreements occupy a position above laws and below the Constitution within the hierarchy of legal sources.[1]

At least two treaties[2] that have been ratified by France are relevant to the topic of germline genome modification. The first one is the Council of Europe's

---

[1]   Constitution du 4 octobre 1958 (Constitution), art. 55: "*Les traités ou accords régulièrement ratifiés ou approuvés ont, dès leur publication, une autorité supérieure à celle des lois, sous réserve, pour chaque accord ou traité, de son application par l'autre partie.*" However, should a treaty conflict with the Constitution, it can be ratified and displace conflicting constitutional provisions only after the Constitution has been amended: "*Si le Conseil constitutionnel, saisi par le Président de la République, par le Premier ministre, par le président de l'une ou l'autre assemblée ou par soixante Députés ou soixante sénateurs, a déclaré qu'un engagement international comporte une clause contraire à la Constitution, l'autorisation de ratifier ou d'approuver l'engagement international en cause ne peut intervenir qu'après révision de la Constitution.*" Constitution, art. 54). On the place of international law within the French legal system, see, in general: I Palaiologou, "The Principle of Supremacy and the Response of Member States' Constitutional Courts" (2018) 8 *Southampton Student L. Rev.* 18, 4; G Carcassonne, "France" (2005) *European Constitutional L. Rev.* 1, 293; L Haopei, "A Comparative Study of the Internal Application of Treaties" (1987) 15 *Korean J. Comp. L.* 159, 174–76; A Cassese, "Constitutional Safeguards Of Compliance With International Treaties" (1985) 192 *Recueil des Cours* 394, 399.

[2]   An additional international instrument that is worth mentioning is the Universal Declaration on the Human Genome and Human Rights of November 11, 1997. Its Article 24 provides: "The International Bioethics Committee of UNESCO should contribute to the dissemination of the principles set out in this Declaration and to the further examination of issues raised by their applications and by the evolution of the technologies in question. It should organize appropriate consultations with parties concerned, such as vulnerable groups. It should make recommendations, in accordance with UNESCO's statutory procedures, addressed to the General Conference and give advice concerning the follow-up of this Declaration, in particular regarding the identification of practices that could be contrary to human dignity, such as germline interventions." This declaration is not subject to ratification and, as such, it cannot be made binding upon Member States. However, as explained by UNESCO, "in view of the greater solemnity and significance of a "declaration," it may be considered to impact, on behalf of the organ adopting it, a strong expectation that Members of the international community will abide by it. Consequently, in so far as the expectation is gradually justified by State practice, a declaration may by custom become recognized as laying down Rules binding upon States." Universal Declaration of Human Rights (adopted 10 December 1948 UNGA

Convention for the Protection of Human Rights and Dignity of the Human Being with Regard to the Application of Biology and Medicine (Oviedo Convention).[3] France ratified it on December 13, 2011.[4] The Oviedo Convention poses rigid constraints to interventions altering the human germline. In particular, Article 13 of the Convention states that "[a]n intervention seeking to modify the human genome may only be undertaken for preventive, diagnostic, or therapeutic purposes and only if its aim is not to introduce any modification in the genome of any descendants." Article 18.2 of the same convention forbids the creation of human embryos solely for research purposes.

A second important treaty is the United Nations Covenant on Economic, Social and Cultural Rights (Covenant).[5] France acceded to it on November 4, 1980.[6] France also ratified the Optional Protocol to the Covenant giving individuals the possibility of raising violations of their rights under the Covenant before the UN Committee on Economic, Social, and Cultural Rights (UNCESCR) on March 18, 2015.[7]

Although by virtue of Article 55 of the Constitution duly ratified treaties are part of French law and prevail over conflicting domestic laws, there are exceptions. For example, in 2005, in a case brought by an organization advocating on behalf of persons with disabilities, the Council of State ruled that Article 12 of the Covenant has no direct effect in the French legal order.[8] This article presents the most comprehensive account of the right to health in international law.[9] It states that every human being has a right to the "highest

---

Res 217 A(III) (UDHR); see United Nations Educational Scientific and Cultural Organization, "General introduction to the standard-setting instruments of UNESCO," http://portal.unesco .org/en/ev.php-URL_ID=23772&URL_DO=DO_TOPIC&URL_SECTION=201.html accessed September 5, 2018.

[3] Convention for the Protection of Human Rights and Dignity of the Human Being with Regard to the Application of Biology and Medicine: Convention on Human Rights and Biomedicine (opened for signatures on April 4, 1997, entered into force January 12, 1999) ETS No. 164 (Oviedo Convention). On the Oviedo Convention, see, in this book, Chapter 6.

[4] Council of Europe, "Chart of signatures and ratifications" https://rm.coe.int/inf-2017–7-rev-etat -sign-ratif-reserves/168077dd22 accessed September 5, 2018.

[5] International Covenant on Economic, Social and Cultural Rights (adopted 16 December 1966, entered into force January 3, 1976) 993 UNTS 3 (ICESCR). On the Covenant, see, in this book, Chapter 2.

[6] UN, "Status of Treaties" https://treaties.un.org/pages/ViewDetails.aspx?src=IND&mtdsg_ no=IV-3&chapter=4&clang=_en, accessed September 5, 2018.

[7] Optional Protocol to the International Covenant on Economic, Social and Cultural Rights, (adopted 10 December 2008, entered into force May 5, 2013), A/RES/63/117.

[8] Conseil d'Etat (1ère et 6ème sous-sections réunies) of September 26, 2005, *Association Collectif contre l'Handiphobie*, Lebon, no 248357, 3rd consideration.

[9] UNCESCR, "General Comment 14: The Right to the Highest Attainable Standard of Health (Art. 12) – Adopted at the Twenty-Second Session of the Committee on Economic, Social and Cultural Rights" (August 11, 2000) UN Doc E/C.12/2000/4, 1–2.

attainable standard of physical and mental health." In 2000, the UNCESCR set out to clarify that the right to health is not to be interpreted as a "right to be healthy" but rather as a set of freedoms and entitlements, including "the right to control one's health and body, including sexual and reproductive freedom, and the right to be free from interference, such as [...] non-consensual medical treatment and experimentation."[10] Entitlements include instead the right to a system of health protection which provides equality of opportunity for people to enjoy the highest possible level of health. In particular, Article 12.2.a prescribes that state parties to the Covenant promote reduction of stillbirth and infant mortality. This should be read as including access to "sexual and reproductive health services, including access to family planning, pre- and post-natal care [...] as well as to resources necessary to act on that information."[11] The right to health as it is formulated in the Covenant thus touches upon ethically controversial issues (e.g. medically assisted reproduction) that, as we explain below, bear directly onto the possibility of technical interventions on the human germline. It is therefore not entirely surprising that the Council of State reclaimed sovereignty over national laws and regulations concerning those subject matters.

The question of the direct applicability of the so-called rights to science (Arts. 15.1.b and 15.2–4 of the Covenant) in the French legal system is less clear.[12] All in all, French law seems rather consistent with both the right *to* science and the right *of* science. As we will see, in France, numerous experimental practices that are considered controversial in other countries (e.g. research on supernumerary embryos) are tightly regulated but permitted. Still, French law poses some limits to what is legitimate for science to attempt, as it bans, for instance, cloning or the creation of a human embryo for research purposes. Whether such restrictions are in accord with the right to science as expressed in the Covenant is a matter of debate.

As a member of the European Union and by virtue of Articles 55 and 88.1–7 of its Constitution, France is bound by all EU legislation.[13] EU legal instruments relevant to the topic of this chapter include the following:

---

[10] Ibid. para 8.

[11] Ibid. para 14

[12] On the "Right to Science" and the "Rights of Science," see in this book, Chapters 2 and 22.

[13] Case C-26/62 *Van Gend en Loos* [1963] ECR 1; Case C-6/64 *Costa v ENEL* [1964] ECR 585. On EU law, see in this book, Chapter 6. On the place of EU law within the French legal system, see, in general: S Boyron, "The New French Constitution and the European Union" (2009), 11 *Cambridge Y.B. Eur. Legal Stud.* 321; L Azoulai and F Ronkes Agerbeek, "Conseil Constitutionnel (French Constitutional Court), Decision No. 2004 – 505 DC of 19 November 2004, on the Treaty Establishing a Constitution for Europe" (2005) 42 *Common Market L. Rev.* 871.

i. Directive 2001/83/EC, on medicinal products for human use.[14]

ii. Regulation (EC) No 726/2004 on EU procedures for the authorization and supervision of medicinal products for human and veterinary use and establishing a European Medicines Agency.

iii. Regulation 1394/2007,[15] on advanced therapy medicinal products, which classifies gene therapy as a medicine for human use.

iv. Regulation 536/2014,[16] on the implementation of good clinical practice in the conduct of clinical trials on medicinal products for human use. Its Article 90 contains provisions that limit clinical research potentially leading to germline genome modifications.[17]

b Laws and Codes

The history of French legislative activities in the domain of the life sciences stretches back to 1983 with the creation of the National Consultative Ethics Committee for health and life sciences (*Comité Consultatif National d'Éthique pour les sciences de la vie et de la santé* – CCNE).[18] A few years later, in 1988, drawing on indications previously formulated also by the CCNE, a study by the Council of State titled "From Ethics to Law: Life Sciences" (*De l'éthique au droit: sciences de la vie*), also known as "Braibant Report" (named after its author, state councillor Guy Braibant), suggested that law-making activities on potentially divisive public policy issues caused by rapid developments in biomedicine and biotechnology should be informed by public debates aimed at soliciting and collecting the views of laypersons and civil society.[19] As we will see shortly, this position has been the defining trait of French legislative activities around life sciences since.

[14]   European Parliament and the Council of the European Union Directive 2001/83/EC of November 6, 2001 on the Community code relating to medicinal products for human use [2001] OJ L 311/67.

[15]   European Parliament and the Council of the European Union Regulation (EC) 1394/2007 of November 13, 2007 on advanced therapy medicinal products and amending Directive 2001/83/EC and Regulation (EC) No 726/2004 [2007] OJ L 324/121.

[16]   European Parliament and the Council of the European Union Regulation (EU) 536/2014 of April 16, 2014 on clinical trials on medicinal products for human use, and repealing Directive 2001/20/EC [2014] OJ L 158/1.

[17]   Ibid., art. 90: "No gene therapy clinical trials may be carried out which result in modifications to the subject's germ line genetic identity."

[18]   Decree no 83–132 of February 23, 1983 establishing a National Advisory Committee on Ethics for Life and Health Sciences [*Décret n. 83–132 du 23 février 1983 portant création d'un Comité consultatif national d'éthique pour les sciences de la vie et de la santé*].

[19]   Conseil d'Etat, *De l'éthique au droit: sciences de la vie: Notes et études documentaires* (Paris La Documentation française, 4855, 1988).

In 1994, three key laws entered into force:

i. Law 94–548 of July 1, 1994 on the treatment of data for research in the health sector;[20]
ii. Law 94–653 of July 29, 1994 on respect for the human body;[21]
iii. Law 94–654 of July 29, 1994 on donation and use of elements and products of the human body, on medically assisted reproduction and on prenatal diagnosis.[22]

Together with the law on the protection of human subjects in biomedical research of 1988,[23] also known as the Huriet Law (*la Loi Huriet*), these four laws constitute the set of legislative provisions known in France as the "Bioethics Laws" (*lois de bioéthique*). Among a variety of other provisions, the Bioethics Laws forbid the creation as well as the use of human embryos for research purposes.

Law 94–654 provided for a mandatory revision, five years after its entry into force, by the Parliamentary Office for the Evaluation of Scientific and Technological Options (*Office Parlementaire d'Évaluation des Choix Scientifiques et Technologiques* – OPECST).[24] However, in 1999, prompted by the mounting global debate caused by the cloning of Dolly the sheep,[25] the prime minister charged the Council of State with the task of reviewing the totality of the Bioethics Laws and suggesting new legislative initiatives as appropriate.[26]

---

[20] *Loi n. 94–548 du 1er juillet 1994 relative au traitement de données nominatives ayant pour fin la recherche dans le domaine de la santé et modifiant la loi n. 78–17 du 6 janvier 1978 relative à l'informatique, aux fichiers et aux libertés.*

[21] *Loi n. 94–653 du 29 juillet 1994 relative au respect du corps humain.*

[22] *Loi n. 94–654 du 29 juillet 1994 relative au don et à l'utilisation des éléments et produits du corps humain, à l'assistance médicale à la procréation et au diagnostic prénatal.*

[23] *Loi n. 88–1138 du 20 décembre 1988 relative à la protection des personnes qui se prêtent à des recherches biomédicales.*

[24] Loi n. 94–654, art. 21; The OPECST, which was created by Law 83–609 of July 8, 1983, informs the Parliament about the consequences of scientific and technological choices to facilitate its decision-making activities. To this end, the OPECST collects information, launches study programs and carries out assessments. See Sénat, "Présentation de l'Office Parlementaire d'Évaluation des Choix Scientifiques et Technologiques (OPECST)." www.senat.fr/opecst/presentation.html accessed September 5, 2018.

[25] I Wilmut and others, "Viable Offspring Derived from Fetal and Adult Mammalian Cells" (1997) 385 *Nature* 810.

[26] T Jean-Francois and others, *Les lois de bioéthique: cinq ans après*, (La Documentation française 1999); see also Sénat, "Projet de loi relatif à la bioéthique – B. Une réflexion préparatoire: l'étude du conseil d'état sur les lois bioéthiques" www.senat.fr/rap/l02-128/l02-1281.html accessed September 5, 2018.

The Bioethics Laws underwent three further rounds of amendment in the subsequent years. Notably, in 2004, a new provision upheld the ban on using human embryos for research, as well as on creating them ad hoc for research, but introduced a temporary (five-year) set of exceptions allowing research activities on supernumerary embryos, created in the context of medically assisted reproduction, upon authorization of the Biomedicine Agency (*Agence de la Biomédecine* – ABM), a public body created by the same law.[27]

Consistent with the spirit of the 1988 report by the Council of State, a law adopted on July 7, 2011 (also called the "2011 Law") vested the CCNE with the authority to organize periodic public engagement initiatives, called the Estates General of Bioethics (*États généraux de la bioéthique*).[28] The Estates General are held whenever changes to the Bioethics Laws are proposed (as in the case for instance of new technical developments in biomedicine or biotechnology) or, in the absence of new legislative initiatives, every five years.[29] Taking into account the conclusions reached within the Estates General, the CCNE drafts legislative proposals that are sent to the OPECST before being eventually submitted to parliamentary discussions.

The 2011 law also upheld the 2004 provisions forbidding all research on human embryos, but extended the possibility of exceptions indefinitely. However, in 2013, the prohibition-exception model was eventually abandoned when research on human embryos was made permissible upon authorization of the ABM and under tightly regulated conditions.[30] We will dwell more closely on these conditions and other provisions in the second part of the chapter.

The Bioethics Laws amended the Public Health Code (*Code de la santé publique*) on issues regarding medically assisted reproduction and research on human embryos and human embryonic stem cells (hESCs).[31]

---

[27] *Loi n. 2004–800 du 6 août 2004 relative à la bioéthique.*

[28] *Loi n. 2011–814 du 7 juillet 2011 relative à la bioéthique*, art. 46, introducing Public Health Code (CSP), arts. L1412-1-1 and L1412-3-1.

[29] CSP, art. L1412-1-1. The 2018 Estates General of Bioethics was convened on January 18, 2018. See (n 31).

[30] *Loi n. 2013–715 du 6 août 2013 tendant à modifier la loi n. 2011–814 du 7 juillet 2011 relative à la bioéthique en autorisant sous certaines conditions la recherche sur l'embryon et les cellules souches embryonnaires*; CSP, arts. L2151-1 to -8.

[31] CSP. In particular, Part II "Santé Sexuelle et Reproductive, Droits de la Femme et Protection de la Santé de l'Enfant, de l'Adolescent et du Jeune Adulte," Book I "Protection et Promotion de la Santé Maternelle et Infantile," Title IV "Assistance Médicale et la Procréation," especially Articles L2141-2, L2141-3, L2141-8, and L2141-11-1, and Title V "Recherche Sur

Several provisions of the Criminal Code (*Code pénal*) make certain activities connected with using human embryos and/or modifying the human genome punishable with pecuniary fines as well as with imprisonment and other kinds of criminal penalties.[32] In particular, Articles 214–1, 214–3, and 214–4 punish all eugenic practices as well as human cloning.[33] Articles 511–15 to 511–25 on the Protection of Human Embryos (*Protection de l'Embryon Humain*) punish those who buy embryos[34]. Anyone who creates embryos for commercial, industrial, or research purposes (including by means of the so-called therapeutic cloning) is punished with imprisonment and pecuniary fines.[35] According to Article 511–19, undertaking a research study on human embryos without donors' consent and authorization by the Biomedicine Agency, or violating the terms of such authorization, is legally punishable with imprisonment and pecuniary fines.[36] If the study involves hESCs, the Criminal Code foresees additional penalties.[37] Other punishable activities include: ceding hESCs to institutes that are not authorized to conduct research with them, or ceding hESCs without informing the Biomedicine Agency; importing and exporting tissues and cells derived from embryos or fetuses;[38] and importing embryos to France (and in all its administered territories) or exporting them without authorization.[39] Finally, Article 511–24 punishes practicing medically assisted reproduction beyond the legally

l'Embryon et les Cellules Souches Embryonnaires," Articles L2151-2, L2151-3, L2151-5, L2151-6, L2151-7, and L2151-7–1.

[32] Criminal Code (C. pén.) available at: Legifrance, "Code pénal" www.legifrance.gouv.fr/affichCode.do;jsessionid=8B8A75640F2469CA258F38BCC907CA13.tplgfr27s_1?cidTexte=LEGITEXT000006070719&dateTexte=20180509 accessed September 5, 2018. C. pén, Book II, "Des Crimes et Délits Contre les Personnes," Title I "Des Crimes Contre l'Humanité et Contre l'Espèce Humaine," Subtitle II "Des Crimes Contre l'Espèce Humaine," Chapter 1 "Des Crimes d'Eugénisme et de Clonage Reproductif," arts. 214–1, 214–3, and 214–4; Ibid., Chapter II "Dispositions Communes," arts. 215–1, 215–2, and 215–3; Ibid., Book V "Des Autres Crimes et Délits," Title I "Des Infractions En Matière De Santé Publique," Chapter I "Des Infractions En Matière d'Éthique Biomédicale," s 3 "De La Protection de l'Embryon Humain," arts. 511–15 to 511–25.

[33] C. pén., arts. 214–1, 214–3, and 214–4. The penalty amounts to thirty years of incarceration (and life sentence if it results from organized group activities) and a sanction of €7.5 million. Additional penalties for physical people who commit such crimes include banning from public office and loss of civic, civil, and family-related rights.

[34] Ibid., arts. 511–15, al.1. The law foresees seven years of prison and a fine of €100,000.

[35] Ibid., arts. 511–17.

[36] Ibid., arts. 511–19. The punishment amounts to seven years of prison and a sanction of €100,000.

[37] Ibid. Two years of prison and a fine of €30,000.

[38] Ibid., arts. 511-19-2. These activities are punished with two years of prison and a fine of €30,000.

[39] Ibid., arts. 511–23. These activities are punished with three years of incarceration and a €45,000 sanction.

mandated scope of this procedure,[40] as well as importing or exporting gametes or other germinal tissues without authorization or with aims that differ from those established by the law.[41]

c Guidelines and Policy Documents

In recent years, several French organizations, such as public research funders, scientific societies, and ethics committees have issued a number of opinions and reports that concern, directly or indirectly, research on modification of the germline of human embryos.

In particular, the Ethics Committee of the French National Institute for Health and Medical Research (*Institut National de Santé et Recherche Médicale* – INSERM) has issued two reports so far. In 2016, in a note concerning the development of CRISPR-related technologies, the committee recommended that basic research should continue and be encouraged to overcome current technical limitations, and that, in line with existing regulations, for the time being, no modification of germline nuclear genome should be attempted with a reproductive intent.[42] However, a few months later, in another statement, the same committee proposed to move beyond the rigid restrictions set forth by the Oviedo Convention and to adopt a case-by-case approach to attempts at modifying the human germline in the case of genetic diseases such as Huntington.[43] Moreover, the committee proposed the creation of a European Steering Committee to assess the risks of off-target mutations in clinical applications of CRISPR-based genome editing, as well as a European Advisory Committee to promote a public debate around controversial societal aspects.[44]

In April 2016, the National Academy of Medicine (*Académie Nationale de Médecine* – ANM) issued a report on the Modification of the Genome of

---

[40]   Ibid., arts. 511–24. This leads to five years of prison and €75,000 of fine.

[41]   Ibid., arts. 511-25-1. This is punished with two years of prison and a sanction of €30,000.

[42]   INSERM, *Saisine concernant les questions liées au développement de la technologie CRISPR (clustered regularly interspaced short palindromic repeat) -Cas9* (Note du Comité d'éthique, février 2016) 14.

[43]   INSERM, *Appendix no 14: Déclaration d'experts européens et français du comité d'éthique de l'INSERM, Promouvoir la recherche responsable avec les technologies de modification ciblée du génome: une perspective européenne* (novembre 2016), *Les enjeux économiques, environnementaux, sanitaires et éthiques des biotechnologies à la lumière des nouvelles pistes de recherche*, Tome II : *Comptes rendus et annexes*, 2017, www2.assemblee-nationale.fr/documents/notice/14/rap-off/i4618-tII/(index)/rapports#P2658_1161355 accessed September 5, 2018.

[44]   Ibid.

Human Germ Cells and Embryos (*Les modifications du génome des cellules germinales et de l'embryon humains*), recommending to uphold existing legislation forbidding all interventions on DNA causing modifications in the genome of descendants; the development of research enabling the use of genome editing, including uses on germ cells and human embryos; relaxing existing prohibitions on the creation of transgenic embryos for research purposes; and, finally, promoting of a pluridisciplinary debate on such controversial issues.[45]

The OPECST has undertaken an extensive set of activities around CRISPR technologies – including expert auditions and visits to scientific laboratories around the world – that have resulted in the publication of a two-volume report in 2017 on the economic, environmental, medical, and ethical challenges of novel biotechnologies.[46]

In 2015, the French Society of Human Genetics (*Société Française de Génétique Humaine*) together with the French Society of Gene and Cell Therapy (*Société Française de Thérapie Cellulaire et Génique*) created a working group on genome editing.[47] The following year, the ethics committee of the French National Institute of Health and Medical Research (INSERM) also established a working group on ethical issues raised by novel techniques of genomic editing.[48]

In January 2018, as required by Law 2011–814 of July 7, 2011, the CCNE convened the Estates General of Bioethics. In early June of the same year, the CCNE published a Summary Report (*Rapport de Synthèse*) describing the results of the public consultation.[49] This document forms the basis for guidelines and legislative proposals by the CCNE.

---

[45]   ANM, *Modifications du génome des cellules germinales et de l'embryon humains – Rapport* (2016) www.academie-medecine.fr/modifications-du-genome-des-cellules-germinales-et-de-lembryon-humains-2/

[46]   OPECST, *Les enjeux économiques, environnementaux, sanitaires et éthiques des biotechnologies à la lumière des nouvelles pistes de recherche*, Tome 1: *Rapport* (2017) www.senat.fr/rap/r16-507-1/r16-507-11.pdf and Tome 2 www2.assemblee-nationale.fr/documents/notice/14/rap-off/i4618-tII/(index)/rapports#P2658_1161355 accessed September 5, 2018.

[47]   See A Blasimme and others, "Genome Editing and Dialogic Responsibility: 'What's in a Name'" (2015) 15:12 *Am. J. Bioeth* 54.

[48]   INSERM, *Saisine concernant les questions liées au développement de la technologie CRISPR (clustered regularly interspaced short palindromic repeat)-Cas9* (INSERM, 2016) www.inserm.fr/sites/default/files/2017-10/Inserm_Saisine_ComiteEthique_Crispr-Cas9_Fevrier2016.pdf accessed September 5, 2018.

[49]   CCNE, *Rapport de Synthèse du Comité consultatif national d'éthique – Opinions du Comité Citoyen*, (Bio Éthique, June 2018) https://etatsgenerauxdelabioethique.fr/media/default/0001/01/013928888b8655e9c41fac63a51385185d5860c8.pdf accessed September 5, 2018.

Finally, in 2018, the ABM issued a report on the application of the Bioethics Laws.[50] This text, issued in preparation for the Estates General of Bioethics, does not set new guidelines for germline modification, but it calls attention – among other things – to the need to discuss genome editing technologies and the creation of gametes in vitro from precursor germ cells or through reprogramming of somatic cells.

2 Definitions

a Embryos

No formal legal definition of "embryo" appears in the French legal system. This is the result of a deliberate choice. Back in 1986, the CCNE stated that, given the lack of scientific consensus on a specific definition of "embryo," it is preferable to adopt a general one, encompassing all stages of development of the zygote before the fetal stage, which is reached at the eighth week of pregnancy.[51] It should be noted that two years earlier, the CCNE had declared that the human embryo (and fetus) ought to be recognized as a potential human being (*personne humaine potentielle*), deserving respect.[52] This crucial statement has been reiterated several times ever since.[53]

b Transgenic and Chimeric Embryos

Although the 2011 Law forbids the creation of transgenic and chimeric embryos, it does not define those terms.[54] In its 2018 report, the ABM

---

[50]   ABM, *Rapport sur l'application de la loi de bioéthique* (Rapport 2018) www.agence-biomedecine.fr/Rapport-sur-l-application-de-la,917?lang=fr> accessed September 5, 2018.

[51]   CCNE, Avis no. 8 du 15 décembre 1986 *relatif aux recherches et utilisation des embryons humains in vitro à des fins médicales et scientifiques – Rapport*: "*Il apparaît donc préférable de retenir une définition générale de l'embryon humain. Suivant en cela la proposition du Conseil de l'Europe (CAHBI, 1986) on désignera par embryon tous les stades de développement du zygote, avant le stade fœtal qui est atteint à la huitième semaine de la grossesse.*"

[52]   CCNE, Avis no 1 du 22 mai 1984, *sur les prélèvements de tissus d'embryons et de fœtus humains morts, à des fins thérapeutiques, diagnostiques et scientifiques – Rapport*, www.ccne-ethique.fr/sites/default/files/publications/avis001.pdf accessed September 5, 2018.

[53]   See eg CCNE, Avis n. 112 du 21 octobre 2010, *Une réflexion éthique sur la recherche sur les cellules d'origine embryonnaire humaine, et la recherche sur l'embryon humain* in vitro, www.ccne-ethique.fr/sites/default/files/publications/avis_112.pdf accessed September 5, 2018.

[54]   *Loi n. 2011-814, art. L. 2151-2, al. 2.*

highlighted the need to clarify these terms and the related provisions,[55] but it remains to be seen whether a definition will be adopted in law.

c   Stem Cells and hESCs

The ABM provided a definition of "stem cells" in its 2018 report. It defined them as cells that, due to their proliferative capacities and due to their multipotency or pluripotency, can reconstitute a tissue in vivo. The term "hESCs" is commonly used to refer to cell lines established from inner cell mass of the blastocyst. Induced pluripotent stem cells are defined as embryo-like stem cells that are created through the reprogramming of the nucleus of differentiated adult cells. They share the properties of hESCs, but are not considered to be embryos.[56]

d   Medically Assisted Reproduction

The Public Health Code defines "medically assisted reproduction" as any clinical and biological practice involving in vitro fertilization, preservation of gametes, germ cells, and embryos, artificial insemination and the implantation of embryos.[57] The PHC also specifies that medically assisted reproduction aims at limiting number of stored embryos.[58]

e   Cloning

French law does not define "cloning" as such, but refers to cloning as any procedure aimed at creating a human being that is genetically identical to

---

[55]   ABM, Rapport 2018 (n 50): *"(l)'instruction de dossiers de demandes d'autorisation peut conduire l'Agence à s'interroger sur l'interprétation de ces notions. Pour des raisons tant éthiques que de sécurité juridique, il conviendrait de clarifier ce que recouvrent ces notions et ce que le législateur entend prohiber."*

[56]   Ibid.

[57]   CSP, art. L-2141-1, al. 1: *"pratiques cliniques et biologiques permettant la conception in vitro, la conservation des gamètes, des tissus germinaux et des embryons, le transfert d'embryons et l'insémination artificielle. La liste des procédés biologiques utilisés en assistance médicale à la procréation est fixée par arrêté du ministre chargé de la santé après avis de l'Agence de la biomédecine."*

[58]   CSP, Art. L-2141-1, al. 5: *"(l)a mise en œuvre de l'assistance médicale à la procréation privilégie les pratiques et procédés qui permettent de limiter le nombre des embryons conservés. L'Agence de la biomédecine rend compte, dans son rapport annuel, des méthodes utilisées et des résultats obtenus"* and al. 6 *"(l)a stimulation ovarienne, y compris lorsqu'elle est mise en œuvre indépendamment d'une technique d'assistance médicale à la procréation, est soumise à des règles de bonnes pratiques fixées par arrêté du ministre chargé de la santé."*

another.[59] The OPECST defined cloning as asexual reproduction – starting from a cell or organism   aimed at the creation of entities that are genetically identical to the initial cell or organism by means of embryo splitting or somatic cell nuclear transfer.[60]

3 Oversight Bodies

Research and experiments on embryos, hESCs, or human gametes could potentially lead to the modification of the human germline. The already-mentioned 2004 law created a specific oversight body – the ABM – and charged it with the task of reviewing authorization requests for research protocols employing supernumerary human embryos and hESCs, derived either directly or from already-established cell lines.[61] The ABM licensing system applies to both academic and private-sector research activities. The agency has an ad hoc advisory panel of experts (called *conseil d'orientation*) that reviews each application and formulates recommendations to the agency. The final decision is then communicated to the ministries in charge of health and research.

Research on human gametes – either derived from donors or developed in vitro from precursor cells or from reprogrammed somatic cells – does not fall under the remit of the ABM. Yet, so far, the only way to test the functionality of in vitro derived gametes is to attempt to use them to create an embryo. As we will discuss further below, at present the ABM does not seem to have the possibility to authorize such procedures.[62]

---

[59]   C. pén., art. 214–2.
[60]   OPECST, *Rapport sur le clonage, la thérapie cellulaire et l'utilisation thérapeutique des cellules embryonnaires* (1999): "(l)e clonage vise à la production asexuée, à partir d'une cellule ou d'un organisme, d'entités biologiques génétiquement identiques à cette cellule ou à cet organisme." It also describes the related processes: "*Deux procédés sont utilisables pour parvenir à cette duplication génétique. (1) Le clonage par scission d'embryon – Ce procédé est le plus facile à mettre en œuvre et, sans doute, le plus efficace. Il consiste à déclencher artificiellement in vitro ce qui se produit à l'état naturel chez les mammifères en cas de gémellité vraie (jumeaux monozygotes). Lorsque l'embryon fécondé se divise en deux cellules, on sépare celles-ci de façon que chacune d'elles produise à son tour un embryon. (...) (2) Le clonage par transfert nucléaire – Il consiste à introduire, dans le cytoplasme d'un ovule non fécondé dont on a retiré le matériel nucléaire, le noyau d'une cellule provenant d'un embryon, d'un fœtus ou d'un organisme adulte. On cherche à leurrer le cytoplasme de l'ovocyte qui tente alors d'organiser le nouveau noyau pour lui redonner ses caractéristiques embryonnaires.*"
[61]   Loi n. 2004–800.
[62]   See, in this chapter, Section II.1.b. "Laws and Codes."

4 *Funding*

In general, public research in France is funded by three types of institutions: universities, the National Center for Scientific Research (*Centre National de la Recherche Scientifique* – CNRS), established in 1939 and funding research in all disciplines, and the INSERM, established in 1964 and employing more than 9,000 researchers (with a total budget of almost €1 billion per year). Individual laboratories and research groups are sometimes affiliated to more than one such institution. Specific project funding can be provided by the National Research Agency (*Agence Nationale de la Recherche* – ANR), which has been in operation since 2005 and is focused on flagship national projects. The ANR research budget started at €710 million in 2005 and continued to grow until 2008, just to start decreasing in the aftermath of the global financial crisis of that year.[63]

Within the limits set forth by the law, all these institutions can fund research activities on embryos, hESCs, and gametes. Funding for similar research can be obtained from the European Union (e.g. the framework program of the European Commission, now called Horizon 2020) and from the European Research Council. Specific EU regulations apply to research involving human embryos, hESCs, and gametes.[64]

III SUBSTANTIVE PROVISIONS

Scientific research and clinical activities on human embryos, hESCs, and gametes can lead to the modification of the human germline. These modifications could become heritable if the modified embryos are implanted in utero and lead to a successful pregnancy. In this section, we will therefore cover research and clinical activities employing any of these human biological materials.

As far as gametes are concerned, we consider sperm and ova derived from adult donors, gametes derived from in vitro maturation of spermatogonial stem cells, and germinal vesicle oocytes, as well as gametes obtained through in vitro reprogramming of mature somatic cells. The last two methods, coupled with genome modification techniques, are starting to offer new possibilities for altering the genetic makeup of human gametes and, therefore, the germline of the resulting offspring.[65] However, we will not consider the

---

[63] S Bos, "La recherche fait face à un manque criant de financements" (Les Echos, August 21, 2017) www.lesechos.fr/21/08/2017/LesEchos/22511-008-ECH_la-recherche-fait-face-a-un-manque-criant-de-financements.htm accessed September 5, 2018.

[64] See in this book, Chapter 6.

[65] R Vassena and others, "Genome Engineering through CRISPR/Cas9 Technology in the Human Germline and Pluripotent Stem Cells" (2016) 22:4 *Hum. Reprod. Update* 411, 411–419.

genetic modification of gametes in vivo (i.e. directly in adult human beings) as this is not a practical method, and, given the available alternatives, there do not seem to be valid technical reasons for experiments of that sort.

1 Regulation of Basic Research

In this subsection, we provide a detailed account of current legal provisions on the use of embryos and hESCs (a) and gametes (b), in the context of basic research. By basic research we mean fundamental research aimed at elucidating human pathophysiological processes and mechanisms that are relevant to medical knowledge or medical practice. We distinguish fundamental research from clinical research. The former aims at producing scientific knowledge regarding the causes and consequences of diseases as well as suggesting possible strategies to prevent and to treat them. The latter (which we discuss below in point (3) refers to testing the safety and efficacy of drugs and other therapeutic interventions in human research subjects (typically, although not exclusively, through clinical trials).

a Human Embryos and hESCs

French laws on the use of human embryos in research do not distinguish between zygotes, pre-embryos (or preimplantation embryos)[66] and embryos. As mentioned above, in the French legal system, notwithstanding the lack of a clear definition, "embryos" are the product of natural or technically assisted fecundation of a human ovum by a human sperm cell.[67] In France, in vitro fertilization (IVF) embryos are cryopreserved within seven days after fertilization.[68] According to the Public Health Code, creating a human embryo through in vitro fertilization with the specific purpose of conducting

---

[66] D Jones and B Telfer, "Before I Was an Embryo, I Was a Pre-embryo: Or Was I?" (1995) 9:1 *Bioethics* 32, 32–49.

[67] See (n 51).

[68] Beyond this time point, embryos are no longer able to implant in utero. As a consequence, supernumerary embryos donated by couples for research uses are not older than seven days. However, it is nowadays technically possible to keep them in culture for as long as thirteen days, and possibly more (S Reardon, "Human embryos grown in lab for longer than ever before – Embryos cultured for up to 13 days after fertilization open a window into early development" (*Nature*, May 4, 2016) www.nature.com/news/human-embryos-grown-in-lab-for-longer-than-ever-before-1.19847 accessed September 5, 2018). That being said, French law has not established temporal limits to in vitro culture of human embryos (See ABM, Rapport 2018 (n 50) 58). International consensus holds that embryos should not be kept in culture beyond gastrulation. At the moment of gastrulation, that is, at Day 14 after fertilization, twinning is not possible anymore. Since this physiological step determines an embryo's

research on it is prohibited.[69] The creation of transgenic or chimeric embryos and the creation of an embryo that is genetically identical to an existing human being (i.e. cloning through somatic cell nuclear transfer or embryo splitting) are also forbidden.[70] Another important normative provision in this domain is the prescription of the Civil Code (*Code civil*) forbidding eugenic practices,[71] that is to say any organized attempt aimed at biologically selecting people according to genetic, physical, or mental traits.[72] As we said earlier, eugenic practices are punished by the Criminal Code.[73]

As a general principle, research activities on embryos and hESCs are only possible upon authorization by the ABM. Since the creation of embryos for research purposes is forbidden, scientists can only work with supernumerary human embryos (i.e. embryos that have been created in vitro in the context of a medically assisted reproduction procedure and that are no longer the object of a reproductive project) and with the written informed consent of the couple.[74] This applies also to procedures undertaken with the aim of deriving hESCs from an embryo. However, such cells can also be imported or obtained from already established cell lines. Even in these cases, an authorization by the ABM is necessary.

To obtain ABM authorization, a research protocol needs to fulfill four conditions:

1. It must be scientifically appropriate;
2. It must have a medical aim and have medical relevance;
3. The study cannot be done without using embryos or hESCs; and, finally,
4. It must conform to the ethical principles governing the type of research in question.[75]

Since the research ban was repealed back in 2004, the ABM has taken 311 decisions regarding research protocols, conservation requests, or importation from abroad of human embryos for research purposes. Of these, 280 resulted in

---

biological individuation, the fourteen-day limit is generally considered an ethical threshold for research on human embryos.

[69]   CSP, art. L2151-2.

[70]   Ibid., arts. L2151-2 and L21511-4; C. civ., art. 16-4.

[71]   C. civ., art. 16-4, al. 2.

[72]   See C. pén., art. 225-1 dealing with discrimination, and art. 214-2 dealing with cloning.

[73]   The penalty amounts to thirty years of incarceration (and life sentence if it results from organized group activities) and a sanction of €7.5 million. Additional penalties for physical people who commit such crimes include banning from public office and loss of civic, civil, and family-related rights. C. pén., arts. 214-1, 214-3, and 214-4.

[74]   CSP, art. L2151-5.

[75]   Ibid.

authorizations and 14 in refusals, while 17 were withdrawn. In total, 88 research protocols, including 18 regarding embryos, have been authorized, whereas 8 research protocol authorizations have been denied. According to the ABM, as of December 31, 2015, out of the 20,000 embryos offered by couples for research, less than 10 percent were made available for research.[76]

So far, neither the French legislator nor any other public body has issued any specific restriction or guideline concerning the experimental techniques that can or cannot be employed in research protocols concerning embryos and hESCs. As a consequence, there are no formal prohibitions on the use of genome modification technologies to alter a human embryo or hESCs in the research context (as long as they do not lead to a pregnancy). It is therefore up to the ABM to authorize research protocols using these techniques. At present, we are not aware of any such protocol in France.

That being said, the Civil Code forbids, in its Article 16–4, any activity that could damage the *integrity of the human species.*[77] Of course, whether an alteration of the human genome through germline modification amounts to a violation of the Civil Code is subject to judicial interpretation. At any rate, embryos that have been used in research cannot be transferred in utero to establish a pregnancy.[78] However, embryos conceived in vitro can be the object of research – before or after implantation – if the research pertains the domain of medically assisted reproduction and if both members of the couple consent to it.[79] Moreover, the French Civil Code, while forbidding genetic modifications intended to alter the progeny of a person, foresees an exception for research activities aimed at preventing or treating genetic diseases.[80] It is not clear, however, whether such research activities can stretch to include the implantation of a genetically modified embryo.

Modifications of the genome of an embryo under circumstances that allow the modifications to pass on to future generations (i.e. through a successful pregnancy) could be construed as an "eugenic attempt" at selecting certain allegedly desirable traits, which is forbidden by both the Civil and the Criminal Code.[81] Likewise, such interventions would also be forbidden if

---

[76] ABM, Rapport 2018 (n 50) 54, 58.

[77] C. civ., art. 16-4, al. 1.

[78] CSP, art. L2151-5, point IV: "*IV.- Les embryons sur lesquels une recherche a été conduite ne peuvent être transférés à des fins de gestation.*"

[79] Ibid., art. L2151-5, point V: "*V.- Sans préjudice du titre IV du présent livre Ier, des recherches biomédicales menées dans le cadre de l'assistance médicale à la procréation peuvent être réalisées sur des gamètes destinés à constituer un embryon ou sur l'embryon in vitro avant ou après son transfert à des fins de gestation, si chaque membre du couple y consent. Ces recherches sont conduites dans les conditions fixées au titre II du livre Ier de la première partie.*"

[80] C. civ., art. 16-4.

[81] Ibid.

their intended objective was that of altering the offspring of a person.[82] How the aim of a research protocol introducing a germline modification should be assessed is not obvious. For instance, if in the context of an experimental procedure CRISPR-based genome editing is used to remove a seriously deleterious mutation from an embryo's genome, this intervention will most likely result in germline modifications in both the resulting person and in his or her offspring. Yet, the *aim* of the intervention is to study whether that technique could be used to prevent a genetic disease, and not to alter the descendants of the treated embryo. It seems therefore that cases in which the aim of research activities is primarily therapeutic, and in particular when that aim is to treat or to prevent the transmission of a genetic disease, the prohibition of altering the germline of future generations may no longer hold. If that is correct, then the use of genome editing technologies on human embryos that will likely result in germline modifications may not be a priori forbidden in France. At least in the case of research undertaken in the context of assisted reproduction[83] and research aimed at preventing or treating a genetic disease,[84] French law does not ban outright research that could potentially lead to human germline genome modifications.

Still, it is dangerous to jump to conclusions. The French regulatory framework was originally laid down when genome editing was nowhere near its current stage of technical development. The existing framework certainly poses considerable limits to the scope of research activities that can be conducted on in vitro–conceived embryos. Yet, interestingly, it does not entirely close the door to germline-altering interventions. This does not mean that the law explicitly allows them, nor that they would be authorized. It will eventually be up to the ABM and the courts to provide viable interpretations of the regulatory framework. Such interpretations will also need to take into account the Oviedo Convention, which France has ratified and which forbids germline-altering interventions.[85] To complicate things further, the text of the Oviedo Convention also leaves room for interpretation:

> An intervention seeking to modify the human genome may only be undertaken for preventive, diagnostic or therapeutic purposes and only if its aim is not to introduce any modification in the genome of any descendants.[86]

---

[82] Ibid.
[83] CSP, art. L2151-5.
[84] C. civ., art. 16–4.
[85] Oviedo Convention (n 3) art. 13.
[86] Ibid.

Although it is uncontested that this article denies the possibility of undertaking germline-altering interventions, it allows altering the genome of human embryos in the context of research, provided the embryo is not implanted. Moreover, the mentioned preventive and therapeutic interventions, if undertaken on embryos in vivo, would most likely result in germline modifications. In its current formulation, since embryos are not mentioned by the article, it is not clear whether this kind of procedure would be forbidden or not.

The ABM and French courts will also have to interpret the regulatory framework for research modifying human germline and embryo genome in light of the Covenant, which France has also ratified, taking into consideration the obligations France has, as a matter of international law, to fulfill the right to health and the right to science.[87]

Given such uncertainties, it is also possible that the upcoming reform of the Bioethics Laws will clarify the interpretation of such provisions or enforce a more explicit ban on germline genome editing by amending them.

b Gametes

Scientists who want to conduct research on human gametes can only do so on gametes donated for assisted reproduction purposes. A specific section of the Public Health Code regulates the use of gametes and stored germ cells.[88] These provisions state that donors who have provided gametes for purposes of assisted reproduction – either for themselves or for other couples – are contacted once a year in writing in order to ask them if they want their unused cells to be stored further. If they no longer want their gametes to be stored, they have three options: (i) ask for the destruction of the material; (ii) make it available to other couples who need it for reproductive purposes; or (iii) consent to the material being used for research purposes.[89] In this latter case, just as any other tissue or cell of human origin, donated gametes can only be used by laboratories in authorized research institutes.[90]

Gametes created in vitro through reprogramming of somatic cells could constitute a novel source of reproductive cells for research uses. At present, these procedures are unregulated, and they could in principle be conducted

---

[87]   ICESCR (n 5) arts. 12, 15.1.b, and 15.2-4.
[88]   CSP, arts. R2141-17 to R2141-23.
[89]   Ibid., art. R2141-17.
[90]   Ibid., art. L1243-3.

by any laboratory in France following the same regulations that pertain to the use of other primary human material in research.[91]

No specific legal provisions regulate which experimental tools and techniques can or cannot be employed in the context of basic research activities on human gametes in France. Research limited to the domain of medically assisted reproduction can be undertaken on human gametes that are destined to be used for in vitro fertilization and implantation.[92] This kind of research could in principle involve genome modifications that could be passed on to the germline of the fetus, and hence to the resulting human being and to his or her offspring – should the modified gametes be used to give rise to an embryo, and should the embryo be implanted. It is therefore imaginable that gametes that are modified using genome editing technologies are used in the context of research on medically assisted reproduction, and in principle could be used to establish a pregnancy – provided the above-specified conditions apply.

The already-mentioned Article 16–4 of the Civil Code states that, with the exception of research aimed at preventing or treating genetic disorders, *the aim* of activities that would result in a modification of genetic traits (*caractères génétiques*) of a person – like the one just mentioned – shall not be modifying the offspring of the person in question.[93] However, it remains to be seen whether altering the DNA of gametes can be interpreted as transforming the genetic traits of a person. Given that there is no "person" when the modification of the DNA of the gametes occurs, from a legal point of view this provision could be considered not pertinent to research activities on human gametes.

2 Regulation of Preclinical Research on Nonhuman Animals

In France, there are no specific regulations restricting research activities that cause alterations to the germline of nonhuman animals. As in most scientifically advanced countries, these types of procedures are standard practice in scientific laboratories throughout France. General research involving nonhuman animals is regulated through the relevant provisions of law,[94] in conformity with European Directive 2010/63/EU on the protection of animals

---

91 Ibid., art. L1243-4.
92 Ibid., art. L2151-5.V.
93 C. civ., art. 16–4. This article in fine states: "*Sans préjudice des recherches tendant à la prévention et au traitement des maladies génétiques, aucune transformation ne peut être apportée aux caractères génétiques dans le but de modifier la descendance de la personne.*"
94 Rural Code (C. rur.) available at: Legifrance, "Code rural et de la pêche maritime" www.legifrance.gouv.fr/affichCode.do?cidTexte=LEGITEXT000006071367&dateText e=20180925 accessed September 5, 2018.

used for scientific purposes,[95] which also contains special limitations for research with primates and great apes.

Currently, in France, the use of gene drive technologies (i.e. genetic engineering *technologies* that can propagate a particular suite of genes throughout a population) is being debated in the context of new genome-editing technologies.[96] This discussion, however, is focusing neither on ethical considerations regarding animal rights and animal welfare nor on the integrity of the genome of nonhuman species. Rather, it is centered on the environmental consequences of testing such techniques in a research context (e.g. containment issues) and of employing them in the wild on an ecosystemic scale (e.g. preservation of biodiversity).

3 Regulation of Clinical Research

By "clinical research" we understand research activities aimed at collecting evidence regarding the safety and efficacy of a drug, therapy, or intervention by conducting clinical trials on human research subjects. Genome editing techniques could be suited to develop therapies for genetic diseases (gene therapy or cell therapy treatments).[97] These therapies would need to be clinically validated through clinical trials. In many countries, some of these innovative therapies (including gene therapies targeting somatic cells) have already been approved for marketing and have now been in clinical use for some time.[98] CRISPR-based genome editing could prove more efficient and precise than other forms of genome editing currently in use, namely zinc-finger nuclease (ZFN) and transcription activator–like

---

[95] European Parliament and the Council of the European Union Directive 2010/63/EU of September 22, 2010 on the protection of animals used for scientific purposes, OJ L276/33.

[96] OPECST, *Les enjeux économiques, environnementaux, sanitaires et éthiques des biotechnologies à la lumière des nouvelles pistes de recherche*, Tome I (2017) www.senat.fr/rap/r16-507-1/r16-507-11.pdf accessed September 5, 2018; Ibid., Tome 2 (2017) www2.assemblee-nationale.fr/do cuments/notice/14/rap-off/i4618-tII/(index)/rapports#P2658_1161355 accessed September 5, 2018; INSERM, *Saisine concernant les questions liées au développement de la technologie CRISPR (clustered regularly interspaced short palindromic repeat) -Cas9* (n 42) 14.

[97] It should be noted that gene and cell therapy can also rely on other means such as viral vectors harboring a DNA cassette that delivers the desired therapeutic modifcation (MA Kay and others, "Viral vectors for gene therapy: the art of turning infectious agents into vehicles of therapeutics" (2001) 7 *Nature Medicine* 33, 33–40; CE Thomas, "Progress and problems with the use of viral vectors for gene therapy" (2003) 4 *Nature Reviews Genetics* 346, 346–358. This is at the moment the most well-established way of performing gene therapy, but the clinical promise of genome editing is that of producing more precise and controllable interventions in the patients' genome.

[98] C Dunbar, "Gene therapy comes of age" (2018) 359:6372 *Science* http://science .sciencemag.org/content/359/6372/eaan4672 accessed September 5, 2018.

effector nucleases (TALENS).[99] Should that be the case, its clinical use would become widespread. As we explained, in France using any such genome editing techniques on gametes and embryos is not forbidden, provided these are not used to establish a pregnancy – in which case the abovementioned limitations apply.[100]

Gene and cell therapies are normally employed on adult patients to correct disease-causing genetic abnormalities,[101] or in the context of cancer immunotherapy.[102] In neither case do they affect the patient's germline. The clinical potential of CRISPR-based genome editing for gene and cell therapy is being studied in laboratories worldwide, but currently clinical trials on human subjects are not underway in France, although in theory they could be.[103] However, should a gene therapy produce germline modifications, for instance by altering the genetic material of gametes or embryos, it could not be tested in a clinical trial because Article 90 of the European Clinical Trial Regulation states: "No gene therapy clinical trials may be carried out which result in modifications to the subject's germ line genetic identity."[104]

For parents who know they are at risk of transmitting a genetic disorder to their progeny, preimplantation genetic diagnosis (PGD) on IVF embryos and embryo selection is currently standard practice and France offers clear regulatory provisions in this domain.[105] Yet, not all parents may be aware of such risks and de novo mutations are also always possible. In such cases, and in the

---

[99]   T Gaj and others, "ZFN, TALEN and CRISPR/Cas-based Methods for Genome Engineering"(2013) 31:7 *Trends Biotechnology* 397, 397–405.

[100] See in this chapter, Section III.1(a) and Section III.4(a).

[101] M Maeder, and C Gersbach, "Genome-editing Technologies for Gene and Cell Therapy" (2016) 24:3 *Molecular Therapy* 430, 430–446.

[102] L Poirot and others, "Multiplex Genome-Edited T-cell Manufacturing Platform for "Off-the-Shelf" Adoptive T-cell Immunotherapies" (2015) 75:18 *Cancer Res* 3853; J Ren and others, "Multiplex Genome Editing to Generate Universal CAR T Cells Resistant to PD1 Inhibition" (2017) 23:9 *Clin Cancer Res* 2256.

[103] To the best of our knowledge, clinical trials testing CRISPR-based therapies in humans are currently only taking place in China (P Rana, AD Marcus, and W Fan, "China, Unhampered by Rules, Races Ahead in Gene-Editing Trials" (*The Wall Street Journal*, January 21, 2018) www.wsj .com/articles/china-unhampered-by-rules-races-ahead-in-gene-editing-trials-1516562360 accessed September 5, 2018). One company has received approval to start the first CRISPR-based clinical trial on β-thalassemia in Europe (K Hignett, "Breakthrough CRISPR Gene Editing Trial Set to Begin This Year" (*Newsweek*, April 16, 2018) www.newsweek.com/crispr-therapeutics-crispr-cas9-gene-editing-beta-thalassaemia-887051 accessed September 5, 2018) and has filed an application for another study on sickle-cell anemia in the US, which is currently under review by the FDA (NP Taylor, "FDA Puts Vertex, CRISPR Sickle Cell Trial on Hold" (*Fierce Biotech*, May 31, 2018) www.fiercebiotech.com/biotech/fda-puts-vertex-crispr-sickle-cell-trial-hold accessed September 5, 2018).

[104] EU Regulation 536/2014.

[105] CSP, art. L2131-1 to L2131-5.

case of people who object to IVF and PGD on religious grounds, editing the genome of embryos in vivo could be an option. This practice is not explicitly forbidden by French law. However, European law forbids clinical trials aimed at testing the safety and efficacy of such procedures.

In November 2018, the alleged birth in China of two girls from genetically modified embryos has been announced. In France, at the moment, clinical research protocols of this kind, potentially resulting in germline alterations, would face the same rigid limitations we described above regarding basic research on embryos. Additionally, they would have to follow specific rules regarding the preparation of gene and cell therapy products,[106] as well as European laws regarding clinical trials and gene therapy.[107] At any rate, given existing rules on clinical research in France, while not prohibited, the approval of a clinical research protocol involving the implantation of an embryo (whether genetically edited or not) is hard to imagine.[108]

In sum, while clinical research resulting in germline alterations is forbidden by European law, clinical research involving edited embryos is not explicitly forbidden. Yet, there seems to exist very limited room in French law to perform clinical research involving genetically edited embryos (either in vivo or in vitro) or embryos resulting from genetically edited gametes. The upcoming reform of the Bioethics Laws may well clarify such matters and introduce more precisely defined limitations on clinical activities around genome editing.

4 Regulation of Clinical Applications

In order to actually produce a germline modification, gene-edited gametes and embryos should be used to establish a pregnancy. By "clinical application" of germline genome editing, we thus mean any attempt at implanting genetically modified embryos (either obtained through modified gametes or resulting from direct genome editing in vitro) in the womb of a woman outside the remit of a research protocol – be it basic or clinical research. Clinical

---

[106]   CSP, art. L5426-1 à L5426-4; Ibid., arts. R5121-207 – R5121-208: Title IV, Chapter III: "*Préparation, conservation, distribution et cession des tissus, de leurs dérivés, des cellules et des préparations de thérapie cellulaire;* Section 16: *Conditions générales d'autorisation des préparations de thérapie génique et des préparations de thérapie cellulaire xénogénique.*" See also EC Regulation 1394/2007.

[107]   EU Regulation 536/2014.

[108]   CSP, arts. L1121-1 to L1126-12, Title II: *Recherches impliquant la personne humaine.*

applications of genome editing hence refer to uses of the technique on embryos in vitro in the course of assisted reproduction.[109]

a Pregnancy with Edited Gametes

As we have seen already, French law does not exclude the possibility that gametes that underwent research activities in the domain of assisted reproduction may be used to establish a pregnancy.[110] However, other kinds of attempts beyond assisted reproduction research, including attempts at introducing genetic traits not linked to a known medical condition (i.e. enhancements), are arguably forbidden.[111] The legislator did not specify whether altering the genome of the gametes through gene editing is permissible or not. However, the *vacuum legis* should not be interpreted to mean that such attempts can be authorized – let alone that they will be in the imminent future. Scientific considerations about safety, as well as legal considerations about whether one can legitimately subsume an entirely novel technique like genome editing under a provision that was created before the technique was available, would definitely play a role in authorizing decisions. This leads us to conclude that under the current regulatory regime, the use of genome-edited gametes to create an embryo and then to implant it will not be authorized in France.

b Pregnancy with Edited Embryos

Analogous considerations apply to clinical uses of gene editing aimed at establishing a pregnancy through a genetically modified embryo. While the law forbids establishing a pregnancy with embryos that have previously been used for research,[112] it admits an exception in the case of research on medically assisted reproduction.[113] Such embryos can be implanted if both members of the couple consent to that. Also Article 16–4 of the Civil Code could be taken into account and interpreted to allow the implantation of embryos used in

---

[109] As we said above, in vivo genome editing of a naturally fertilized embryo is not impossible in principle, but, from a technical point of view, it seems unreasonably demanding.

[110] CSP, art. L2151-5 V.

[111] See (n 71) and (n 73).

[112] CSP, art. L2151-5 IV.

[113] Ibid., art. L2151-5 V.

research aimed at preventing or treating genetic diseases.[114] Yet, also in this case, it could be argued that since the regulatory framework was never intended to cover the application of genome editing, it should not be construed as authorizing such procedures. Specific guidelines or amendments are needed to clarify such matters in light of recent progress in genome editing technologies.

IV CURRENT PERSPECTIVES AND FUTURE POSSIBILITIES

As foreseen by the law, the CCNE organized the Estates General of Bioethics in 2018. At the end of a nationwide public consultation, the CCNE released a document highlighting the orientations that emerged from the consultation phase and suggesting legislative action on wide array of controversial topics.[115] This document also includes recommendations regarding novel genome-editing techniques, and, more generally, activities involving human embryos and pluripotent stem cells.

The CCNE recognizes that the use of CRISPR/Cas9 genome editing on human embryos can potentially represent a threat to their integrity, but is favorable to its use in the context of research, provided that edited embryos are not introduced in a woman's womb.[116] Therefore, legislators are encouraged to introduce clearer provisions allowing genome-editing research on embryos, while retaining a prohibition to implant genetically modified human embryos.[117]

The CCNE questions the opportunity of treating within the same legal framework embryos and hES cells. The latter, argues the CCNE, while deriving from an embryo, do not carry the symbolic connotation of "potential persons" that embryos possess.[118] This makes their use less controversial from an ethical point of view. As a consequence, embryos and hES cells should not be assimilated for juridical purposes.[119] The CCNE therefore suggests to pass ad hoc legislation for pluripotent cells, either of embryonic origin (hES

---

[114]   C. civ., art. 16–4. This article in fine states: "*Sans préjudice des recherches tendant à la prévention et au traitement des maladies génétiques, aucune transformation ne peut être apportée aux caractères génétiques dans le but de modifier la descendance de la personne.*"

[115]   CCNE, Avis 129, Contribution du Comité consultatif national d'éthique à la révision de la loi de bioéthique 2018–2019, as adopted on September 18, 2018.

[116]   Ibid. 51.

[117]   Ibid. 60.

[118]   Ibid. 52.

[119]   Ibid. 58. The need to retain the existing prohibition on creating embryos for research purposes is restated, but suggests to create an exemption for assisted reproduction research aimed at studying early embrogenesis.

TABLE 14.1 *France: list of key legal instruments*

Key applicable legislation and guidelines	Specific parts
1. International instruments	
Council of Europe – Convention for the Protection of Human Rights and Dignity of the Human Being with regard to the Application of Biology and Medicine (Oviedo Convention) – April 4, 1997.	Article 13 and Article 18
UN General Assembly – International Covenant on Economic, Social and Cultural Rights – December 16, 1966.	Article 12 and Article 15
UN General Assembly – Optional Protocol to the International Covenant on Economic, Social and Cultural Rights – 2009.	All
EU – Regulation (EU) No 536/2014 of the European Parliament and of the Council of 16 April 2014 on clinical trials in medicinal products for human use and repealing Directive 2001/20/EC – 2014.	Article 90
EU – Directive 2010/63/EU of the European Parliament and of the Council of 22 September 2010 on the protection of animals used for scientific purposes – 2010.	All
EU – Regulation (EC) No 1394/2007 on advanced therapy medicinal products – 2007.	All
EC – Directive 2001/83/EC of 6 November 2001 on the Community code relating to medicinal products for human use – 2001.	All
2. Constitution	
Constitution – October 4, 1958	Article 54, Article 55, and Article 88
3. Applicable laws[120]	
Civil Code (C. civ.)	Chapter II *"Du Respect du Corps Humain"*

---

[120] Most of the articles included in the *"Specific Parts"* listed in this table were created or modified by the French Bioethics Laws. These laws are as follows:

(1) The Law n° 94–548 of July 1, 1994 on the treatment of data for research in the health sector (*"relative au traitement de données nominatives ayant pour fin la recherche dans le domaine de la santé et modifiant la loi n° 78–17 du 6 janvier 1978 relative à l'informatique, aux fichiers et aux libertés,"* and the two laws of July 29, 1994, that is, the Law n° 94–653 on respect for the human body (*"relative au respect du corps humain"*) and the Law n° 94–654 on donation and use of elements and products of the human body, on medically assisted reproduction and on prenatal diagnosis (*"relative au don et à l'utilisation des éléments et*

TABLE 14.1 *(continued)*

Key applicable legislation and guidelines	Specific parts
Code of Public Health (CSP)	Title IV "*Assistance Médicale et la Procréation*" Title V "*Recherche Sur l'Embryon et les Cellules Souches Embryonnaires*"
Criminal Code (C. pén.)	Chapter I "*Des Crimes d'Eugénisme et de Clonage Reproductif*" Chapter II "*Dispositions Communes*" Chapter I "*Des Infractions en Matière d'Éthique Biomédicale,*" Section 3 "*De La Protection de l'Embryon Humain*"
Rural and Sea Fisheries Code (C. rur.)	Chapter IV "*La Protection des Animaux*"

4. Guidelines

Council of State ("*Conseil d'État*") – *De l'éthique au droit: sciences de la vie – Braibant Report* – 1988.	All
French National Institute for Health and Medical Research ("*Institut National de Santé et Recherche Médicale*") (INSERM):	All

produits du corps humain, à l'assistance médicale à la procréation et au diagnostic prénatal") – These laws essentially created new articles or modified existing articles in the Civil Code, the Code of Public Health, and the Criminal Code. Together with the Law n° 88–1138 of 20 December 1988 on the protection of human subjects in biomedical research ("*relative à la protection des personnes qui se prêtent à des recherches biomédicales*" – "*Loi Huriet*"), these laws constitute the set of legislative provisions known in France as the Bioethics Laws ("*lois de bioéthique*").

(2) The Law n° 2004-800 of August 6, 2004 on Bioethics ("*relative à la bioéthique*"), which created new articles or modified existing articles in the Civil Code, the Code of Public Health and the Criminal Code;

(3) The Law n° 2011-814 of July 7, 2011 on Bioethics ("*relative à la bioéthique*"), which created new articles or modified existing articles in the Civil Code, the Code of Public Health and the Criminal Code and introduced the Estates General of Bioethics; and

(4) The Law n° 2013-715 of August 6, 2013 allowing research on embryos and stem cells under certain conditions ("*tendant à modifier la loi n° 2011-814 du 7 juillet 2011 relative à la bioéthique en autorisant sous certaines conditions la recherche sur l'embryon et les cellules souches embryonnaires*"), hence modifying Art. L2151-5 of the Code of Public Health.

TABLE 14.1 *(continued)*

Key applicable legislation and guidelines	Specific parts
– Déclaration d'experts européens et français du Comité d'éthique de l'INSERM – *Promouvoir la Recherche Responsable avec les Technologies de Modification Ciblée du Génome: une Perspective Européenne (Déclaration)* – November 2016; – *Saisine Concernant les Questions Liées au Développement de la Technologie CRISPR (Clustered Regularly Interspaced Short Palindromic Repeat) – Cas9 (Note du Comité d'Éthique)* – February 2016.	
French National Academy of Medicine (*"Académie Nationale de Médecine"*) (ANM) – *Les Modifications du Génome des Cellules Germinales et de l'Embryon Humains – Rapport* – 2016.	All
French Parliamentary Office for Scientific and Technological Assessment (*"Office Parlementaire d'Évaluation des Choix Scientifiques et Technologiques"*) (OPECST):	All
– *Les Enjeux Économiques, Environnementaux, Sanitaires et Éthiques des Biotechnologies à la Lumière des Nouvelles Pistes de Recherche (Rapport, T. I and Rapport, T. II, Comptes Rendus et Annexes)* – 2017; – *Rapport sur le clonage, la thérapie cellulaire et l'utilisation thérapeutique des cellules embryonnaires* – 2000.	
French National Consultative Ethics Committee *"Comité Consultatif National d'Éthique"* (CCNE):	All
– *Rapport de Synthèse du Comité Consultatif National d'Éthique – Opinions du Comité Citoyen* – 2018; – *Avis n° 112 du 21 octobre 2010, Une réflexion éthique sur la recherche sur les cellules d'origine embryonnaire humaine, et la recherche sur l'embryon humain in vitro* – 2010; – *Avis n° 8 du 15 décembre 1986 relatif aux recherches et utilisation des embryons humains in vitro à des fins médicales et scientifiques – Rapport* – 1986; – *Avis n° 1 du 22 mai 1984 sur les prélèvements de tissus d'embryons et de fœtus humains morts, à des fins thérapeutiques, diagnostiques et scientifiques – Rapport* – 1984.	
French Biomedicine Agency (*"Agence de la Biomédecine"*) (ABM) – *Rapport sur l'application de la loi de bioéthique (Rapport 2018)* – 2018.	All

cells) or reprogrammed from adult cells (iPS cells), to regulate them according to the existing provisions regarding the use and distribution of elements of the human body, thereby dispensing them from the authorization-based model that is currently in place for both embryos and hES cells. Moreover, the CCNE proposes to introduce specific provisions regulating possible controversial uses of pluripotent cells, such as their differentiation into gametes, or their use in vivo for the development of human germ or neuronal cells into chimeric animals.[121] Regarding research on chimeras, the CCNE recommends that the law set specific provisions to explicitly allow chimeric embryo complementation assays (one of the assays used to test stem cell pluripotency) and research aimed at creating human organs in large mammals.[122] An ad hoc multidisciplinary body could, according to the CCNE, oversee these types of experiments.

The CCNE also advises the removal of two of the existing conditions to research on embryos and hES cells: the absence of alternative means and the need for research to demonstrate a medical aim. It also suggests the introduction of a clear temporal limit for in vitro culture of human embryos and to specify information and consent procedures for donors involved in research on embryos and pluripotent cells.

While it is not possible to predict whether and how the above recommendations will be taken into account by the French Parliament, the report of the CCNE signals some degree of openness in the French civil society regarding the use of genetic engineering and genome editing in the context of research. On the other hand, France is bound to a rather restrictive approach toward human germline modification – as stated through both internal and international law – and does not seem keen on revising its position on such matters in the near future.

[121]   Ibid. 53.
[122]   Ibid. 57.

15

The Regulation of Human Germline Genome Modification in Switzerland

Alessandro Blasimme, Dorothée Caminiti, and Effy Vayena

I INTRODUCTION

The Swiss Confederation (*Schweizerische Eidgenossenschaft/Confédération suisse/Confederazione Svizzera/Confederaziun svizra*) is composed of twenty-six Cantons. The federal government, called Federal Council and based in Berne, possesses executive power. Its members (the seven Federal Councilors – one for each department) are elected for a period of four years by the bicameral Swiss Parliament.[1]

While the parliament has legislative power, Switzerland is known for its frequent use of popular referenda (both legislative and constitutional) as instruments of direct democracy. The Cantons enjoy legislative and organizational autonomy in a number of key policy areas, from taxation to healthcare. The latter in particular is organized in an entirely Canton-based manner. Switzerland does not have a federal government–sponsored healthcare system, and each Canton is responsible for healthcare, which is offered by a mix of public and private providers. However, all residents in the Swiss territory have to buy a healthcare coverage plan from a private insurer recognized by the Federal Department of Home Affairs.

The use of genetic technologies for reproductive, farming, agricultural, and scientific purposes has long been a matter of public concern in Switzerland. As a result of a series of legislative initiatives at the federal level, as well as of popular referenda, the country developed one of the most restrictive regulatory environments in Europe for research potentially leading to human germline genome modification. In particular, any genetic manipulation of reproductive cells or embryos is strictly forbidden, regardless its intended purpose.

---

[1] In general, in a Confederation, the federal government is accountable to the member states, which have ultimate authority and are the sovereigns. Conversely, in a Federation, the federal government is the sovereign and it has ultimate authority over the member states.

In such a restrictive regulatory context, it is highly unlikely that recent technical advances in genetic engineering and genome editing will be employed to produce germline genome modifications for either medical or purely scientific purposes. Furthermore, while the Swiss National Advisory Commission for Biomedical Ethics has recently expressed partial support for basic research possibly involving the genetic modification of human embryos, there are currently no indications that legislative initiatives will be undertaken to ease current regulations on such controversial matters.

II THE REGULATORY ENVIRONMENT

All medical and scientific activities involving reproductive cells and embryos are tightly regulated, and indeed rigidly limited, by a number of constitutional provisions and federal laws. In particular, provisions relevant to human germline genome modification are to be found in legal texts regarding reproductive medicine, gene technology, transplant medicine, genetic testing, research involving human subjects, and stem cell research. Those relevant to germline modification technologies in nonhumans are included in laws on animal protection and animal experimentation.

1 *Regulatory Framework*

a Legislative Sources

The Federal Constitution of the Swiss Confederation[2] was adopted by popular vote on April 18, 1999,[3] and replaced and updated the previous 1874 Constitution.[4] Famously, popular referenda are a characteristic direct democracy tool of the Swiss political systems. Whenever the Swiss Parliament proposes amendments to the constitutional text, a mandatory popular referendum is held. The Federal Constitution calls attention to matters related to human germline modification in its Article 119.c on "Reproductive Medicine and Gene Technology Involving Human Beings,"[5] covering cloning as well as

---

[2] *Constitution fédérale de la Confédération suisse du* 18 avril 1999 (Federal Constitution), RS 101.
[3] *Arrêté fédéral relatif à une mise à jour de la Constitution fédérale du* 18 décembre 1998.
[4] *Constitution fédérale de la Confédération Suisse du* 29 mai 1874, RS 101.
[5] Federal Constitution, RS 101, art. 119 "Reproductive Medicine and Gene Technology Involving Human Beings." This article is included in Title 3 "Confederation, Cantons and Communes," Chapter 2 "Powers," Section 8 "Housing, Employment, Social Security and Health." Art. 119a on "Transplant Medicine" – adopted by the popular vote on February 7, 1999 (Federal Decree of June 26, 1998, Federal Council Decree of March 23, 1999; AS 1999 1341; BBl 1997 III 653, 1998 3473, 1999 2912 8768) is also relevant, albeit to a lesser degree, as it covers also

forms of technical interventions on the human genetic material of human reproductive cells and human embryos. This article was added to the Federal Constitution by popular vote on June 14, 2015 and it has been in force since then.[6] Article 80 of the Federal Constitution, on the protection of animals,[7] and Article 120, on nonhuman gene technology,[8] are also relevant when discussing experimentation with germline genome modifying techniques on animals.

In 1992, Switzerland ratified the International Covenant on Economic, Social and Cultural Rights (henceforth, the Covenant), first adopted by the United Nations' General Assembly in 1966.[9] However, Switzerland has not signed the 2008 Optional Protocol to the Covenant,[10] establishing mechanisms for individual complaints regarding the rights specified by the Covenant. Some provisions of the Covenant are of relevance to the issue of germline genome modification. These include the right to health specifying the right to

cell transplants, that is, one of the possible way of introducing genetic modifications in a human being.

[6] *Arrêté fédéral concernant la modification de l'article constitutionnel relatif à la procréation médicalement assistée et au génie génétique dans le domaine humain du 12 décembre 2014.*

[7] "The Confederation shall legislate on the protection of animals. It shall in particular regulate: a. the keeping and care of animals; b. experiments on animals and procedures carried out on living animals; c. the use of animals; d. the import of animals and animal products; e. the trade in animals and the transport of animals; f. the killing of animals. The enforcement of the regulations is the responsibility of the Cantons, except where the law reserves this to the Confederation." Federal Constitution, RS 101, Art. 80 "Protection of Animals." This article is included in Title 3 "Confederation, Cantons and Communes," Chapter 2 "Powers," Section 4 "Environment and Spatial Planning," Article 80 "Protection of Animals."

[8] "Human beings and their environment shall be protected against the misuse of gene technology. The Confederation shall legislate on the use of reproductive and genetic material from animals, plants and other organisms. In doing so, it shall take account of the dignity of living beings as well as the safety of human beings, animals and the environment, and shall protect the genetic diversity of animal and plant species." Federal Constitution, RS 101, art. 120 "Non-Human Gene Technology." This article is included in Title 3 "Confederation, Cantons and Communes," Chapter 2 "Powers," Section 8 "Housing, Employment, Social Security and Health."

[9] International Covenant on Economic, Social and Cultural Rights (adopted December 16, 1966, entered into force January 3, 1976) 993 UNTS 3 (ICESCR).

[10] Optional Protocol to the International Covenant on Economic, Social and Cultural Rights (adopted December 10, 2008, entered into force May 5, 2013), A/RES/63/117. In 2008, the Human Rights Council approved the text of the draft Optional Protocol. In December, the UN General Assembly adopted it, and the NGO Coalition launched an official campaign for ratification. In 2009, the Optional Protocol opened for signature and ratification, with immediate signature by twenty-nine States. On May 5, 2013, and in accordance with Article 18(1), the Optional Protocol entered into force, following submission of the tenth ratification by Uruguay. So far, the Optional Protocol has been signed by forty-five countries and ratified by twenty-three countries.

the "highest attainable standard of physical and mental health" (Art. 12.1), which implies that states parties shall strive to improve standards of perinatal health (Art. 12.2.a) and to reduce the burden of disease (Art. 12.2.c). Article 15, on the other hand, spells out the so-called right to science and right of science, specifying that everyone has a right to "enjoy the benefits of scientific progress and its applications" (Art. 15.1.b) and that states parties should protect and promote scientific freedom (Art. 15.2). Both Article 12 and Article 15, therefore, may in principle implicate that new techniques, such as genome editing, must be made widely available for both medical and research purposes.

However, the applicability of international norms within the Swiss legal system follows the so-called *jurisprudence Schubert* (or *Schubert-Praxis* in German), according to which, while international law has immediate internal validity, the national legislator can intentionally derogate from international norms (with the exception of those contained in the European Convention on Human Rights).[11] Switzerland's second and third United Nations' reports on the implementation of the Covenant specify that, while the Federal Constitution establishes scientific freedom, this freedom is not unlimited, and restrictions are in place in particular in areas such as assisted reproduction, embryo research, and stem cell research.[12] Such restrictions, according to the *jurisprudence Schubert*, have precedence over the obligations contained in the Covenant.

In 2008, Switzerland ratified the Council of Europe's Convention for the Protection of Human Rights and Dignity of the Human Being with Regard to the Application of Biology and Medicine (henceforth, Oviedo Convention).[13] Article 13 of the Oviedo Convention notoriously states that interventions "seeking to modify the human genome may only be undertaken for preventive, diagnostic or therapeutic purposes and only if its aim is not to introduce any

---

[11]   Conseil Fédéral Suisse, *La relation entre droit international et droit interne – Rapport du Conseil fédéral En réponse au postulat 07.3764 de la Commission des affaires juridiques du Conseil des États du 16 octobre 2007 et au postulat 08.3765 de la Commission des institutions politiques du Conseil national du 20 novembre 2008*, FF 2010 2067.

[12]   ICESCR, *Implementation of the International Covenant on Economic, Social and Cultural Rights: Second and third periodic reports of States parties under articles 16 and 17 of the Covenant – Switzerland*, E/C.12/CHE/2–3 (July 17, 2009) para 599.

[13]   Convention for the Protection of Human Rights and Dignity of the Human Being with Regard to the Application of Biology and Medicine: Convention on Human Rights and Biomedicine (opened for signatures on April 4, 1997, entered into force January 12, 1999) ETS No. 164 (Oviedo Convention). Switzerland signed the Oviedo Convention on May 7, 1999, and ratified it on July 24, 2008. The Oviedo Convention entered into force in Switzerland on November 1, 2008.

modification in the genome of any descendants." This provision, while leaving a door open for genetic engineering of somatic cells in adults, is consistent with the restrictive legislation that Switzerland has adopted on scientific and medical interventions on embryos, germ cells, and their genetic material.

The Federal Act on Medically Assisted Reproduction (*Loi fédérale sur la procréation médicalement assistée*) of December 18, 1998[14] contains provisions relevant to human germline modification in Articles 3 to 5b,[15] in Articles 8 to 12,[16] and in Articles 15 to 17 of Section 3 on "Handling of Reproductive Material."[17] Articles 29 to 38, in Chapter 4 of the same act, contain criminal provisions regarding violation of the rules established by such act.[18]

In Switzerland, in order to perform research involving animals or human subjects, study protocols need to be reviewed and approved by one of Switzerland's seven local ethics committees (each serving one or more Canton(s) in a given linguistic region) coordinated by the Federal Office of Public Health (see Section I.3).

The use of human embryonic stem cells (hESCs) is regulated through the Federal Act on Research Involving Embryonic Stem Cells (*Loi fédérale relative à la recherche sur les cellules souches embryonnaires*) of December 19, 2003.[19] The act sets the conditions for the derivation of hESCs from surplus embryos from in vitro fertilization (IVF) and for their research use. Article 3 specifies which practices are forbidden,[20] whereas Article 4 forbids the commercialization of both embryos and hESCs.[21] In Section 2 on "Derivation of Embryonic Stem Cells from Surplus Embryos," Articles 7 to 10 explain the licensing requirements for derivation, use, and storage of hESCs.[22] Section 3 on "Management of Embryonic Stem Cells" regulates, among other things, the scientific and ethical requirements for hESCs research (Art. 12),[23] while Sections 4 and 5 (Arts. 24 to 26) contain, respectively, enforcement and criminal provisions.[24] This law does not cover the use of hESCs in the context of regenerative medicine clinical trials.[25]

---

[14]   *Loi fédérale sur la procréation médicalement assistée (LPMA) du* 18 décembre 1998, RS 810.11.
[15]   Ibid., c 2 "Techniques of Medically Assisted Reproduction," s 1 "Principles."
[16]   Ibid., s 2a "Evaluation."
[17]   Ibid., s 3 "Handling of Reproductive Material."
[18]   Ibid., c 4 "Criminal Provisions."
[19]   *Loi fédérale relative à la recherche sur les cellules souches embryonnaires (LRCS) du* 19 décembre 2003, RS 810.31.
[20]   Ibid., art. 3. See Section II.1.b for details.
[21]   Ibid., art. 4.
[22]   Ibid., arts. 7–10.
[23]   Ibid., art. 12.
[24]   Ibid., arts. 17–23 and arts. 24–26.
[25]   Ibid., art. 1, al. 1 and 3.

Another relevant text for research that could lead to altering the human germline genome is the Federal Act on Human Genetic Testing (*Loi fédérale sur l'analyse génétique humaine*) of 2004.[26] Its broad remit embraces the medical, employment, and insurance contexts as well as issues of liability. In particular, Article 11 on prenatal tests forbids genetic testing of embryos and fetuses for nonmedical reasons.[27]

The Federal Act on the Transplantation of Organs, Tissues and Cells of 2004 (*Loi fédérale sur la transplantation d'organes, de tissus et de cellules*) regulates the handling of organs, tissues, and cells of both animal and human origin.[28] The act does not apply to germ cells, impregnated ova, and embryos in the context of medically assisted reproduction.[29] However, it is relevant to the handling of embryonic or fetal human tissues or cells.[30] Chapter 6 of this act contains criminal provisions.[31]

Provisions regarding research on embryos and fetuses are also contained in Articles 25 and 26 of the Federal Act on Research involving Human Beings (*Loi fédérale relative à la recherche sur l'être humain*) of 2011,[32] which is the text that regulates all research activities on human subjects in medicine as well as in any other discipline. This text also disciplines research on embryos and fetuses from induced and spontaneous abortions or stillbirths,[33] and contains criminal sanctions and jurisdictional dispositions.[34]

Over the last decade or so, Switzerland has developed comprehensive legislation around animal welfare and protection both in the context of farming and industrial use and in the context of scientific research. The Federal Act on Non-Human Gene Technology (*Loi fédérale sur l'application du génie génétique au domaine non humain*) of March 21, 2003 applies to the handling of all genetically modified plants, animals, and organisms.[35] Its aim is "to protect human beings, animals and the environment form abuses of gene technology,"[36] and "to serve the welfare of human beings, animals and the

[26]   *Loi fédérale sur l'analyse génétique humaine* (LAGH) du 8 octobre 2004, RS 810.12.

[27]   Ibid., art. 11.

[28]   *Loi fédérale sur la transplantation d'organes, de tissus et de cellules (Loi sur la transplantation)* du 8 octobre 2004, RS 810.21, art. 1.1.

[29]   Ibid., art. 2, al. 2.

[30]   Ibid., s 9, art. 37 to art. 42.

[31]   Ibid., arts. 69–70.

[32]   *Loi fédérale relative à la recherche sur l'être humain (LRH) du* 30 septembre 2011, RS 810.30, arts. 25–26.

[33]   Ibid., arts. 39–40.

[34]   Ibid., arts. 62–64.

[35]   *Loi fédérale sur l'application du génie génétique au domaine non humain* (LGG) du 21 mars 2003, RS 814.91, art. 3, al. 1.

[36]   Ibid., art. 1, al. 1, a.

environment in the application of gene technology."[37] Its provisions apply also to scientific experiments and experimental release of genetically modified organisms.

Chapter 3 of the Federal Act on the Transplantation of Organs, Tissues and Cells (*Loi fédérale sur la transplantation d'organes, de tissus et de cellules*) of 2004 regulates the transfer of organs, tissues, or cells of animal origin into humans,[38] both in the context of a clinical trial[39] or for medical treatment.[40] This kind of procedures may give rise to germline modifications through the creation of interspecific chimeric embryos, which, however, is currently not allowed in Switzerland.[41]

Finally, a cornerstone of animal legislation in Switzerland is the 2005 Animal Welfare Act (*Loi fédérale sur la protection des animaux*).[42] Section 2 specifies the conditions for the breeding and production of animals obtained through genetic engineering.[43] Sections 6 and 6a cover animal experimentation.[44] Criminal provisions are included in Chapter 5, Articles 26 to 31.[45]

In Switzerland, violating legal rules regarding the use of human gametes, embryos, and embryonic stem cells in the clinical and research contexts gives rise to criminal penalties. According to the Federal Act on Medically Assisted Reproduction, anyone who creates an embryo by impregnation without the intention of implanting it in utero or using an embryo for purposes other than establishing a pregnancy is liable to imprisonment for up to three years or to a pecuniary penalty.[46] The same applies to storing an impregnated ovum or an embryo for purposes other than establishing a pregnancy.[47]

Up to three years of prison or a fine are foreseen for anyone who develops an embryo in vitro beyond the point at which it is still possible to implant it.[48] Using reproductive material obtained from an embryo or fetus to attempt to create or develop another embryo, or selling and purchasing such material, is

---

[37] Ibid., art. 1, al. 1, b.
[38] *Loi sur la transplantation*, RS 810.21, art. 43.1.
[39] Ibid., art. 43.2.
[40] Ibid., art. 43.3.
[41] LRCS, RS 810.31, art. 3, al. 1, c.
[42] *Loi fédérale sur la protection des animaux (LPA) du 16 décembre 2005*, RS 455.
[43] Ibid., arts. 10–12.
[44] Ibid., arts. 17–20.e.
[45] Ibid., arts. 26–31.
[46] LPMA, RS 810.11, art. 29, al. 1.
[47] Ibid., art. 29, al. 2.
[48] Ibid., art. 30, al.1.

also punished with up to three years of prison or a fine.[49] The same penalty applies to those who analyze the genetic material of sperm cells, ova, or embryos and select them according to sex or other features – except when this is needed to maximize the chances of successful medically assisted reproduction, or to avoid the transmission of a genetic disease.[50]

Germline modifications are specifically addressed in Article 35 of the Federal Act on Medically Assisted Reproduction, which provides the following: "Any person who genetically modifies a germline cell or an embryonic cell shall be liable to a custodial sentence not exceeding three years or to a monetary penalty. The same penalty shall apply to any person who uses a genetically modified reproductive cell for impregnation or uses a similarly modified impregnated ovum for further development into an embryo."[51] Interestingly, the legislator specifies that the sanctions in Article 35 do not apply to 'the case of unintentional germline modifications occurring as a consequence of chemotherapy, radiotherapy, or other treatments.[52]

The Federal Act on Research Involving Embryonic Stem Cells also addresses issues of relevance for germline modification in its criminal provisions. According to Article 24, the following can be punished with imprisonment: deriving hESCs from an embryo created for research purposes, or from a genetically modified embryo, or from a clone, chimera, hybrid, or parthenote; using a surplus embryo for purposes other than the derivation of hESCs; deriving stem cells from an embryo beyond the seventh day of development; and implanting a surplus embryo used for the derivation of stem cells.[53]

Criminal provisions are attached to the legislative sources that regulate activities on human reproductive cells, embryos, and hESCs. They create rigid limits around the possibility of using such material for research or clinical activities that could lead to germline modification. However, it should be noted that the penalties in question are not particularly heavy, probably indicating that the legislator did not consider violations to be likely to occur.

b Binding Guidelines

In the Swiss legal system, federal ordinances issued by the Federal Council or, occasionally, by the Parliament are binding normative provisions sitting right below laws in the hierarchy of legal sources. Let us briefly describe those

---

[49] Ibid., art. 32.
[50] Ibid., art. 33.
[51] Ibid., art. 35, al. 1 and 2.
[52] Ibid., art. 35, al. 3.
[53] LRCS, RS 810.31, art. 3.

ordinances that are directly attached to the acts mentioned so far and that specify requirements and procedures that need to be followed to fulfill their provisions.

The Reproductive Medicine Ordinance (*Ordonnance sur la procréation médicalement assistée*) of 2000 sets guidance for licensing requirements and monitoring activities concerning the practice of medically assisted reproduction.[54]

The Ordinance on Research involving Embryonic Stem Cells (*Ordonnance relative à la recherche sur les cellules souches embryonnaires*) of 2005[55] contains guidance on the determination of the surplus status of embryos that are no longer needed for the purpose of assisted reproduction,[56] specification of informed consent requirements for donating surplus embryos for research,[57] and provisions related to licensing for the derivation of hESCs from donated embryos as well as for storing surplus embryos and importing and exporting hESCs lines.[58] The ORCS also contains guidelines regarding the assessment and authorization of research projects by competent ethics committees.[59]

The Ordinance on Human Genetic Testing (*Ordonnance sur l'Analyse Génétique Humaine*) of 2007 sets the conditions for the authorization to perform cytogenetic and molecular analyses as well as screenings on humans.[60]

The 2007 Ordinance on Transplantation of Human Organs, Tissues and Cells (*Ordonnance sur la transplantation d'organes, de tissus et de cellules d'origine humaine*) disciplines, among other matters, the transplantation of tissues or cells derived from embryos or fetuses, as well as the conditions for the release of the authorization to perform such activities.[61]

As the Federal Act on Research involving Human Beings, the Human Research Ordinance (*Ordonnance relative à la recherche sur l'être humain à l'exception des essais cliniques*)[62] applies to scientific research involving humans, but with the exception of clinical trials. It contains guidance on

---

[54] *Ordonnance sur la procréation médicalement assistée (OPMA) du 4 décembre 2000*, RS 810.112.2.
[55] *Ordonnance relative à la recherche sur les cellules souches embryonnaires (ORCS) du 2 février 2005*, RS 810.311.
[56] Ibid., art. 1.
[57] Ibid., arts. 2–4.
[58] Ibid., ss 2–6.
[59] Ibid., arts. 17–21.
[60] *Ordonnance sur l'analyse génétique humaine (OAGH) du 14 février 2007*, RS 810.122.1.
[61] *Ordonnance sur la transplantation d'organes, de tissus et de cellules d'origine humaine (Ordonnance sur la transplantation) du 16 mars 2007*, RS 810.211, c 6.
[62] *Ordonnance relative à la recherche sur l'être humain à l'exception des essais cliniques (ORH) du 20 septembre 2013*, RS 810.301.

research on human embryos derived from both induced and spontaneous abortions.[63]

Clinical trials guidance is contained in the 2013 Ordinance on Clinical Trials in Human Research (*Ordonnance sur les essais cliniques dans le cadre de la recherche sur l'être humain*).[64] It sets special provisions for clinical trials consisting in the transplantation of embryonic or fetal tissues and cells into human subjects.[65]

Relevant guidance on how to treat genetically modified organisms can also be found in the Ordinance on Handling Organisms in Contained Systems (*Ordonnance sur l'utilisation des organismes en milieu confiné*).[66]

In application of both the law on the protection of animals and the Federal Act on Non-Human Gene Technology, the Animal Welfare Ordinance (*Ordonnance sur la Protection des Animaux*)[67] issued in 2008 contains guidance on animals whose germline has been intentionally modified. In particular, Chapter 6 on "Animal Experiments, Genetically Modified Animals and Mutants that Have a Significant Clinical Pathological Phenotype" takes into account the well-being of genetically modified animals, including animals bearing pathologic mutations.[68]

In 2010, the Federal Food Safety and Veterinary Office (*Office Fédéral de la sécurité alimentaire et des affaires vétérinaires* – OSAV) issued a guidance called the Ordinance Regarding Detention of Test Animals, the Production of Genetically Modified Animals and the Methods Used in Animal Experimentation (*Ordonnance de l'OSAV Concernant la Détention des Animaux d'Expérience, la Production d'Animaux Génétiquement Modifiés et les Méthodes Utilisées dans l'Expérimentation Animale*).[69] While it lists recognized methods for the production of genetically modified animals,[70] it does not mention any specific mode of genome modification such as recombinant DNA or genome editing tools.

---

[63]   Ibid., c 5, arts. 44–46.
[64]   *Ordonnance sur les essais cliniques dans le cadre de la recherche sur l'être humain (OClin) du 20 septembre 2013, RS 810.305.*
[65]   Ibid., art. 56.
[66]   *Ordonnance sur l'utilisation des organismes en milieu confiné (OUC) du 9 mai 2012, RS 814.912.*
[67]   *Ordonnance sur la protection des animaux (OPAn) du 23 avril 2008, RS 455.1.*
[68]   Ibid., s 3, arts. 123–127.
[69]   *Ordonnance de l'OSAV concernant la détention des animaux d'expérience, la production d'animaux génétiquement modifiés et les méthodes utilisées dans l'expérimentation animale (Ordonnance sur l'expérimentation animale) du 12 avril 2010, RS 455.163.*
[70]   Ibid., art. 9, al. 1, referring to Annex 1.

The authorizations to perform scientific experiments on animals and the mandatory reporting on such experiments are handled through an informatics system, as specified in the Ordinance on Informatics Systems to Manage Experiments of Animals (*Ordonnance sur le Système Informatique de Gestion des Expériences sur Animaux*).[71]

c Nonbinding Guidelines

The Swiss National Advisory Commission on Bioethics (*Commission nationale d'éthique dans le domaine de la médecine humaine*) has recently issued a position document on gene editing on human embryos.[72] In the document, the Commission recommends an open public debate on the theme and states that, in the current legal landscape, it is in principle forbidden to intervene in the genetic makeup of human reproductive cells and human embryos.[73] The document also states that while some members of the Commission are supportive of this ban, some others are ready to accept basic research on the embryonic germline.[74]

2 Definitions

Swiss laws and ordinances provide quite an extensive collection of definitions related to our topic.

In Article 2, the Federal Act on Medically Assisted Reproduction defines, inter alia:

Assisted reproductive techniques as "methods of establishing a pregnancy without sexual intercourse
- in particular, insemination, *in vitro* fertilization with embryo transfer and gamete transfer";[75]
- *Insemination* (insémination) as "the introduction, by means of instruments, of sperm cells into the female reproductive organs";[76]

---

[71]   *Ordonnance sur le système informatique de gestion des expériences sur animaux (O-SIGEXPA)* du 1er septembre 2010, RS 455.61.
[72]   NEK / CNE, *Gene editing sur les embryons humains – un état des lieux: Prise de Position no. 25/ 2016* (NEK CNE, 2016) www.nek-cne.admin.ch/inhalte/Themen/Stellungnahmen/fr/DEF_NEK_Kurzstellungnahme_GeneEditing_FR.pdf accessed September 4, 2018.
[73]   The National Advisory Commission on Bioethics cites Articles 119, al. 2, let. A of the Federal Constitution and Article 35 of the Federal Act on Medically Assisted Reproduction in support of its legal interpretation – Ibid., see "Les Conclusions de la CNE."
[74]   Ibid.
[75]   LPMA, RS 810.11, art. 2.
[76]   Ibid.

- *In vitro fertilization* as "the bringing together of an ovum and sperm cells outside the woman's body";[77]
- *Germline cells* as "reproductive cells (including their precursor cells), impregnated ova and embryonic cells whose genetic material can be passed on to offspring";[78]
- *Impregnated ovum* as "the fertilized ovum before pronuclear fusion";[79]
- *Embryo* as "the developing offspring from the time of pronuclear fusion until the end of organogenesis";[80]
- *Fetus* as "the developing offspring from the end of organogenesis until birth";[81]
- *Cloning* as "the artificial production of genetically identical organisms";[82]
- *Chimera formation* as "the fusion of totipotent cells from two or more genetically different embryos";[83]
- *Hybrid formation* as occurring when "a non-human sperm cell ... penetrate[s] into a human ovum, or a human sperm cell into a non-human ovum."[84]

The Federal Act on Research Involving Embryonic Stem Cells offers the following definitions in Article 2:[85]

- *Surplus embryo* means an embryo produced in the course of an in vitro fertilization (IVF) procedure that cannot be used to establish a pregnancy and therefore has no prospect of survival;
- *Embryonic stem cell* means a cell from an IVF embryo with the ability to differentiate into the various cell types, but not to develop into a human being, and the cell line derived therefrom;
- *Parthenote* means an organism derived from an unfertilized oocyte.

The definitions found in Article 5 of the Federal Act on Non-Human Gene Technology include:[86]

[77] Ibid.
[78] Ibid.
[79] Ibid.
[80] Ibid.
[81] Ibid.
[82] Ibid.
[83] Ibid.
[84] Ibid.
[85] LRCS, RS 810.31.
[86] LGG, RS 814.91.

– *Genetically modified organisms* are organisms in which the genetic material has been altered in a way that does not occur under natural conditions by crossing or natural recombination.

The Animal Welfare Ordinance mentions, inter alia, the following definition of genetically modified animals in Article 2:[87]

– *Genetically modified animals* are animals whose genetic material has been modified in germ cells by genetic modification techniques, within the meaning of Annex 1 of the Ordinance on Handling Organisms in Contained Systems of May 9, 2012, in a manner that does not occur by crossing under natural conditions or by natural recombination.[88]

The Ordinance on Handling Organisms in Contained Systems offers, in Article 3, the definition of *gene technology methods* as follows:

a. recombinant nucleic acid techniques, in which nucleic acid molecules synthesized outside an organism are inserted into viruses, bacterial plasmids or other vector systems to produce novel combinations of genetic material, which are then transferred to a recipient organism in which they do not naturally occur but are capable of continued propagation;

b. techniques in which genetic material produced outside the organism is inserted directly into an organism, in particular by micro-injection, macro-injection and micro-encapsulation, electroporation or on micro-projectiles;

c. cell fusion or hybridization techniques in which cells with novel combinations of genetic material are produced by the fusion of two or more cells through processes that do not occur under natural conditions.[89]

3 Oversight Bodies

The National Advisory Commission on Biomedical Ethics, whose mandate is described in the Ordinance of the National Commission of Ethics in the Field of Human Medicine (*Ordonnance sur la Commission nationale d'éthique dans*

---

[87]   OPAn, RS 455.1.

[88]   Original: *"les animaux dont le matériel génétique a été modifié dans les cellules germinales par des techniques de modification génétique au sens de l'annexe 1 de l'ordonnance du 9 mai 2012 sur l'utilisation confinée d'une manière qui ne se produit pas par croisement dans des conditions naturelles ou par recombinaison naturelle."*

[89]   OUC, RS 814.912.

422 Alessandro Blasimme, Dorothée Caminiti, and Effy Vayena

le domaine de la médicine humaine), monitors scientific developments and their applications in the areas of human health and disease. Furthermore, it takes position and issues statements on the social, scientific, and legal issues related to such activities without any specific oversight prerogative.[90] Switzerland also has a Federal Ethics Committee on Non-Human Biotechnology that advises the Federal Council and other federal authorities on legislative matters and informs federal and cantonal bodies on activities concerning the implementation of federal provisions.[91] A further consultative body that may be of relevance to germline modification is the Commission of Experts for Human Genetic Testing (*Commission d'Experts pour l'Analyse Génétique Humaine*).[92] It produces technical updates and recommendations to the Federal Council and to federal authorities.[93]

Proper oversight, authorization, or licensing of research activities involving human reproductive cells, embryos, and hESCs or potential genetic modification of humans and nonhuman organisms are performed by a variety of public bodies and committees. Cantonal ethics committees have the power to assess and authorize research protocols. The Swiss Agency for Therapeutic Products (*Swissmedic, Institut suisse des produits thérapeutiques*)[94] authorizes gene therapy clinical trials and clinical trials with genetically modified organisms with advice from the Swiss Expert Committee for Biosafety (*Commission fédérale d'experts pour la sécurité biologique*),[95] the Federal Office for the Environment (*Office fédéral de l'environnement*),[96] and from the Federal Office of Public Health (*Office fédéral de la santé publique*, OFSP).[97]

The OFSP decides on requests regarding the transplantation of organs, tissues, or genetically modified cells after having consulted the Federal

90 *Ordonnance sur la Commission nationale d'éthique dans le domaine de la médecine humaine (OCNE) du* 4 décembre 2000, RS 810.113.
91 Federal Ethics Committee on Non-Human Biotechnology (ECNH) www.ekah.admin.ch/en/ homepage/ accessed September 4, 2018.
92 Office federal de la santé publique OFSP, "Commission d'experts pour l'analyse génétique humaine (CEAGH)" www.bag.admin.ch/bag/fr/home/das-bag/organisation/ausserparlamen tarische-kommissionen/expertenkommission-fuer-genetische-untersuchungen-beim- menschen.html accessed September 4, 2018.
93 OAGH, RS 810.122.1, c 6.
94 Swissmedic, "Swissmedic, Swiss Agency for Therapeutic Products" (2017) www.swissmedic .ch/swissmedic/en/home/about-us/swissmedic–swiss-agency-for-therapeutic-products.html accessed September 4, 2018.
95 "Swiss Expert Committee for Biosafety (SECB)" www.efbs.admin.ch/en/homepage/ accessed September 4, 2018.
96 "Federal Office for the Environment (FOEN)" www.bafu.admin.ch/bafu/en/home.html accessed September 4, 2018.
97 OClin, RS 810.305, art. 35; Federal Office of Public Health (FOPH) www.bag.admin.ch/bag/ en/home.html accessed September 4, 2018.

TABLE 15.1 *Switzerland: list of key legal instruments*

Key applicable legislation and guidelines	Specific parts
1. International instruments	
Council of Europe – Convention for the Protection of Human Rights and Dignity of the Human Being with regard to the Application of Biology and Medicine of April 4, 1997	Article 13
UN General Assembly – International Covenant on Economic, Social and Cultural Rights – December 16, 1966	Article 12 and Article 15
2. Constitution	
The Federal Constitution of the Swiss Confederation of April 18, 1999, CC 101	Section 4 "Environment and Spatial Planning" Section 8 "Housing, Employment, Social Security and Health"
3. Applicable laws	
(i) Humans	
Federal Act on Medically Assisted Reproduction (LPMA) of December 18, 1998, CC 810.11	Chapter 2 "Techniques of Medically Assisted Reproduction" Chapter 4 "Criminal Provisions"
Federal Act on Research Involving Embryonic Stem Cells (LRCS) of December 19, 2003, CC 810.31	Section 1 "General Provisions"
Federal Act on Human Genetic Testing (LAGH) of October 8, 2004, CC 810.12	Chapter 3 "Genetic Testing in a Medical Context"
Federal Act on the Transplantation of Organs, Tissues and Cells (Loi sur la transplantation) of October 8, 2004, CC 810.21	Chapter 1 "General Provisions" Section 9 "Handling Embryonic or Fetal Human Tissues or Cells" Chapter 3 "Organs, Tissues and Cells of Animal Origin" Chapter 6 "Criminal Provisions"
Federal Act on Research involving Human Beings (LRH) of September 30, 2011, CC 810.30	Chapter 3 "Additional Requirements for Research involving Particularly Vulnerable Persons" – Section 2 "Research Involving Pregnant Women and Embryos and Fetuses In Vivo" Chapter 6 "Research involving Embryos and Fetuses from Induced Abortions and from Spontaneous Abortions including Stillbirths" Chapter 11 "Criminal Provisions"

(continued)

Alessandro Blasimme, Dorothée Caminiti, and Effy Vayena

TABLE 15.1 *(continued)*

Key applicable legislation and guidelines	Specific parts
Federal Act on the Promotion of Research and Innovation (LERI) of December 14, 2012, CC 420.1	All
Federal Act on Medicinal Products and Medical Devices (LPTh) of December 15, 2000, CC 812.21	All
(ii) Nonhumans	
Federal Act on the Transplantation of Organs, Tissues and Cells (Loi sur la transplantation) of October 8, 2004, CC 810.21	Chapter 3 "Organs, Tissues and Cells of Animal Origin"
Federal Act on Non-Human Gene Technology (LGG) of March 21, 2003, CC 814.91	All
Animal Welfare Act (LPA) of December 16, 2005, CC 455 (French version)	Section 2 "Élevage d'Animaux et Modifications Obtenues par Génie Génétique" Section 6 "Expérimentation Animale" Section 6a "Système d'Information dans le Domaine de l'Expérimentation Animale" Chapter 5 "Dispositions Pénales"
4. Applicable binding guidelines	
(i) Humans	
Reproductive Medicine Ordinance (OPMA) of December 4, 2000, CC 810.112.2	All
Ordinance on Research involving Embryonic Stem Cells (ORCS) of February 2, 2005, CC 810.311	All
Ordinance on Human Genetic Testing (OAGH) of February 14, 2007, CC 810.122.1 (French version)	All
Ordinance on the Transplantation of Human Organs, Tissues and Cells (Ordonnance sur la transplantation) of March 16, 2007, CC 810.211 (French version)	All
Ordinance on Human Research with the Exception of Clinical Trials (ORH) of September 20, 2013, CC 810.301	Chapter 5 Research involving Embryos and Fetuses from Induced Abortions and from Spontaneous Abortions including Stillbirths

TABLE 15.1 *(continued)*

Key applicable legislation and guidelines	Specific parts
Ordinance on Clinical Trials in Human Research (OClin) of September 20, 2013, CC 810.305	Section 3 "Procedure for the FOPH"
Ordinance of the National Commission of Ethics in the Field of Human Medicine (OCNE) of December 4, 2000, CC 810.113 (French version)	All
Ordinance related to the Federal Act on the Promotion of Research and Innovation (O-LERI) of November 29, 2013, CC 420.11 (French version)	All
(ii) Nonhumans	
Ordinance on Handling Organisms in Contained Systems (OUC) of May 9, 2012, CC 814.912	All
Animal Welfare Ordinance (OPAn) of April 23, 2008, CC 455.1 (French version)	Chapter 6 "Expérimentation Animale, Animaux Génétiquement Modifiés et Mutants Présentant un Phénotype Invalidant," Section 1 "Champ d'Application, Dérogations Admises"
	Section 3 "Détention, Élevage et Commerce d'Animaux Génétiquement Modifiés et de Mutants Présentant un Phénotype Invalidant"
	Section 4 "Exécution d'Expériences sur Animaux"
Ordinance Regarding Detention of Test Animals, the Production of Genetically Modified Animals and the Methods Used in Animal Experimentation (Ordonnance sur l'expérimentation animale) of April 12, 2010, CC 455.163 (French version)	Section 3 "Production, Élevage et Détention d'Animaux d'Expérience Génétiquement Modifiés et de Mutants Présentant un Phénotype Invalidant"
Ordinance on Informatics Systems to Manage Experiments of Animals (O-SIGEXPA) of September 1, 2010, CC 455.61 (French version)	All

Ethics Committee on Non-Human Biotechnology, the Federal Office for the Environment, and the Swiss Expert Committee for Biosafety.[98]

Cantonal health departments are responsible for licensing and supervision of reproductive medicine activities, but Cantons may nominate other ad hoc authority with the needed expertise.[99] Authorizations regarding experiments with genetically modified organisms in contained systems are handled by the Federal Coordination Center for Biotechnology.[100]

4 Funding

Scientific research in Switzerland is sustained both by public and by private funding. Investment in research and development activities amounts to 3% of the gross domestic product, totaling about 16 billion Swiss francs yearly (about US$16 billion).[101] Two-thirds of this cost is borne by the private sector, but the Federal Act on the Promotion of Research and Innovation (*Loi fédérale sur l'encouragement de la recherche et de l'innovation*) ensures sustained public support to research activities.[102] The Swiss National Science Foundation (*Fonds national Suisse de la recherche scientifique*), established in 1952, is the country's main public grant agency and has a research budget of almost 1 billion Swiss francs (about US$1 billion) a year.[103]

Although Switzerland is not a member of the European Union, thanks to its Associated Country status, it has been a full participant to the European Union's research funding schemes (namely *Horizon 2020*) since 2016, and Swiss researches can tap those funds too. That being said, funding for research that could in principle lead to human germline modification is available only as long as the rigid regulatory constraints that characterize embryo-related and genetic research in Switzerland are complied with. As we explain below, research on surplus human embryos is forbidden in Switzerland. Surplus

[98]   Ordonnance sur la transplantation, RS 810.211, art. 39.

[99]   OPMA, RS 810.112.2, art. 8.

[100] OUC, RS 814.912, arts. 11–17; Federal Office for the Environment (FOEN), "Federal Coordination Center for Biotechnology" www.bafu.admin.ch/bafu/en/home/topics/biotech nology/info-specialists/activities-in-contained-use/federal-coordination-centre-for-biotechnology.html accessed September 4, 2018.

[101] State Secretariat for Education, Research and Innovation, "Research and Innovation in Switzerland" www.sbfi.admin.ch/sbfi/en/home.html accessed September 4, 2018.

[102] *Loi fédérale sur l'encouragement de la recherche et de l'innovation (LERI) du* 14 décembre 2012, RS 420.1. See *also Ordonnance relative à la loi fédérale sur l'encouragement de la recherche et de l'innovation (O-LERI) du* 29 novembre 2013, RS 420.11.

[103] Swiss National Science Foundation, "Ongoing SNSF projects" (2017) www.snf.ch/SiteColl ectionDocuments/profil/2017/SNF-Profil-2017–2018-en-Statistiken-Kurzversion.pdf accessed September 4, 2018.

embryos can be used only to derive hESCs, which in turn can be used for research purposes. However, research on hESCs, while not illegal, is considered controversial, and it is hard to obtain public funding in this area.[104] Back in 2017, there were only fifty-six imported hESCs lines and four lines derived from surplus embryos available for research in Switzerland.[105] On June 15, 2017, the Federal Office of Public Health listed a total of twenty-four hESCs projects, fourteen of which were finished while the remaining ten were still ongoing.[106] Currently, no projects for the derivation of hESCs from surplus embryos are underway in Switzerland.[107]

III SUBSTANTIVE PROVISIONS

Research activities and clinical applications that can result in altering the genetic material of the germline in humans and animals presuppose interventions on gametes, embryos, or embryonic stem cells. Gametes can be obtained directly from adult humans or animals or derived in vitro from progenitors of sperm and ova, or through in vitro reprograming of somatic cells and subsequent differentiation into reproductive cells. Genome editing can thus rely on a variety of different routes to produce a germline mutation, but each presents its own technical and regulatory challenges. In animals, the use of some of these tools is already well established. However, regulations have long been in place in countries like Switzerland to prevent the birth of children harboring a germline genetic alteration. In what follows, we will review how Swiss regulations play out with respect to germline modification in the context of (1) basic research, (2) preclinical research, (3) clinical research, and (4) clinical applications.

1 *Regulation of Basic Research*

This section offers insight into Swiss regulation concerning the use of (i) embryos, (ii) hESCs, and (iii) gametes in the context of basic research. By "basic research," we mean fundamental research on human pathophysiological processes and mechanisms. While basic

---

[104] See Federation of European Academies of Medicine, *Human Genome Editing in the EU – Report of a Workshop Held on 28th April 2016 at the French Academy of Medicine* (2016) 22.

[105] OFSP, "Research Projects and Stem Cell Lines" www.bag.admin.ch/bag/en/home/medizin-und-forschung/forschung-an-humanan-embryonalen-stammzellen/stammzellenforschung .html accessed September 4, 2018.

[106] Ibid.

[107] Ibid.

research is of relevance to medical knowledge or medical practice, it should be distinguished from clinical research. Basic research has to do with the elucidation of the causes and consequences of diseases as well as with suggesting possible preventive or therapeutic strategies. Clinical research, on the other hand, encompasses those activities aimed at testing the safety and efficacy of drugs and other therapeutic interventions in human research subjects, for instance through interventional clinical trials (see below, point 3).

From a general perspective, Switzerland has adopted a restrictive regulatory stance toward research activities that could lead to germline modification. Article 119 of the Federal Constitution, for instance, sets out to protect human beings "against the misuse of reproductive medicine and gene technology."[108] Framed this way, the medical or scientific benefits of such activities are de-emphasized and the risks that scientific practices can produce, if left unchecked by the law, are instead stressed.[109] Similarly, the Federal Act on Medically Assisted Reproduction, in Article 1.2, declares that it aims at protecting "human dignity, personality and the family and prohibits misuses of biotechnology and gene technology."[110] In a similar vein, the purpose of the Federal Act on Research Involving Embryonic Stem Cells is "to prevent the misuse of surplus embryos and embryonic stem cells, and to protect human dignity."[111] The same stance is present in Article 1 of the Federal Act on Non-Human Gene Technology stating that its purpose is "to protect human beings, animals and the environment from abuses of gene technology."[112]

The precautionary approach adopted by the Swiss legislator is not a consequence of the attribution of moral or legal status to the human embryo.

---

[108] Federal Constitution, RS 101, art. 119, al. 1.

[109] It has to be noticed that such framing was already present in the original form of the article, introduced in 1992 by popular vote and then amended in 2015 through another referendum that left that part unchanged. *Arrêté fédéral concernant la modification de l'article constitutionnel relatif à la procréation médicalement assistée et au génie génétique dans le domaine humain.* The referendum of June 5, 2015 introduced the possibility of creating in vitro all the embryos that could be employed for reproduction, and not anymore only those that could be immediately implanted. According to the commentators, this change has not only made assisted reproduction less burdensome and more efficient, but should have also resulted in an increased number of surplus embryos available for research.

[110] LPMA, RS 810.11, art. 1 al. 2.

[111] LRCS, RS 810.31, art. 1 al. 2.

[112] LGG, RS 814.91, art. 1 al. 1, a.

Rather, it is based on a so-called dignitarian approach,[113] one that posits that legal subjects deserve protection against the risk of alteration of those essential features upon which their dignity is based.[114] The dignitarian approach makes the notion of "dignity" the source of legal recognition, and it applies to humans as well as to animals, as we explain in detail below, in subsection 2 (Regulation of Pre-Clinical Research on Non-Humans).

a Human Embryos

Article 119 of the Swiss Federal Constitution forbids cloning as well as any interference with the genetic material of human reproductive cells and embryos.[115] According to the Federal Act on Medically Assisted Reproduction, embryos can only be created in vitro with the purpose of being used in assisted reproductive procedures (to overcome infertility or to avoid the transmission of a serious disease that cannot be avoided otherwise) and cannot be created in vitro solely for the purpose of research.[116] Research activities involving in vitro human embryos are not allowed.[117] However, it is possible for scientists to access donated surplus embryos and to use them, but only to derive hESCs.[118] Donated embryos cannot be kept in culture in vitro for more than seven days.[119] Even the genetic testing of embryos for nonmedical reasons is prohibited.[120] Given such rules, it is not possible to perform genome editing of embryos in Switzerland for basic research purposes.

Both the Federal Act on Medically Assisted Reproduction and the Federal Act on Research Involving Embryonic Stem Cells forbid the creation of clones, chimera, hybrids, and parthenotes – embryonic entities that in recent years have been the object of both scientific interest and public controversy.[121] Import and export of such types of embryos as well as of embryos solely created for research, of surplus embryos, and of genetically modified embryos are also prohibited.[122]

---

[113] R Brownsword, "Bioethics today, bioethics tomorrow: Stem cell research and the dignitarian alliance" (2003) 17 *Notre Dame J. Law Ethics Public Policy*, 15, 15–51. See also R Ashcroft, "Making sense of dignity" (2005) 31 *J. Med. Ethics*, 679, 679–682.

[114] See (n 141).

[115] Federal Constitution, RS 101, art. 119.

[116] LPMA, RS 810.11, art. 29 al. 1. The prohibition to create embryos for research purposes or to derive hES cells is restated in the LRCS, art. 3, al 1, a.

[117] LRCS, RS 810.31, art. 3, al. 2, a.

[118] Ibid., art. 1, al. 1, a.

[119] Ibid., art. 3, al. 2, c.

[120] LAGH, RS 810.12, art. 11.

[121] LPMA, RS 810.11, art. 36 al. 1; LRCS, RS 810.31, art. 3, al.1, c. and 3, al.1, d.

[122] LRCS, RS 810.31, art. 3.

Article 25 of the Federal Act on Research involving Human Beings prohibits research projects involving the modification of embryos' characteristics for nondisease-related reasons.[123] While this formulation seems to imply that other types of disease-related modifications are possible, as we saw, Swiss law does not allow research on in vitro embryos unless they are surplus embryos and their intended use is hESCs derivation. The same article also states that research on embryos in vivo is possible, but only if direct benefits for the pregnant woman and for the embryo outweigh the risk and burdens for them, or, if no direct benefits are expected, if research "entails no more than minimal risks and burdens" and is expected to generate knowledge that can in the future be beneficial to pregnant women or embryos.[124] While these provisions leave a door open to research on embryos, it seems that the stated conditions do not leave room for genome editing interventions in vivo, which would be prohibited anyway under Article 119 of the Federal Constitution.[125]

b Human Embryonic Stem Cells

The Federal Act on Research Involving Embryonic Stem Cells allows the derivation of hESCs from surplus embryos and research on hESCs derived in this way or imported from abroad. However, it is prohibited to create embryos for the sole purpose of research – including for the derivation of hESCs [126]– to derive stem cells from genetically modified embryos,[127] to derive stem cells from clones, chimera, hybrids, and parthenotes,[128] and to derive stem cells from an embryo of more than seven days.[129] Anyone who intends to derive stem cells from a surplus embryo for research purposes needs to receive appropriate license from the Federal Office of Public Health.[130]

Some constraints apply to research projects involving hESCs. Such projects must be approved by a cantonal ethics committee which needs to ensure that the project is scientifically significant, has relevance for human medicine or developmental biology, cannot gain equivalent insights without resorting to hESCs, is of sufficient scientific quality, and is ethically acceptable.[131] There seems to be no formal ban on genetic modification of hESCs for research

---

[123] LRH, RS 810.30, art. 25.
[124] Ibid., art. 26.
[125] Federal Constitution, RS 101, art. 119.
[126] LRCS, RS 810.31, art. 3, al. 1, a.
[127] Ibid., art. 3, al. 1, b.
[128] Ibid., art. 3, al. 1, c. and d.
[129] Ibid., art. 3, al. 2, c.
[130] Ibid., art. 7, al. 1.
[131] Ibid., art. 12.

purposes. However, any modified cell can only be studied in vitro and cannot be inserted in an embryo or otherwise used to give rise to an embryo.

c Gametes

Rigid restrictions also apply to basic research activities on human gametes. The already-cited Article 119 of the Swiss Federal Constitution forbids any "interference with the genetic material of human reproductive cells."[132] Modifying the genetic material of human germ cells is also prohibited by the Federal Act on Medically Assisted Reproduction[133] and by the Federal Act on Research Involving Embryonic Stem Cells.[134] Importing genetically modified germ cells is also unlawful according to Article 3.1.e. of Federal Act on Research Involving Embryonic Stem Cells.[135]

2 Regulation of Preclinical Research on Nonhumans

Overall, the Federal Constitution adopts a precautionary approach toward genetic engineering, stating that "human beings and their environment shall be protected against the misuse of gene technology."[136] This refers also to the use of gene technology in animals, as made explicit in the Federal Act on Non-Human Gene Technology, specifying the meaning of the precautionary principle in Article 2: "Early precautions shall be taken to prevent hazards or harm that may be caused by genetically modified organisms."[137]

Legislating on matters relevant to animal protection is a federal prerogative, as established in Article 80 of the Federal Constitution.[138] The same text, in Article 120 on nonhuman gene technology, states that "the Confederation shall legislate on the use of reproductive and genetic material from animals, plants and other organism."[139]

This approach is grounded in the principle that legislation in this domain "shall take account of the dignity of living beings" as well as human safety and genetic diversity of animal and plant species.[140] This dignitarian approach to animal protection and to the use of biotechnology in animals is restated in

---

[132]   Federal Constitution, RS 101, art. 119.
[133]   LPMA, RS 810.11, art. 35 al. 1.
[134]   LRCS, RS 810.31, art. 3, al. 1, b.
[135]   Ibid., art. 3, al. 1, e.
[136]   Federal Constitution, RS 101, art. 120, al. 1.
[137]   LGG, RS 814.91, art. 2.
[138]   Federal Constitution, RS 101, art. 80.
[139]   Ibid., art.120, al. 2.
[140]   Ibid.

Article 8 of Federal Act on Non-Human Gene Technology on respect for the dignity of living beings:

> In animals and plants, modification of the genetic material by gene technol-ogy must not violate the dignity of living beings. In particular, violation is deemed to have occurred if such modification substantially harms species-specific properties, functions or habits, unless this is justified by overriding legitimate interests. In evaluating the harm, the difference between animals and plants must be taken into consideration.[141]

Which circumstances constitute a violation of the dignity of living beings is not specified a priori. Such determination needs to take into account what the law defines as legitimate interests that justify interventions that would prima facie appear as violations of animals' dignity, that is: human health, food security, reducing environmental harm, preserving and improving of the environment, securing substantial economic, social, and environmental ben-efit for society, and, finally, expanding knowledge.[142] Under special circum-stances, the Federal Council can authorize certain modifications to the genetic material without a weighing of other interests.[143]

We now turn to some specific regulations that directly address genetic modifications undertaken for research purposes on animals (we will not cover provisions regarding animal breeding or handling for farming or com-mercial purposes).[144] The Federal Act on Non-Human Gene Technology of 2003 takes both a precautionary[145] and a dignitarian[146] approach to genetic engineering of non-human animals. The text has a dual purpose: on the one hand, it sets out "to protect human beings, animals and the environment from abuses of gene technology"; on the other, it serves "the welfare of human beings, animals and the environment in the application of gene technology."[147]

Another fundamental aim of legislation in this domain is ensuring biodiver-sity. However, the legislator also acknowledges the significance of scientific research on gene technology for human beings, animals, and the

---

[141] LGG, RS 814.91, Art. 8, al. 1, available at www.admin.ch/opc/fr/classified-compilation/19996 136/index.html, accessed September 4, 2018.

[142] Ibid., art. 8.2.

[143] Ibid., art. 8.3.

[144] We also do not cover general provisions on animal experimentation, that is, provisions that are not specific to genetic engineering. Such norms can be found in LPA, RS 455, especially c 2, s s 6 and 6a, arts. 17–20e and c 3, art. 22.

[145] LGG, RS 814.91, art. 2.

[146] Federal Constitution, RS 101, art. 120, al. 2; LGG, RS 814.91, art. 1, al. 2, c.

[147] LGG, RS 814.91.

environment.[148] This is reflected in Article 9 of the Federal Act on Non-Human Gene Technology, which states that "putting in circulation"[149] genetically modified vertebrates is prohibited unless it is done in the context of research, therapy, or diagnostics in human or veterinary medicine.[150]

Genetically modified organisms can only be released for experimental purposes if certain conditions apply, namely: if release is the only way to attain the research objective; if the release contributes to knowledge in the biosafety of genetically modified organisms; if the organisms in question do not contain engineered genes that confer antibiotic resistance; if the dispersal of the organism can be excluded; and if such release does not endanger human beings, animals, and the environment or reduce biological diversity.[151] A federal authorization is needed for experimental release of genetically modified organisms in the environment.[152]

All genetically modified organisms that do not fulfill the conditions for experimental release should be handled only within contained systems, and only following proper notification or authorization by the Federal Coordination Centre for Biotechnology (within the Federal Office for the Environment).[153]

Germline modification of animals is currently commonplace in Swiss science laboratories. In principle, there is no reason to believe that employing gene-editing techniques to alter the germline of laboratory animals would face any kind of regulatory restriction beyond those that apply to any research activity involving animals. However, genome editing is not yet included in the list of recognized methods for the production of genetically modified animals.[154]

3 Regulation of Clinical Research

We understand "clinical research" as referring to research activities aimed at collecting evidence regarding the safety and efficacy of a drug, therapy, or

---

[148] Ibid., art. 1, al. 2, g.

[149] The LGG defines "putting into circulation" as any supply of organisms to third parties in Switzerland, in particular by sale, exchange, giving as a gift, renting, lending or sending on approval, as well as their import. Supply for activities in contained systems or experimental release does not count as "putting into circulation." Ibid., art. 5.

[150] Ibid., art. 9.

[151] Ibid., art. 6.

[152] Ibid., art. 11.

[153] Ibid., art. 10; OUC, RS 814.912, art. 11.

[154] *Ordonnance sur l'expérimentation animale*, RS 455.163, art. 9.1. and Annex 1.

intervention by conducting trials on human research subjects. Using genome modification techniques in therapeutic applications means undertaking either gene therapy or cell therapy treatments in the course of a clinical trial.[155] Gene and cell therapy treatments are currently being tested on somatic cells in clinical trials in most scientifically advanced countries, including Switzerland, where about sixty gene therapy clinical trials had been conducted until 2014.[156] According to the Swiss Therapeutic Products Act,[157] its related ordinances and guidance by the Swiss Agency for Therapeutic Products,[158] gene therapy and cell therapy trials can only be employed on somatic cells – thereby excluding, for instance, the possibility of modifying a gamete in the context of a clinical trial. In any event, as already discussed, embryos cannot be the object of research practices at all.

Genetically modified somatic cells intended for clinical use are classified as "transplant products." Viral and bacterial vectors containing artificially introduced genetic material and used to deliver gene therapy in humans are classified as "genetically modified organisms." Even if this type of research is permitted on somatic cells, the Swiss Agency for Therapeutic Products has

---

[155] Genome editing is just one of the possible techniques to be employed in the course of gene or cell therapy. Viral vectors are another, and actually more established, way of performing these procedures. (M Kay, J Glorioso, and L Naldini, "Viral vectors for gene therapy: the art of turning infectious agents into vehicles of therapeutics" (*Nature Medicine*, January 1, 2001) www.nature.com/articles/nm0101_33 accessed September 4, 2018; C Thomas, A Ehrhardt, and M Kay, "Progress and problems with the use of viral vectors for gene therapy" (*Nature Reviews*, May 1, 2003) www.nature.com/articles/nrg1066 accessed September 4, 2018). For gene therapy in this chapter, we intend only therapies aimed at inserting, in a human, patient's cells – from a donor or from the patient himself or herself – that underwent some form of genetic modification ex vivo. Yet, genome editing promises more targeted and more easily manageable interventions. See T Gaj, C Gersbach, and C Barbas, "ZFN, TALEN and CRISPR/Cas-based methods for genome engineering" (2013) 31:7 *Trends Biotechnology* 397, 397–405.

[156] A Marti, "Requirements for Clinical Trials with Gene Therapy and Transplant Products in Switzerland" in M Galli and M Serabian (eds.), *Regulatory Aspects of Gene Therapy and Cell Therapy Products. Advances in Experimental Medicine and Biology* (Springer, Cham, vol. 871., 2015) 131–145.

[157] *Loi fédérale sur les médicaments et les dispositifs médicaux (Loi sur les produits thérapeutiques) du* 15 décembre 2000, RS 812.21.

[158] Swissmedic, *Guidance Document for the Compilation of the Documentation on Possible Risks for Humans and the Environment in Support of Applications for the Authorization to Carry Out Clinical Trials of Somatic Gene Therapy and with Medicines Containing Genetically Modified Microorganisms (Environmental Data) in accordance with Art. 22, 35 and Annex 4 KlinV / ClinO* (Swissmedic, 2017), www.swissmedic.ch/dam/swissmedic/en/dokumente/bewilligun gen/i-315/i-315_aa_01-a11dwegleitunggentherapiegvoumweltdaten.pdf.download.pdf/i-315_a a_01-a11eguidancedocumentgenetherapygmoenvironmentaldata.pdf accessed September 4, 2018.

issued technical prescriptions to exclude the risk of inadvertent germline integration.[159]

Some gene and cell therapy products for somatic cells have already received clearance and are currently in use. They are used on patients affected by genetic diseases,[160] or injuries, as well as in cancer immunotherapy.[161] None of such uses entail modifications of the patient's germline. Clinical research is still underway to test whether CRISPR-based genome editing could be a valid alternative to other genome editing tools, namely, zinc-finger nuclease (ZFN) and transcription activator–like effector nucleases (TALENS).

Currently, in Switzerland there are no clinical trials on CRISPR-based genome editing (either in embryos, or on adult subjects). However, *CRISPR Therapeutics*, a company headquartered in Zug, Switzerland, has received approval to start a clinical trial in late 2018. The study will assess the safety and efficacy of a CRISPR gene-edited therapy on up to thirty transfusion-dependent β-thalassemia patients. At the time of this writing, it was still unclear whether patients will be recruited and treated in Switzerland or in other countries in Europe.[162]

In theory, the same procedure could be used to modify gametes and human embryos presenting genetic alterations. This technique would be an alternative to preimplantation genetic diagnosis for parents who disapprove of it for religious reasons, or for couples in which, for instance, one parent is homozygous for a dominant disease-causing mutation, or in which both parents are homozygous for a recessive disease-causing mutation. Yet, as we have observed above, it is forbidden to modify the genetic makeup of gametes in Switzerland.[163] Also, the general prohibition to conduct research

[159]   Ibid.
[160]   M Maeder and C Gersbach, "Genome-editing Technologies for Gene and Cell Therapy" (2016) 24:3 *Molecular Therapy* 430, 430–446.
[161]   L Poirot and others, "Multiplex Genome-Edited T-cell Manufacturing Platform for 'Off-the-Shelf' Adoptive T-cell Immunotherapies" (2015) 75:18 *Cancer Res* 3853; J Ren and others, "Multiplex Genome Editing to Generate Universal CAR T Cells Resistant to PD1 Inhibition" (2017) 23:9 *Clin. Cancer Res* 2256.
[162]   CRISPR Therapeutics, "CRISPR Therapeutics Submits First Clinical Trial Application for a CRISPR Gene-Edited Therapy, CTX001 in β-thalassemia-Phase 1/2 trial in β-thalassemia expected to begin in 2018" (Globe News Wire, 2017) https://globenewswire.com/news-release/2017/12/07/1247360/0/en/CRISPR-Therapeutics-Submits-First-Clinical-Trial-Application-for-a-CRISPR-Gene-Edited-Therapy-CTX001-in-β-thalassemia.html, accessed September 4, 2018; CRISPR Therapeutics, "CRISPR Therapeutics AG: Common Shares, Prospectus Supplement (To prospectus dated December 4, 2017)" (CRISPR Therapeutics, 2017) www.stifel.com/prospectusfiles/PD_3604.pdf accessed September 4, 2018.
[163]   Federal Constitution, RS 101, art. 119.

on embryos forbids the use of genome editing on embryos in the course of a clinical trial.[164]

4 Regulation of Clinical Applications

A clinical application of germline genome editing is any procedure resulting in the implantation of in vitro–modified embryos, or embryos resulting from the combination of one or more genetically modified gametes. We speak of clinical applications only when such procedures take place independently of a basic or clinical research protocol. Herein we consider as clinical applications also those procedures that are not established yet as standard clinical procedures but that are not administered to patients in the course of a clinical trial either. Access to such not-yet-validated procedures (and use of not-yet-authorized drugs) can happen via a number of regulatory pathways subject to a special authorization by Swiss Agency for Therapeutic Products.[165]

Establishing a pregnancy with gametes or embryos that underwent genetic modifications for clinical purposes will result in a person harboring a genetically modified germline, who will in turn pass the modification on to his or her descendants.

a Pregnancy with Genetically Modified Gametes

The constitutional ban on interfering with the genome of reproductive cells – restated, as we saw, also by other pieces of legislation (Art. 35.2 of the Federal Act on Medically Assisted Reproduction) – makes it so that establishing a pregnancy with edited gametes is prohibited in Switzerland.[166]

b Pregnancy with Genetically Modified Embryos

Swiss legislation punishes anyone who tries to establish a pregnancy using an embryo that results from genetically modified gametes or that is by any

---

[164] Ibid.

[165] Swissmedic, "Special Authorizations" (2017). www.swissmedic.ch/swissmedic/en/home/huma narzneimittel/bewilligungen_zertifikate/special-authorisation.html accessed September 4, 2018. Such pathways comprise compassionate use (when a patient receives a treatment that is currently been tested in a clinical trial without the patient being enrolled in it), off-label prescriptions, and special authorizations (importing a drug that is not yet approved in Switzerland).

[166] Federal Constitution, RS 101, art. 119.

other means genetically altered (Art. 35.2 of the Federal Act on Medically Assisted Reproduction).

IV CURRENT PERSPECTIVES AND FUTURE POSSIBILITIES

Both the Swiss media and the Helvetic academic community pay considerable attention to the rapid development of gene-editing technologies. However, Switzerland's legal framework limits, to a high degree, research activities that may lead to human germline modifications. Such a restrictive environment was created well in advance of recent technical progress in genome editing, and, at the moment, nothing indicates that the regulatory framework might be relaxed to accommodate technological developments in the field of genome editing.

Yet, in a recent report, the National Advisory Commission on Biomedical Ethics highlighted that existing provisions on the use of human embryos for research purposes are not coherent. The law allows the destruction of surplus embryos if the aim is deriving stem cells from them.[167] However, the same law forbids any other form of research on human embryos.[168] As the National Advisory Commission on Biomedical Ethics noted, both activities lead to the destruction of the embryo. Treating the two activities differently, so the argument goes, is not justified on moral grounds and, therefore, the law should be amended, perhaps opening the door to other forms of research on human embryos beyond hESCs derivation.[169] At present, however, there are no initiatives underway in the country that could lead to a revision of any law concerning research on human embryos in the direction indicated by the National Advisory Commission on Biomedical Ethics.

---

[167] LRCS, RS 810.31, art. 1, al. 1.

[168] Ibid., art. 3, al. 2, a.

[169] NEK / CNE, "Gene editing sur les embryons humains – un état des lieux" (2016) www.nek-cne.admin.ch/inhalte/Themen/Stellungnahmen/fr/DEF_NEK_Kurzstellungnahme_Gene Editing_FR.pdf accessed September 4, 2018; see Ibid., Title "Les Conclusions de la CNE."

PART III

ASIA

16

The Regulation of Human Germline Genome Modification in Japan

Tetsuya Ishii

I INTRODUCTION

Researchers have developed several techniques to modify the nuclear or mitochondrial genome of the human germline (i.e. germ cells and zygotes). Since 1996, some types of germline mitochondrial genome modification have been clinically used for infertility treatment,[1] but public concerns regarding experimental reproductive medicine led to restrictive or prohibitive policies in the United States and China.[2] Some countries were less hesitant. In 2015, the United Kingdom legally permitted a form of germline genome modification that changes the composition of mitochondrial genome in the eggs or zygotes by nuclear transfer to prevent serious mitochondrial disease in offspring (the so-called mitochondrial replacement technique – MRT).[3]

With the advent of more precise and efficient genome-editing techniques, such as CRISPR-Cas9, germline nuclear genome editing is becoming clinically feasible.[4] Reports on basic research involving human germline genome modification suggest it could be used to correct prenatally genetic diseases to free offspring from them.[5] However, germline genome modification is highly controversial for clinical, ethical, legal and social reasons. In theory, modifying genome in the germline can lead to genetic modifications that can be

---

[1] T Ishii, 'Reproductive medicine involving mitochondrial DNA modification: evolution, legality, and ethics' (2018) 4:1 *EMJ Repro Health* 88, 88–99.
[2] Ibid.
[3] UK Human Fertilization and Embryology (Mitochondrial Donation) Regulations 2015.
[4] M Araki and T Ishii, 'International regulatory landscape and integration of corrective genome editing into in vitro fertilization' (2014) 12: 108 *Reprod Biol Endocrinol* 1.
[5] P Liang and others, 'CRISPR/Cas9-mediated gene editing in human tripronuclear zygotes' (2015) 6:5 *Protein Cell* 363; L Tang and others, 'CRISPR/Cas9-mediated gene editing in human zygotes using Cas9 protein' (2017) 292:3 *Mol Genet Genomics* 525, 525–533; H Ma and others, 'Correction of a pathogenic gene mutation in human embryos' (2017) 548:7668 *Nature* 413, 419.

transferred from the offspring on to the subsequent generations, which could lead to social issues. Unsurprisingly, at this time in history, it is prohibited, by law or guidelines, in 29 out of 39 countries recently surveyed (74%).[6]

Although all developed countries are facing a rising average age of women at first birth, with consequent fertility problems, in Japan the problem is all the more urgent, as its fertility rates are lower than most countries and the population is rapidly ageing. Together with the United States, Japan is the country where in vitro fertilization (IVF) is resorted to most often (368,000 cycles in 2013).[7] Resorting to reproductive medicine, such as IVF, is encouraged to facilitate reproduction in some infertility cases, and, probably as a consequence, Japan's policy of reproductive medicine is permissive, more so than most European countries.[8] It is noteworthy that, in Japan, reproductive medicine is regulated solely by guidelines established by professional societies, such as the Japan Society of Obstetrics and Gynecology (JSOG). There are no laws directly regulating reproductive medicine.

When it comes to research involving human germline genome modification, the Japanese regulatory framework is characterized by gaps and inconsistencies. This chapter will discuss first the overall regulatory framework for research on human germline and the specific regulations affecting research on human germline genome modification. Then, it will discuss some of the reasons why Japan lacks a key law governing the medical use of human germ cells and embryos, and will conclude discussing a possible regulatory reform.

II THE REGULATORY ENVIRONMENT

What are the legal instruments and bodies regulating research involving human germline genome modification, key definitions, funding opportunities and restrictions in Japan?

1 *Regulatory Framework*

Figure 16.1 summarizes the national regulatory framework of human germline genome modification. In short, it consists of one pivotal law (the *Act on*

---

6   M Araki and T Ishii, 'International regulatory landscape and integration of corrective genome editing into in vitro fertilization' (2014) (n 4).
7   ESHRE, 'ART fact sheet' (ESHRE, 2018) www.eshre.eu/Press-Room/Resources.aspx accessed 15 May 2018.
8   ESHRE, 'Regulation and legislation in assisted reproduction' (ESHRE, 2018) www.eshre.eu /Press-Room/Resources.aspx accessed 15 May 2018.

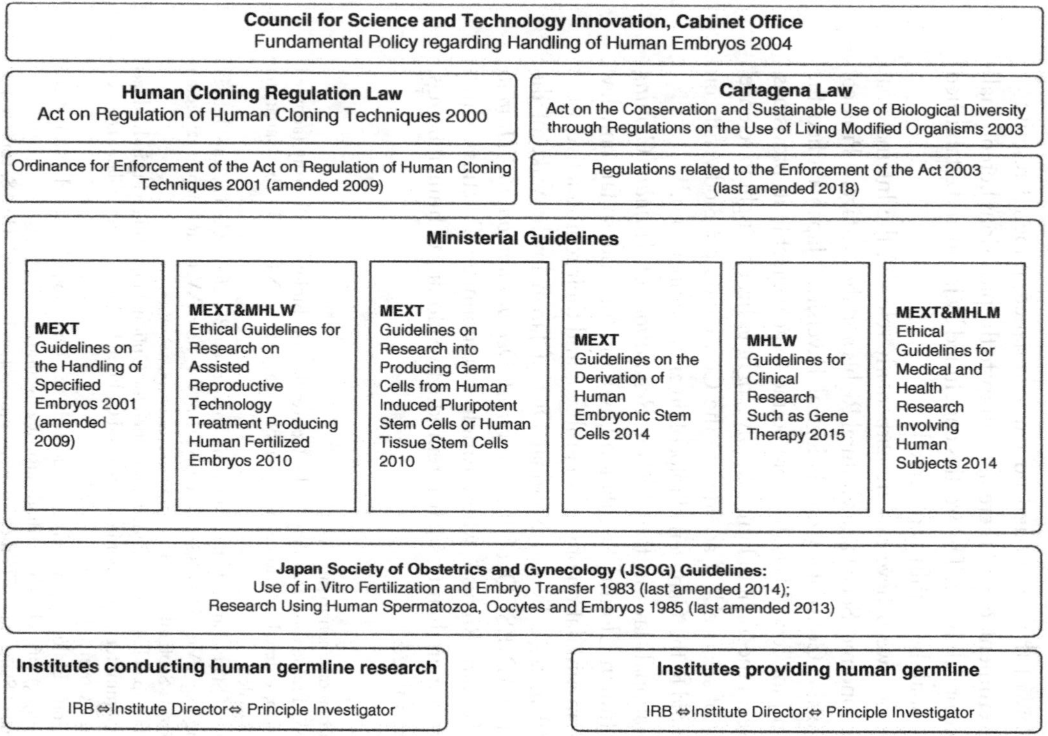

Council for Science and Technology Innovation, Cabinet Office
Fundamental Policy regarding Handling of Human Embryos 2004

Human Cloning Regulation Law
Act on Regulation of Human Cloning Techniques 2000

Cartagena Law
Act on the Conservation and Sustainable Use of Biological Diversity through Regulations on the Use of Living Modified Organisms 2003

Ordinance for Enforcement of the Act on Regulation of Human Cloning Techniques 2001 (amended 2009)

Regulations related to the Enforcement of the Act 2003 (last amended 2018)

Ministerial Guidelines

MEXT
Guidelines on the Handling of Specified Embryos 2001 (amended 2009)

MEXT&MHLW
Ethical Guidelines for Research on Assisted Reproductive Technology Treatment Producing Human Fertilized Embryos 2010

MEXT
Guidelines on Research into Producing Germ Cells from Human Induced Pluripotent Stem Cells or Human Tissue Stem Cells 2010

MEXT
Guidelines on the Derivation of Human Embryonic Stem Cells 2014

MHLW
Guidelines for Clinical Research Such as Gene Therapy 2015

MEXT&MHLM
Ethical Guidelines for Medical and Health Research Involving Human Subjects 2014

Japan Society of Obstetrics and Gynecology (JSOG) Guidelines:
Use of in Vitro Fertilization and Embryo Transfer 1983 (last amended 2014);
Research Using Human Spermatozoa, Oocytes and Embryos 1985 (last amended 2013)

Institutes conducting human germline research

IRB ⇔Institute Director⇔ Principle Investigator

Institutes providing human germline

IRB ⇔Institute Director⇔ Principle Investigator

FIGURE 16.1 The regulatory framework on human germline research in Japan

MHLW: Ministry of Health, Labour and Welfare.

MEXT: Ministry of Education, Culture, Sports, Science, and Technology

Regulation of Human Cloning Techniques), a policy issued by the Cabinet of the prime minister (the *Fundamental Policy Regarding Handling of Human Embryos*), and a series of ministerial and scientific society guidelines. Of course, all of these normative instruments are subordinate to the Constitution, which recites in its Article 98: '1. This Constitution shall be the supreme law of the nation and no law, ordinance, imperial rescript or other act of government, or part thereof, contrary to the provisions hereof, shall have legal force or validity. 2. The treaties concluded by Japan and established laws of nations shall be faithfully observed.'[9]

Amongst treaties, we believe two are particularly relevant: the International Covenant on Economic, Social and Cultural Rights and the Cartagena Protocol on Biosafety to the Convention on Biological Diversity.[10] Japan ratified the Covenant on 21 June 1979.[11] It is therefore bound to give effect to its provisions, including those on the 'right to form a family', 'right to health', 'right to science' and the 'rights of science'.[12] It acceded to the Convention on Biodiversity on 28 May 1993[13] and to the Cartagena Protocol on 21 November 2003.[14]

The key national law is the *Act on Regulation of Human Cloning Techniques* (Human Cloning Regulation Law), the first and, so far, only law addressing the manipulation of human embryos and their reproductive use.[15] It was adopted in 2000 by the National Diet of Japan in the wake of the adoption by the UNESCO of the *Universal Declaration on the Human Genome and Human Rights*,[16] whose Article 11 announces: 'Practices which are contrary to human dignity, such as reproductive cloning of human beings, shall not be permitted.'[17]

[9] The Constitution of Japan 1946.
[10] The Cartagena Protocol on Biosafety to the Convention on Biological Diversity (adopted 29 January 2000, entered into force 11 September 2003) 2226 UNTS 208 (The Cartagena Protocol on Biosafety).
[11] UNTS, 'Status of States Parties to the Cartagena Protocol on Biosafety' (UNTS, 2018) https://treaties.un.org/Pages/ViewDetails.aspx?src=IND&mtdsg_no=XXVII-8-a&chapter=27&clang=_en accessed 12 September 2018.
[12] However, so far Japan has neither signed nor ratified the Optional Protocol giving individuals access to the Committee on Economic, Social and Cultural Rights claiming violation of their rights under the Covenant.
[13] Convention on Biological Diversity (adopted 22 May 1992, entered into force 29 December 1993) 1760 UNTS 79 (Biodiversity Convention).
[14] UNTS, 'Status of States Parties to the Cartagena Protocol on Biosafety' (UNTS, 2018) n 11.
[15] Act on Regulation of Human Cloning Techniques (ヒトに関するクローン技術等の規制に関する法律: Hito-ni-Kansuru-Kurohn-Gijyutsu-tou-no-Kisei-ni-Kansuru-Houritsu) (Act No 146 of 2000).
[16] UNESCO, Universal Declaration on the Human Genome and Human Rights (11 November 1997).
[17] Ibid.

Four years later, in 2004, the Council of Science and Technology Policy in the Cabinet Office[18] issued the *Fundamental Policy Regarding Handling of Human Embryos* (Fundamental Policy).[19] At that time, the prime minister, and therefore chair of the Council of Science and Technology, was Mr Jun-Ichiro Koizumi. The policy is not law, and therefore is not binding (there is no penalty, such as fines and imprisonments, except the limitations of public funding). It is an aspirational document, pointing the way for future guidelines regarding the permissible uses of human embryos for research.

Echoing Article 11 of the UNESCO Declaration, the Fundamental Policy declared the principle of 'human dignity', mandating respect for 'human embryos', as defined by the *Human Cloning Regulation Law*.[20] At the same time, it also concluded that the ethical and legal position of human embryos is different from that of human beings. On this basis, it restricted the creation of and research on human embryos as such. Specifically, the *Fundamental Policy* exceptionally permits the following: (i) the creation and use of human embryos for basic research, but only with the aim to enhance assisted reproductive technology (reproductive medicine), such as IVF;[21] and (ii) the derivation of human embryonic stem cells, but only from surplus IVF, and only for stem cell therapy.[22]

The Council of Science and Technology Policy postponed deciding whether to allow the creation and use of human embryos to perform basic research on congenital intractable diseases, probably because at the time (2004) the feasibility of these kinds of research was only speculative.[23]

Years later, the Ministry of Education, Culture, Sports, Science and Technology (MEXT)[24] and the Ministry of Health, Labour and Welfare (MHLW)[25] adopted ministerial research guidelines (again, not biding law) to implement the above-mentioned points (i) and (ii) of the Fundamental Policy. The first one is the *2010 MEXT & MHLW Ethical Guidelines for Research on Assisted Reproductive Technology Treatment Producing Human*

---

[18] It is currently named Council for Science, Technology and Innovation (総合科学技術・イノ ベーション会議: Sougou-Kagaku-Gijyutsu・inobehsyon-Kaigi).

[19] Council of Science and Technology Policy, *Fundamental Policy Regarding Handling of Human Embryos* (ヒト胚の取扱いに関する基本的考え方: Hito-Hai-no-Toriatsukai-ni-Kansuru-Kihonteki-Kangaekata) (2004).

[20] Ibid. 4–6.

[21] Ibid. 14.

[22] Ibid. 14.

[23] The Cartagena Protocol on Biosafety (n 10) 6–7.

[24] Ministry of Education, Culture, Sports, Science and Technology (文部科学省 (Monbu-Kagaku-Syou)).

[25] Ministry of Health, Labour and Welfare (厚生労働省 (Kousei-Roudou-Syou)).

Fertilized Embryos.[26] The second one is the 2009 MEXT *Guidelines on the Derivation and Distribution of Human Embryonic Stem Cell.*[27] Since then, the MEXT and the MHLW have updated and amended the two guidelines. Their current versions are the 2014 *MEXT & MHLW Guidelines on the Derivation of Human Embryonic Stem Cells*[28] and the 2014 *MEXT Guidelines on the Distribution and Utilization of Human Embryonic Stem Cells.*[29]

Besides these two key guidelines, there are other ministerial guidelines relevant to basic research on human germline. One is the 2010 *MEXT Guidelines on Research into Producing Germ Cells from Human Induced Pluripotent Stem Cells or Human Tissue Stem Cells.*[30] These guidelines regulate gametogenesis research using human induced pluripotent stem (iPS) cells.[31] Japan is one of the few countries in the world that enacts specific regulation for basic research on creating germ cells from stem cells.[32] However, at the same time, the guidelines prohibit researchers from fertilizing induced gametes to respect human embryos' dignity as enshrined in the Fundamental Policy.[33]

The *Human Cloning Regulation Law* and related guidelines (specifically the 2001 MEXT *Guidelines for Handling of Specified Embryos*)[34] prohibit

---

[26]   MEXT and MHLW, *Ethical Guidelines for Research on Assisted Reproductive Technology Treatment Producing Human Fertilized Embryos* (ヒト受精胚の作成を行う生殖補助医療研究に関する倫理指針: Hito-Jyusei-Hai-no-Sakusei-wo-Okonau-Seisyokuhojyoiryou-Kenkyu -ni-Kansuru-Rinrishishin) (Notice No 2, 2010).

[27]   MEXT, *Guidelines on the Derivation and Distribution of Human Embryonic Stem* Cells (ヒトＥＳ細胞の樹立及び分配に関する指針: Hito-ES-Saibou-no-Jyuritsu-oyobi-Bunpai-ni-Kansuru-Shishin) (Notice No 156, 2009).

[28]   MEXT and MEXT, *Guidelines on the Derivation of Human Embryonic Stem Cells* (Notice No 2, 2014).

[29]   MEXT, *Guidelines on the Distribution and Utilization of Human Embryonic Stem Cells* (Notice No 174, 2014).

[30]   MEXT, *Guidelines on Research into Producing Germ Cells from Human Induced Pluripotent Stem Cells or Human Tissue Stem Cells* ヒトiPS細胞又はヒト組織幹細胞からの生殖細胞の作成を行う研究に関する指針: Hito-iPS-Saibou-matawa-Hito-Sosiki-Kansaibou-karano-Seisyoku-Saibou-no-Sakusei-wo-Okonau-Kenkyu-ni-Kansuru-Shishin) (Notice No 88, 2010).

[31]   However, if researchers use human germ cells in the tissues obtained during surgery (excluding reproductive medicine), they shall observe MEXT and MHLW Ethical Guidelines for Medical and Health Research Involving Human Subjects 2014. MEXT and MHLW, *Ethical Guidelines for Medical and Health Research Involving Human Subjects* (人を対象とする医学系研究に関する倫理指針: Hito-wo-Taisyou-tosuru-Igakukei-Kenkyu-ni-Kansuru-Rinrishishin) (Notice No 3, 2014).

[32]   T Ishii and M Saitou, 'Promoting In Vitro Gametogenesis Research with a Social Understanding' (2017) 23:11 *Trends Mol Med* 985, 985–988.

[33]   MEXT, *Guidelines on Research into Producing Germ Cells from Human Induced Pluripotent Stem Cells or Human Tissue Stem Cells* (n 30).

[34]   MEXT, Guidelines for Handling of Specified Embryos (特定胚の取扱いに関する指針: Tokuteihai-no-Toriatsukai-ni-Kansuru-Shishin) (Notice No 173, 2001) (amended by Notice No 83, 2009).

human cloning (reproductive use of a human somatic cell nuclear transfer embryo) and regulate eight other related techniques for the reproduction of humans, animals or their subspecies.[35] They permit basic research on embryos created using human somatic cell nuclear transfer, but only for therapeutic cloning (the derivation of patient-specific embryonic stem cell lines for stem cell therapy) and only within the fourteen-day limit, and the creation of animal-human chimeric embryos for producing human organs in animals.[36]

The 2003 Act on the Conservation and Sustainable Use of Biological Diversity through Regulations on the Use of Living Modified Organisms (also known as the Cartagena Law),[37] which was adopted to implement nationally the *Cartagena Protocol on Biosafety to the Convention on Biological Diversity*.[38] The *2015 Guidelines for Clinical Research Such as Gene Therapy* regard the clinical use of plasmid DNA and virus vectors to transfer a gene.[39] As we will discuss in greater detail below, they prohibit the clinical use of some germline genome editing techniques.

To complete the picture of the regulatory framework of human germline research in Japan, one needs to take into account also some earlier guidelines of the Japan Society of Obstetrics and Gynecology (JSOG).[40] While the Fundamental Policy addresses only basic research, the JSOG guidelines regulate reproductive medicine in Japan, at all stages of research. JSOG also certifies reproductive medicine specialists. Thus, the first JSOG standard to be taken into consideration is the *1983 Guidelines on the Use of In Vitro*

---

[35] Ibid. Nine techniques: Human Split Embryo, Human Embryonic Nuclear Transfer Embryo, Human Somatic Cell Nuclear Transfer Embryo, Human-Human Chimeric Embryo, Human-Animal Hybrid Embryo, Human-Animal Clone Embryo, Human-Animal Chimeric Embryo, Animal-Human Clone Embryo and Animal-Human Chimeric Embryo.

[36] MEXT and MEXT, *Guidelines on the Derivation of Human Embryonic Stem Cells* (n 28).

[37] Act on the Conservation and Sustainable Use of Biological Diversity through Regulations on the Use of Living Modified Organisms (遺伝子組換え生物等の使用等の規制による生物の多様性の確保に関する法律: Idenshi-Kumikae-Seibutsu-tou-no-Shiyou-tou-no-Kisei-niyoru-Seibutsu-no-Tayousei-ni-Kansuru-Houritsu) 2003 (Cartagena Law).

[38] The Cartagena Protocol on Biosafety (n 10).

[39] A 'plasmid' is a small DNA molecule within a cell that is physically separated from a chromosomal DNA and can replicate independently. Plasmids are most commonly found in bacteria; however, they are sometimes present in archaea and eukaryotic organisms. In nature, plasmids often carry genes that may benefit the survival of the organism, for example antibiotic resistance. While the chromosomes are big and contain all the essential genetic information for living under normal conditions, plasmids usually are very small and contain only additional genes that may be useful to the organism under certain situations or particular conditions. Artificial plasmids are widely used as vectors in molecular cloning, serving to drive the replication of recombinant DNA sequences within host organisms.

[40] The Japan Society of Obstetrics and Gynecology (日本産科婦人科学会: Nihon-Sanka-Fujinka-Gakkai).

Fertilization and Embryo Transfer.[41] The second is the 1985 *Guidelines on Research Using Human Spermatozoa, Oocytes and Embryos.*[42] Of course, these guidelines apply only to JSOG members, not to all researchers. Moreover, since the enactment of ministerial guidelines relevant to human germline research (such as *MEXT & MHLW Ethical Guidelines for Research on Assisted Reproductive Technology Treatment Producing Human Fertilized Embryos* and the 2009 *MEXT Guidelines on the Derivation and Distribution of Human Embryonic Stem Cell*), the importance of the JSOG guidelines has relatively diminished. However, they still play a regulatory role in human germline research (e.g. testing a new culture medium product or cryopreservation agent using human sperm, oocytes and embryos) that does not involve the creation of human embryos for research.

In sum, the current regulatory framework on human germline research in Japan is rather elaborate, but gaps and inconsistencies abound. In particular, as we will discuss in Sections III and IV, there appears to be inconsistency between the Fundamental Policy, the Human Cloning Regulation Law, ministerial guidelines and JSOG guidelines, especially on basic research using human germline genome modification of surplus IVF embryos. Still, it is safe to conclude that in Japan it is lawful to: derive human embryonic stem cells from surplus embryos and use them for research; conduct basic research using human germline cells; generate human germ cells from stem cells; and therapeutic cloning (the derivation of patient-specific embryonic stem cells for developing autologous cell therapy); generate animal-human chimeric embryos for producing human organs in animals; and create and use human embryos for basic research to enhance reproductive techniques, such as IVF within the fourteen-day limit.

2 Definitions

The Japanese regulatory system is riddled with inconsistent definitions that are often at odds with scientific understanding. For instance, Article 2.1.i of the *Human Cloning Regulation Law* defines 'human embryo' as: 'A cell (except for a germ cell) or a cell group which has the potential to grow into an

---

[41]   JSOG On Use of in Vitro Fertilization and Embryo Transfer (体外受精・胚移植に関する見解: Taigai-Jyusei・Hai-Isyoku-ni-Kansuru-Kenkai). Available at: www.jsog.or.jp/ethic/taigaijusei_201406.html (in Japanese), accessed 15 May 2018.

[42]   JSOG On Research Using Human Spermatozoa, Oocytes and Embryos 1985 (ヒト精子・卵子・受精卵を取り扱う研究に関する見解: Hito-Seisi・Ranshi・Jyuseiran-wo-Toriatsukau-Kenkyu-ni-Kansuru-Kenkai). Available at: www.jsog.or.jp/modules/statement/index.php?content_id=29 (in Japanese), accessed 21 January 2019.

individual through the process of development in utero of a human or an animal and remains at a stage prior to placental formation.'[43] The definition is echoed in Article 7 of the 2015 *Guidelines for Clinical Research Such as Gene Therapy*,[44] as well as the Fundamental Policy. In other words, no distinction is made between the stages of embryonic development. 'Zygotes' (one-cell-stage embryos) are considered 'embryos'.

The *Human Cloning Regulation Law* defines also other key technical terms, such as 'germ cell', 'unfertilized embryo' and 'human fertilized embryo'. Germ cell is: 'A sperm (including a spermatid and a spermatocyte whose number of chromosomes is equal to the number of chromosomes of the sperm . . .) and an unfertilized egg.'[45] Clearly, the legal definition of 'germ cell' in the *Human Cloning Regulation Law* is not consistent with the scientific definition of this term. The former includes haploid cells, but not diploid cells,[46] while, scientifically, germ cells include diploid cells that produce haploid cells via meiosis.[47] On the other hand, the definition of 'germ cell' in the 2010 *MEXT Guidelines on Research into Producing Germ Cells from Human Induced Pluripotent Stem Cells or Human Tissue Stem Cells* is broader. It includes not only haploid germ cells but also primordial diploid germ cells. The law and the guidelines are obviously inconsistent.

The *Human Cloning Regulation Law* defines 'unfertilized egg' as 'an unfertilized ovum and oocyte (limited to an oocyte whose number of chromosomes is equal to the number of chromosomes of the ovum)'.[48] Moreover, it defines 'human fertilized embryo' as 'an embryo produced by fertilization between a human sperm and a human unfertilized egg (including each embryo which is produced successively by single or multiple splitting of such an embryo and is not a human split embryo)'.[49]

---

[43]   Human Cloning Regulation Law, art. 2.1.i.
[44]   Ministry of Health, Labour and Welfare, *The Guidelines for Clinical Research Such as Gene Therapy* (遺伝子治療等臨床研究に関する指針: Idenshi-Chiryou-tou-Rinsyou-Kenkyu-ni-Kansuru-Shishin) (2015).
[45]   Human Cloning Regulation Law, art. 2.1.ii.
[46]   'Haploid cells' are cells that contain only one complete set of chromosomes. The most common type of haploid cells is gametes, or sex cells. When the haploid cells from the parent donors come together and are fertilized, the offspring has a complete set of chromosomes and becomes a 'diploid cell'. Diploid cells contain the complete set of necessary chromosomes for life to begin, while haploid have only half the number of chromosomes found in the nucleus.
[47]   Haploid cells are produced by 'meiosis'. Meiosis is a type of cell division that results in four daughter cells each with half the number of chromosomes of the parent cell, as in the production of gametes and plant spores.
[48]   Ibid., art. 2.1.iii.
[49]   Ibid., art. 2.1.iv.

Finally, the same law defines 'human embryonic nuclear transfer embryo' as an Embryo produced by [t]usion between a Human Enucleated Egg and a Human Fertilized Embryo or Human Split Embryo at the one-cell stage or an Embryonic Cell with a cell nucleus of a Human Fertilized Embryo, Human Split Embryo or Human-Human Chimeric Embryo'.[50] 'Human embryonic nuclear transfer embryos' can be interpreted as the reconstituted embryo produced using nuclear transfer between human zygotes to reduce mutated mtDNA in the oocytes (one form of MRT: specifically, pronuclear transfer).[51] The legal implications of this definition are discussed further in Section II.

3 *Oversight Bodies*

As it was said, the Ministry of Technology (MEXT) and the Ministry of Health (MHLW) exercise oversight over the creation of human embryos for research, human embryonic stem cell derivation, and in vitro gametogenesis using human iPS cells and tissue stem cells.[52] Human germline genome modification experiments involving the derivation of human ES cells require the approval of both the national authority (MEXT) and the institutional review board (IRB). Researchers need to find first a fertility clinic that can provide surplus IVF embryos with the informed consent of the parents.[53] Then, they need to get IRB approval at the institute where they are doing the research and the one that provided the surplus IVF embryos. Finally, they must submit their research protocol to the MEXT for national review. The MEXT publishes approved protocols on a dedicated website.[54] Researchers must also report their research results to MEXT when they transfer cell lines to other researchers or when they have finished their research protocol. Members of the Japan Society of Obstetrics and Gynecology are also required to register the protocol of human germline research with, and report the research results to, the society.[55] But, all

---

[50] Ibid., art. 2.1.ix.

[51] A Greenfield and others, 'Assisted reproductive technologies to prevent human mitochondrial disease transmission' (2017) 35:11 *Nat Biotechnol* 1059, 1059–1068.

[52] See, in this chapter, Section II.1.

[53] Council of Science and Technology Policy, *Fundamental Policy Regarding Handling of Human Embryos* (n 19).

[54] MEXT, 'A List of Derivation, Distribution and Research Use of Human Embryonic Stem Cells (ヒト ES 細胞の樹立・分配・使用に関する計画一覧: Hito-ES-Saibou-no-Jyuritsu・Bunpai・Shiyou-ni-Kansuru-Keikaku-Ichiran)' (2018) www.lifescience.mext.go.jp/files/pdf/E SC_utilization_plan_180518.pdf accessed 15 May 2018.

[55] MEXT and MHLW, *Ethical Guidelines for Research on Assisted Reproductive Technology Treatment Producing Human Fertilized Embryos* (n 26).

of this is still only theory. As of September 2018, no article on human germline genome modification has been published in Japan.

4 Funding

Human germline genome modification research – whether basic (modified human germline is used only in laboratories and not used for human reproduction) or clinical (modified human germline is used for human reproduction) – done in compliance with law and guidelines becomes eligible to receive public funds from MEXT, MHLW and other public funding agencies. As far as we can tell, in Japan there have been no instances yet of clinical research involving human germline genome modification supported by public funding.

Recently, a fertility clinic in Osaka, which is a JSOG member, performed clinical research on Autologous Germline Mitochondrial Energy Transfer (AUGMENT).[56] AUGMENT is an infertility treatment to boost the egg's energy levels for better embryo development. It uses the energy-producing mitochondria (containing their own DNA) from the same woman's egg precursor cells, which are immature egg cells found in the protective lining of the ovaries, to supplement the existing mitochondria in the egg.[57] However, the research was privately funded.[58]

III SUBSTANTIVE PROVISIONS

Let's now look at the specific substantive provisions relevant for germline genome modification research and its clinical application in Japan.

1 Basic Research

The Fundamental Policy stipulates that basic research on human germline genome modification cannot be done on human embryos that have been created ad hoc. This reflects the principle of recognizing human embryos' 'dignity' enshrined in the UNESCO Declaration and the Fundamental

---

[56] UMIN-CTR, 'Autologous germline mitochondrial energy transfer, UMIN000021387' (ICMJE, 2016) https://upload.umin.ac.jp/cgi-open-bin/ctr/ctr_view.cgi?recptno= R000024309 accessed 14 May 2018.

[57] MH Fakih, MEl Shmoury, J Szeptycki, DB dela Cruz, C Lux, et al., 'The AUGMENT[SM] Treatment: Physician Reported Outcomes of the Initial Global Patient Experience' (2015). *JFIV Reprod Med Genet* 3, 154. doi:10.4172/2375-4508.1000154.

[58] Op. cit. 56.

Policy.[59] According to the *2014 MEXT & MHLW Ethical Guidelines for Medical and Health Research Involving Human Subjects*, human germline genome modification research is permissible only if it is done on embryos that have been created for IVF but that have not been transferred into uterus (i.e. surplus embryos), or on germ cells, such as spermatogonial stem cells (i.e. male germ stem cells taken from the testis).[60] This procedure can only be done with IRB approval (and, if necessary, also JSOG registration). This has already been done in Japan and the research results have been published.[61]

If the research involves embryonic stem cell derivation, researchers must go through an additional regulatory step, one that lengthens the approval process. This is so because they must also observe the *2014 MEXT Guidelines on the Derivation of Human Embryonic Stem Cells*, according to which both research on human germline genome modification involving embryonic stem cells and human embryo research not involving embryonic stem cell derivation end up with the destruction of the embryo. However, the guidelines are inconsistent since the former must undergo a national review in addition to an IRB review, while the latter does not.

The *2010 MEXT & MHLW Ethical Guidelines for Research on Assisted Reproductive Technology Treatment Producing Human Fertilized Embryos* are not quite relevant for research on human germline genome modification. These guidelines are meant only for basic research to enhance existing reproductive medicine, such as IVF,[62] as per directive (i) of the Fundamental Policy. This is why, to date, there has been no instance of approval of research protocols involving human germline genome modification under the 2010 MEXT & MHLW Guidelines. Currently, the first

---

59   UNESCO, Universal Declaration on the Human Genome and Human Rights (n 16); Council of Science and Technology Policy, *Fundamental Policy Regarding Handling of Human Embryos* (n 19).

60   Such basic research is considered legitimate according to the 2014 MEXT & MHLW Guidelines on the Derivation of Human Embryonic Stem Cells and the 2010 MEXT Guidelines on Research into Producing Germ Cells from Human Induced Pluripotent Stem Cells or Human Tissue Stem Cell, which is discussed in Section II.1.

61   For example, K Kawai and others, 'Parental age and gene expression profiles in individual human blastocysts' (2018) 8:1 *Sci Rep* 2380; S Hashimoto and others, 'Quantitative and qualitative changes of mitochondria in human preimplantation embryos' (2017) 34:5 *J Assist Reprod Genet* 573, 573–580; H Kurosawa and others, 'Development of a new clinically applicable device for embryo evaluation which measures embryo oxygen consumption' (2016) 31:10 *Hum Reprod* 2321, 2321–2330.

62   Council of Science and Technology Policy, *Fundamental Policy Regarding Handling of Human Embryos* (n 19).

application of in vitro egg maturation is under review by the national committee.[63]

The 2010 *Guidelines on Research into Producing Germ Cells from Human Induced Pluripotent Stem Cells or Human Tissue Stem Cells* allow the MEXT to approve research protocols involving human germline genome modification research, if it is done on gametes derived from in vitro gametogenesis from human iPS cells or tissue stem cells. However, fertilization of artificial gametes is prohibited.[64] Moreover, the research protocol must be reported to MEXT, which opens the door for more intervention by this ministry. For instance, it could ask that the protocol be re-reviewed by the IRB.[65]

Before initiating human germline genome modification experiments using surplus IVF embryos or spermatogonial stem cells, the 2003 *Act on the Conservation and Sustainable Use of Biological Diversity through Regulations on the Use of Living Modified Organisms* (the Cartagena Law) must also be considered. This law, adopted to implement nationally the Cartagena Protocol on Biosafety to the Convention on Biological Diversity, seeks to protect biological diversity from the potential risks posed by genetically modified organisms resulting from modern biotechnology. Although it is mostly relevant for agriculture and farming, it becomes relevant for a discussion of germline genome editing because of the way the Cartagena Law incorporated the Protocol in Japan's legal system.

The Cartagena Protocol defines 'living organism' as 'any biological entity capable of transferring or replicating genetic material, including sterile organisms, viruses and viroids'.[66] 'Living modified organism' is defined as 'any living organism that possesses a novel combination of genetic material obtained through the use of modern biotechnology'.[67] And 'modern biotechnology' is defined as 'the application of: (a) In vitro nucleic acid techniques, including recombinant deoxyribonucleic acid (DNA) and direct injection of nucleic acid into cells or organelles; or (b) fusion of cells beyond the taxonomic family'.[68]

The Cartagena Law further defines 'organism' as 'a single cell (excluding a single cell forming a cell colony) or a cell colony, excluding human cells, which is stipulated in the ordinance of the competent

---

[63] Health and Science Council, *Minutes of 3rd human embryo research review committee, science and technology division* (MHLW, 2017).

[64] MEXT, *Guidelines on Research into Producing Germ Cells from Human Induced Pluripotent Stem Cells or Human Tissue Stem Cells* (n 30) art. 6.E.

[65] Ibid., art. 13.E.

[66] The Cartagena Protocol on Biosafety (n 10) art. 3.h.

[67] Ibid., art. 3.g.

[68] Ibid., art. 3.i.

ministries as having the capacity to transfer or replicate nucleic acid, and viruses and viroids'.[69] It defines 'living modified organism' as 'an organism that possesses nucleic acid, or a replicated product thereof, obtained through use of any of the following technologies: (i) Those technologies as stipulated in the ordinance of the competent ministries, for the processing of nucleic acid extracellularly.'[70]

In the previous reports on genome editing in human embryos,[71] nuclease mRNA or proteins are injected into the zygotes, along with template DNA (Figure 16.2). Injected nuclease mRNA (nucleic acid) or proteins are soon digested, and the encoded nuclease catalyses gene modification using the introduced template DNA.[72]

Because nuclease mRNA is nucleic acid, its injection falls under the definition contained in Article 2.2 of the Cartagena Law. However, the nuclease in the form of protein can be used. In addition, human cells are excluded from 'organisms' in the ordinance, as mentiomed above. Even if human cells are 'organisms', this way of genetic modification can be not within the scope of the Cartagena Law.

Article 2.i of the 2003 *Regulations Related to the Enforcement of the Cartagena Law*[73] excludes some technologies from the technologies for processing nucleic acid extracellularly. The exempted technologies include: '(i) Technology for processing by using, as nucleic acid to be introduced into cells, only the nucleic acid shown in the following: A. The nucleic acid of living organism belonging to the same species as that of the living organism which the cells originate from.'[74] Thus, embryos that were genetically modified in this way are not 'living modified organism' for the purpose of the Cartagena Law.

---

[69] Cartagena Law (n 37) art. 2.1.

[70] Ibid., art. 2.2.

[71] P Liang and others, 'CRISPR/Cas9-mediated gene editing in human tripronuclear zygotes' (2015) 6:5 *Protein Cell* 363; L Tang and others, 'CRISPR/Cas9-mediated gene editing in human zygotes using Cas9 protein' (2017) 292:3 *Mol Genet Genomics* 525, 525–533; H Ma and others, 'Correction of a pathogenic gene mutation in human embryos' (2017) 548:7668 *Nature* 413, 419.

[72] T Ishii, 'Germline genome-editing research and its socio-ethical implications' (2015) 21:8 *Trends Mol Med* 473, 473–481.

[73] Regulations related to the Enforcement of the Law concerning the Conservation and Sustainable Use of Biological Diversity through Regulations on the Use of Living Modified Organisms (遺伝子組換え生物等の使用等の規制による生物の多様性の確保に関する法律施行規則: Idenshi-Kumikae-Seibutsu-tou-no-Siyou-tou-no-Kisei-niyoru-Seibutsu-no -Tayousei-no-Kakuho-ni-Kansuru-Houritsu-Shikoukisoku) 2003 (Regulations Related to the Enforcement to the Cartagena Law).

[74] The provisions in Japanese are shown below:
一 細胞に移入する核酸として、次に掲げるもののみを用いて加工する技術
イ 当該細胞が由来する生物と同一の分類学上の種に属する生物の核酸

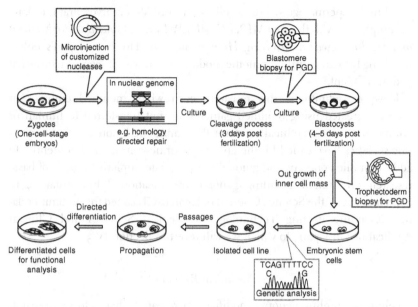

FIGURE 16.2 Example of a protocol for human germline genome-editing research

Note: Researchers can use existing surplus embryos derived from assisted reproductive treatment cycles that were not to be used for future treatment. Otherwise, research begins by creating human embryos after the approval of an institutional review board (IRB) and/or health authority. Genome-editing nucleases are microinjected into the one-cell-stage embryos. The embryos are then cultured, and biopsied embryonic cells are subjected to genetic analyses, including preimplantation genetic diagnosis (PGD), to investigate the gene modification. Subsequently, embryonic stem cells (ESCs) can be established if necessary. Genetic and functional analyses can be conducted using ESC-derived differentiated cells. If spermatogonial stem cells (SSCs) are edited, the developmental potential of the spermatozoa derived from the edited SSCs would then be assessed, thus requiring the creation of human embryos for research if relevant regulation permits. Adapted from Ishii, T. Trends in Molecular Medicine 21, 473–481. 2015

To repeat, neither the Cartagena Protocol nor the Cartagena Law is meant to apply to humans. However, the production of nuclease mRNA or repair template DNA, for example, requires the use of 'living modified' microorganism that is transfected with exogenous DNA. This could make the Cartagena Law potentially relevant at least for a part of human germline genome modification research.

To summarize, under the current regulatory framework, some basic research on human germline genome modification is permissible as long it uses human spermatogonial stem cells or surplus IVF embryos, and it is done in compliance with the 2014 *MEXT & MHLW Ethical Guidelines for Medical and Health Research Involving Human Subjects*. However, Japan's policy regarding human germline genome modification research seems inconsistent and insufficient.

However, admittedly this research, even when modified embryos or embryos created using modified gametes are not transferred to human or animal uterus, raises ethical concerns that cannot be ignored.[75] Some scientific societies have called for the enactment of new ministerial guidelines to address public concerns and guide the appropriate implementations of basic research on human germline genome modification.[76] For instance, in September 2017, the Science Council of Japan (SCJ) issued the recommendation 'Genome Editing Technology in Medical Sciences and Clinical Applications in Japan', to which we will revert in Section IV.3.

2 Preclinical Research

Preclinical germline genome modification research, that is to say research involving animals, is regulated by the *Act on Welfare and Management of Animals* and related guidelines.[77] In Japan, animal experimentation is managed in each research organization. There is no need for a general license for animal research. Institutional animal care and use committees review protocols for research on animal experiments, keeping in mind the so-called 3 R Principles: Replacement of animal uses in experiments such as cell culture experiments, Reduction of number of animal uses, and Refinement of procedures applied to animals to avoid animals' pains and distress.[78]

Research using animal germ cells is also subject to the Cartagena Law. As we said earlier, when nuclease mRNA (nucleic acid) or proteins are injected into animal zygotes or germ cells, along with template DNA, to produce genetically modified animals, the resulting genetic modification is a 'living modified' animal, as defined by Article 2.2 of the Cartagena Law.

---

[75] T Ishii, 'The ethics of creating genetically modified children using genome editing' (2017) 24:6 *Curr Opin Endocrinol Diabetes Obes* 418, 418–423.

[76] Japan Society of Gene and Cell Therapy, *Joint Position Statement on Human Genomic Editing* (2015).

[77] Fundamental Guidelines for Proper Conduct of Animal Experiment and Related Activities in Academic Research Institutions 2006 (Notice No 71).

[78] Ibid., arts. 4.1.1.i, ii and iii.

3 *Clinical Research*

As mentioned before, currently the main regulations relevant to clinical research on human germline genome modification are the *Human Cloning Regulation Law*, the *2001 MEXT Guidelines for Handling of Specified Embryos* and the *2015 MHLW Guidelines for Clinical Research Such as Gene Therapy*. The combined effect of these regulations is that clinical research using germline genome editing is largely, but not completely, prohibited in Japan.

Mitochondrial Replacement Therapy (MRT), also known as 'mitochondrial donation' in the UK, is a special form of IVF in which the future baby's mitochondrial DNA (mtDNA) comes mostly from a third party. This technique is used in cases when the mitochondria of the mother has pathogenic mutations (acquired or inherited) in the mtDNA that code for mitochondrial components. They may also be the result of acquired mitochondrial dysfunction due to adverse effects of drugs, infections or other environmental causes. At this point in history, the two most common techniques to replace mitochondria are the 'pronuclear transfer' and 'maternal spindle transfer' (Figure 16.3).[79]

The clinical introduction of these techniques has raised concerns about the adverse effects they may have on resultant embryos and offspring as well as ethical concerns, since the offspring will carry genomes of three parents, not just two (the third one being the donor of the mitochondria). Some countries prohibit it, while a few allow it. For instance, as we said before, MRT was made legal in the United Kingdom in 2015,[80] and regulations put in place the following year cleared the way for procedures to begin.[81]

In the case of Japan, the language of the *2001 MEXT Guidelines for Handling of Specified Embryos*, implementing the *Human Cloning Regulation Law*, seems to entirely prohibit pronuclear transfer. It provides that: 'For the time being, specified embryos (including human embryonic nuclear transfer embryo) not prescribed in the provisions of Article 3 of the Act may not be transferred to human or animal uterus.'[82] Also, the same law stipulates that: 'For the time being, of specified embryos, the types of embryos that may be produced shall be limited to human somatic cell nuclear transfer (hSCNT) embryos and animal-human chimeric embryos.'[83]

---

[79] 'Polar body' transfer is an additional technique.
[80] UK Human Fertilization and Embryology (Mitochondrial Donation) Regulations 2015 (n 3).
[81] J Gallagher, 'Babies made from three people approved in UK' *BBC News* (15 December 2016).
[82] Guidelines for Handling of Specified Embryos (n 34) art. 7.
[83] Ibid, art. 2.

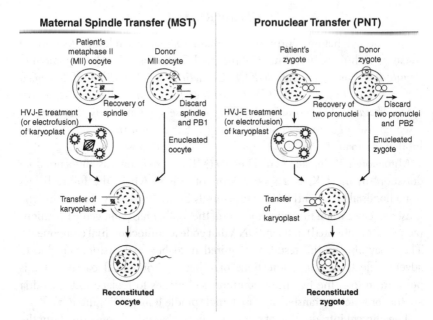

FIGURE 16.3 The procedures of two mitochondrial replacement techniques
PB: polar body; HVJ-E: Hemagglutinating Virus of Japan Envelope

In pronuclear transfer, a karyoplast (a cellular nucleus together with a plasma membrane containing a small amount of cytoplasm) is taken from a patient's zygote (fertilized egg) and is transferred to a zygote coming from a donor or a zygote created using a patient's egg and emptied of its nucleus (enucleated). Under the *Human Cloning Regulation Law*, a 'human enucleated egg' is 'an enucleated human unfertilized egg or an enucleated human fertilized embryo or human split embryo at the one-cell stage'.[84] 'Human embryonic nuclear transfer embryo' is defined as 'an embryo produced by fusion between a human enucleated egg and a *human fertilized embryo* or human split embryo at the one-cell stage, or an embryonic cell with *a cell nucleus of a human fertilized embryo,* human split embryo or human-human chimeric embryo'.[85] It follows that, in the case of a reconstituted embryo generated via pronuclear transfer, the enucleated zygote is the 'human fertilized embryo' to which the *Human Cloning Regulation Law* refers. The 'cell nucleus of a human fertilized embryo'

---

[84] Human Cloning Regulation Law, art. 2.1.xxiii.
[85] Human Cloning Regulation Law, art. 2.1.ix (emphasis added).

is a karyoplast. Therefore, the *Human Cloning Regulation Law* prohibits pronuclear transfer.

That being said, the *Human Cloning Regulation Law* does not prohibit all clinical research using germline genome modification. Indeed, the statute has no language that bars the other major MRT: maternal spindle transfer. Probably, this is because the *Human Cloning Regulation Law* focused on regulating use of embryos from humans, animals or their subspecies, not their germ cells.

Crucially for the purposes of this volume, the 2015 *MHLW Guidelines for Clinical Research Such as Gene Therapy* do not seem to rule out germline genome editing either. On the one hand, Article 7 of these guidelines seems to prohibit entirely clinical research involving germline genome editing and MRTs: 'Clinical research that intentionally conducts or may conduct genetic modification of human germ cells or embryos (a cell or a cell group which has the potential to grow into an individual through the process of development in utero of a human or an animal and remains at a stage prior to placental formation) is prohibited.'[86] However, the same guidelines define 'gene therapy' as 'the administration of a gene or cells into which a gene was transferred for the purposes of treatment and prohibition of a disease'.[87] If germline genome editing, for instance of a human embryo, is done by way of a plasmid,[88] harbouring a nuclease gene and/or a gene of template DNA, then this conduct falls under the prohibition contained in Article 7. However, the nuclease – the enzyme that cleaves the chains of nucleotides in nucleic acids into smaller units – can also be deployed in the form of mRNA or protein.[89] If this can be done without using template DNA, then the germline genome editing falls outside the scope of Article 7 of the 2015 Guideline.

Consider the clinical research of AUGMENT. As women age, mitochondria, which is the 'battery pack' of a cell, tend to wear out and not be as efficient as they were in producing energy. AUGMENT intends to boost the egg's energy levels by complementing mitochondria. Mitochondria is extracted from the same woman's egg precursor cells, which are immature egg cells found in the protective lining of the ovaries, to add the mitochondria to the egg.[90] By adding

---

[86]   Guidelines for Clinical Research Such as Gene Therapy (n 34) art. 7.

[87]   Ibid., art. 2.1.

[88]   Plasmid is a genetic structure in a cell that can replicate independently of the chromosomes, typically a small circular DNA strand in the cytoplasm of a bacterium or protozoan. Plasmids are often used in the laboratory manipulation of genes.

[89]   M Araki and T Ishii, 'International regulatory landscape and integration of corrective genome editing into in vitro fertilization' (n 4).

[90]   Op. cit. 57.

battery packs, egg quality is expected to be improved.[91] However, there have been concerns about the safety and efficacy of AUGMENT.[92] When Japanese researchers applied to the MHLW to obtain authorization for clinical research on AUGMENT, the MHLW answered that AUGMENT falls outside the prohibition contained in Article 7 of the 2015 Guidelines, and can be used in clinical research according to 2014 *MEXT & MHLW Ethical Guidelines for Medical and Health Research Involving Human Subjects*.[93] As a result, the clinical research of AUGMENT needed to be approved only by the IRB of the clinic where the research was to be carried out.[94] However, subsequently, a committee in the SCJ[95] criticized this regulatory response by the MHLW by pointing out that the European Society of Human Reproduction and Embryology and the British Fertility Society had raised doubts about the safety as well as efficacy of AUGMENT.[96]

In sum, the *Human Cloning Regulation Law* prohibits the clinical research of pronuclear transfer but not of maternal spindle transfer. The *2015 MHLW Guidelines for Clinical Research Such as Gene Therapy* ban clinical research of germline genome editing, but only if it meets the statutory definition of 'gene therapy', whereas the prohibition contained in the 2015 Guidelines does not apply to the clinical research of germline genome editing that uses only nucleases in the form of mRNA or protein. There is no clear rationale for these distinctions. They seem to be loopholes created by vaguely and not well-enough-drafted regulations and legislation rather than intentional distinctions supported by a clear rationale.

4　Clinical Applications

The *Human Cloning Regulation Law* applies also to clinical applications of germline genome modification. The law punishes anyone who 'creates

---

[91]　K Oktay and others, 'Precursor Cell-Derived Autologous Mitochondria Injection to Improve Outcomes in Women With Multiple IVF Failures Due to Low Oocyte Quality: A Clinical Translation' (2015) 22:12 *Reprod Sci* 1612, 1612–1617.

[92]　B Heindryckx and others, Opinion Statement: The Use of Mitochondrial Transfer to Improve ART Outcome (ESHRE, 2015) www.eshre.eu/-/media/sitecore-files/SIGs/Stem-Cells/SIG-Stem-Cells-opinion-16102015.pdf?la=en&hash=8C8820B5C2AA60CCAEC4D1B6D51519B6A251C350 accessed 15 May 2018.

[93]　Science Forest, "Oocyte rejuvenation" hypothesis application concern expectation for infertility treatment, the influence to the next generation is unknown' *Mainichi Shimbun* (Tokyo, 20 October 2016) (in Japanese).

[94]　MEX and MHLW, *Ethical Guidelines for Medical and Health Research Involving Human Subjects* (n 31).

[95]　Science Council of Japan (日本学術会議: Nihon-Gakujyutsu-Kaigi).

[96]　Op. cit. 92.

human fertilized embryos' without obtaining MEXT approval 'with imprison-ment with work for not more than one year or a fine of not more than one million yen' (about USD 8,900).[97] By contrast, it is unclear how Article 7 of the 2015 Guidelines applies to the clinical applications of germline genome editing, which can be performed with only artificial nuclease (pro-tein, neither gene nor DNA), since the instrument is meant to primarily provide clinical guidance only for somatic gene therapy that involves gene transfer. Moreover, they contain no criminal sanctions in case of violation.[98] If physicians or researchers violate Article 7 of the 2015 Guidelines ('Clinical research that intentionally conducts or may conduct genetic modification of human germ cells or embryos . . . is prohibited'), the only sanction they face is loss of public research funds for one to ten years and the reimbursement of public funds already used.[99] Obviously, similar penalties matter only for grantees of public research funds, not for those doing research using private funds. In the above-mentioned AUGMENT case, research was carried out using private funds only.[100]

IV CURRENT PERSPECTIVES AND FUTURE POSSIBILITIES

1 Current Perspectives on Human Germline Genome Modification Research

As we saw, basic research on human germline genome modification is per-missible in Japan as long as it is done on human spermatogonial stem cells or surplus IVF embryos. However, the regulatory framework inconsistently requires that research involving embryonic stem cells undergoes a national review in addition to an IRB review, whereas a national review is not manda-tory in human germline genome modification research involving embryonic stem cells derivation.

With regard to the clinical use of germline genome modification, the *Human Cloning Regulation Law* and the related 2003 *Guidelines* prohibit pronuclear transfer. However, another MRT technique, maternal spindle transfer, is not regulated. The 2015 *MHLW Guidelines for Clinical Research Such as Gene Therapy* forbid the clinical use of germline genome editing if it involves gene transfer. However, other clinical uses of germline genome

---

[97]   Human Cloning Regulation Law, art. 17.
[98]   No provisions, including the miscellaneous provisions, address penalty.
[99]   Japan's Ministries, *Guidelines for Proper Use of Competitive Public Research Funds* (2005).
[100]   Op. cit. 56.

editing, such as the injection of site-directed nucleases in the form of protein for clinical practice (not clinical research), are not prohibited.

These inconsistencies in the regulatory framework have been laid bare by technical advancements. They are probably due to uncoordinated drafting, and inconsistent interpretation and implementation of the law. Consider, again, the case of clinical research of AUGMENT performed by a clinic in Osaka. When a journalist asked the MHLW Minister, Mr Yasuhisa Shiozaki, about the ethical and safety concerns about AUGMENT, he answered that the MHLW had consulted with the JSOG and concluded that AUGMENT research is not subject to the ban in Article 7 of 2015 *MHLW Guidelines for Clinical Research Such as Gene Therapy*. Therefore, it could be performed as clinical research according to the 2014 *MEXT & MHLW Ethical Guidelines for Medical and Health Research Involving Human Subjects*.[101] After the approval by the competent IRB, the clinical research was carried out from March 2016 to August 2017.[102]

However, Minister Shiozaki could have been simply wrong. In the AUGMENT research, human embryos had actually been created, by injecting autologous mitochondria, for the research purpose of 'improvement of the developmental competence of human mature oocytes via AUGMENT'.[103] Clinical research that includes human embryo basic research should have been subject to review, at least partly, according to the 2010 *MEXT & MHLW Ethical Guidelines for Research on Assisted Reproductive Technology Treatment Producing Human Fertilized Embryos*. The inconsistent regulatory framework led the MHLW to make an incorrect regulatory decision on the medical use of human germline genome modification.

2 *The Fundamental Policy Regarding Handling of Human Embryos*

The Fundamental Policy Regarding Handling of Human Embryos was the first national policy to address the medical use of human embryos. However, as many pioneer policies, it was both incomplete and tentative. It was clear that it was an insufficient legal instrument already at the time of its adoption, in 2004. To begin

---

[101]   MHLW, 'Shiozaki Minister's Interview Summary from the Minister Press Conference' (MHLW, 7 October 2016) www.mhlw.go.jp/stf/kaiken/daijin/0000139301.html accessed 15 May 2018 (in Japanese).

[102]   MEXT and MHLW, *Ethical Guidelines for Medical and Health Research Involving Human Subjects* (人を対象とする医学系研究に関する倫理指針: Hito-wo-Taisyou-tosuru-Igakukei-Kenkyu-ni-Kansuru-Rinrishishin) (Notice No 3, 2014) www.lifescience.mext.go.jp/files/pdf/n1500_01.pdf accessed 21 January 2019.

[103]   Op. cit. 57.

with, as we saw, it was not fully consistent with the existing regulatory framework, including the *Human Cloning Regulation Law*, which predated the Fundamental Policy by four years. Second, it had a narrower scope than needed, since it addressed only embryonic stem cell derivation and the creation of human embryos for reproductive medicine. However, it did not fully consider the ethical aspects of reproductive techniques other than IVF, such as preimplantation genetic diagnosis (PGD). A JSOG guideline has permitted PGD as clinical research to avoid childbirths with serious genetic disease and screen embryos with a type of chromosomal abnormality (balanced translocation) through two-step ethics reviews (first, JSOG review, then clinic review).[104]

Third, it was already outdated even before it was adopted. For instance, the Council of Science and Technology Policy in the Cabinet Office[105] did not consider sufficiently the medical use of genome modification in germ cells, and it was not because germline genome modification had not become a reality yet. Indeed, ooplasmic transfer (also called cytoplasmic transfer), an experimental fertility technique that involves injecting a small amount of ooplasm from eggs (germ cells, not embryos) of fertile women into eggs of women whose fertility is compromised and which causes a modification of the germline genome, had been discussed in Japan since at least 1997.[106]

Fourth, the ethically difficult but necessary decision whether to allow the use of human embryos in basic research on congenital intractable diseases was postponed to a day yet to come.[107] However, the Fundamental Policy has not been revised since it was adopted in 2004. Even so, it was lacking already at the outset and it is becoming increasingly inadequate as it cannot keep up with developments in the biomedical field. What Japan needs is a coherent, up-to-date, fundamental law that governs both basic research and medical use of human germ cells as well as embryos, one that is discussed in and approved by the Diet, Japan's bicameral legislature, instead of by a Cabinet Committee, to ensure broad social understanding of, and support for, scientifically important research on human germline.[108]

[104]   JSOG, *Guidelines on Preimplantation Diagnosis* (1998, last amended 2018).
[105]   Council of Science and Technology Policy, *Fundamental Policy Regarding Handling of Human Embryos* (n 19).
[106]   T Ishii and Y Hibino, 'Mitochondrial manipulation in fertility clinics: Regulation and responsibility' (2018) Reproductive Biomedicine & Society Online https://doi.org/10.1016/j.rbms.2018.01.002.
[107]   Council of Science and Technology Policy, *Fundamental Policy Regarding Handling of Human Embryos*, 7.
[108]   For the examples, see NME Fogarty and others, 'Genome editing reveals a role for OCT4 in human embryogenesis' (2017) 550:7674 Nature 67, 67–73.

3 *Future Possibilities of Human Germline Genome Modification Research in Japan*

In July 2016, a committee of the SCJ initiated a discussion of the scientific, clinical, socio-ethical and regulatory issues surrounding human genome editing.[109] In September 2017, it issued a document entitled: *Recommendation on Genome Editing Technology in Medical Sciences and Clinical Applications in Japan*.[110] In the document, the SCJ stressed the pressing need for appropriate regulations on human germline genome modification. It urged the government to adopt guidelines to prohibit germline genome editing clinical trials and applications for the time being. In addition, the SCJ asked the government to carefully consider the need for laws to regulate experimental manipulation of human germ cells and embryos, including germline genome editing and mitochondrial replacement techniques. At the same time, the SCJ recognized the usefulness of germline genome editing to obtain fundamental knowledge of human reproduction and development. To help research advance, it called for the adoption of appropriate regulation to promote basic germline genome editing research while taking into account social values and transparency.

Meanwhile, at the governmental level, a bioethics investigation committee in the Council for Science, Technology and Innovation (CSTI)[111] started discussing ethical and regulatory issues of human germline genome editing in June 2015. In April 2016, it published an interim report that barred the transfer of human embryos whose genome has been edited to uterus.[112] In July 2017, it established a subcommittee called 'Task Force for Reviewing Fundamental Policy Regarding Handling of Human Embryos'.[113] In March 2018, the task force issued its first full report, which was later endorsed at the CSTI, urging MEXT and MHLW to establish guidelines that prohibit

---

[109]   The Science Council of Japan (SCJ) is a representative organization of Japanese scholars and scientists in all fields of sciences, including humanities, social sciences, life sciences, natural sciences and engineering. As of 2015, it consists of 210 elected members (appointed by the prime minister) and 2,000 associate members.

[110]   SCJ Committee on Genome Editing Technology in Medical Sciences and Clinical Applications, *Recommendation Summary on Genome Editing Technology in Medical Sciences and Clinical Applications in Japan* (SCJ, 2017).

[111]   Council for Science, Technology and Innovation (総合科学技術・イノベーション会議: Sougou-Kagaku-Gijyutsu・inobehsyon-Kaigi) (n 18).

[112]   Bioethics Investigation Committee, *CSTI Interim Report* (Cabinet Office, 2016).

[113]   Task Force for Reviewing Fundamental Policy Regarding Handling of Human Embryos (ヒ ト 胚の取扱いに関する基本的考え方　見直し等に係るタスク・フォース: 「Hito-Hai-no-Toriatsukai-ni-Kansuru-Kihonteki-Kangaekata」 Minaoshi-tou-ni-Kakaru-Tasuku・fohsu); CSTI, 'Task Force for Reviewing Fundamental Policy Regarding Handling of Human Embryos' (2017) www8.cao.go.jp/cstp/tyousakai/life/tf/tfmain.html accessed 12 September 2018.

clinical use of human germline genome editing for any reasons but for the promotion of basic research on human germline genome modification to help improve reproductive medicine.[114] The task force report ruled out the creation of human embryos for research, and postponed the consideration of the legal regulation of experimental manipulation of human germline, SCJ recommendation to that end notwithstanding. It did not revise the Fundamental Policy.[115]

As the CSTI Task Force Report suggested, the MEXT and MHLW have recently prepared draft guidelines for basic research using the techniques to modify genetic information in human embryos.[116] They were opened for public comments for a one-month period starting 17 October 2018. Under the draft guidelines, researchers can genetically modify surplus human embryos within fourteen days of fertilization to obtain fundamental knowledge on the development, growth and implantation of human embryos, the techniques to cryopreserve human embryos, and methods to improve reproductive medicine.[117] Importantly, the draft guidelines explicitly prohibit the transfer of human embryos used for the research to the human or animal uterus.[118] However, even new research guidelines will not be enough. A national law is needed, one with extraterritorial reach. Guidelines can regulate only the work of researchers in Japan. However, in the era of cross-border reproductive medicine, some prospective parents might choose to go abroad to seek germline genome modification as the last-resort remedy for their infertility problems, or to treat a genetic disease in their offspring.[119]

4 Why Japan Lacks Laws Governing the Medical Use of Human Germline Genome Modification

Given that Japan is a world leader in scientific research and is one of the countries in the world where artificial reproductive technology is most frequently resorted to, the lack of a law governing the medical use of human

---

[114]   CSTI, *First Report on Review of Fundamental Policy Regarding Handling of Human Embryos (on use of genome editing etc. for contributing assisted reproductive medicine)* (2017).

[115]   Ibid.

[116]   'Public Comment on the Proposal for Establishment of Ethical Guidelines Concerning Research Using Genetic Information Alteration Technology etc. in Human Fertilized Embryo' (E.Gov Japan, 2018) http://search.e-gov.go.jp/servlet/Public?CLASSNAME=PCM MSTDETAIL&id=495180206&Mode=0, accessed 18 October 2018 (in Japanese).

[117]   MEXT and MHLW, *Draft Ethical Guidelines on Research on Using Genetic Information Modification Technology etc. on Human Fertilized Embryos* (2018).

[118]   Ibid., art. 3.

[119]   Op. cit. 106.

germline genome modification is befuddling. Some Japanese politicians repeatedly made efforts to send a bill for assisted reproductive medicine involving a third party to the Diet, but in vain.[120] More recently, in the Health, Labour and Welfare Committee of the National Diet (House of Councilors) on 6 December 2018, one of the committee members asked the MHLW to legally prohibit the clinical use of germline genome editing; however, the MHLW denied the request because they considered ministerial guidelines sufficiently effective.[121]

Reproductive medicine is unique amongst medical disciplines because it does not involve just one person, but at least two (the parents), if not three or four (a gamete donor and/or a surrogate mother), plus the future human being that might result from the procedure. The complexity deriving from the need to balance the rights of multiple parties in reproductive medicine has generally made it difficult to discuss its appropriate use. To complicate matters further, religious beliefs are part of the discussion about human reproduction and family-building, too. Some religious groups largely accept reproductive medicine, while others condemn some or all types of reproductive medicine.[122] Often, religious beliefs shape the discussion on the acceptable use of reproductive medicine, but not so much in Japan since there is no state religion. Article 20 of the Constitution of Japan stipulates: 'Freedom of religion is guaranteed to all. No religious organization shall receive any privileges from the State, nor exercise any political authority.'[123] Moreover, a majority of Japanese are not affiliated with religion. The Pew Research Center Religion and Public Life Project 2010 showed that 57% of Japanese respondents answered that they were unaffiliated with a religion, whereas 36.2% of them answered that they are Buddhists.[124] Nearly all forms of Buddhism currently support reproductive medicine.[125]

---

[120] T Furukawa, 'As the chair of project team on assisted reproductive medicine, why rapid legislation is needed?' (2014) 129 *Chuo-Kouron* 28, 28–33.

[121] Minutes No. 7 of the Health, Labour and Welfare Committee of House of Councillors on 6 December 2018 (in Japanese). http://kokkai.ndl.go.jp/SENTAKU/sangiin/197/0062/197120 60062008.pdf accessed 21 January 2019.

[122] HN Sallam and NH Sallam, 'Religious aspects of assisted reproduction' (2016) 28:8 *Facts Views Vis Obgyn* 33, 33–48.

[123] The Constitution of Japan 1946, art. 20.

[124] Pew Research Center, 'Religion & Public Life Project 2010: Japan' (Pew-Templeton Global Religious Futures, 2010) www.globalreligiousfutures.org/countries/japan#/?affiliations_religi on_id=0&affiliations_year=2010&region_name=All%20Countries&restrictions_year=2015 accessed 15 May 2018.

[125] HN Sallam, NH Sallam, 'Religious aspects of assisted reproduction' (2016). *Facts Views Vis Obgyn* 8, 33–48.

However, if religion is not a major factor shaping discussion on these matters in Japan, ethics and moral are. 'Morals' is a vague concept in Japanese, which is one of the reasons why there are no legal regulations on reproductive medicine in Japan. The English words 'morals' and 'morality', which derive from Latin 'mores', meaning 'custom', 'habits', translates to the Japanese word 道徳 (*doutoku*). However, *doutoku* also has philosophical roots in ancient Chinese Confucianism and Taoism.[126] Those concepts are similar but not the same. Ancient Chinese Confucianism and Taoism put great value on 'ideology' and respect for authority, whereas European 'morals' value traits like freedom, equality and philanthropy.[127] While in the West 'ethical concerns' are the soul of many biomedical regulations, in Japan it is a concern with safety and the impact that activities might have on society that is paramount.

General social and philosophical considerations aside, the hastened implementation of research on AUGMENT underscores the pressing need for legal regulations on experimental reproductive medicine involving germline genome modification. The well-being of future generations is at stake.

As a closing remark, let us consider four provisions of the Constitution of Japan. As we said at the outset, the Constitution is the supreme law of the nation and no law, ordinance, imperial rescript or other act of government, contrary to the provisions of the Constitution, has legal force or validity.[128] The day the Japanese legislator finally drafts a fundamental law regulating human genome editing, it will have to keep in mind at least these four articles of the Constitution.

Article 11: 'The people shall not be prevented from enjoying any of the fundamental human rights. These fundamental human rights guaranteed to the people by this Constitution shall be conferred upon the people of this and future generations as eternal and inviolate rights.'[129]

Article 12: 'The freedoms and rights guaranteed to the people by this Constitution shall be maintained by the constant endeavor of the people, who shall refrain from any abuse of these freedoms and rights and shall always be responsible for utilizing them for the public welfare.'[130]

---

[126] T Yuuki, 'Moral Education and Religion – The Inseparability of Religious Elements in Moral Education' (2010) 44 *Journal of Ibaraki Christian University* 13, 13–24.
[127] Ibid.
[128] Constitution of Japan, art. 98.1.
[129] Ibid.
[130] Ibid.

Article 13: 'All of the people shall be respected as individuals. Their right to life, liberty, and the pursuit of happiness shall, to the extent that it does not interfere with the public welfare, be the supreme consideration in legislation and in other governmental affairs.'[131]

And, finally, Article 98.2: 'The treaties concluded by Japan and established laws of nations shall be faithfully observed.'[132]

These provisions in the Constitution demand a national law that addresses human germline genome modification, particularly in the clinical context, that protects human rights and is consistent with international legal standards.

---

[131]   Ibid.
[132]   Ibid.

The Regulation of Human Germline Genome Modification in the People's Republic of China

Lingqiao Song and Rosario Isasi

I INTRODUCTION

Recently, the development of precise gene editing technologies, such as CRISPR-Cas 9, has captured significant attention worldwide. The publication in 2016 of an article describing an attempt to modify the human germline genome by a team of Chinese scientists ignited further local and global ethical and policy debates.[1] This was followed in late November 2018 by a still unconfirmed report of the first live birth of twin girls following IVF and gene-editing techniques, placing China once again at the epicenter of controversy.[2]

Overall, China's regulatory framework of genome modification activities is permissive, more so than many other countries surveyed in this volume. However, as all other countries, it prohibits human cloning, research on human embryos fourteen days after fertilization and genetic manipulation of human gametes, zygotes, and embryos for reproductive purposes.

China's advances in the field and the size of its scientific community and resources position it at the forefront of biotechnological and gene editing research. This makes understanding the Chinese regulatory framework and the strength of its governance to address the vast scientific, social, ethical, and political global implications of germline genome modification paramount. This chapter explores how the People's Republic of China (PRC) legal system regulates human gene editing with particular focus on germline applications. It further outlines existing governance frameworks and addresses the possibility of policy convergence by contrasting Chinese approaches to those adopted worldwide.

---

[1] X Kang and others, "Introducing precise genetic modifications into human 3PN embryos by CRISPR/Cas-mediated germline editing" (2016) 33:5 *J Assit Reprod Genetic*, 581.

[2] D Cyranoski, "CRISPR-gene-editing tested in a person for the first time" (2016) 539:7630 *Nature* 479.

II THE REGULATORY ENVIRONMENT IN THE PRC

1 *The General Political and Legal System*

In China, the supreme political authority rests in the Communist Party, whose centrality to the People's Republic of China is guaranteed by the Constitution. The government of the PRC is made up of several organs:

- The National People's Congress (NPC) and its Standing Committee (legislature);[3]
- The president and the vice president;
- The State Council (SC), whose premier (prime minister) is the head of government (executive).[4] It is the central executive body of the PRC with supreme administrative power;
- The Central Military Commission, whose chairman is the commander-in-chief of the People's Liberation Army (PLA);
- The National Supervisory Commission: It is an agency created by Constitutional amendment at the first session of the Thirteenth National People's Congress (2018). It is endowed with the equivalent legal position as the Supreme Court and the Supreme Procuratorate, and mandated to be the supreme anti-corruption body in the PRC as well as to oversee Communist officials;
- The Supreme People's Court and the Supreme People's Procuratorate (judiciary).

Of these, the primary organs of state power are the NPC, the president, and the State Council. The State Council directly oversees the various subordinate People's governments in the provinces and is composed by the top levels of the Communist Party of China. Currently, the State Council has thirty-five members. It is chaired by the premier and includes the heads of each cabinet-level executive departments (ministries).

The Chinese legal system follows the Civil Law tradition. It has at least four levels of legal sources.[5] The Constitution of the People's Republic of China (1982) is the fundamental law.[6] The Constitution and its twenty-one

---

[3] National People's Congress and its Standing Committee 全国人民代表大会及其常务委员会 (*Quánguó Rénmín Dài biǎo Dàhuì Chángwù Wěiyuányuánhuì*).

[4] National Assembly of the PRC 中华人民共和国国务院 (*Zhōnghuá Rénmín Gònghéguó Guówùyuàn*).

[5] Lehman, Lee & Xu, "What are the sources of legal rules in China?" http://spanish .lehmanlaw.com/resource-centre/faqs/china-legal-services/what-are-the-sources-of-legal-rules-in-china.html accessed July 31, 2018.

[6] Constitution of the PRC 1982, Preamble.

amendments have established a series of fundamental human rights, including the right to education and the right to work.[7] The integration in the Constitution in 2004 of a provision explicitly stating that "the State respects and preserves human rights"[8] was a major milestone in the Chinese system of human rights protections.

In 2001, the Standing Committee of the NPC authorized the SC to ratify the International Covenant on Economic, Social, and Cultural Rights (ICESCR).[9] Although China is legally bound externally to comply with the ICESCR, it does not have direct effect domestically. The rights enshrined in the ICESCR have legal force in China only to the extent that they are also protected by the Constitution or are included in legislation.

The right to science and the right to health are not codified in the Constitution. The right to health is protected by law (法律 – fǎlǜ), which is enacted either by the NPC or by its Standing Committee. As such, it occupies second-rank hierarchy in the Chinese legal system, below the Constitution. Both the Torts Liability Law (2009)[10] and the General Rules of the Civil Law (2017)[11] recognize the right to health, a basic civil right equivalent to the right to life, the right to bodily integrity, and the right to privacy.[12]

All states that have ratified the ICESCR have an obligation to submit a periodic report to the Committee on Economic, Social, and Cultural Rights detailing what they have done to implement its provisions, including on the right to health and the right to science.[13] Since ratification, China has produced two reports.[14] According to the 2010 report, China has enacted a variety of policies and governmental activities to ensure the implementation

---

[7] Constitution of the PRC 1982, art. 46.

[8] Constitution of the PRC 1982, art. 33.

[9] International Covenant on Economic, Social and Cultural Rights (adopted December 16, 1966, entered into force January 3, 1976) 993 UNTS 3 (ICESCR). China ratified the ICESCR on March 27, 2001. However, so far, the PRC has not ratified the Optional Protocol giving individuals the possibility to raise violations of the Covenant before the Committee on Economic, Social and Cultural Rights. Optional Protocol to the International Covenant on Economic, Social and Cultural Rights (adopted December 10, 2008, entered into force May 5, 2013), A/RES/63/117.

[10] Torts Liability Law of the PRC 2009 (侵权责任法).

[11] General Rules of the Civil Law of the PRC 2017 (民法总则).

[12] Torts Liability Law of the PRC 2009, art. 2; General Rules of the Civil Law of the PRC 2017, art. 111.

[13] ICESCR, arts. 16–17.

[14] PRC "Second Periodic Report under Articles 16 and 17 of the Covenant, Implementation of the International Covenant on Economic, Social and Cultural Rights" (June 30, 2010) UN Doc E/C.12/CHN/2 (PRC July 2010); PRC "Initial Report submitted by States parties under articles 16 and 17 of the Covenant" (June 27, 2003) UN Doc E/1990/5/Add.594.

of both the right to health and the right to science.[15] The right to health has been implemented by a series of reforms, including the creation of a basic health insurance system to improve the quality of medical services provided to citizens and the promotion of a basic medical and healthcare delivery system, and by conferring special protections to ensure that women and children's right to health is respected.[16]

The right to science is being implemented by means of the Popularization of Science and Technology Act (2002)[17] and the amended Science and Technology Progress Act (2008).[18] These laws specify that the state protects the freedom of scientific and technological research and development, encourages scientific exploration and technological innovation, and protects the legitimate rights and interests of science and technology personnel.[19] In the 2010 ICESCR periodic report, statistics provided regarding the state budget allocated to scientific research and development showcase the importance of the scientific development in China.[20]

Under the Constitution, "The ministries and commissions issue orders, directives and regulations within the jurisdiction of their respective departments and in accordance with the statutes and the administrative rules and regulations, decisions and orders issued by the State Council."[21] In other words, the State Council issues national "administrative regulations" (*xíngzhèng fǎguī* – 行政法规). These regulations are the framework within which the various ministries enact the regulations and laws within their administrative mandates called "administrative rules" (*xíngzhèng guīzhāng* – 行政规章).[22] Ministries also have the power to issue "normative instruments" (*guīfànxìng wénjiàn* – 规范性文件). The most obvious distinction between "administrative rules" and "normative instruments" is that for administrative rules to enter into force, they must be reported to the State Council, whereas it is not the case for the normative instruments. In addition, administrative rules are usually issued in the form of "decree" (*lìng* – 令), while the normative instruments are released in a less formal manner usually taking the form of

[15]   PRC July 2010 (n 14) 44 and 58.
[16]   Ibid. 46.
[17]   Popularization of Science and Technology Act of the PRC 2002 (中华人民共和国科技普及法).
[18]   Science and Technology Progress Act of the PRC 2008 (中华人民共和国科技进步法).
[19]   Popularization of Science and Technology Act of 2002 (n 17) art. 5; Science and Technology Progress of 2008 (n 18) art. 3.
[20]   PRC July 2010 (n 14) 58.
[21]   Constitution of PRC 1982, art. 90.
[22]   Y Zhu, *Concise Chinese Law* (China Law Press, 2007) 10–23.

departmental documents.[23] Even though normative instruments are not considered a legal source in the Chinese legal system, they are enforced by the ministries with binding effect.

2 The Regulatory Framework Governing Human Germline Genome Modification in the PRC

a Regulatory Agencies

Currently, three agencies represent the main national authorities governing human germline genome modification: the National Health Commission (NHC), the Drug Inspection Administration (DIA), and the Ministry of Science and Technology (MoST). The NHC and the DIA were created in 2018 to replace the now-defunct Ministry of Health (MoH) and the Chinese Food and Drug Administration (CFDA). Indeed, over the past decades, the State Council has passed several institutional reforms to improve the quality and efficiency of its administrative powers, as well as realizing the varying social development missions given to it by the Communist Party at different stages in history. These reforms have created some confusion with respect to the roles and responsibilities of agencies mandated to regulate gene technologies in research and clinical applications, including human germline genome modification techniques. Let us clarify.

Until 2013, the main body governing human germline genome modification was the Ministry of Health (MoH),[24] acting under the leadership of the State Council. However, in 2013, it was integrated with the National Population and Family Planning Commission and its mandate was expanded. As a result of these reforms, a new ministry called the Health and Family Planning Commission (HFC) was created.[25] In March 2018, it became the NHC.[26] Despite some minor modifications to its mission, the Commission remains an important ministry, with a mandate that includes drafting population health policy, promoting healthcare system reform as well as drawing up

---

[23] L Huang, "The Legal Scope and Effect of Normative Instruments (规范性文件的法律界限和效力)" (2014), *Law Science*, Issue 7, available at http://b.38zhubao.net/KCMS/detail/detail.aspx?QueryID=0&CurRec=1&recid=&filename=FXZZ201407003&dbname=CJFD2014&dbcode=CJFQ&yx=&pr=&URLID=&forcenew=no; accessed 21 January 2019.

[24] Ministry of Health 中华人民共和国卫生部 (*Zhōnghuá Rénmín Gònghéguó Wèishēngbù*).

[25] National Health and Family Planning Commission of the PRC 中华人民共和国国家卫生与计划生育委员会 (*Zhōnghuá Rénmín Gònghéguó Guójiā Wèishēng Hé Jìhuà Shēngyù Wěiyuánhuì*).

[26] National Health and Wellness Commission of the PRC 中华人民共和国国家卫生健康委员会(*Zhōnghuá Rénmín Gònghéguó Guójiā Wèishēng Jiànkāng Wěiyuánhuì*).

the basic national drug regulations, and overseeing public health, medical care, and health emergency services.[27] As such, it is within the mandate of the NHC to regulate human germline genome modification.

Another relevant institutional change that took place in 2018 is the dissolution of the CFDA.[28] A new agency, the DIA, was created to replace it.[29] It retains some responsibilities of the former CFDA pertaining to the "registration and oversight administration of drugs, cosmetic products, and medical devices" under the supervision of a brand new ministry: the State Administration for Market Regulation (SAMR).[30] In the latest announcement issued by the Communist Party, it specifies the responsibility of the DIA. The DIA is mandated to regulate the licensing, oversight, and sanctions process of the research and development activities pertaining to pharmaceutical products, medical devices, and cosmetic products.[31]

Unlike the MoH and the CFDA, the MoST has remained unchanged over the years, retaining its core mandate.[32] The MoST assists socioeconomic growth by coordinating research on basic and advanced technology and social service technology, together with key and common technology. It is tasked with the elaboration and implementation of the science and technology development plans and policies, including the enactment of related laws and regulations.[33]

The MoST is responsible for the National Basic Research Program,[34] the National High-tech Research and Development (R&D) Program,[35] and the

---

[27] Communist Party of PRC, *Decision on Deepening Reform of Party and State Institutions* (深化党和国家机构改革方案) (March 21, 2018) s 38.

   Communist Party of PRC, *Decision about the Functional Allocation, Internal Organization, and Staffing of the National Health and Wellness Commission of the PRC* (国家卫生健康委员会职能配置、内设机构和人员编制规定) (September 10, 2018) art. 3.

[28] China Food and Drug Administration 中华人民共和国食品药品监督管理局 (*Zhōnghuárénmíngònghéguó Shípǐnyàopǐn Jiāndūguǎnlǐjú*).

[29] Drug Administration 药品监督管理局 (*Yàopin Jiāndu Guanliju*); Communist Party of PRC, *Decision on Deepening Reform of Party and State Institutions* (n 27) s 34.

[30] Ibid.

[31] Communist Party of PRC, *Decision about the functional allocation, internal organization, and staffing of the Drug Inspection Administration of the PRC* (国家药品管理局职能配置、内设机构和人员编制规定) (2018) art. 3.

[32] Ministry of Science and Technology 中华人民共和国科技部 (*Zhōnghuárénmíngònghéguó Kējìbù*).

[33] Ministry of Science and Technology, "Functions of the Ministry of Science and Technology" (General Office of State Council, 2008) www.most.gov.cn/zzjg/kjbzn/ accessed May 7, 2018 (Functions of the Ministry of Science and Technology).

[34] National Basic Research Program 国家重点基础研究计划 (*Guójiā Zhòngdiǎ jīchǔyánjiū Jìhuá*).

[35] National High Technology Research and Development Program 国家高技术研究发展计划 (*Guójiāgāojìshùyánjiufāzhǎn Jìhuá*).

Science and Technology Enabling Program.[36] Given its mission, issues regarding gene engineering, biotechnology, and high-tech research and development of gene technology are under the governance of the MoST, which includes the basic research on genomic modification.

In addition, the MoST has a leading role in advancing the Chinese Precision Medicine Initiative (PMI).[37] The PMI is listed on the State's Thirteenth Five-Year Technology and Innovation Plan,[38] which aims to build a multiple-layered knowledge system and a national biomedical data sharing platform to promote precision medicine.[39] As an indispensable component of personalized medicine, research on human germline genome modification will certainly become a primary focus of the MoST and scientists.

In sum, nowadays the NHC and the MoST are the main regulatory authorities of preclinical research, clinical trials, and clinical applications of gene editing. They (or their predecessors) have been vested with the legislative power to draft regulations to achieve their mandates.[40] The ministries also have the authority to draft a joint administrative rule to perform their missions,[41] as the MoH and the MoST did when they jointly published the *Interim Administrative Measures for Human Genetic Resources* (1998).[42]

The MoST's main task is to outline and regulate the PRC's significant research and development projects, including human germline genome modification research projects. The NHC, in turn, is the ministry with the principal authority to regulate matters regarding public health and the health system in terms of medical technology, clinical research, and (potential) clinical applications of human germline modification. Finally, the newly established DIA has taken over some duties of the former CFDA, such as administrating

---

[36]   Science and Technology Support Plan 科技支撑计划 (*Kējì Zhīchēng Jìhuà*); Functions of the Ministry of Science and Technology (n 34) art. 2.

[37]   Precision Medicine Plan 精准医疗计划 (*Jīngzhǔn Yīliáo Jìhuà*).

[38]   Ministry of Science and Technology, "The Initial Experts Conference of Personalized Medicine" (精准医疗第一次专家会议) (2015) www.most.gov.cn/ accessed May 17, 2018 (The Initial Experts Conference of Personalized Medicine).

[39]   State Council, *The State's 13th Five-Year Technology and Innovation Plan* (十三五"科技创新规划) (2016).

[40]   Legislation Law of the PRC 2015 (立法法), art. 80.

[41]   Ibid., art. 81; As stipulated in Article 81 of the *Legislation Law of the PRC*, these institutions are entitled to jointly formulate rules for matters that are within the scope of authority of two or more of them.

[42]   Interim Administrative Measures for Human Genetic Resources of the PRC 1998 (人类遗传资源管理暂行规定) art. 25.

the drug market and giving technical guidance for human germline genomic modification.

However, the exact scope of its authority still needs to be determined.[43]

b Regulations and Rules

As we saw, the overall regulatory framework in the PRC is made of the Constitution, laws, administrative regulations, administrative rules, and normative instruments. When it comes to human germline genome modification, a number of department regulations and rules drafted by the three main responsible agencies form the basic regulatory framework. They range widely from regulating the basic research on gene technology, clinical applications of stem cell research, to clinical applications of gene therapy.

In Table 17.1, included at the end of this chapter, we have listed all relevant instruments and specified the legal nature of each (i.e. administrative regulation, rule, or normative instrument). It should be noted that none of them specifically addresses human germline genome modification.

The most relevant ones are the following: the *Ethical Guiding Principles for Research on the Human Embryonic Stem Cell*, an administrative regulation issued by the MoST and MoH in 2003;[44] the *Technical Norms on Assisted Reproduction*, an administrative rule issued by the MoH in 2003;[45] and the *Administrative Measures for the Clinical Application of Medical Technology*, an administrative rule issued by the MoH in 2009.[46] The substantive provisions contained in these instruments are discussed below. Moreover, in 2003, the CFDA published the *Guiding Principles for Human Gene Therapy Research and Preparation Quality Control*, in which gene therapy refers to one type of medical treatment based on modifying the cellular genetic material and is currently merely applicable to somatic cells.[47] All these instruments have been issued and enacted by the abovementioned three main functional departments in this field.

Chinese bioethics took a big step forward in 2016 when the HFC drafted and fully implemented the *Ethical Review Guidelines on Biomedical Research*

---

43 The DIA has not yet taken on its full duties so far. Given the notice announced by the SAMR, the official website for the new agency, DIA, is still under construction. Instead, the website of CFDA is still functioning and serving the main duties in this transitional phase.

44 Ethical Guiding Principles for Research on the Human Embryonic Stem Cell of the PRC 2003 (人胚胎干细胞研究伦理指导原则).

45 Technical Norms on Assisted Reproduction of the PRC 2003 (人类辅助生殖技术规范).

46 Administrative Measures for the Clinical Application of Medical Technology of the PRC 2009 (医疗技术临床应用管理办法).

47 Guiding Principles for Human Gene Therapy Research and Quality Control of Preparation (Trial) of the PRC 2003 (干细胞制剂质量控制及临床前研究指导原则(试行)) Introduction, 1.

TABLE 17.1 *People's Republic of China: key regulatory instruments regarding human germline genome modification*

Title	Relevant Provisions (in Chinese)	Relevant Provisions (Authors' Translation Into English)
Decision to Ratify the Covenant on Economic. Social and Cultural Rights (CESCR) (2001, the Standing Committee of the National People's Congress) 全国人民代表大会常务委员会关于批准《经济、社会及文化权利国际公约》的决定 (2001, 2, 28)	N/A	The 9th Standing Committee of the People's Congress authorized the State Council to sign the CESCR in 1997. Four years later, in 2001, the Standing Committee of the People's Congress ratified it. The CESCR is applicable in Hong Kong and Macao. However, based on the *Legislative Law of PRC* (2015), international agreements and conventions are only enforceable if they are implemented through a domestic law, which means rights stipulated in the CESCR can only be enforced if a domestic law has incorporated them.
Administrative Measures for Gene Engineering Safety (1993, MoST) 基因工程安全管理办法 (1993) [Administrative Rule]	第二条: 本办法所称基因工程,包括利用载体系统的重组体DNA技术,以及利用物理或者化学方法把异源DNA直接导入有机体的技术.但不包括下列遗传操作: (一)细胞融合技术 (二)传统杂交繁殖技术(三)诱变技术,体外受精技术,细胞培养或者胚胎培养技术 第三条: 本办法适用于在中华人民共和国境内进行的一切基因工程工作,包括实验研究,中间试验,工业化生产以及遗传工程体释放和遗传工程产品使用等.从国外进口遗传工程体,在中国境内进行基因工程工作的,应当遵守本办法. 第六条 按照潜在危险程度,将基因工程工作分为四个安全等级:	Article 2. Genetic engineering includes the use of recombinant DNA technology of vector systems and the use of physical and chemical methods for the direct introduction of heterologous DNA into organisms. However, the following genetic manipulations are not included: (i) cell fusion technology, (ii) traditional crossbreeding techniques, (iii) mutagenesis techniques, in vitro fertilization techniques, and cell culture and embryo culture techniques. Article 3. The Measures apply to all gene-engineered projects undertaken within the territory of the PRC, including experimental research, intermediate tests, industrial production, and release of gene-engineered body, and utilization of genetic engineering products.

TABLE 17.1 *(continued)*

Title	Relevant Provisions (in Chinese)	Relevant Provisions (Authors' Translation Into English)
	安全等级I, 该类基因工程工作对人类健康和生态环境尚不存在危险. 安全等级II, 该类基因工程工作对人类健康和生态环境具有底度危险. 安全等级III, 该类基因工程工作对人类健康和生态环境具有中度危险. 安全等级IV, 该类基因工程工作对人类健康和生态环境具有高度危险.	In the case where genetic engineering body is imported abroad but genetic engineering work is performed in China, the Measures will apply. Article 6. The genetic engineering project are ranked in four levels of safety depending on the potential risk: Level I This kind of genetic engineering work poses no risk to human health and ecological environment. Level II This kind of genetic engineering work poses low risk to human health and ecological environment. Level III This kind of genetic engineering work poses intermediate risk to human health and ecological environment. Level IV This kind of genetic engineering work poses significant risk to human health and ecological environment.
Interim Administrative Measures for Human Genetic Resources (1998, MoH and MoST) 人类遗传资源管理暂行规定 [Administrative Regulation]	第二条 本办法所称人类遗传资源是指含有人体基因组、基因及其产物的器官、组织、细胞、血液、制备物、重组脱氧核糖核酸（DNA）构建体等遗传材料及相关的信息资料。 第三条 凡从事涉及我国人类遗传资源的采集、收集、研究、开发、买卖、出口、出境等活动, 必须遵守本办法。 第四条 国家对重要遗传家系和特定地区遗传资源实行申报制度。发现和持有重要遗传家系和特定地区遗传资源的单位或个人, 应及时向有关部门报	Article 2. Genetic resources refer to human tissues, cells, blood specimens, preparations and recombinant DNA constructs that contain the human genome, genes, or gene products, and information concerning such material. Article 3. Whoever involved in such activities in China as sampling, collecting, researching, developing, trading, or exporting human genetic resources or taking such resources outside the territory of the People's Republic of China shall abide by the Measures.

Article 4 The state adopts a reporting and registration system on important pedigrees and genetic resources in specified regions. Any institution or individual who discovers or holds important pedigrees and genetic resources in the specified regions shall immediately report to the relevant departments. No institution or individual may sample, collect, trade, and export human genetic resources or take them outside the territory of the People's Republic of China, or provide them to other countries in any form without permission.

Article 5 Where the human genetic resources and the relevant information or data are classified as state scientific or technological secrets, the Rules for the Protection of State Secrets in Science and Technology shall be observed.

Chapter Five: Rewards and Penalties

Article 20 Any institution or individual that discovers and reports important human genetic resources shall be praised and rewarded: whoever exposes illegal activities shall be rewarded and protected.

Article 21 If any Chinese institution or individual, in violation of the provisions stipulated in the Measures, exports the human genetic materials without authorization by hand carrying, mailing, or transporting, the human genetic materials shall be confiscated by the Chinese Customs and the institution or individual shall be punished ranging from administrative sanctions to prosecution by the judicial department according to the seriousness of the circumstances: if anyone, in violation of the

告。未经许可，任何单位和个人不得擅自采集、收集、买卖、出口、出境或以其他形式对外提供。

第五条 人类遗传资源及有关信息、资料，属于国家科学技术秘密的，必须遵守《科学技术保密规定》。

第五章 奖励与处罚

第二十条 对于发现和报告重要遗传家系和资源信息的单位或个人，给予表彰和奖励；对于揭发违法行为的，给予奖励和保护。

第二十一条 我国单位和个人违反本办法的规定，未经批准，私自携带、邮寄、运输人类遗传资源材料出口、出境的，由海关没收其携带、邮寄、运输的人类遗传资源材料，视情节轻重，给予行政处罚直至移送司法机关处理；给予行政处罚并处以罚款；情节严重的，给予行政处罚直至提供人类遗传资源材料的，没收所提供的人类遗传资源材料，并处以罚款；情节严重的，依照我国有关法律追究其法律责任。

第二十二条 国（境）外单位和个人违反本办法的规定，未经批准，私自采集、收集、买卖我国人类遗传资源材料，并处以罚款；情节严重的，依照我国有关法律追究其法律责任。私自携带、邮寄、运输我国人类遗传资源材料出口、出境的，由海关没收其携带、邮寄、运输的人类遗传资源

TABLE 17.1 *(continued)*

Title	Relevant Provisions (in Chinese)	Relevant Provisions (Authors' Translation Into English)
	材料，视情节轻重，给予处罚或移送司法机关处理。 第二十三条 管理部门的工作人员和参与审核的专家负有为申报者保守技术秘密的责任。玩忽职守、徇私舞弊，造成技术秘密泄露或人类遗传资源流失的，视情节给予行政处罚直至追究法律责任。	provisions stipulated in the Measures provides human genetic materials to foreign institutions or individuals without permission, the human genetic materials shall be confiscated and the institution or individual shall be fined; if the circumstances are serious, he shall be investigated for legal responsibility according to the Chinese law. If anyone exports the human genetic resources outside China by hand carrying, mailing, or transporting without authorization, the human genetic materials shall be confiscated by the Chinese Customs and he shall be punished or put under the prosecution of the judicial department according to the seriousness of the circumstances. Article 23 Any staff member of the administrative department or expert engaging in the examination shall have the duty to keep technological secret for the applicants. Whoever causes the exposure of technological secrets or loss of the human genetic resources of China due to negligence in his duty or malpractice for personal gains shall be imposed with a punishment ranging from administrative sanctions to being investigated for legal responsibility.
Guiding Principles for Human Gene Therapy Research and Quality Control of Preparation (2003, CFDA)	一 引言 基因治疗是指改变细胞遗传物质为基础的医学治疗。目前仅限于体细胞。 二，研究内容和制品质量控制。 （八）伦理学考虑	Introduction: Gene therapy refers to medical treatment based on changes in cellular genetic material. Currently, it is only limited to somatic gene therapy. Chapter 2 (8):Ethical Considerations.

人基因治疗研究和制剂质量控制技术指导原则 (2003) [Normative Instrument]

必须充分重视伦理学的原则，并具体按国家药品监督管理局GCP规定的要求严格实施。包括在实施本方案前，须向病人说明该治疗方案属试验阶段，它可能的有效性及可能发生的风险，同时保证病人有权选择该方案治疗或中止该方案治疗，以及保证一旦中止治疗能得到其它治疗的权利。严格保护病人及家属的隐私。在病人及家属充分理解并签字后才能开始治疗。

Ethical Guiding Principles for Research on the Human Embryonic Stem Cells (2003, Former MoH and MoST)

人胚胎干细胞研究伦理指导原则 (2003) [Administrative Rule]

第二条 本指导原则所称的人胚胎干细胞包括人胚胎来源的干细胞、生殖细胞起源的干细胞和通过移植核获得的干细胞。

第四 禁止进行生殖性克隆的任何行动

第五 用于研究的人胚胎干细胞只能通过下列方式获得：

(一) 体外受精时多余的配子或胎儿细胞
(二) 自然或自愿选择流产的胎儿细胞
(三) 体细胞核移植技术本部所获得的囊胚和单性分裂囊胚
(四) 自愿捐献的生殖细胞

第六条 进行人胚胎干细胞研究必须遵守以下行为规范

Researchers must pay full attention to the principles of ethics and implement them in strict accordance with the requirements of the Good Clinical Practice (GCP) provided by the State Drug Administration. The patient must be informed that the treatment is in its trial phase and of its possible effectiveness and possible risks. Researchers must also ensure that the patient has the right to choose or stop the treatment program and that he or she can still retain other treatment rights when the treatment is discontinued. Researchers must strictly protect the privacy of patients. The treatment can only start after the patient and family members have fully understood and given consent.

Article 2. The human embryonic stem cell shall include stem cells originated from the human embryo, from germ cells, and stem cells obtained from the transplant of nucleus.

Article 4. Any research on reproductive cloning is prohibited.

Article 5. The human embryonic stem cell used for research may only be obtained through the following methods:

1. The unwanted gametes or blastula from in vitro fertilization;
2. The fetal cells from natural abortion or voluntary abortion;

TABLE 17.1 (continued)

Title	Relevant Provisions (in Chinese)	Relevant Provisions (Authors' Translation Into English)
		3. The blastula and asexual split blastula obtained through the technology of body nucleolus transplant; or 4. The germ cell donated on one's own initiative.
	（一）利用体外受精，体细胞核移植，单性复制技术或遗传修饰获得的囊胚，其体外培养期限自受精或移植开始不得超过14天。 （二）不得将前款中获得的已用于研究的人囊胚植入人或任何其他动物的生殖系统 （三）不得将人的生殖细胞与其他物种的生殖细胞结合。	Article 6. The following criteria must be observed for performing research on the human embryonic stem cell: 1. For the blastulas obtained from in vitro fertilization, transplant of nucleolus body, asexual reproduction technology, or genetic modification, the time for their in vitro breeding shall not exceed fourteen days from the date of fertilization or nucleus transplant. 2. No human blastula obtained as mentioned in the preceding paragraph that has been used for research purposes shall be transplanted into the genital system of human or any other creatures. 3. No human germ cell may be combined with the germ cell of other species.
Management Ordinance for the Quality of Drug Clinical Trial (2003, CFDA) 药物临床试验质量管理规范 (2003) [Administrative Rule]	第三条 药物临床试验必须符合《世界医学大会赫尔辛基宣言》原则，受试者的权益和安全是临床试验考虑的首要因素，并高于对科学和社会获益的考虑。伦理委员会与知情同意书是保障受试者权益的主要措施。	Article 3. Clinical trials of drugs must meet the principles of the World Medical Association Declaration of Helsinki. The subjects' rights and safety are the primary considerations in clinical trials and are higher than the considerations of scientific and social benefits. The ethics committee and informed consent form are the main measures to protect the subjects' rights.
Technical Norms on Human Assisted Reproduction (2003, MoH) 人类辅助生殖技术规范 (2003) [Administrative Rule]	三、实施技术人员的行为准则 （九）禁止以生殖为目的对人类配子、合子和胚胎进行基因操作； （十五）禁止克隆人。	Chapter 3: Code of Conduct (9) Genetic manipulation of human gametes, zygotes, and embryos for reproduction purposes is prohibited; (15) Human cloning is prohibited.

Document	Chinese	English
Administrative Measures for the Clinical Application of Medical Technology (2009, MoH) 医疗技术临床应用管理办法 (2009) [Administrative Rule]	异种干细胞治疗技术、异种基因治疗技术、人类体细胞克隆技术等医疗技术暂不得应用于临床。 第三类医疗技术目录 一、涉及重大伦理问题、安全性、有效性尚需经规范的临床试验研究进一步验证的医疗技术：克隆治疗技术、自体干细胞和免疫细胞治疗技术、基因治疗技术、中枢神经系统手术戒毒、立体定向手术治疗精神病技术、异基因干细胞移植技术、瘤苗治疗技术等。	Medical technologies such as xenogeneic stem cell therapy, heterologous gene therapy, and human somatic cell cloning are temporarily not allowed to be used clinically. Catalogue for the third level of medical technology First, medical technology that involves major ethical issues or the safety and efficacy need to be further validated by standardized clinical trials: technology such as cloning therapy, autologous stem cell and immune cell therapy, gene therapy, central nervous system surgery, detoxification, targeted surgery for psychotic techniques, allogeneic stem cell transplantation techniques, tumor vaccine treatment, and so on.
Guiding Principles for Quality Control and Preclinical Research of Stem Cell-Based Medical Products Formulation and (Trial) (for public deliberation) (CFDA, 2013) 《干细胞制剂质量控制及临床前研究指导原则（试行）》(2013) [Normative Instrument]	名词解释：胚胎干细胞（Embryonic stem cell）：源自第5~7天的胚胎中内细胞团的初始（未分化）细胞，可在体外非分化状态下"无限制地"自我更新，并且具有向三个胚层所有细胞分化的潜力，但不具有形成胚外组织（如胎盘）的能力。 胚胎干细胞系（Embryonic stem cell line）：在体外培养的条件下，可保持未分化状态连续增殖的胚胎干细胞。	Definitions: Embryonic stem cells: Initial (undifferentiated) cells of the inner cell mass in embryos from Days 5–7, which can be "unrestrictedly" self-renewed in an undifferentiated state in vitro and have all three germ layers that have the potential for cell differentiation but do not have the ability to form extra-embryonic tissues (such as the placenta). Embryonic stem cell line: embryonic stem cell that can maintain continuous growth in an undifferentiated state under in vitro culture conditions.
Ethical Review Guidelines for Biomedical Research Involving Humans (2016, HFC) 涉及人的生物医学研究伦理审查办法 (2016)	第二条 本办法适用于各级各类医疗卫生机构开展涉及人的生物医学研究伦理审查工作。	Article 2. These guidelines apply to the ethical review of biomedical research involving human subjects conducted in the medical institutions of all tiers.

TABLE 17.1 (continued)

Title	Relevant Provisions (in Chinese)	Relevant Provisions (Authors' Translation Into English)
[Administrive Rule] Administrative Safety Measures for Biotechnology Research and Development (2017, MoST) 生物技术研究开发安全管理办法 (2017) [Administrative Rule]	生物技术研究开发活动风险分级 高风险等级: 7. 涉及存在重大风险的人类基因编辑等基因工程的研究开发活动 较高风险等级: 4. 涉及存在较大风险的人类基因编辑等基因工程的研究开发活动 一般风险等级: 4 涉及存在一般风险的人类基因编辑等基因工程的研究开发活动	Risk ranking for biotechnology research and development activity Significant risk … 7. Research and development activities involving genetic engineering, such as gene editing, with significant risk. Substantial risk: … 4. Research and development activities involving genetic engineering, such as human gene editing, with substantial risk. General Risk: …4. Research and development activities involving genetic engineering, such as human genetic editing, with general risk.

Involving Human Subjects.[48] These *Guidelines* outline the ethical principles applicable in China by all scientific institutions and medical centers conducting biomedical research that involves human subjects. They incorporate the main international standards for biomedical research on human subjects, such as the World Health Organization's *International Ethical Guidelines for Biomedical Research Involving Human Subjects* and the World Medical Association's *Declaration of Helsinki.*[49] Despite the status of *Guidelines* as administrative rules, they contain criminal and civil sanctions (e.g. fines) that can be imposed at the discretion of the issuing Department.

These *Guidelines* contain six main ethical principles governing biomedical research on human subjects: informed consent; balance of benefits and risks; free participation; privacy protections; and indemnity for the necessary costs as well as compensation for damages.[50] They require medical institutions of all tiers engaging in biomedical research to establish independent institutional ethical review boards (IRBs).[51] That means that research on human germline genome modification is also subject to oversight by an institutional review board.[52] IRBs are mandated to undertake ethical review of research projects including the initial and follow-up review in order to protect the legitimate rights of the subjects, ensure their dignity as well as promote biomedical research.[53]

The scope of the *Guidelines* is not limited to research conducted with human embryos or with human embryonic stem cells. There are no specialized oversight bodies similar to the Embryonic Stem Cell Research Oversight Committee (ESCRO) in the United States. In China, research with human embryos or with embryonic stem cell falls under the general supervision of an IRB as with any other area of biomedical research.[54] Chapter 6 of the *Guidelines* sets up legal responsibilities and sanctions for the health institutions and their staffs. For example, a penalty of 30,000 yuan (about US$5,000) is levied for the medical institution which conducts biomedical research

---

[48] Ethical Review Guidelines on Biomedical Research Involving Human Subjects of the PRC 2016 (涉及人的生物医学研究伦理审查办法).

[49] Health and Family Planning Committee, *Interpretation of the Ethical Review Guidelines for Biomedical Research involving Human Subjects* (关于《涉及人的生物医学研究伦理审查办法 的解读) (2016), s 2 (*Ethical Review Guidelines for Biomedical Research involving Human Subjects*).

[50] Ethical Review Guidelines on Biomedical Research Involving Human Subjects (n 50) art. 18.

[51] Ethics Review Committee伦理审查委员会 (*Lúnlǐ Shěnchá Wěiyuánhuì*); Ethical Review Guidelines on Biomedical Research Involving Human Subjects (n 50) art. 7.

[52] Ibid., art. 3.

[53] Ibid., art. 8.

[54] Ibid., art. 9.

without constituting an ethical review board in their institutions.[55] Civil or criminal liabilities are triggered when illegal behavior causes severe bodily injury or financial loss.[56]

Except for the sanctions stipulated in the *Guidelines*, sanctions rarely appear in regulatory instruments. This raises the issue of their actual implementation.

3 Funding

Although so far there are no specific funding mechanisms for research or clinical application of human germline genome modification, there are a number of funding initiatives that could be tapped for research of this kind.

In 2016, Prime Minister Li Keqiang announced the State Council's "Thirteenth Five-Year Plan" agenda, in which the Precision Medicine Initiative (PMI) occupies a central position.[57] Under the plan, China will invest a total of 60 billion yuan (more than US$9 billion) by 2030 for research on precision medicine.[58] Although nearly forty countries have their own version of a precision medicine initiative, China's is by far the largest.[59] For instance, the United States Precision Medicine Initiative started at $215 million in 2016.[60]

As part of the initiative on precision medicine, in 2018, the MoST released a plan called the *Major Research Projects on Reproductive Health and Prevention and Control of Major Birth Defects of 2018*, with approximately 90 million yuan funding (about US$13 million), mostly aimed at mitochondrial gene editing technology.[61] Section 2.2 of the plan specifies the development of innovative assisted reproductive technology targeting mitochondrial genetic diseases, which some authors suggest is a form of germline genetic

---

55 Ibid., art. 45.
56 Ibid., art. 49.
57 The Initial Experts Conference of Personalized Medicine (n 39).
58 G Dana, "3 ways China is leading the precision medicine" (World Economic Forum, November 2, 2017) www.weforum.org/agenda/2017/11/3-ways-china-is-leading-the-way-in-precision-medicine/ accessed August 6, 2018; P Liu, "China Initiative Would Pour Billions into Precision Medicine" (BioWorld) www.bioworld.com/content/china-initiative-would-pour-billions-precision-medicine-0 accessed July 31, 2018.
59 Ibid.
60 Ibid.
61 Ministry of Science and Technology, *Major Research Projects on Reproductive Health and Prevention and Control of Major Birth Defects of 2018 plan* (生殖健康及重大出生缺陷防控研究重点专项2018年度项目申报指南) s 2.2.

modification.[62] It further explains that the primary purpose is to build safe and efficient mitochondrial genetic editing practices and improve the gene editing of stem cells with antiaging, anti-apoptosis, and anti-tumorigenicity objectives. The research should clarify the efficacy of the technology, enhancing safety of treatments, performing mitochondrial gene editing of germ cells, and further elucidating the feasibility of treating mitochondrial genetic disease through gene editing.[63]

Moreover, in the same year, MoST announced the *National Major Research & Development Agenda,* in which stem cell and its translational research is an important part of the plan.[64] According to the official guidelines on project applications, twenty projects will be funded by a budget of around 630 million yuan (about US$100 million).[65] One of the supported topics is preclinical evaluation of stem cells using animal models. The funded projects will be evaluated based on their capability to successfully gene edit or modify (such as gene deletion or knock-in) human pluripotent stem cell lines.[66] These large amounts of research funding in gene editing suggest that the government is taking active measures to promote gene-editing research.

Finally, another source of funding supporting biomedical research in China is the National Natural Science Foundation.[67] During 2017, the National Natural Science Foundation funded forty-eight projects, with amounts ranging from US$40,000 to US$100,000 to exploit CRISPR/Case 9 in the area of medical science.[68] Most of the projects are basic research using animal models. However, there are still some projects on the treatment of human genetic diseases, such as using CRISPR/Cas 9 to correct the gene mutation of maternal hereditary deafness. According to the released funding

---

[62] AJ Newson and A Wrigley, "Is Mitochondrial Donation Germ-Line Gene Therapy? Classifications and Ethical Implications" (2017) 31:1 *Bioethics* 55.

[63] Major Research Projects on Reproductive Health and Prevention and Control of Major Birth Defects 2018 plan (n 62) s 2.2.

[64] Ministry of Science and Technology, "National Major Research & Development Agenda (国家重大研发计划)" (National Science and Technology Information System Public Service Platform) http://service.most.gov.cn/sbylb2018zy/index_2.html accessed May 17, 2018.

[65] Ministry of Science and Technology, *Project Application Guidelines for the Stem Cell and its Translational Study* (2017) http://service.most.gov.cn/u/cms/static/201710/16145441jur5.pdf accessed August 16, 2018.

[66] Ministry of Science and Technology, "National Major Research & Development Agenda" (n 65).

[67] National Natural Science Foundation 国家自然科学基金 (*Guójiāzìránkēxué Jījīn*).

[68] National Natural Science Foundation (国家自然科学基金网络信息系统), "Project Comprehensive Inquiry" (Internet-based Science Information System) https://isisn.nsfc.gov.cn/egrantindex/funcindex/prjsearch-list accessed July 31, 2018.

guideline, there is no prohibition on funding research on human germline genome modification.

III SUBSTANTIVE PROVISIONS

1 *Definitions*

A remarkable feature of the Chinese regulatory instruments pertaining to human germline modification is the absence of clear definitions of key terms. For instance, there is no precise definition – at least not as precise as in other countries – of what constitutes a *human embryo, zygote,* or *gamete*. Determining the legal status of these three is essential to grasp the full breadth and scope of existing policy.[69] For now, they are simply inferred from the definitions provided in policy governing distinct technologies.[70]

2 *Basic Research*

In the PRC, basic research on human embryos and germ cells, including their genetic modification, is permitted and regulated by a series of ministerial regulations and rules. These instruments have established the baseline requirements for procuring embryonic cells or gametes, conducting human genome editing, and preventing gene modification of gametes, zygote, or embryo for reproductive purposes.

The only legitimate sources of human embryonic cells or gametes in China are spare gametes or embryonic cells from in vitro fertilization, embryonic cells from voluntary or natural abortion, blastulas and cleavage of single blastulas from the nucleus transfer of the somatic cell, and donated productive cell.

The fourteen-day rule is recognized by these instruments indicating that in vitro culture period of the blastulas must not exceed fourteen days.[71] Reproductive cloning is also strictly forbidden.[72]

---

[69] RM Isasi and BM Knoppers, "Mind the Gap: Policy Approaches to Embryonic Stem Cell and Cloning Research in 50 Countries" (2006) 13 *European Journal of Health Law*, 9.

[70] Guiding Principles for Quality Control and Preclinical Research of the Stem Cell-based Medical Products Formulation (for the public deliberation) of the PRC 2013 (干细胞制剂质量控制及临床前研究指导原则（试行）) Section of Terminology, 11.

[71] Ethical Guiding Principles for Research on the Human Embryonic Stem Cell of the PRC 2003 (n 45) art. 2.

[72] Technical Norms on Assisted Reproduction of the PRC 2003 (n 46) art. 3.

a 1993 Administrative Measures for Gene Engineering
Safety (MoST)

The *Administrative Measures for Gene Engineering Safety*, an administrative rule issued by the Ministry of Science and Technology, regulates genetic engineering technologies at large. It applies to all genetic engineering activities including laboratory research, intermediate experiments, industrialized production, release of genetic engineering body, and usage of genetic engineered products.[73] This might suggest that the *Measures* apply widely to plants, animals, and probably humans involving genetic engineering activities. It defines genetic engineering activities as interventions that "use recombinant DNA technology of vector systems and the use of physical or chemical methods for the direct introduction of heterologous DNA into organisms."[74] However, genetic manipulations of cell fusion technology, traditional crossbreeding, mutagenesis, and fertilization techniques together with in vitro, cell culture, and embryo culture techniques are excluded from the *Measures*.[75] For these areas, the *Measures* do not indicate which agency is responsible or what regulation is applicable.

The *Measures* rank gene engineering projects on four levels based on the risk they pose to human health and the environment and stipulate different requirements for the projects' application and review process and safety control measures accordingly.[76] The *Measures* require any institution that plans to undertake gene engineering research to proceed with the application and approval process.[77]

Given that the *Measures* were adopted more than twenty years ago, it is not surprising that they do not explicitly mention human germline genome modification. Although their application to human gene editing remains unclear, human gene editing may be understood as one category of gene engineering, making the *Measures* difficult to ignore when discussing human germline genome modification. After all, the *Measures* continue to be China's official policy guiding gene engineering projects.

---

[73] Administrative Measures for Gene Engineering Safety of the PRC 1993 (基因工程管理办法), art. 3.

[74] Ibid., art. 2.

[75] Ibid., art. 2.

[76] Ibid., art. 6.

[77] Ibid., art. 13.

490 *Lingqiao Song and Rosario Isasi*

b 1998 Interim Administrative Measures for Human Genetic Resources (MoST and MoH)

Second, in 1998, the MoST and the MoH released the overarching administrative regulation in the field of human genetic resources: the *Interim Administrative Measures for Human Genetic Resources*.[78] The *Interim Measures* impose the requirement of an administrative license for the collection, storage, and export of human genetic materials.[79] "Human genetic materials" are defined broadly to include human tissues, cells, blood specimens, preparations or recombinant DNA constructs that contain the human genome, genes, or gene products, and information concerning such material, including genomic data.[80] The licensing agency is Human Genetic Resources Management Office (HGRMO).[81] Notably, a license by the HGRMO is required not just for projects conducted in China but also for projects done abroad that involve Chinese human genetic samples and data derived from them.[82] Chapter 5 of the *Interim Administrative Measures* imposes administrative and criminal responsibilities for the unauthorized export, shipment, and trading of human genetic resources.[83] Even though the *Interim Administrative Measures* were initially intended to be temporary, they remain, to date, the only effective rule regulating the Chinese human genetic resources. In the future, the *Ordinance of Human Genetic Resources* may replace the *Interim Administrative Measures*.[84]

c 2003 Guiding Principles for Human Gene Therapy Research and Preparation Quality Control (CFDA)

The CFDA 2003 *Guiding Principles for Human Gene Therapy Research and Preparation Quality Control* complete the regulatory framework.[85] Even

[78] Interim Administrative Measures for Human Genetic Resources of the PRC 1998 (n 43); Another MoST regulation, the *Human Genetic Resources Ordinance*, drafted to meet the challenges of the ever-changing field of genomics, has been stuck at the public deliberation stage since 2015, raising doubts about its possible implementation in the future; Human Genetic Resources Ordinance of PRC 2012 (人类遗传资源管理条例).
[79] Interim Administrative Measures for Human Genetic Resources of the PRC 1998 (n 43) art. 3.
[80] Ibid., art. 2.
[81] Human Genetic Resources Management Office 人类遗传资源管理办公室 (*Rénlèi yíchuán zīyuán guǎnlǐ Bàngōngshì*); Interim Administrative Measures for Human Genetic Resources of the PRC 1998 (n 43) art. 7.
[82] Interim Administrative Measures for Human Genetic Resources of the PRC 1998 (n 43) art. 3.
[83] Ibid., c 5.
[84] Ordinance of Human Genetic Resources of the PRC 2012 (Drafted) (人类遗传资源管理条例).
[85] Guiding Principles for Human Gene Therapy Research and Preparation Quality Control of the PRC 2003 (人类基因治疗研究和制剂质量控制技术指导原则).

though they are department-level normative instruments, the *Guiding Principles* are strictly enforced by the CFDA. Their overall purpose is promoting gene therapy research and reinforcing support for innovative research. According to the *Guiding Principles*, China allows human gene therapy only for somatic cells, not germline cells.[86] The *Guiding Principles* include measures directed at ensuring the safety and effectiveness of gene therapy, anticipating future risks, and taking corresponding measures.[87] They further contain specific requirements for research projects and quality control of the DNA recombinant or construction of gene introduction system, the establishment of cell bank and microbe banks, the preparation of gene therapy products, and the production process.[88] Special conditions are required for human gene therapy clinical trials (for somatic cells) including a Good Manufacturing Practices (GMPs) certification from the clinical institution conducting the trial, and informed consent provisions.[89] Even though they outline the general principles to be followed for human gene therapy, they still leave room for further regulations to identify the more specific measures needed to ensure their implementation.

The *Guiding Principles* distinguish between two modes of gene therapy: ex vivo and in vivo. In ex vivo therapy, genes are artificially introduced into a human cell, and then the cell is injected into the human body thereby producing a transformed cell with an exogenous gene. Under the *Guiding Principles*, ex vivo gene therapy can only be performed in medical institutions that are GMP compliant and with specialized technical personnel. In vivo gene therapy involves introducing a gene directly in the human being, through an introduction system, whether using viral vectors or not. Its product is the gene-modified virus and the recombinant or composite DNA. Different requirements are established for in vivo and ex vivo gene therapy in terms of the preparation and production process of therapy products, their quality control, validity tests of the gene therapy, and safety tests of the therapy.[90]

Finally, the *Guiding Principles* outline the ethical requirements for gene therapy research. These include informed consent, voluntary participation, withdrawal of consent, and privacy protection.[91] A notable feature that sets China's approach on these matters apart from other countries is that not only must consent be obtained from the patient, but also that his/her family

---

[86]   Ibid., Preface.
[87]   Ibid.
[88]   Ibid., 2–7.
[89]   Ibid., 3.
[90]   Ibid., 3–10.
[91]   Ibid., 7.

members should fully understand the research before accepting any type of treatment.[92] The *Guiding Principles* do not explain whether family consent must be obtained in addition to individual consent and what happens when it is not given, or whether families must simply be informed. The *Guiding Principles* do not explain either the process for obtaining such consent,[93] nor do they mention sanctions for violation of the family consent requirement. However, lack of family consent will most likely cause the denial of the administrative permit for exploiting Chinese human genetic resources as stipulated in the *Interim Administrative Measures*.[94] Rooted in the paternalism and collectivism tradition embedded in the Chinese society,[95] family consent is not foreign to Chinese bioethics. In clinical practice, close family members are often involved in the process and participate in the consent of the patient.

d 2003 Ethical Guiding Principles for Research on the Human Embryonic Stem Cells (MoH and MoST)

In 2003, the MoH and the MoST adopted jointly the administrative rules called *Ethical Guiding Principles for Research on Human Embryonic Stem Cells*. They regulate any research activities involving human embryonic stem cell within the territory of the PRC. No sanctions are mentioned in the text, which raises doubts about their actual enforceability.

The *Principles* prohibit research on reproductive cloning.[96] Research on human embryonic stem cells is regulated. "Human embryonic stem cells" are defined as stem cells originated from the human embryo, from germ cells, and stem cells obtained from the transplant of nucleus.[97] The ad hoc creation of human embryonic stem cells for research is prohibited. They can only be sourced from germ cells that are donated, surplus embryos from in vitro fertilization, fetal cells from natural abortion or voluntary abortion, and the "blastula and asexual split blastula obtained through the technology of body nucleolus transplant."[98]

Three prohibitions limit research using human embryonic stem cells:

---

92   Ibid.
93   Ibid., c 8.
94   Interim Administrative Measures for Human Genetic Resources of the PRC 1998 (n 43) art. 13.
95   H Chi and others, "Ethical Reflection on Informed Consent of Close Relatives in the Research and Application of Assisted Reproduction Techniques (ART) (人类辅助生殖技术实施中关于患者近亲属知情权的伦理思考)" (2010), 23:5 Chinese Medical Ethics 20, 20–21.
96   Ethical Guiding Principles for Research on the Human Embryonic Stem Cells of the PRC 2003 (n 45) art. 4.
97   Ibid., art. 2.
98   Ibid., art. 5.1–4.

"1. For the blastulas obtained from in vitro fertilization, transplant of nucleolus body, asexual reproduction technology or genetic modification, the time for their in vitro breeding shall not exceed 14 days from the date of fertilization or nucleus transplant.

2. No human blastula obtained as mentioned in the preceding paragraph that has been used for research purposes shall be transplanted into the genital system of human or any other creatures.

3. No human germ cell may be combined with the germ cell of other species."[99]

These three limitations are consistent with those imposed on research on human embryos and germ cells in all nations surveyed in this volume.

e 2017 Administrative Safety Measures for Biotechnology Research and Development (MoST)

In 2017, the MoST issued an administrative rule called the *Administrative Safety Measures for Biotechnology Research and Development*. Its main purpose is to strengthen biosafety in biotechnology research and development activities.[100] In response to rising concerns about biosafety, the *Safety Measures* include a supplementary document that classifies biotechnology research and development activities into three levels of risk: significant risk, substantial risk, and general risk.[101] These three levels are classified on the basis of the potential risk posed to humans and animals.[102] The *Safety Measures* authorize the responsible departments of the central government or the provincial counterparts to impose sanctions for individuals and legal entities who breach the safety requirements and cause biosafety accidents.[103]

Interestingly, human gene editing is identified as an example in all levels of risks without further explanation, making it difficult to understand what level of risk has been assigned to it.[104] Moreover, the *Safety Measures* fail to differentiate the level of risk assigned to gene editing on somatic and on

---

[99] Ibid., art. 6.

[100] Administrative Safety Measures for the Biotechnology Research and Development of the PRC 2017 (生物技术研究开发安全管理办法), art. 1.

[101] Ministry of Science and Technology, "One Chart to Understand the Administrative Security Measures for the Biotechnology Research and Development (一张图看懂生物技术研究开发管理办法)" (2017) www.most.gov.cn/kjbgz/201708/P020170803514013436923.pdf accessed August 8, 2018.

[102] Administrative Safety Measures for the Biotechnology Research and Development (n 101) art. 4.

[103] Ibid., art. 11.

[104] Ibid., Annex.

germline cells. The uncertainty and opaqueness of the document creates significant hurdles for compliance and implementation. Fortunately, the *Safety Measures* created an expert committee, called the National Biotechnology Research and Development Security Management Experts Committee, charged to propose more specific risk lists regarding research and development biotechnology activities.[105] The Committee was set up in 2017, and it remains to be seen what role it will play in further defining and delineating the risks of human gene editing that are associated with various types of applications.

3 *Clinical Research and Applications*

When it comes to human germline genome modification, the PRC distinguishes between basic research on the one hand, and clinical research and clinical applications on the other hand. Preclinical research is not regulated as such, although there are rules regulating research on animals, which we will not discuss here for sake of brevity.

The *Guiding Principles for Human Gene Therapy Research and Quality Control of Preparation*, a normative instrument released by the CFDA in 2003, allow only genetic therapy on somatic cells, not on germline cells.[106] However, it is unclear whether the *Guiding Principles* allow gene therapy on human embryos.

The MoH administrative measures for *Clinical Applications Technical Norms on Assisted Reproduction (2003)* comprise three parts. They establish the baseline requirements respectively for in vitro fertilization and embryo transfer (Part I), artificial insemination (Part II), and the *Code of Conduct for All Staff Conducting Assisted Reproductive Technology* (Part III), ethical including issues such as informed consent and privacy protections.[107] The *Technical Norms* prohibit a wide range of activities, including human oocyte cytoplasmic or nuclear transfer for the purpose of treating infertility.[108] They further prohibit the hybridization between humans and heterologous gametes, transplantation of heterologous gametes, zygotes and embryos in humans, and transplantation of human gametes, zygotes, and embryos in xenogeneic bodies (that is to say in nonhuman species).[109] Human clone and human chimeric embryo experiments are also

---

[105]   Ibid., art. 6.
[106]   Guiding Principles for Human Gene Therapy Research and Quality Control of Preparation of the PRC 2003 (n 48).
[107]   Technical Norms on Assisted Reproduction of the PRC 2003 (n 46) pts I, II, and III.
[108]   Ibid., pt III, para 7.
[109]   Ibid., pt III, para 8.

explicitly forbidden. Crucially, these administrative measures explicitly prohibit "gene manipulation of human gametes, zygotes or embryos for reproductive purposes," thereby outright banning germline gene editing in clinical applications.[110] However, it is unknown whether germline gene editing research or gene editing for nonreproductive purposes is allowed according to this provision.[111]

In 2009, the MoH published an administrative rule entitled the *Administrative Measures for the Clinical Application of Medical Technology.*[112] This rule classifies gene therapy as a "third-level medical technology," one that carries significant ethical concerns. Unlike the abovementioned risk levels established by the *Safety Measures*, which measure the risk posed to humans and the environment, the *Administrative Measures* focus on the ethical concerns, reliability, and safety of medical technologies. Third-level technologies require stringent clinical trials to validate their safety and efficacy.

For now, but without specifying until when, the *Administrative Measures* ban xenogeneic gene therapy, xenogeneic stem cell therapy, and human somatic cell cloning.[113] Although the *Administrative Measures* do not give reasons why xenogeneic gene therapy is prohibited, the presupposition is that this technology carries significant ethical concerns and that there are considerable doubts about its clinical utility. The *Administrative Measures* further stipulate sanctions for health institutions and their staffs that engage in medical practices that are outside the authorized scope.[114]

Moreover, in 2013, the State Council issued the *Decision on Canceling Non-Administrative Licensing Approval Items*, an administrative rule eliminating the need to obtain an administrative license from the local health administration for the clinical application of the third-level medical technologies, including heterologous gene therapy.[115] This shifted responsibility for the clinical application of a medical technology from the local health administration to the individual medical institutions, which now bear full responsibility of overseeing compliance with legal and ethical

---

[110]   Ibid., pt III, para 9.
[111]   L Jiang, "Human gene editing is calling for a governing law (基因编辑应当有法可依)" (Chinese Social Sciences Net, February 2018) www.cssn.cn/fx/201802/t20180207_3842890.shtml accessed September 11, 2018.
[112]   Administrative Measures for the Clinical Application of Medical Technology of the PRC 2009 (医疗技术临床应用管理办法).
[113]   Ibid., Annex, para 3.
[114]   Ibid., Sanctions.
[115]   State Council, *Decision on Canceling Non-Administrative License* (国家卫生计生委关于取消第三类医疗技术临床应用准入审批有关工作的通知) (2015), available at www.gov.cn/zhengce/content/2015-05/14/content_9749.htm, accessed May 17, 2018.

requirements.[116] This policy move suggests that the PRC is lessening its centralized control of gene therapy, thus leaving more discretion to the local health institutions.

Finally, the surprising announcement of the birth in China of twin girls whose CCR5 genes had been edited by Dr. He Jiankui, a professor at the Southern University of Science and Technology,[117] was swiftly followed by strong condemnation from a wide range of Chinese institutions, including from Dr. He's own university[118] as well as from Chinese regulatory and funding agencies,[119] and professional societies.[120] All these statements converged on a strong denunciation of Dr. He's experiment as an intervention contrary to the law, regulation, and medical ethics of China. They further reiterated that it was contrary to "internationally accepted ethical principles regulating human experimentation and human rights law," which China's regulatory framework endorses.[121] Mirroring policy approaches adopted around the world,[122] the Academic Division of the Chinese Academy of Sciences concluded that "under current circumstances, gene editing in human embryos still involves various unresolved technical issues, might lead to unforeseen risks, and violates the consensus of the international scientific community."[123] It further opposed carrying out human gene editing for reproductive or clinical purposes.

[116] Health and Family Planning Commission, "Notice on Canceling the License for the Third Type of Medical Technology" (2015), (卫生与计划生育委员会关于取消第三类医疗技术行政许可的通知).

[117] D Normile, "Shock greets claim of CRISPR-edited babies" (2018) 362:6418 *Science* 978, 978–979.

[118] Southern University of Science and Technology, "Statement on the Gene Editing of Human Embryos Conducted by Dr. Jiankui He" (November 26, 2018) http://sustc.edu.cn/en/news_e vents_/2871 accessed December 11, 2018.

[119] Academic Division of the Chinese Academy of Sciences (CASAD), "Statement About CCR5 Gene-edited Babies by the Enforcement of Scientific Ethics Committee" (CASAD, November 27, 2018) http://english.casad.cas.cn/bb/201811/t20181130_201704.html accessed December 11, 2018 (Statement About CCR5 Gene-edited Babies by the Enforcement of Scientific Ethics Committee).

[120] Chinese Society for Cell Biology (CSCB), "Official Statement from CSCB & GSC Condemning the reproductive application of gene editing on human germline" (CSCB, November 27, 2018) www.cscb.org.cn/news/20181127/2988.html accessed December 11, 2018.

[121] Committee of Genome Editing, Genetics Society of China and Chinese Society for Stem Cell Research, Statement "Condemning the Reproductive Application of Gene Editing on Human Germline," November 27, 2018. www.cscb.org.cn/news/20181127/2988.html, accessed December 11, 2018.

[122] KE Ormond and others, "Human Germline Genome Editing" (2017) 101:2 *Am J Hum Genet* 167, 167–176.

[123] Statement About CCR5 Gene-edited Babies by the Enforcement of Scientific Ethics Committee (n 120).

IV CURRENT PERSPECTIVES AND FUTURE POSSIBILITIES

As in many countries around the world, the Chinese regulatory and governance framework of human germline genome modification is fragmented and mostly outdated.[124] Frequent institutional reforms have paved a rugged way for building a coherent regulatory framework. Numerous policies have been adopted in a collaborative effort by the MoH, MoST, and the CFDA, particularly regarding genetic engineering.[125] However, the division of powers between the three agencies with respect to the regulation of human germline genome modification remains unclear. What is more, the current framework still relies on a heterogeneous set of departmental regulatory instruments, which are low in China's legal hierarchy.

Dr. He's case represents an illustrative example of the insufficiency of existing oversight mechanisms and the lack of accountability not only in China, but also around the globe, which is caused in part by loopholes in respective legal frameworks.[126] Indeed, in China, obscurity in the breadth and scope of normative instruments, paired with blurred jurisdictional boundaries between governmental actors, have created what seems to be an unstable regulatory environment where accountability is uncertain, with chilling effects on research. As Dr. Zhou Qi, a leading scientist at the Chinese Academy of Sciences, eloquently summarized:

> As a key gene technology, research and development and application of gene editing is still in disorder. Problems such as weak ethics oversight, inadequate supervision and management mechanisms, and groundless regulatory frameworks are urgently needed to be resolved.[127]

Shortly after Dr. He's announcement, the Chinese government moved to immediately shut down his laboratory and halted all of his activities. It further started conducting a throughout investigation of his practices promising stiff sanctions for all individuals and institutions involved in Dr. He's unethical and illegal experiment. In addition, the Chinese Clinical Trial Registry with-

---

[124] R Isasi, E Kleiderman, and BM Knoppers, "Editing Policy to Fit the Genome? Framing Genome Editing Policy Requires Setting Thresholds of Acceptability" (2016) 351:6271 *Science* 337.

[125] Q Zhou, "Gene editing in China (基因编辑在中国)" (Chinese Academy of Science, 2016) www.casisd.cn/zkcg/zjsd/201610/t20161017_4679428.html accessed May 17, 2018.

[126] Y Peng, "The Morality and Ethics Governing CRISPR-Cas 9 Patents in China" (2016) 34:6 *Nature Biotechnology* 616.

[127] Q Zhou, "Gene editing in China (基因编辑在中国)" (Chinese Academy of Science, 2016) (n 126).

drew Dr. He's postdated registry.[128] While many aspects of Dr. He's story remain obscure at the time of writing this chapter, the message sent by the Chinese government has been unequivocal: genetic modification of the human germline for reproductive purposes is strictly forbidden under the current regulatory framework.

Even before Dr. He's debacle, there have been calls in China for improved collaborative efforts between regulatory agencies and the scientific community to establish robust and explicit ethical principles that will guide future clinical applications involving human germline genome modifications.[129] The Chinese society seems to have less qualms with genetic modification of human cells than the Chinese government and surely less than Western governments and their public.[130] Historical, sociocultural, and religious factors and beliefs might help explain these differences.[131] Yet, condemnation of the first human germline gene editing clinical trial as highly unethical has been virtually unanimous at all levels across Eastern and Western societies. The Chinese regulatory system is permissive with respect to preclinical and clinical applications of modified somatic cells. Although the legal framework applicable to germline gene editing or embryo modification prohibits transfer in utero of genetically modified embryos, it is considerably liberal if compared to most of the world.

However, China's bold approach to disruptive genetic research does not lack detractors, internally and internationally.[132] For instance, after the publication of a study involving human germline modification,[133] the Chinese bioethics community and the international community voiced concerns[134] which have been magnified in the aftermath of Dr. He's affair. In particular, issues of adequate informed consent, appropriate balance of benefits and risks, potential for genetic

[128] Chinese Clinical Trial Registry, *Been withdrawn with the reason of the original applicants cannot provide the individual participants data for reviewing Safety and validity evaluation of HIV immune gene CCR5 gene editing in human embryos, Registry no ChiCTR1800019378* (Chinese Clinical Trial Registry, 2018).

[129] Ibid.

[130] JB Nie, "He Jiankui's Genetic Misadventure: Why Him? Why China?" (The Hasting Center, 2018) www.thehastingscenter.org/jiankuis-genetic-misadventure-china/ accessed December 11, 2018.

[131] Y Peng, "The Morality and Ethics Governing CRISPR-Cas 9 Patents in China" (2016) (n 127).

[132] Ibid.

[133] D Cressey and D Cyranoski, "Human-Embryo Editing Poses Challenges for Journals" (2015) 520:7549 *Nature* 520, 594.

[134] P Liang and others, "CRISPR/Cas9-mediated gene editing in human tripronuclear zygotes" (2015) 6:5 Protein & Cell 363, 363, 372.

discrimination, and even eugenic applications were raised, both in China[135] and abroad.[136] Yet, leading Chinese bioethicists pointed out that studies conducted in China have followed international accepted ethical principles, such as respecting the "14-day rule,"[137] obtaining informed consent, and approval by an institutional ethics review board.[138] Similar claims of ethics probity have been raised by Dr. He to further justify his actions.[139]

Despite harsh global criticism and skepticism, which, at times, has led major scientific journals to even decline publishing results,[140] the Chinese scientific community has not lost enthusiasm for pushing boundaries, although not going as far as Dr. He's experiments.[141] In 2015 and 2018, China and Hong Kong took the initiative to cohost the first and second international summits on gene editing to improve the understanding of the legal and ethical frameworks necessary to support research.[142] History will tell whether the summits and the ensuing international debate they channeled had any real-world effects.

We can only speculate what the next chapter in the book of human history will bring. Reformulating the scientific, legal, and ethical debates, to instill them with some guiding principles, such as respect for human rights and dignity, transparency, and democratic participation of those affected and the scientific community, might turn out to be difficult but it is imperative.

---

[135] H Wang, "Ethical Inquiries about CRISPR/Cas9-Mediated Gene Editing in Human Tripronuclear Zygotes (利用CRISPR/Cas 9介导基因编辑人类三核体受精卵伦理问题探讨)" 37:7A *Medicine and Philosophy*, 552.

[136] DK Tatlow, "A Scientific Ethical Divide between China and West" *The New York Times* (New York, June 29, 2015).

[137] Ethical Guiding Principles for Research on the Human Embryonic Stem Cell of the PRC 2003 (n 45) art. 6.

[138] R Qiu, "Debating Ethical Issues in Genome Editing Technology" (2016) 8:4 *Asian Bioethics Review* December 307.

[139] H Jiankui and others, "Draft Ethical Principles for Therapeutic Assisted Reproductive Technologies" (2018) *The CRISPR Journal*.

[140] D Cyranoski and S Reardon, "Chinese Scientists Genetically Modify Human Embryos" (2015) *Nature* www.nature.com/news/chinese-scientists-genetically-modify-human-embryos-1.17378 accessed August 17, 2018.

[141] NME Fogarty and others, "Genome Editing Reveals a Role for OCT4 in Human Embryogenesis" (2017) 550 *Nature* 67.

[142] S Reardon, "Global Summit Reveals Divergent Views on Human Gene Editing" (2015) 528:7581 *Nature* 173.

18

The Regulation of Human Germline Genome Modification in the Republic of Korea

Hannah Kim and Yann Joly

I INTRODUCTION

For as long as genetics existed, there has been a debate about whether humankind should or should not be allowed to manipulate, directly or indirectly, the human genome. The theoretical foundations of genetics were laid during the second half of the 1800s, with Gregor Mendel's discovery of the rules of heredity and Friedrich Miescher's isolation of nucleic acid, paving the way for the identification of DNA as the carrier of inheritance.[1]

The Korean government gradually developed the country's genomic policies over the second half of the twentieth century. Earlier in the twentieth century, genetics, combined with radical evolutionary theories, was used to justify eugenics and racist policies in countries such as Germany and Japan.[2] The Japanese policies, such as the 1940 National Eugenics Law and the 1948 Eugenic Protection Law, influenced the development of eugenics policies in the postwar Korean society.[3] In particular, although the Criminal Act prohibits induced abortion, the Mother and Child Act of 1973 introduced a eugenic or genetic exception, in cases "where a woman or her spouse suffers from any genetic mental disability or physical disease prescribed by Presidential Decree."[4] The Genetic Engineering Promotion Act of 1984 was Korea's first

---

[1] JH Kim, *History's Turning Point, 2nd volume: Discovery of DNA Double Helix* (21st century Books 2012).

[2] M Landler, "Results of secret Nazi breeding program: Ordinary folks" *The New York Times* (New York, November 7, 2006); JC Lee, "Korea's health care policy of the twentieth century" (1999) 8:2 *Korean J Med Hist* 138, 146; YJ Shin, "The characteristics of Korea's Eugenic Movement in the Colonial Period represented in the Bulletin, *Woosaeng*" (2006) 15:2 *Korean J Med Hist* 133, 155.

[3] YJ Shin, "The characteristics of Korea's Eugenic Movement in the Colonial Period represented in the Bulletin, *Woosaeng*" (2006) 15:2 *Korean J Med Hist* 133, 155.

[4] Criminal Act (형법) 2018, art. 269–1; Mother and Child Health Act (모자보건법) 2018 art. 14–1-1.

attempt to introduce a real bioethical perspective in the regulation of genetic engineering science and technology, stipulating guidelines for genetic experiments and precautionary measures for biohazardous and ethical problems.[5] The Genetic Engineering Promotion Act was later revised, and became the Biotechnology Support Act in 1995.[6] In the early 2000s, with rapid developments in the biomedical field and human embryo and stem cell research, the Republic of Korea (ROK) consolidated its ethical and legal positions on this research in different economic, social, and political contexts.[7] The adoption of the Bioethics and Safety Act (BSA) in 2004 was an important milestone in this endeavor.[8] Notably, the BSA prohibits human reproductive cloning and the fusion of human embryos with those of other species.[9]

However, the rapid development of genome editing has generated new regulatory issues, from research using human embryos to the use of genome editing.[10] Recent breakthroughs in genome editing, such as CRISPR/Cas9, enables researchers to use highly accessible programming tools for gene alteration that can perform gene engineering of any organism and cell type including human embryos and germline cells.[11] The ROK is one of the global leaders in genome engineering, including the CRISPR system,[12] and Korean scientists have collaborated in genome editing research on human embryos in other countries, such as the United States and the United Kingdom.[13] Yet, to date, no clinical genome editing research using human embryos has been undertaken in the ROK. This gap between science and the law has accelerated the debate about whether human germline genome modification research should be allowed to proceed and how far.

This chapter provides an overview of the ethical, legal, and policy framework of the ROK, applicable to research on human germline genome modification and of the political and social fundaments behind it. It will start with the introduction of

---

[5] Genetic Engineering Promotion Act (유전공학육성법) 1993.

[6] Biotechnology Support Act (생명공학육성법) 2017.

[7] SH Kim, "Science, technology and the imaginaries of development in South Korea" (2017) 46:2 *Development and Society* 341, 371.

[8] Bioethics and Safety Act (생명윤리 및 안전에 관한 법률) 2018 (BSA 2018).

[9] BSA 2018, arts. 20–21.

[10] C Fellmann and others, "Cornerstones of CRISPR-Cas in drug discovery and therapy" (2016) 16 *Nature Reviews Drug Discovery* 89, 100.

[11] Ibid.

[12] SW Cho and others, "Targeted genome engineering in human cells with the Cas9 RNA-guided endonuclease" (2013) 31 *Nature Biotechnology* 230, 232; H Kim and JS Kim, "A guide to genome engineering with programmable nucleases" (2014) 15 *Nature Review Genetics* 321, 334.

[13] H Ma and others, "Correction of a pathogenic gene mutation in human embryos" (2017) 548 *Nature* 413, 419; NME Fogarty and others, "Genome editing reveals a role for OCT4 in human embryogenesis" (2017) 550 *Nature* 67, 73.

the legal system, and the assessment of specific laws and policies related to the research on human embryos and the legal framework applicable to the modification of the human genome will subsequently be presented. The final part will critically review the implications of technological advancements in this field for legal reform including the possibility of integrating solutions from international human rights instruments to develop a more progressive, yet responsible, national regulatory framework to human germline genome modification.

II THE REGULATORY ENVIRONMENT

The regulatory environment plays a pivotal role in research on human germline genome modification in the ROK. As research advances, it needs to conform to an increasingly complex legal and ethical framework. The first section of this chapter presents the relevant constitutional, regulatory, and ethical norms associated with research on human germline genome modification taking place in the ROK.

1 *Regulatory Framework*

a Constitution

Since its adoption in 1952, the Constitution of the Republic of Korea is the supreme law of the country.[14] It has been in force for three decades, from the last amendment of the Constitution in 1987, as a historical product of democratization of the ROK. The Constitution establishes the fundamental rights of the people, such as the rights to education, work, life, housing, and health. These rights are also protected through ordinary laws enforced by a judicial review system of the Constitutional Court of Korea.

No provision in the Constitution explicitly mentions the human embryo or human embryo research, including human genome editing. However, decisions of the Constitutional Court regarding the right to life have put forward key principles that would eventually be applied to research using human embryos. The right to life is not explicitly stipulated in the Constitution, but is supported by the Constitution through protecting the sanctity of human life and human dignity of all citizens and guaranteeing the fundamental and inviolable human rights of all citizens.[15] In 1996, in a decision on capital punishment, the Constitutional Court elaborated the concept of the right to

---

[14] Constitution of the ROK (대한민국 헌법) 1987.
[15] Constitution of the ROK 1987, art. 10.

life.[16] The court began by stating that human life is "the basis for human existence which is precious and cannot be replaced by any worldly thing," and the right to life is "the most fundamental right and the precondition to all the basic rights set forth in the Constitution," which is "derived from intuitive and natural law based on human instinct for survival and on the purposes of human existence."[17] In 2008, in a case on stillborn fetuses' right to claim for damages, the court reflected on whether the right to life also applies to a human fetus.[18] This landmark decision recognized that the right to life applies to a human fetus as it does to a human being and specified the obligation of the state to protect the life of a fetus.[19]

However, the Constitutional Court also established that the capacity of an unborn fetus to claim damages through the Civil Act of 1960 and Constitutional rights is not absolute.[20] It depends, retrospectively, on the future viability and/or practical capacity of the child to be to exercise said rights after its birth. However, the Constitutional Court found that preimplanted human embryos do not possess fundamental rights, including the right to life.[21] According to the Constitutional Court, in a case on bioethics regarding embryo research in 2010: "(1) It is hard to affirm the continuity of the entity from embryo to an independent human being, unless they are implanted into a mother's womb or the embryological primitive streaks appear. (2) Given the contemporary level of technological development, an embryo is expected to develop into a human entity only after being implanted into a woman's womb. (3) Moreover, there seems to be no social recognition that such early human embryos are regarded or should be treated as human entities."[22]

In the 2010 case on bioethics regarding embryo research, the Constitutional Court reaffirmed that the constitutional right to life is not applicable to a human embryo that is generated by in vitro fertilization (IVF) and is in the early development stage.[23] In this case, the Constitutional Court also found that regulating human embryo research with measures to protect the human embryo or germline cells through the oversight process established by the BSA did not undermine the constitutional order.[24] The BSA permits the storage of

---

[16]   95 Hun-Ba1 (Constitutional Court of Korea, November 28, 1996).
[17]   Ibid.
[18]   2004 Hun-Ba81 (Constitutional Court of Korea, July 31, 2008).
[19]   Ibid.
[20]   Civil Act (민법) 2018.
[21]   2005 Hun-Ma346 (Constitutional Court of Korea, May 27, 2010).
[22]   Ibid.
[23]   Ibid.
[24]   Ibid.

"embryos up to 5 years, or less than 5 years, as authorized by the patients," and "for the embryo-producing medical institutions, to dispose of all embryos which do not use for the purpose of research at the end of their period of storage."[25]

Regretfully, the "right to science," defined as the right "to enjoy the benefits of scientific progress and its applications" found expressed in the International Covenant on Economic, Social, and, Cultural Rights (ICESCR), remains underused in the Constitutional system of the ROK. This is especially noteworthy since the Korean Constitution stipulates: "Treaties duly concluded and promulgated under the Constitution and the generally recognized rules of international law shall have the same effect as the domestic laws of the Republic of Korea."[26] The ROK acceded to the ICESCR on April 10, 1990.[27] However, while the right to science in the ICESCR implies a right for everyone to benefit from scientific advances, including scientists and citizens, the provision included in the Constitution so far solely focuses on the rights of creators.[28] The ICESCR has mainly influenced the development of the ROK's civil and political rights among the constitutional rights through its influence on the decisions of the Constitutional Court.[29]

The right to science encompasses two distinct but interrelated sets of rights: the right of everyone to benefit from advancements in science and technology, which, in the case of biomedicine, is related to the right to health, and the so-called rights of science.[30] These entail the scientists' rights and duties to do research and to benefit from their scientific and technological discoveries (rights of science).[31] To facilitate the realization of the "rights of science," member countries have to provide an enabling environment for scientific research,[32] to encourage international contacts and cooperation to take steps to allow for the benefits of scientific progress to become available and accessible

---

[25] Ibid.

[26] Constitution of the ROK 1987, art. 6–1.

[27] International Covenant on Economic, Social and Cultural Rights (adopted December 16, 1966, entered into force January 3, 1976) 993 UNTS 3 (ICESCR); Optional Protocol to the International Covenant on Economic, Social and Cultural Rights, (adopted December 10, 2008, entered into force May 5, 2013), A/RES/63/117. The ROK has not ratified yet the Optional Protocol to the Covenant giving individuals the possibility to bring claims to the Committee on Economic, Social and Cultural Rights.

[28] Constitution of the ROK 1987, art. 22–2.

[29] JY Lee, "The development of the international covenant on economic, social and cultural rights and its national implications" (2016) 61:2 *Korean Journal of International Law* 125, 157.

[30] On the "right to science" and the "rights of science" see, in this volume, Part 2, Chapter 2, Section III.

[31] ICESCR, arts. 15–1-2, 3.

[32] Ibid., art. 15–2.

to everyone,[33] and to respect the freedom for scientific and technological developments.[34]

b Bioethics and Safety Act (BSA)

The BSA is the key legislation regulating research on human embryos in the ROK. The BSA attempts to reconcile scientific innovation with the ethical requirement to protect individuals from the exploitation that may happen in a laboratory setting. The BSA has been amended several times since 2005. The most significant amendments are those of 2008,[35] 2012,[36] and 2015,[37] which have established rules regarding treatment of human embryos and human embryo research, including genome therapy research.

The BSA has provided special protective measures limiting research on human sperm, eggs, and embryos through specific prohibitions.[38] It defines key terms such as "embryo," "gene therapy," and "somatic cell nuclear transfer." An "embryo" is defined as "a fertilized human ovum or a group of [segmented] cells divided during a period from the moment of fertilization time at which all organs of the given organism have developed embryologically."[39] A "residual embryo" is "an embryo remaining after embryos [which are] created through IVF procedures are used for pregnancy."[40] "Somatic cell nucleus transfer" is defined as "a transfer of a human somatic nucleus into a human ovum from which the nucleus has been removed."[41] "Parthenogenesis" is defined as "a process through which a human ovum is divided into cells aside from the process of fertilization."[42] "Gene therapy" refers to "a series of procedures to alter genes in the body for the purpose of preventing or treating a disease or to transfer hereditary substances or cells to which hereditary substances are introduced to the body."[43]

The overall approach to scientific research using human embryos remains cautious and conservative in the ROK. Under the BSA, "no one shall produce

---

[33] Ibid., art. 15–4.
[34] Ibid., art. 15–3.
[35] Bioethics and Safety Act 2008 (Law no 9100).
[36] Bioethics and Safety Act 2012 (Law no 11250).
[37] Bioethics and Safety Act 2015 (Law no 13651).
[38] BSA 2018, c 4.
[39] Ibid., art. 2–3.
[40] Ibid., art. 2–4.
[41] Ibid., art. 2–6.
[42] Ibid., art. 2–7.
[43] Ibid., art. 2–16.

embryos other than for the purpose of pregnancy."[44] Furthermore, when an
embryo is produced for the purpose of pregnancy, the following actions are
prohibited: "fertilizing an oocyte, when the oocyte and/or sperm have been
specially selected for the purpose of producing offspring of a particular
gender";[45] "fertilizing an oocyte, when the oocyte and/or sperm are those of
a non-living human";[46] and "fertilizing an oocyte, when the oocyte and/or
sperm are those of an underage human [...] this (last action) shall be allowed
when married under-aged parents wish to conceive a child."[47] The "inducement or assistance in providing or utilizing sperm or oocytes for the purpose of
receiving financial reward, property, or any other personal benefits" is also
prohibited.[48]

Human cloning, including cloning via somatic cell nuclear transfer or
parthenogenic embryo (except under special circumstances), and implantation of any of cloned embryos are prohibited.[49] Human and animal
hybrid experiments, fusing or fertilizing human and animal embryos, as
well as implanting any human-animal hybrids in the uterus of a human
or animal are also prohibited.[50] The BSA prohibits "gene therapy"
applied to human embryos, ovum, sperm, or fetuses.[51] In addition, the
institutions carrying out research using human embryos must be registered embryo research institutions with the Ministry of Health and
Welfare (MOHW).[52]

The BSA imposes criminal sanctions in case of illicit generation and
use of human embryos. First, any person who unlawfully uses residual
embryos shall be sentenced to imprisonment for up to three years or
given a fine of up to 50 million South Korean won (about US$45,000).[53]
Second, any person who illegally carries out research on gene therapy,
or provides gene therapy as part of his/her clinical practice shall be
sentenced to imprisonment for up to two years or given a maximum
50 million Won fine.[54]

---

44   Ibid., art. 23–1.
45   Ibid., art. 23–2–1.
46   Ibid., art. 23–2–2.
47   Ibid., art. 23–2–3.
48   Ibid., art. 23–3.
49   Ibid., art. 20.
50   Ibid., art. 20.
51   Ibid., art. 47–3.
52   Ibid., art. 29–2.
53   Ibid., art. 66–2.
54   Ibid., art. 67–1–5.

c Pharmaceutical Affairs Act, the Regulation on Approval/Inquiry
of Biological Agent, and the Guidelines on the Evaluation
of Nonclinical Trials of Genetic Therapy Medicinal Products

It is also worth mentioning three additional statutes that are relevant to the regulation of human germline genome modifications. In the ROK, the *Pharmaceutical Affairs Act* (PAA) of 1954 is a general law that regulates clinical trials.[55] According to the PAA, a "non-[human] clinical trial" means a test conducted by using animals, plants, microorganisms, a physical or chemical medium, or a composite thereof under the same conditions as those in a laboratory, so as to obtain various data on the nature or safety of a test material which influences the health of humans.[56] A "clinical trial" is a test to confirm pharmacokinetic, pharmacodynamic, pharmacologic, and clinical effects of a drug on humans to prove its safety and validity and study adverse reaction.[57]

In comparison to the BSA, which considers gene therapy as a practice, the Regulation on Review and Authorization of Biological Products, the Public Notification under the Enforcement Rule on the Safety of the Pharmaceutical Product of the PAA, and so on, approach gene therapy as a product that is the subject of a clinical trial.[58] In the Regulation, a gene therapy product means "a medicinal product containing one of the following: a genetic material to be administered in the human body for the purpose of influencing genetic expression; or a cell which is genetically modified or administered genetic material."[59] In other words, the regulation under the PAA classifies gene therapy products as: (1) in vivo gene therapy product that administers genetic material directly into the human body; and (2) ex vivo gene therapy product that inserts genetically modified cells in vitro into the human body.[60]

The PAA also regulates permissions to manufacture, sell, and manage drugs, including gene therapy products. These products should be approved by the minister of the Ministry of Drug and Food Safety (MDFS).[61] Under the Enforcement Rule of the PAA, the minister of the MDFS "does not permit the

[55] Pharmaceutical Affairs Act (약사법) 2018 (PAA 2018).
[56] Ibid., art. 2–16.
[57] Ibid., art. 2–15.
[58] Ministry of Food and Drug Safety, Regulation on Review and Authorization of Biological Products (생물학적제제 등의 품목허가·심사 규정) 2018.
[59] Regulation on Review and Authorization of Biological Products 2018, art. 2–15.
[60] National Institute of Food and Drug Safety Evaluation, *Guidelines on the Evaluation of Nonclinical Trials of Genetic Therapy Medicinal Products – Public Guidelines* (유전자치료제 비임상시험 평가 가이드라인 – 민원인 안내서) (No 0819–01, 2017).
[61] PAA arts. 31, 35, 42 and 76.

manufacture of any gene therapy product which raises ethical concerns, including treatment through genetic modification of human germline cells."[62] However, interestingly, there is no prohibition on nonclinical or clinical studies on human germline genome modification under the PAA. In a 2016 report, the MDFS put forth the possibility that the existing guidelines (i.e. the 2008 Guidelines on the Evaluation of Nonclinical Trials of Genetic Therapy Medicinal Products) could be used to regulate research on human germline genome modification.[63]

In sum, there remains a degree of confusion about research studies on human germline genome modification even though genome editing tools such as CPISPR/Cas are considered as gene therapy products in the ROK. The genetic modification of a human embryo can be understood as a series of procedures on an embryo in its initial step to grow into a human child, conceived yet unborn, under the BSA,[64] or as a medicinal product under the PAA.[65]

d Laboratory Animal Act

Generally, nonclinical trials are regulated by the PAA, while the ethical aspects related to the usage and handling of laboratory animals are regulated by the Laboratory Animal Act of 2009 (LAA).[66] Any other animal research activities that are not prescribed in the LAA are regulated by the Animal Protection Act of 1991.[67] The MFDS oversees the compliance with the PAA and the LAA. According to the LAA, *animal testing* refers to "testing conducted for laboratory animals or the scientific procedures, such as education, testing, research and production of biological medicines,"[68] and *laboratory animal* is defined as a "vertebrate used or raised for the purpose of animal testing."[69]

The LAA mandates the creation of a Laboratory Animal Management Committee in each research institution,[70] and requires advanced reporting

---

[62]   Rule on Manufacturing, Sale and Management of Biological Products (생물학적 제제 등의 제조·판매관리 규칙) 2017, art. 3–3.

[63]   National Institute of Food and Drug Safety Evaluation, *Report on R&D Trend on Gene Editing Technology* (2017) 5.

[64]   BSA 2018, art. 2–16.

[65]   Regulation on Review and Authorization of Biological Products 2018, art. 2–15.

[66]   Laboratory Animal Act (실험동물에 관한 법률) 2018, art. 1.

[67]   Laboratory Animal Act 2018, art. 4.

[68]   Ibid., art. 2–2.

[69]   Ibid., art. 2–3.

[70]   Ibid., art. 7–1.

to the head of the MDFS when using "biohazardous substances" on laboratory animals,[71] including substances of genetic recombination "that can cause harm to human dignity."[72] Nevertheless, these laws do not apply to genome editing or the CRISPR system on animal embryo.

2 Governance

a Governmental Bodies That Oversee Human Germline Modification Research

The BSA assigns to the MOHW, the National Bioethics Committee, and the Institutional Bioethics Committees the task of determining policies regarding research involving human embryos. These bodies are also responsible for overseeing this research.

The National Bioethics Committee is an advisory body reporting directly to the president of the ROK. It reviews issues concerning bioethics and biosafety in the life sciences and biotechnologies, including "policies concerning national bioethics and biosafety issues; the permissive range of research involving residual embryos; the permissive range of research involving somatic cell nucleus transfer; and other issues of social or moral significance concerning the research, development, and utilization of life sciences and biotechnologies."[73]

The BSA requires that embryo-generating medical institutions, embryo research institutions, somatic cell nucleus transferred embryo research institutions, biobanks, and, more generally, any other institution that researches, develops, or utilizes life sciences and biotechnologies that may have significant moral or social consequences establish their own Institutional Bioethics Committee.[74]

Research plans must first be submitted to the Institutional Bioethics Committee.[75] Then, the research plan and the Institutional Bioethics Committee's research approval are submitted to the minister of health for final approval.[76] The approval of the minister is also needed to change the research plan, including the purpose, duration, number of residual embryos to be used in the research, embryo-generating medical institution that provides the residual embryos, and principal investigator of the research.[77]

---

[71]   Ibid., art. 19.
[72]   Enforcement Decree of the Biotechnology Support Act 2017, art. 15-2-2.
[73]   BSA 2018, arts. 7-1-1, 5, 6 and 10.
[74]   Ibid., art. 10-1.
[75]   Ibid., art. 31-2.
[76]   Ibid., art. 30-1.
[77]   Ibid., arts. 30-1 and 4; Enforcement Decree of the Bioethics and Safety Act 2017, art. 13.

b Funding Agencies

The Institute for Basic Research (IBS) is the major public research agency funding genome editing research in the ROK.[78] It was established in 2011 by the Special Act on the Establishment of and Support for International Science and Business Belt.[79]

The Ministry of Science and Information and Communications Technology is the body supporting the IBS under the Special Act. According to the Special Act, the president of the IBS "shall formulate a five-year plan [...] [including] the budget of the IBS and report the five-year plan to the Minister of Science."[80] Then, the Special Act stipulates "the State shall provide the funding necessary for the implementation of the five-year plan in a stable manner."[81] The Special Act also guarantees the financial contribution under the Act on the Management of Public Institutions from "the State, local governments, or public institutions, universities, enterprises, foreign governments, etc." to the IBS "to help it cover expenses incurred in the establishment, research, and operation of the IBS."[82] The approximate governmental funding for the year 2016 was US$7 billion.[83]

III SUBSTANTIVE PROVISIONS

1 *Regulation of Basic Research Using the Human Embryo*

As mentioned earlier, gene therapy on embryos, ovum, sperm, or fetuses is prohibited in the ROK.[84] However, the 2015 amendment of the BSA was a turning point for research on genetic modification. First, this amendment modified the definition of "gene therapy" in the BSA to "a series of procedures to alter genes in the body for preventing or treating a disease or to transfer hereditary substances or cells through which hereditary substances are introduced to the body." It is worth noting that this change does not clarify whether the concept of gene therapy includes genome editing.

---

[78] Institute for Basic Research, "Center for Genome Engineering" (2018) www.ibs.re.kr/eng/su bo2_06_02.do accessed July 12, 2018.

[79] Special Act on Establishment of and Support for International Science and Business Belt (국제과학비즈니스벨트 조성 및 지원에 관한 특별법) 2018, art. 14.

[80] Ibid., art. 21–1.

[81] Ibid., art. 21–2.

[82] Ibid., art. 22–1.

[83] KH Kang and SR Kim, "I spared the tenure, but I only wanted to research without concerns of money" *Joongang Ilbo* (Seoul, April 20, 2016) http://news.joins.com/article/19912416 accessed July 12, 2018.

[84] BSA 2018, art. 47–3.

Second, the prohibition of "gene therapy" on embryos, ovum, sperm, or fetuses notwithstanding, the BSA left the door open to gene therapy research.[85] Research on gene therapy is allowed in cases that meet the following two conditions: "(1) a research study to treat or cure genetic disorders, cancer, Acquired Immune Deficiency Syndrome (AIDS), and other life-threatening or seriously [damaging] disabling diseases; and (2) a research study to treat diseases for which there currently is no cure or when the effect of gene therapy expects to outweigh those of other therapies."[86]

Third, the BSA allows human embryo research under specific conditions. Human embryo research can use residual embryos (i.e. supernumerary embryos left over after IVF) only. Research using human embryos must be performed before the appearance of the primitive streak in the embryonic development process.[87] As aforementioned, the appearance of the primitive streak is recognized as a significant point in the process of becoming a human entity from an embryo in the ROK.[88] The BSA stipulates the purpose of human embryo research as: "developing infertility treatments and contraception"; "curing rare or incurable diseases such as muscular dystrophy, as stipulated by the Presidential decree"; or "when approved by the Presidential Decree after being reviewed by the National [Bioethics] Committee."[89] The consent of donors is needed for embryos that have been stored in cryopreservation for less than five years for using them for such research purposes.[90]

However, the BSA does not provide for an explicit distinction between somatic or germline gene therapy research, allowing research using residual human embryos in parallel. This gap seems to imply that the BSA allows for genome editing research, somatic or germline, as long as the research meets either of the two abovementioned conditions.[91] This confusion had the consequence of preventing human germline genome editing research in the ROK. Even if the BSA was interpreted as allowing human germline genome editing research, research using residual embryos for genome modification must target one of the rare or incurable diseases under the BSA and the Enforcement Decree of the BSA.[92] Rare diseases include multiple sclerosis, Huntington's disease, hereditary ataxia,

---

[85]   Ibid., art. 47.
[86]   Ibid., art. 47–1-1, 2.
[87]   Ibid., art. 29–1.
[88]   2005 Hun-Ma346 (Constitutional Court of Korea, May 27, 2010).
[89]   BSA 2018, arts. 29–1-1, 2, 3.
[90]   Ibid., arts. 24 and 25–1, 2.
[91]   BO Jun, "Ethical issues on the human embryonic genome editing" (2016) 16:2 *J Korean Bioethics Assoc*, 17, 29; H Kim, SH Kim, S Kim, "Issues of genetic modification in human embryo and its future orientation" (2015) 23:2 *Korean Journal of Medicine and Law* 211, 224.
[92]   Enforcement Decree of the BSA 2017, art. 12–1-1, 2; BSA 2018, art. 29–2.

amyotrophic lateral sclerosis, cerebral palsy, spinal cord injury, congenital immunodeficiency syndromes, aplastic anemia, leukemia, osteochondrodysplasia, adrenoleukodystrophy, metachromatic leukodystrophy, Krabbe's disease, and muscular dystrophy. Incurable diseases include myocardial infarction, liver cirrhosis, Parkinson's disease, stroke, Alzheimer's disease, optic nerve damage, diabetes mellitus, and acquired immunodeficiency syndrome (AIDS).

Such a conservative regulatory approach that allows only specific types of research to be carried out risks making the ROK lag behind in the areas of technology and therapy development and potentially miss breakthrough opportunities. For instance, although research carried out at Sun Yat-sen University, a major Chinese public research university located in Guangdong, has shown promise for treating β-thalassemia through human embryo genome editing, the same research could not be conducted in the ROK because this disease is not among those listed in the BSA and the Enforcement Decree of the BSA.

2 *Regulation of Preclinical Research in Animals*

As described above, the LAA regulates research involving animals, but there is no specific law or policy regulating genome editing in an animal germline cell and implanting the modified cell in the womb of an animal. Therefore, the Institutional Bioethics Committee may require genome editing researchers to follow relevant professional standards. For instance, a recent scientific research followed the Statement for the Use of Animals in Ophthalmic and Vision Research of the Association for Research in Vision and Ophthalmology for animal research.[93]

The BSA prohibits "implanting a somatic cell embryo or parthenogenic embryo in a human or animal womb, keeping such an embryo implanted, or bearing a child therefrom."[94] The BSA also prohibits implanting a human embryo in an animal womb or an animal embryo in a human womb.[95] Furthermore, the BSA prohibits the following activities: "fertilizing a human ovum with an animal spermatozoon or an animal ovum with a human spermatozoon, except for medical tests for examining the activity of human spermatozoa";[96] "implanting an animal somatic nucleus into a human ovum with its nucleus removed or a human somatic nucleus into

---

[93] T Koo and others, "CRISPR-LbCpf1 prevents choroidal neovascularization in a mouse model of age-related macular degeneration" (2018) 9:1 *Nature Communications* 1855.

[94] BSA 2018, art. 20-1.

[95] Ibid., art. 21-1.

[96] Ibid., art. 21-2-1.

an animal ovum with its nucleus removed";[97] and "fusing a human embryo with an animal embryo."[98] Implanting a manipulated human embryo from any procedure referred to in the Article 21–2 in an animal womb or a human womb is also prohibited.[99] However, those regulations would not prevent the new types of chimeric research, such as implanting a human pluripotent stem cell into an animal embryo.[100]

3 Regulation of Clinical Research and Application

The BSA prohibits gene therapy on human embryos and fetuses and prohibits the implantation in utero of an embryo whose genome has been modified. However, the BSA does not close the door to practices that are outright banned in many other states around the world. Indeed, the BSA does not regulate genome editing for non-therapeutic purposes. Although the BSA bans "gene therapy" on germline cells, it does not ban the modification of their genome for ˌnontherapeutic goals, such as enhancement for appearance, physical ability, or intellect, for example.

IV CURRENT PERSPECTIVES AND FUTURE POSSIBILITIES

Recently, the stakeholders in the ROK have promoted legislative changes that endeavor to strike a balance between relaxing the regulations to allow responsible research in gene therapy and eliminating ethical loopholes. However, so far, the legal framework regarding human germline genome modification remains prohibitory and strict, although less so than several other developed countries. To recapitulate, under the BSA: (1) the permission for research on human embryos that causes germline genome modification is unclear; (2) the implantation of a genetically manipulated human embryo in utero is prohibited; and (3) creating a human embryo ad hoc for research is prohibited.

As we saw, the overall concern of the BSA is the protection of human dignity rather than that of human rights flowing from this concept, including the right to health and the right to science.[101] The logical underpinning of this choice is

---

97   Ibid., art. 21–2-2.
98   Ibid., art. 21–2-3.
99   Ibid., art. 21–3.
100 J Wu and others, "Interspecies chimerism with mammalian pluripotent stem cells" (2017) 168:3 *Cell* 473, 486.
101 HC Kim and others, *A Study of the Korean Legal Institutionalization of the UNESCO Universal Declaration on Bioethics and Human Rights* (Korea Legislation Research Institute, 2016) 50–51.

that human rights are not applicable to unborn entities while human dignity is.[102] A human rights approach might not be sufficient to control actions that aim to manipulate human germline cells, while putting human dignity front and center is deemed to better protect life that is conceived but still unborn.[103]

However, the "human dignity approach" is becoming an obstacle for studies on human germline genome modification. Recent scientific developments have raised the question of further extending the possibility of researching on embryos beyond fourteen days.[104] Until 2016, culturing human embryos in vitro had never gone beyond nine days.[105] In 2016, human embryos were sustained in vitro for twelve or thirteen days.[106] As the fourteen-day limit is approached, scientists in the ROK and around the world are increasingly questioning the reasonability of this limit because it prevents them from doing a range of research, including studying epigenetic expressions.[107]

Perhaps it is time to recalibrate the overall ethical and legal approach to the ROK gene therapy framework. Human rights, such as the right to health and the right to science, could be used, in addition to dignity, as a compass to help Korean policy makers, legal scholars, and scientists better navigate the ethical and legal hurdles they face. The right to health enshrined in the Constitution entitles all Koreans to "a life worthy of a human being" and "the health of all citizens."[108] Fulfilling the right to health inevitably depends on improving health care through scientific research and development. While the Constitution does not grant the right to life to a child conceived but yet unborn, the ICESCR protects both the right "to enjoy the benefits of scientific progress and its applications" and the right to health, and duly ratified treaties have the force of law in the ROK.[109] This could be a sound legal basis to advocate for research on germline genome modification beyond the appearance of the primitive streak.

[102]   Ibid.

[103]   Ibid.

[104]   A Deglincerti and others, "Self-organization of the in vitro attached human embryo" (2016) 533:7602 *Nature* 251, 254; MN Shahbazi and others, "Self-organization of the human embryo in the absence of maternal tissues" (2016) 18:6 *Nat Cell Biol* 700, 708; J Harris, "It's time to extend the 14-day limit for embryo research" *The Guardian* (May 6, 2016).

[105]   Ibid.

[106]   I Hyun, A Wilkerson, and J Johnston, "Embryology policy: Revisit the 14-day rule" 533:7602 *Nature* 169, 171.

[107]   FS Collins and S Gottlieb, "The next phase of human gene-therapy oversight" (2018) 379 *N Engl J Med* 1393, 1395; G Cavaliere, "A 14-day limit for bioethics: the debate over human embryo research" (2017) 18 *BMC Medical Ethics* 38.

[108]   Constitution of the Republic of Korea 1987, arts. 34–1 and 36–3.

[109]   2005 Hun-Ma346 (Constitutional Court of Korea, May 27, 2010); Constitution of the ROK 1987, art. 22–2.

A "human rights approach" could also help bypassing the current legislative hurdles regarding preimplantation genetic diagnosis (PGD). The BSA permits PGD only for the sixty-three genetic diseases in the list of rare and incurable diseases.[110] The right to health and the right to science support an expansion of opportunities for PGD for patients who carry genetic diseases or mutations or want to detect severe or life threatening genetic diseases and mutations in an embryo before its implantation. These suggestions show that the ROK has arrived at a crossroads in the way it addresses some of the more substantive applications of genomic research. It is hoped that the legislator will chose the "human rights" path and move toward a more progressive, yet reflexive, approach, and toward this type of research and clinical practice and the ethical issues it raises.

---

[110]   Enforcement Decree of the BSA 2017, art. 12–1-1, 2; BSA 2018, art. 29–2.

19

The Regulation of Human Germline Genome Modification in Singapore

Calvin W. L. Ho

I INTRODUCTION

This chapter provides an analysis of the regulatory framework for genome editing technologies for human biomedical research and cell therapies (somatic and germline) in Singapore. By 'genome editing technologies' I refer to a group of techniques that include CRISPR/Cas9, a tool that allows the creation of a specific genetic modification at a precise locus in the genome. These technologies can modify the genetic composition of any living organism, including human cells. Application of these techniques to human somatic and germline cells could unlock therapeutic opportunities to treat severe hereditary diseases by repairing the disease-causing gene(s). At the time of this writing, genome editing technologies remain technically challenging and complex to implement safely and effectively, mainly because of fundamental limitations in the ability to control precisely how genetic material can be introduced into cells. Apart from concerns about physical well-being, genome editing technologies have also raised concerns that relate to eugenics, as they can be applied for non-therapeutic purposes, or for the purposes of genetic enhancement.

This chapter discusses the legal responses in Singapore to these and related concerns. It is a case study of legal governance of an emergent biomedical technology. By 'regulatory framework' I refer not only to statutory laws and regulations, but also to ethical principles, recommendations and good practices that either have regulatory effect or represent the legal or regulatory ends of sanction-backed requirements that apply to genome editing technologies. Under the heading 'regulatory framework', this chapter further discusses the two major actors that have developed research governance in Singapore: the Ministry of Health (MOH) and the Bioethics Advisory Committee (BAC). While the MOH remains primarily responsible for the governance of clinical care and certain innovative medical technologies that apply to reproductive medicine, in recent years it has assumed responsibility for the governance of human biomedical research as well. Both the MOH and the BAC have been instrumental in setting up a research ethics governance framework, primarily administered through Institutional Review Boards (IRBs). The BAC was established by the Singapore government as an expert body to provide it with advice on the ethical, legal and social implications of emergent technologies in human biomedical research. While the BAC is not itself a regulatory body, its ethical recommendations and guidelines have varying degrees of regulatory force, most evidently in that IRBs are expected to give effect to them.

In Singapore, genome editing techniques may be applied in research but not for therapeutic (or clinical) purposes. This conclusion is based on an analysis of the legal, regulatory and ethical provisions that apply to genetic research and clinical genetics. Consequently, it is necessary to consider how these provisions, which apply more generally to emergent genetic technologies, are applicable to genome editing techniques. In this respect, the issues to be considered include: (a) applications that are prohibited and those that are permitted; (b) for applications that are permitted, the procedures that must be observed (in order to obtain a specific license, for instance); and (c) gaps and ambiguities that arise.

Broadly speaking, the legal, regulatory and ethical provisions that apply to genome editing are drawn from five (at times overlapping) regulatory regimes: (i) research involving human embryos and stem cells; (ii) research involving assisted reproduction centres (and assisted reproduction services); (iii) research involving non-human animals; (iv) human biomedical research other than clinical trials; and (v) clinical trials.

In Section II, a short introduction to Singapore's political and legal system is provided. This is followed by an overview of the first two of the aforementioned regulatory regimes in Section III. Essentially, the regulatory regime for research involving human embryos and stem cells makes it clear that a human embryo can be created solely for the purposes of research (to study a genome editing technique, for instance), provided certain conditions are met. It is also clear that any such research should not extend beyond fourteen days from the time that the embryo was created, whether specifically for research or initially for assisted reproductive purposes. The regulatory requirements that must be observed, and approvals that must be obtained for the use of human gametes and embryos in research, are set out in the second of these regulatory regimes, on assisted reproduction centres and assisted reproduction services. These requirements and approvals apply *mutatis mutandis* to the use of genome editing techniques to modify the genome of human oocytes and embryos. This section concludes with an overview of the regulation of research involving non-human animals. The requirements under this regime apply to the use of genome editing techniques on non-human animals or to biological materials that are obtained from them. If such materials are used in combination with human materials (such as cytoplasmic hybrid embryos), the requirements of this regime, as well as those stipulated in the fourth regime, apply.

Section IV provides an overview of the requirements under the fourth regulatory regime: human biomedical research other than clinical trials. It begins by setting out the general requirements, such as when and how approval is to be obtained from an IRB, informed consent, procurement of tissue and establishment of tissue banks or repositories. These requirements apply also to the genetic modification of human biological materials for research purposes. This section further elaborates on additional regulatory requirements and approvals that apply to certain types of 'restricted research', to be further explained below. For the purposes of this volume, genome editing techniques that are applied to human oocytes and embryos or to human-animal combinations (such as cytoplasmic hybrid embryos) are considered as, and are regulated as, 'restricted research'. The chapter then considers, in Section V, the regulatory framework that applies to the use of genome editing technologies in clinical trials. Depending on the research design and the intervention concerned, professional governance might also be applicable, although this topic is beyond the scope of the chapter.

This chapter concludes with a discussion, in Section VI, on how the regulatory framework in Singapore is likely to evolve in the foreseeable future. Owing to a moratorium imposed by the BAC, human germline genetic modification is not allowed in the clinical setting. To the extent procedures such as mitochondrial replacement therapy are considered a form of genome editing, they cannot be applied by assisted reproduction centres in Singapore. However, it is less clear if such procedures could be applied in a clinical trial. For reasons that will be explained in this section, clinical trial of mitochondrial replacement therapy is probably prohibited at this time. However, this might change in the foreseeable future, and further guidance from the BAC is expected.

II A BRIEF OVERVIEW OF SINGAPORE'S LEGAL AND POLITICAL SYSTEM

The Republic of Singapore is an island and a sovereign city-state in Southeast Asia, at the southern tip of the Malay Peninsula. Singapore was founded in 1819 as a trading post of the British East India Company, and eventually became a crown colony. It gained full independence from Britain in 1963 to form Malaysia, only to separate two years later to form its sovereign state. Despite its small size (smaller than the city of New York), lacking natural resources and a hinterland, Singapore transitioned from 'third world to first world' in a single generation.

Politically, Singapore is a parliamentary representative democratic republic with a multiparty system. Although in Singapore there are three separate branches of government (legislature, executive and judiciary), it applies the British Westminster political tradition in which the legislature and the executive branches are intermingled. The People's Action Party (PAP) has been the main political party since the 1959 general election. Yet, despite lack of political change, Singapore has consistently been rated as the least-corrupt country in Asia and amongst the top-ten cleanest in the world. Meritocracy and multiracialism are key governing principles, and English has served as the common language to integrate its immigrant society.

As many former British colonies, Singapore's legal system follows the common law tradition. At the apex of its legal system, there is a triad of documents: the 1963 Constitution of the State of Singapore;[1] the Republic of Singapore Independence Act 1965;[2] and certain provisions of the Federal Constitution of Malaysia made applicable to Singapore by the Independence Act itself.[3]

As to the interface with International Law, Singapore is largely dualist. The Constitution is very protective of the country's sovereignty, which should be no surprise given its small size and a long history of colonization. Part III of the Constitution, entitled 'Protection of the Sovereignty of the Republic of Singapore', on the one hand prohibits the 'surrender of sovereignty or relinquishment of control over the Police Force or the Armed Forces except by referendum'.[4] But, at the same time, it allows for 'participation in co-operative international schemes which are beneficial to Singapore, including joining international organizations',[5] and entering into treaties that are 'or appear to be, beneficial or advantageous to Singapore in any way'.[6] The Singapore Constitution does not contain a formal provision regulating the reception of

---

[1] The Constitution of Singapore Amendment Act 1965 (Act No 8, amending the State Constitution). It changed procedure required for constitutional amendment.

[2] Republic of Singapore Independence Act 1965 (Act No 9). It vested the powers relinquished by the Constitution and Malaysia Singapore Amendment Act in the legislative branch of government.

[3] The Constitution and Malaysia Singapore Amendment Act 1965 (Act No 53). It transferred all legislative and executive powers to the new government of Singapore.

[4] 1963 Constitution, art. 6.

[5] Ibid., art. 7.a.

[6] Ibid., art. 7.b.

international law or establishing the hierarchical ordering of international and domestic law. Treaties are not self-executing as the dualist system treats international and municipal law as distinct systems of law; international treaties and agreements have no domestic legal effect until they are incorporated by a subsequent act of Parliament.[7]

Constitutionally sanctioned rights in Singapore are the due process rights, freedom of religion, freedom of expression, assembly and association. For different historical and political reasons, Singapore has not ratified the two pillars of the so-called international Bill of Rights: the Covenant on Civil and Political Rights,[8] and the Covenant on Economic, Social and Cultural Rights.[9] It does not explicitly recognize, therefore, as a matter of international law, the 'right to health', 'right to science' and the 'rights of science'. That being said, Singapore is member of the Association of South-East Asian Nations (ASEAN). In 2012, the ten-member regional organization adopted the ASEAN Declaration of Human Rights.[10] In its Preamble, the ASEAN member states reaffirm their 'commitment to the Universal Declaration of Human Rights, the Charter of the United Nations, the Vienna Declaration and Programme of Action, and other international human rights instruments to which ASEAN Member States are parties'. Under Article 26, the ASEAN members 'affirm all the economic, social and cultural rights in the Universal Declaration of Human Rights'. Specifically, ASEAN Member States affirm the following: 'Every person has the right to the enjoyment of the highest attainable standard of physical, mental and reproductive health, to basic and affordable health-care services, and to have access to medical facilities.'[11] 'Every person has the right, individually or in association with others, to freely take part in cultural life, to enjoy the arts and the benefits of scientific progress and its applications and to benefit from the protection of the moral and material interests resulting from any scientific, literary or appropriate artistic production of which one is the author.'[12]

Moreover, while treaty norms have to be statutorily incorporated to be given domestic legal effect, the approach towards customary international law

---

[7] L Thio, 'Reception and Resistance: Globalisation, International Law and the Singapore Constitution' (2009) 4 *NTU L Rev* 335, 349.

[8] International Covenant on Civil and Political Rights (adopted 16 December 1966, entered into force 23 March 1976) 999 UNTS 171 (ICCPR).

[9] International Covenant on Economic, Social and Cultural Rights (adopted 16 December 1966, entered into force 3 January 1976) 993 UNTS 3 (ICESCR).

[10] ASEAN Human Rights Declaration, adopted by Member States of the Association of Southeast Asian Nations (2012).

[11] Ibid., art. 29.1.

[12] Ibid., art. 32.

follows the practice in English law and is monistic, in that it becomes part of Singapore law without a further act of incorporation.[13] Thus, to the extent certain human rights have become norms of customary international law, they are applicable in Singapore.

III REGULATION OF RESEARCH INVOLVING A HUMAN EMBRYO, ASSISTED REPRODUCTION CENTRES AND NON-HUMAN ANIMALS

1 *Research Involving a Human Embryo*

The first set of recommendations published by the BAC concerns human embryos and stem cell research.[14] These were issued in 2002. Particularly relevant for this volume is the pronouncement that the creation of a human embryo specifically for research is permissible as long as: (a) there is strong scientific merit in, and potential medical benefit from, such research; (b) no acceptable alternative exists; and (c) specific regulatory approval is obtained on a highly selective, case-by-case basis.[15] Conceivably, a human embryo may be created for research in relation to genome editing if these justifications are satisfied.

Other recommendations of the BAC include proposals for stringent regulation of human embryonic stem cell research in Singapore and the legal prohibition of reproductive cloning.[16] These recommendations were endorsed by the legislature with the enactment in 2004 of the *Human Cloning and Other Prohibited Practices Act* (Human Cloning Act).[17] As most states that have significant research activities on human embryos, the Human Cloning Act fixes at fourteen days after fertilization the limit within which research involving a human embryo can take place.[18] Since public consultation showed that there was no consensus among the main religious groups in Singapore as to when 'personhood' could be said to begin, it was decided to adopt the fourteen-day limit as a compromise.[19] Around fourteen days after fertilization

---

[13] L Thio, 'Reception and Resistance: Globalisation, International Law and the Singapore Constitution' (n 7) 369.

[14] Bioethics Advisory Committee, *Ethical, Legal and Social Issues in Human Stem Cell Research, Reproductive and Therapeutic Cloning* (2002).

[15] Ibid., 30, Recommendation 5.

[16] Ibid., 31, Recommendation 7.

[17] Human Cloning and Other Prohibited Practices Act 2004 (revised 2005) c 131B (Human Cloning Act).

[18] Human Cloning Act, s 8.

[19] CWL Ho, B Capps, and TC Voo, 'Stem Cell Science and its Public: The Case of Singapore' (2010) 4:1 *East Asian Science, Technology and Society: An International Journal* 7.

the 'primitive streak' (i.e. a transient structure whose formation marks the start of the process in which the inner cell mass is converted into the three germ layers: ectoderm, mesoderm and endoderm) becomes evident. The BAC considers the development of the primitive streak a significant threshold, as, once it has been crossed, it signifies the creation of a unique human being. The appearance of the 'primitive streak' signals that 'the individuality of an embryo is assured'.[20] However, the law might place the beginning of life at other points depending on the context. For instance, in Singapore a pregnant woman has the discretion to terminate her pregnancy up to twenty-four weeks from conception.[21] Similarly, the common law relating to inheritance and Singapore's Penal Code attribute legal 'personhood' at a much later stage of fetal development.[22]

2 Assisted Reproduction Centres and Assisted Reproduction Services

In Singapore, artificial reproductive technology (ART) is regulated by the MOH through licensing requirements for Assisted Reproduction Centres (ARCs). Where research is done involving artificial reproductive technologies, the licensing terms and conditions provide:

> [ARCs] shall ensure that no research on oocytes (including those obtained from excised ovarian tissue) or on embryos shall be carried out without the prior written approval of the Director of Medical Services. Approval from the Director of Medical Services is required for the release of human oocytes (including those obtained from excised ovarian tissue) and/or embryos to other research centres.[23]

The director of Medical Services is the chief medical officer and is a senior position in the MOH, primarily responsible for regulating the medical profession and medical practice in Singapore.[24] Research protocols must be reviewed and approved by an appropriate IRB before they are submitted to the director of Medical Services for consideration. The research that ARCs engage in tends to be of an applied nature and is more concerned with the safety and/or efficacy of an assisted reproductive technique rather than solely

---

20   LifeMap Discovery, *Primitive Streak* http://discovery.lifemapsc.com/in-vivo-development/pri mitive-streak accessed 8 November 2018.
21   Termination of Pregnancy Act 1974 (revised 1985), c 324.
22   T Kaan, 'At the Beginning of Life' (2010) 22 *Singapore Academy of Law Journal* 883.
23   Ministry of Health, *Licensing Terms and Conditions on Assisted Reproduction Services* (2011), cl 9.1.
24   Medical Registration Act 1997 (revised 2014), c 174.

directed at the production of knowledge. The scope of the research conducted by ARCs is also likely to be limited by the fact that in Singapore assisted reproduction services are only available to heterosexual couples who are legally married. For this reason, the donation of a surplus embryo for research must be supported by the informed consent of both the wife and her husband.

ARCs are required to ensure that all prospective oocyte donors (i.e. patients who come primarily to donate their oocytes for research and not as part of fertility treatment) are assessed by an independent panel.[25] This panel may be part of the hospital's ethics committee, and consists of a layperson and two medical practitioners, one of whom is an authorized ART practitioner.[26] The panel interviews the prospective donor before the commencement of the oocyte donation procedure and must be satisfied that the prospective donor: (a) is of sound mind; (b) has a clear understanding of the nature and consequences of the oocyte donation; and (c) has given express consent for the donation (freely and without coercion or inducement), before allowing any procedure leading to the donation to proceed.[27] In addition, the panel must take into consideration the public interest and community values when assessing an application for donation of oocytes for research. For consent purposes, ARCs are required to ensure that prospective donors of oocytes or embryos for research are provided with all relevant material information relating to the proposed donation in an understandable form. This includes, at the minimum: (i) the purpose and nature of the research; (ii) the risks of the donation procedure; (iii) whether the donated oocytes/embryos will be destroyed after being used for research; (iv) whether the oocytes/embryos, or the derived cells, will be kept/stored for future research; and (v) the right to withdraw consent or vary the terms of consent at any time before their oocytes/embryos are actually used in research, how to withdraw consent, and the implications of a withdrawal.[28] Consent for research on oocytes or embryos must be obtained separately from any consent for ART treatment.[29] If a potential oocyte donor for research is also a woman undergoing fertility treatment, her consent for oocyte donation is taken independently of the treatment team. The potential donor (if an oocyte) or donors (if an embryo) must confirm in writing that the full implications of the

---

[25] Ministry of Health, *Licensing Terms and Conditions on Assisted Reproduction Services* (n 23) cl 9.9.
[26] Ibid.
[27] Ibid.
[28] Ministry of Health, *Licensing Terms and Conditions on Assisted Reproduction Services* (n 23) cl 9.10.
[29] Ibid.

donation have been explained and that these oocytes or embryos (previously provided for ART treatment) are no longer required for future reproductive use.

The regulatory regime for research involving ARCs will be likely the one applicable to germline genetic modification activities, such as mitochondrial replacement therapy, the day they are allowed in Singapore.[30] At the time that this chapter was written, the therapeutic application of germline genetic modification in any form was not allowed in Singapore as a result of a moratorium issued by the BAC in 2005.[31] More recently, the BAC observed that any intervention that alters the germline of an individual leading to a change in the genetic make-up of that individual's descendants raises serious ethical and moral concerns. It further observed that there is insufficient knowledge of the potential long-term health consequences of such interventions, as they are still in the experimental stage, and that many countries have laws that prohibit germline modification.[32] However, the BAC appears to be open to certain types of genetic germline modification technologies, such as assisted reproductive techniques to prevent the transmission of mitochondrial diseases, including ooplasmic transfer, pronuclear transfer and maternal spindle transfer, provided these techniques are shown to be sufficiently safe and effective.[33] In 2014, the BAC appointed a Germline Modification Working Group to review these developments, and, at the time of this writing, recommendations are being formulated.

3 *Research Involving Non-Human Animals*

Research involving non-human animals is regulated by the Agri-Food and Veterinary Authority (AVA) of Singapore. The AVA is the licensing authority of research facilities that use animals for scientific purposes.[34] All research facilities working with non-human animals must comply with guidelines and directives issued by the AVA.[35]

---

[30] The United Kingdom has recently allowed the therapeutic application of mitochondrial replacement therapy on a strictly licensed basis. No other forms of germline genetic modification are permissible in the United Kingdom. See Human Fertilisation & Embryology Authority, 'Mitochondrial donation treatment' www.hfea.gov.uk/treatments/embryo-testing-and-treatments-for-disease/mitochondrial-donation-treatment/ accessed 20 August 2018.

[31] Bioethics Advisory Committee, *Genetic Testing and Genetic Research* (2005) paras 4.51 and 4.52, Recommendation 11, paras 4.53–4.58, Recommendation 12.

[32] Bioethics Advisory Committee, *Ethical Guidelines for Human Biomedical Research* (2015) para 6.4.

[33] Ibid., para 6.5.

[34] Animals and Birds (Care and Use of Animals for Scientific Purposes) Rules 2004 (revised 2007), GN No S 668/2004.

[35] Ibid., r 18.

Of relevance to this volume are the guidelines issued by the National Advisory Committee for Laboratory Animal Research (NACLAR) for the proper care and use of animals for scientific purposes in 2004 (NACLAR Guidelines).[36] These guidelines apply to all types of use of animals for scientific purposes, including their use in teaching, field trials, environmental studies, research, diagnosis, product testing and the production of biological products. Three main principles underscore the NACLAR Guidelines that apply to the research use of animals:[37] (a) replacement of animals with alternative methods; (b) reduction of the number of animals used; and (c) refinement of projects and techniques used to minimize impact on animals. The overarching aim of these principles is to promote humane and responsible care and use of animals. The NACLAR Guidelines further require any research facility that houses and uses animals for scientific purposes to establish an Institutional Animal Care and Use Committee (IACUC).[38] The IACUC has the responsibility of overseeing and evaluating animal care and use programmes in the facility, and in ensuring that appropriate animal experimental procedures are complied with. Where the users of animals for research are concerned, the NACLAR Guidelines require that they undergo appropriate training before carrying out any experiments involving animals.[39]

IV LEGAL AND REGULATORY REQUIREMENTS THAT APPLY TO HUMAN BIOMEDICAL RESEARCH INVOLVING GENOME EDITING IN A NON-CLINICAL TRIAL SETTING

Genome editing techniques used in human biomedical research other than clinical trials must comply with the regulatory regime that comprises the Human Biomedical Research Act (HBRA) and supporting regulations and ethical provisions, some of which have binding effect. If the research is categorized as 'restricted' (explained later in this section) under the HBRA, additional requirements, such as approval by the MOH, apply.

---

[36] National Advisory Committee for Laboratory Animal Research, *Guidelines on the Care and Use of Animals for Scientific Purposes* (2004). Available at www.ava.gov.sg/docs/default-source/tools-and-resources/resources-for-businesses/Attach3_AnimalsforScientificPurposes.PDF, accessed on 20 August 2018.

[37] Ibid., c 2.

[38] Ibid., c 7.

[39] Ibid., c 8, pt 8.5.

1 *IRB Approval*

The *Human Biomedical Research Act* (HBRA), enacted by Parliament on 18 August 2015, establishes the legislative framework for human biomedical research.[40] In many ways it is an amalgamation of critical features of a prototype legislation that was proposed by the MOH in 2003[41] and the BAC-MOH framework implemented through IRBs. The explanatory statement describes the goals of the HBRA as: (a) regulating the conduct of human biomedical research, regulating tissue banks and tissue banking activities; (b) prohibiting certain types of human biomedical research; and (c) prohibiting the commercial trading of human tissue, most of the legislative provisions are concerned with goals (a) and (b).[42]

The human biomedical research activities that are prohibited under the HBRA are those that involve the following:[43]

(1) Development of cytoplasmic hybrid embryos or human-animal combination embryos created in vitro beyond fourteen days or the appearance of the primitive streak, whichever is earlier;

(2) Implantation of any human–animal combination embryo into the uterus of an animal or a human;

(3) Introduction of human stem cells (including induced pluripotent stem cells (iPSC)) or human neural cells into the brain of living great apes whether prenatal or postnatal; and

(4) Breeding of animals that have had any kind of human pluripotent stem cells (including iPSCs) introduced into them.

Essentially, all human biomedical research that is not explicitly prohibited and that falls within the scope of the HBRA must be approved (unless exempted) by an appropriate IRB before it is carried out. To determine whether a research falls within the scope of the legislation, the HBRA applies an 'inclusion-exclusion' two-stage test. The first test is the 'inclusion'. It assesses the nature of the research to determine whether it involves biological material deemed ethically, culturally or religiously sensitive,[44] or the intention to:

---

[40] Human Biomedical Research Act 2015 (No. 29 of 2015) (Human Biomedical Research Act).

[41] Ministry of Health, 'Draft Bill to Regulate Research On Human Stem Cells and Tissues' (30 October 2003).

[42] Ministry of Health, *Human Biomedical Research Act: Reflection on Transition* (2 March 2018).

[43] Human Biomedical Research Act, sch 3.

[44] Ibid., sub-s 3(3). This sub-section identifies these materials as human gametes or human embryos, cytoplasmic hybrid embryos, human–animal combination embryo and human stem cells or neural cells.

(a) Prevent, prognosticate, diagnose or alleviate of any disease, disorder or injury affecting the human body;
(b) Restore, maintain or promote the aesthetic appearance of human individuals through clinical procedures or techniques; or
(c) Enhance the performance or endurance of human individuals.[45]

If it turns out that the research has one of these goals, a second step is required to ascertain if it is 'excluded'. Does the research:

(a) have a temporary or permanent physical, mental or physiological effect on the research participant?
(b) use any individually identifiable human biological material? or
(c) use any individually identifiable health information?[46]

Thus, for instance, clinical trials of medicinal products will satisfy the inclusion criteria, but fall within the exclusion criteria.[47]

While the HBRA did not significantly modify the existing ethics review infrastructure, this rearticulation was necessary to redraw the scope of application of regulatory oversight so that the HBRA could be applied uniformly to all human biomedical researchers and research institutions.[48]

The HBRA also defines a 'research institution' as an entity composed of two or more persons and that has managerial control over human biomedical research that is conducted in Singapore. In order to be legally recognized, a 'research institution' is required to notify the MOH and submit a declaration of compliance before it commences operation. Once recognized, it must appoint an IRB to review all research under its supervision and control, and to report any serious adverse events as defined in the legislation. The HBRA further requires a close relationship between the research institution and its IRB, primarily because the research institution assumes responsibility for research that is reviewed and approved by its IRB. The legal responsibilities of an IRB are essentially similar to its role within the ethical framework: the protection of the safety, dignity and welfare of human research participants.

---

[45] Ibid., sub-ss 3(2)(a)-(c).
[46] Ibid., sub-ss 3(2)(i)-(iii).
[47] Ibid., sch 2, paras 1, 2, 6, 7. Other research excluded from the statutory definition of human biomedical research includes minimal risk studies and tests that relate to normal human psychological responses and behaviours or measurement of human intelligence, as well as activities that are already governed under a different statutory regime.
[48] Ibid., s 3(5).

528 *Calvin W. L. Ho*

In 2015, the BAC consolidated the ethical principles, recommendations and guidelines published prior to that year.[49] Five ethical principles were identified as foundational to the ethics governance of human biomedical research in Singapore. They are as follows: (i) respect for persons; (ii) solidarity; (iii) justice; (iv) proportionality; (v) and sustainability. As far as genome editing technologies are concerned, the BAC explanation of the principle of sustainability is pertinent:

> The research process should be sustainable, in the sense that it should not jeopardise the welfare of later generations. For example, research leading to permanent change to the human genome might not be considered ethical, even if immediately beneficial, because the unforeseeable, potentially harmful long-term implications outweigh the immediate benefits of the research.[50]

In addition, the BAC highlighted that research institutions have a responsibility to ensure that research integrity, that is to say the integrity and validity of the research process, is observed and maintained, and IRBs have a responsibility to check that it has been considered.[51] It is on the basis of this ethical premise that the BAC sets out the substantive responsibilities of research institutions, IRBs and researchers. Their ethical responsibilities have been codified in subsequent regulations and are binding and enforceable under the HBRA.[52]

2 *Appropriate Consent*

In giving effect to the ethical principle of respect of persons, the HBRA places considerable emphasis on obtaining appropriate consent from research participants. For consent to be appropriate, it must be in writing, obtained from the research participant after certain prescribed information has been provided, and it must have been obtained in the presence of a witness.[53] Where the research involves a minor (defined as a person under the age of twenty-one) or an adult without decision-making capacity, additional requirements and

---

49 Bioethics Advisory Committee, *Ethical Guidelines for Human Biomedical Research* (n 32) para 2.2.
50 Ibid., para 2.11.
51 Ibid., para 2.13. As the BAC explains, the principle of beneficence is not considered to be distinct from the principle of respect for persons for many research endeavours and, hence, is not set apart as a standalone principle.
52 Human Biomedical Research (Restricted Research) Regulations 2017, S 622/2017 (Human Biomedical Research Regulation).
53 Human Biomedical Research Act, s 6.

procedures have been set out.[54] The information to be provided to the participant for the purposes of consent-taking includes all of the following:[55]

(a) Nature of the research;

(b) Purpose of the research;

(c) Reasonably foreseeable risks, discomforts or inconveniences to a living research subject arising from the research;

(d) Benefits that the research participant may reasonably expect from the research;

(e) Where applicable, whether there are any alternative procedures or treatments available, and the potential benefits and risks of such alternatives;

(f) Any compensation and treatment available in the event of injury;

(g) Any anticipated expenses the research participant is likely to incur;

(h) The extent to which information identifying the research participant will be kept confidential;

(i) Whether individually identifiable information obtained from the research participant will be used for future research;

(j) Whether any biological material taken from the research participant will be destroyed, discarded or stored for future research;

(k) Whether the research involves information in individually identifiable form;

(l) The circumstances, if any, under which, the research participant will be contacted for further consent, including but not limited to changes in the proposed research, and serious adverse events that would lead to a change in the proposed research;

(m) Whether the research participant would wish to be reidentified in the case of an incidental finding if the proposed research expressly provides for such reidentification;

(n) The right to withdraw consent and the limitations to such withdrawal;

(o) The person or persons to contact to obtain further information on the research and to provide feedback; and

(p) Any other information that the Institutional Review Board may require.

Apart from appropriate consent, the HBRA further requires participation to be voluntary (i.e. free from coercion or intimidation, or deception or

---

[54] Ibid., ss 7 and 8.

[55] Ibid., s 12(1).

misrepresentation),[56] and for privacy and confidentiality to be respected or safeguarded, as the case may be.[57]

3 Human Tissue

As noted above, another important component of the HBRA is the regulation of the collection and use of human tissue. These provisions are set out in Part 6, entitled 'Regulation of Human Tissues Activities and Tissue Banking'. The definition of 'human tissue' is broad.[58] It encompasses any human biological material except those specified in the First Schedule of the legislation (i.e. essentially those that have limited scientific value, such as hair shaft, nail plate and naturally excreted bodily fluids and waste products) or materials that have been substantially manipulated and rendered non-individually identifiable.

The provisions on human tissue give effect to three objectives that were articulated during the public consultation leading to the adoption of the HBRA: protect the safety and welfare of tissue donors; prohibit commercial trading of human tissue; and ensure human tissue used in biomedical research is obtained only through altruistic donations.[59] The second and third of these objectives are relatively straightforward, and take the form of prohibitions of commercial trading of human tissue,[60] advertisements relating to such trade,[61] and compelling a person to provide tissue by means of coercion or intimidation, or by deception or misrepresentation.[62] The first objective is more intricate, and entails consent requirements and restrictions on certain activities of the tissue banks or the IRB. A tissue banking activity includes the collection, storage, procurement, importation, supply, provision or export of human tissue for purposes that are not limited to research, but may be for reasons of public health or epidemiological.[63] For all of these purposes, 'appropriate consent' must be taken. Procedurally, the requirements are as follows: obtaining consent in writing;[64] from the tissue donor personally or otherwise in accordance with additional requirements that apply to adults who

---

[56]   Ibid., s 26.
[57]   Ibid., ss 27–29.
[58]   Ibid., s 2.
[59]   Ministry of Health, *Human Biomedical Research Bill: Public Consultation Nov-Dec* 2014 (2014).
[60]   Human Biomedical Research Act, s 32.
[61]   Ibid., s 33.
[62]   Ibid., s 38.
[63]   Ibid., s 2.
[64]   Ibid., s 6.

lack mental capacity for decision-making,[65] minors[66] and deceased persons;[67] after certain information has been provided and explained, including a list of possible concerns;[68] and in the presence of a witness.

Crucially, the legislation empowers an IRB to waive the requirement of 'appropriate consent' where research involving human biological material (or health information) is concerned. To grant a waiver, the IRB must be satisfied of the following: (a) the proposed research on the individually identifiable human biological material may not practically be carried out unless there is a waiver; (b) the use of such material involves no more than minimal risk to the research subject or donor; (c) the waiver will not adversely affect the rights and welfare of the research subject or donor; and (d) the research would reasonably be considered to contribute to the greater public good.[69] This statutory provision addresses a long-standing legal lacuna relating to the research use of legacy tissue that was first highlighted within a public policy forum held by the BAC more than a decade ago.[70]

4 Additional Requirements for Restricted Research

The Fourth Schedule addresses ethically contentious types of research such as those that involve human embryonic stem cells and human–animal combinations, and is particularly relevant for the purposes of this volume. The types of research for which additional regulatory requirements apply are those that involve:

(a) Human eggs or human embryos;

(b) Human–animal combination embryos, specifically cytoplasmic hybrid embryos, human–animal combination embryos created by the incorporation of human stem cells (including iPSC) and human–animal combination embryos created in vitro by using human gametes and animal gametes, or one human pro-nucleus and one animal pro-nucleus;

(c) Introduction of human stem cells (including iPSC) into a prenatal animal fetus or animal embryo;

(d) Introduction of human pluripotent stem cells (including iPSC) into a living postnatal animal;

---

[65] Ibid., s 9.
[66] Ibid., s 10.
[67] Ibid., s 11.
[68] Ibid., s 12(2).
[69] Ibid., sch 5, pt 2.
[70] Bioethics Advisory Committee, *Human Tissue Research* (2002), para 9.6.

(e) Introduction of human stem cells (including iPSC) or human neural cells into the brain of a living postnatal animal; and

(f) Any entity created as a result of the process referred to in sub-paragraphs (c), (d) and (e) above.

To be sure, 'restricted research' may only be conducted after the requirements that apply to all human biomedical research under the HBRA are satisfied, as well as the additional provisions that are specific to this category of research. These include notification to be provided to MOH; IRB review; appropriate consent having been obtained from the research subject; and conduct of the research only by certain specific persons, at certain specific premises and in a specific manner.[71]

a Procedures that Apply to Restricted Research

Any 'restricted research' can only be conducted after it has been approved: (1) by an IRB of the research institution where the research is to be carried out; (2) where applicable, by an institutional animal care and use committee where any living postnatal animal, living animal fetus or a living animal embryo is to be used; and (3) by the MOH, acting through its director or an authorized designate.[72] Application to conduct restricted research to the MOH must be done through an online process known as Tissue and Research Application System (TIARAS), and detailed information about the research protocol and researcher are to be provided.[73] The MOH has the power to impose certain conditions for the conduct of restricted research. If the restricted research is approved, researchers are required to provide periodic reports to MOH and to ensure that there is no deviation from the approved application.[74] Importantly, appropriate consent for participation in restricted research (including the use of human tissue) must be obtained from the research participant himself or herself, and not from another person.[75] This means that a legal minor who is unable to provide consent or an adult without decision-making capacity will not be able to participate or otherwise contribute identifiable tissue or information to a restricted research. Similarly, tissue from a deceased

---

[71] Human Biomedical Research Act, s 31.
[72] Human Biomedical Research Regulations, s 3(2).
[73] Ministry of Health, 'A guide to TIARAS (Tissue and Research Application System)' (2017) www.moh.gov.sg/content/dam/moh_web/Legislations/Legislation%20And%20Guidelines/TI ARAS%20screenshots%20for%20-%20notifications.pdf accessed 20 August 2018.
[74] Human Biomedical Research Regulation, ss 6–8.
[75] Ibid., s 5.

person cannot be used in a restricted research unless the donor has provided consent prior to her or his demise. Be that as it may, while the HBRA gives the MOH the power to prescribe additional requirements to be satisfied for restricted human biomedical research, it is questionable whether the scope of the legislation can be narrowed, for instance by excluding certain individuals as research participants, through secondary legislation.[76]

b Additional Requirements that Apply to Research Involving Human Oocytes, Embryos and Human–Animal Combinations

In the case of research on human oocytes and embryos, there are additional requirements for appropriate consent-taking. A woman is allowed to donate oocytes purely for research use only when not undergoing infertility treatment.[77] However, if she is undergoing such treatment, donation is possible but additional regulatory requirements must be observed to ensure that the oocytes originally intended for clinical use are no longer needed for therapeutic purposes. Regardless of whether a woman is undergoing infertility treatment or not, the additional regulatory requirements seek to ensure that the donation of oocytes is voluntary and free of undue influence, and that the donor is clearly aware that her donation to research is neither for artificial reproduction nor other therapeutic treatment.[78]

Practically, the potential donor of an oocyte must provide consent in writing, after she has been informed of the full implications of the donation, and she must state that she does not want to have her oocyte set aside for future reproductive use. Where an embryo is concerned, consent in writing is required from both the donor and her husband after they have been informed of the full implications of the donation and confirmed that they no longer require the embryo for their future reproductive use. Additionally, unless the donation relates to immature oocytes or non-viable embryos, a cooling-off period of eight days must be allowed between when the donors are given the requisite information and the donation is done.[79] For oocytes or embryos obtained outside of Singapore, documentary evidence must be produced to show that consent has been obtained in accordance with the legal or ethical requirements of the source jurisdiction.[80]

---

[76] Ibid., s 31(1).
[77] Bioethics Advisory Committee, *Donation of Human Eggs for Research* (2008), para 4.14.
[78] Human Biomedical Research Regulation, ss 11 and 12.
[79] Ibid., s 13.
[80] Ibid., s 9.

In any event, as it was said, regardless of where and how the embryos are obtained, no research is allowed after fourteen days from the time of their creation, excluding any period when their development is suspended.[81] These requirements are reiterated in the regulations in terms of specific responsibilities of an IRB in reviewing a research proposal that involves the use of oocytes or embryos.[82] Crucially, besides approval by the IRB, the MOH could also subject, at its discretion, research protocols to scientific review, and the MOH review could also encompass ethical issues and concerns at the wider societal level.[83]

The creation and use of cytoplasmic hybrid embryos and animal chimeras for research are permitted in Singapore but are strictly regulated. In the consolidated guidelines published in June 2015, the BAC indicated that the main 'ethical hazard lies in the possibility of inadvertently creating an animal with human characteristics, especially, but not exclusively, mental attributes'.[84] While carrying out ethics review, the BAC takes into consideration six relevant factors:[85] (i) proportion or ratio of human to animal cells in the animal's brain; (ii) age of the animal; (iii) recipient species; (iv) brain size of the animal involved; (v) state of integration of human neural cells; (vi) and presence of pathologies in the host animal. The BAC further indicates that research using established pluripotent stem cell lines and confined to cell culture or research that involves routine and standard research practice with laboratory animals is exempted from IRB review.[86] The other ethical requirements set out by the BAC follow international best practices, although clinical and research personnel maintain the right to 'conscientious objection' to research involving human–animal combinations. No one will be put at a disadvantage only because of her or his objection.[87]

V REGULATION OF CLINICAL RESEARCH

In Singapore, genome editing techniques are generally considered to be 'medicinal products' and, thus, fall under the scope of the Medicines

---

[81] Ibid., s 10.

[82] Ibid., s 14.

[83] Ibid., s 15.

[84] Cf. Bioethics Advisory Committee, *Ethical Guidelines for Human Biomedical Research* (n 32) para 7.22.

[85] Ibid. These considerations have been drawn from M Greene and others, 'Moral Issues of Human-Non-Human Primate Neural Grafting' (2005) 309 *Science* 385.

[86] Cf. Bioethics Advisory Committee, *Ethical Guidelines for Human Biomedical Research* (n 32) para 7.24.

[87] Ibid., para 7.31.

Act.[88] A 'medicinal product' is defined as any substance or article (not being an instrument, apparatus or appliance) which is manufactured, sold, supplied, imported or exported for use wholly or mainly in either or both of the following ways:

> (a) by being administered to one or more human beings or animals for a medicinal purpose;
>
> (b) ... as an ingredient in the preparation of a substance or article which is to be administered to one or more human beings or animals for a medicinal purpose, which means treating or preventing disease, diagnosing disease or ascertaining the existence, degree or extent of a physiological condition, contraception, inducing anaesthesia, or otherwise preventing or interfering with the normal operation of a physiological function, whether permanently or temporarily, and whether by way of terminating, reducing or postponing, or increasing or accelerating, the operation of that function or in any other way.[89]

However, 'medicinal product' does not include any substance or article which is manufactured for use wholly or mainly by one or more human beings or animals, where it is to be administered to them:

> (a) in the course of the business of the manufacturer or on behalf of the manufacturer in the course of the business of a laboratory or research establishment carried on by another person;
>
> (b) solely by way of a test for ascertaining what effects it has when so administered; and
>
> (c) in circumstances where the manufacturer has no knowledge of any evidence that those effects are likely to be beneficial to those human beings, or beneficial to, or otherwise advantageous in relation to, those animals, as the case may be, and which (having been so manufactured) is not sold, supplied or exported for use wholly or mainly in any way not fulfilling all the conditions specified in paragraphs (a), (b) and (c).[90]

These requirements are set out in the Medicines Act, the Medicines (Clinical Trials) Regulations,[91] the Medicines (Medicinal Products as Clinical Research Materials) Regulations 2016[92] and regulatory guidance

---

[88]   Medicines Act 1975 (revised 1985).
[89]   Ibid, s 3.
[90]   Ibid.
[91]   Medicines (Clinical Trials) Regulations 2016.
[92]   Medicines (Medicinal Products as Clinical Research Materials) Regulations 2016.

documents issued by the Health Sciences Authority (HSA).[93] The HSA is the principal regulator of clinical trials in Singapore and it reports to the MOH. For all clinical trials of 'medicinal products', prior approval must be obtained from the HSA (in the form of a Clinical Trial Certificate) and from an IRB. An application for approval can be submitted concurrently to HSA and the relevant IRB.

The requirements for IRB approval do not substantially differ from those discussed in Section IV. Crucially, the IRB must ensure that: (1) the proposal is reviewed and approved by a scientific review committee with the relevant expertise; (2) there is strong evidence of safety and efficacy from preclinical studies; (3) the research participants have been provided with sufficient information, especially in relation to the nature and risks of the research, and, where applicable, the source of the cells; and (4) appropriate and informed consent has been obtained, without any inducement, coercion or undue influence.[94] If the clinical trial participant is a legal minor (i.e. a person under twenty-one years of age), an adult lacking decision-making capacity or if the clinical trial is conducted in an emergency situation, additional requirements relating to consent will apply.[95]

Before recruiting for a clinical trial, a Clinical Trial Certificate must be obtained from the HSA. A certificate is issued if the proposed clinical trial is conducted in accordance with requirements set out in the *Medicines (Clinical Trials) Regulation and the Good Clinical Practice (GCP) Guidelines* (ICH E6 Guidelines).[96] The ICH E6 GCP Guidelines have been amended to encourage research institutions to implement quality management systems to ensure protection of human subjects participating in trials and clinical trial data integrity. Following are the regulatory goals of the HSA: (1) facilitating access to gene therapy medicinal products for use in clinical research through a simplified and harmonized regulatory notification system; (2) requiring imported or locally manufactured gene therapy medicinal products to be of sufficient quality; (3) restricting supply of imported or locally manufactured gene therapy medicinal products to regulated trials or IRB-approved clinical research; (4) requiring traceability and accountability of gene therapy

---

[93] For Regulatory guidance, see Health Sciences Authority, 'Health Products Regulation' (Singapore, 2018) www.hsa.gov.sg/content/hsa/en/Health_Products_Regulation/Clinical_Tri als/Overview/Regulatory_Guidelines.html accessed 20 August 2018.

[94] Ibid.

[95] HSA, *Regulatory Guidance: Safeguards and Consent Requirements in Vulnerable Subjects* (2 May 2017).

[96] HSA, *Regulatory Guidance* (Singapore, 2018) www.ich.org/products/guidelines/efficacy/arti cle/efficacy-guidelines.html#6–2 accessed 20 August 2018.

medicinal products through record-keeping; (5) requiring disposal or export of imported or locally manufactured gene therapy medicinal products after the research or trial ends; (6) requiring appropriate gene therapy medicinal products labelling; and (7) requiring reporting of unexpected serious adverse drug reactions related to the use of gene therapy medicinal products.[97] The processing time to obtain a Clinical Trial Certificate is set at thirty working days.[98] Regulatory requirements are set out in a guidance document,[99] and the regulatory roadmap for conducting clinical trials of medicinal products is depicted in Table 19.1.[100]

VI CURRENT PERSPECTIVES AND FUTURE POSSIBILITIES

Genome editing techniques could be used in basic research, to effect somatic gene modification, or to alter the germline genome. Earlier sections of this chapter have discussed how existing laws and regulations could apply in the governance of such techniques in Singapore, and I conclude in this section with a summary of activities or applications that are prohibited or permissible on a restricted basis, and the procedures that need to be observed. Where pertinent, gaps or ambiguities in the existing regimes will be noted.

The application of genome editing techniques in basic research falls within the scope of the HBRA. Approval from an appropriate IRB is legally required, unless an exemption applies. The requirements relating to consent, privacy and confidentiality, as well as those that apply to the collection and use of human tissue, must be observed. The research can be conducted only once IRB approval has been obtained following expedited or full review (depending on whether the nature of the research is sensitive or if it poses significant risks of harm to potential participants), unless it is exempted from review. It is important to note that beyond what is legally required, a research institution is at liberty to require research protocols that fall outside the remit of the HBRA, and submit them nevertheless for review and approval by its IRB.

Research involving human oocytes, embryos and human–animal combinations (as defined in the Fourth Schedule of the HBRA) is 'restricted research',

---

[97] HSA, *Regulatory Framework* (Singapore, 2018) www.hsa.gov.sg/content/hsa/en/Health_Product s_Regulation/Clinical_Trials/Overview/Regulatory_Framework.html accessed 20 August 2018.

[98] HSA, *Target Processing Timelines* (Singapore, 2018) www.hsa.gov.sg/content/hsa/en/Health_Pro ducts_Regulation/Useful_Information_for_Applicants/Target_Processing_Timelines.html accessed 20 August 2018.

[99] Cf. HSA, *Regulatory Guidance: Safeguards and Consent Requirements in Vulnerable Subjects* (n 95).

[100] HSA, *Overview* (Singapore, 2018) www.hsa.gov.sg/content/hsa/en/Health_Products_Regulat ion/Clinical_Trials/Overview.html accessed 20 August 2018.

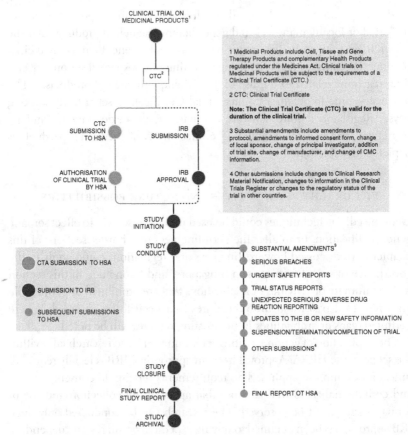

CLINICAL TRIAL ON
MEDICINAL PRODUCTS[1]

CTC[2]

1 Medicinal Products include Cell, Tissue and Gene
Therapy Products and complementary Health Products
regulated under the Medicines Act. Clinical trials on
Medicinal Products will be subject to the requirements of a
Clinical Trial Certificate (CTC.)

2 CTC: Clinical Trial Certificate

**Note: The Clinical Trial Certificate (CTC) is valid for the
duration of the clinical trial.**

CTC
SUBMISSION
TO HSA

IRB
SUBMISSION

3 Substantial amendments include amendments to
protocol, amendments to informed consent form, change
of local sponsor, change of principal investigator, addition
of trial site, change of manufacturer, and change of CMC
information.

AUTHORISATION
OF CLINICAL TRIAL
BY HSA

IRB
APPROVAL

4 Other submissions include changes to Clinical Research
Material Notification, changes to information in the Clinical
Trials Register or changes to the regulatory status of the
trial in other countries.

STUDY
INITIATION

STUDY
CONDUCT

SUBSTANTIAL AMENDMENTS[3]

CTA SUBMISSION TO HSA

SERIOUS BREACHES

URGENT SAFETY REPORTS

SUBMISSION TO IRB

TRIAL STATUS REPORTS

UNEXPECTED SERIOUS ADVERSE DRUG
REACTION REPORTING

SUBSEQUENT SUBMISSIONS
TO HSA

UPDATES TO THE IB OR NEW SAFETY INFORMATION

SUSPENSION/TERMINATION/COMPLETION OF TRIAL

OTHER SUBMISSIONS[4]

STUDY
CLOSURE

FINAL CLINICAL
STUDY REPORT

FINAL REPORT OT HSA

STUDY
ARCHIVAL

FIGURE 19.1 Singapore: regulatory road map for conducting clinical trials of
medicinal products
Source: Health Sciences Authority, www.hsa.gov.sg/content/hsa/en/health_pro
ducts_regulation/clinical_trials/overview.html accessed 21 January 2019.

and the protocol will require full review by the appropriate IRB. Where legal
and regulatory provisions on human oocytes are concerned, a purposive reading of the legislation suggests that these requirements under the HBRA will
similarly apply to any oocyte that has been genetically modified or reconstituted as an artificial gamete. Once approval has been obtained from the IRB,
approval from MOH must also be obtained. These approvals can only be
obtained if all the ethical and regulatory requirements are observed, including
those that apply specifically to consent-taking and the procurement of these
ethically sensitive biological materials. As noted above, it is unclear what
standard the MOH applies in its review, although it is clear that it has the

power to impose additional requirements as it considers necessary. Where the research involves the application of genome editing techniques to a human embryo, it is strictly restricted to the first fourteen days of embryonic development (excluding any period of suspended growth). In addition, human oocytes and embryos that have been subject to research manipulation cannot be used for reproductive purposes. Any person who is convicted for contravening any of these legal requirements could be subject to a fine of not exceeding S$100,000 (about US$72,000) or to imprisonment for a term not exceeding ten years, or to both.[101]

Genome editing of somatic cells is likely to be regarded as a research intervention and either falls within the scope of the HBRA or is otherwise regulated as a clinical trial, if applied therapeutically. If a genome editing technique is to be applied in a clinical trial setting, approvals of an appropriate IRB and the HSA are required before the trial may be conducted. Comprehensibly, there will be greater focus on ensuring the safety of trial participants. The principles of good clinical practice must be observed in addition to ethical and regulatory provisions. If trial participants are 'vulnerable', as defined by the HSA's regulatory guidance document, additional requirements will apply to obtain the requisite approvals.

Genome editing of germline cells and embryos that is conducted as basic research is 'restricted research' under the HBRA. Consequently, the legal and regulatory requirements pertaining to the HBRA, as well as the ethical provisions of the BAC, will be applicable. At this point in history, germline genome editing cannot be done for therapeutic purposes. As noted earlier, in 2005 a moratorium was imposed against germline genetic modification. The practical implication is that no IRB in Singapore will approve any research protocol of this nature. It is less clear if certain types of genetic modification technologies, like ooplasmic transfer,[102] which can help in treating mitochondrial diseases, could be permitted on a case-by-case basis, and regulated directly by the MOH as an ART, or otherwise in the context of a clinical trial. Given that the BAC is currently deliberating on this matter and it is expected to provide recommendations in due course, it is unlikely that the MOH will approve any germline genetic modification applications until the BAC has pronounced on the subject.

---

[101] Human Biomedical Research Act, s 31(4).
[102] Ooplasmic transfer, also known as cytoplasmic transfer, is an experimental fertility technique that involves injecting a small amount of ooplasm from eggs of fertile women into eggs of women whose fertility is compromised.

It is probably premature to discuss the possible use of genome editing techniques for non-therapeutic enhancement of physical or social traits as the BAC has considered such genetic interventions to be unethical and, at this time, socially unacceptable. In relation to preimplantation genetic diagnosis (PGD), the BAC prohibits its use for the selection of desired traits or sex for non-medical reasons.[103] The same can be logically extended to genome editing techniques, should they eventually near clinical trial. In other words, genome editing technology could only be used to treat serious genetic disorders, and not simply any medical conditions, or for sex selection, or indeed any 'lifestyle' reasons. The BAC does not consider sex selection done for non-medical reasons (i.e. to balance the gender ratio in the family, personal preferences or certain social, cultural, religious or economic motivations) to be ethically and socially acceptable.[104] The regulation that applies to ARCs in Singapore reflects this stance. Services of ARCs are limited to specific medical indications, and PGD can be done only to screen for specific serious medical conditions.[105]

Moving forward, it is likely there will be significant developments in the regulatory landscape for genetic-related activities in Singapore. Regulatory focus has been on the inevitable translation of whole-genome and whole-exome sequencing to clinical practice,[106] and the existing regulatory framework could well be expanded to include the clinical application of certain types of germline genetic modification techniques. In addition, it remains to be seen whether the MOH will establish an independent regulatory agency to oversee human biomedical research, much like the HSA where clinical trials are concerned. Meanwhile, further systematization and refinement of review standards and practices is expected, as the government of Singapore moves towards achieving greater transparency and efficiency, among other values, in the regulation of human biomedical research and clinical trials.

---

[103] Cf. Bioethics Advisory Committee, *Genetic Testing and Genetic Research* (n 31) paras 4.47–4.50, Recommendation 10.

[104] Ibid., paras 4.44–4.46.

[105] Cf. Ministry of Health, *Licensing Terms and Conditions on Assisted Reproduction Services* (n 23), cls 13–14.

[106] Sequencing (i.e. determining the order of DNA building blocks (nucleotides) in an individual's genetic code) is one technique used to test for genetic disorders. Two methods (whole exome sequencing and whole genome sequencing) are increasingly being used in healthcare and research to identify genetic variations. Both methods rely on new technologies that allow rapid sequencing of large amounts of DNA. These approaches are known as 'next-generation sequencing'.

PART IV

OTHER OECD COUNTRIES

20

The Regulation of Human Germline Genome Modification in Australia

Dianne Nicol

I INTRODUCTION

Australia has a long history of engagement with innovative reproductive technologies. Indeed, the development of modern artificial reproductive technologies (ART) was pioneered in Australia in the late 1960s and early 1970s.[1] In parallel, some of the states of Australia were early adopters of legislation aimed at regulating certain forms of ART and prohibiting others.[2] Since then, a number of other statutes have been enacted across Australia that have direct bearing on the practice of human germline genome modification and research involving human embryos.

Developments in the late 1990s, including the creation of Dolly the sheep through reproductive cloning and the discovery of the pluripotency of embryonic stem cells, indicated that urgent and sharp attention needed to be paid to the adequacy of existing legislation, both in Australia and elsewhere. In the Australian context, the Australian Health Ethics Committee,[3] one of the principal committees of the National Health and Medical Research Council

---

[1] J Leeton, 'The Early History of IVF in Australia and Its Contribution to the World (1970–1990)' (2004) 44 *Australian and New Zealand Journal of Obstetrics and Gynaecology* 495, 495–501.

[2] The world's first ART legislation was the *Infertility (Medical Procedures) Act 1984* (Vic) introduced in the state of Victoria. That act was later replaced by the *Infertility Treatment Act 1995* (Vic), and, subsequently, the *Assisted Reproductive Treatment Act 2008* (Vic). Similar, though not identical, legislation was introduced in South Australia (*Assisted Reproductive Treatment Act 1988* (SA)) and Western Australia (*Human Reproductive Technology Act 1991* (WA)) and, some time later, in New South Wales (*Assisted Reproductive Technology Act 2007* (NSW)). This legislative history is discussed in: D Nicol, D Chalmers and B Gogarty, 'Regulating Biomedical Advances: Embryonic Stem Cell Research' (2002) 2 *Macquarie Law Journal* 3.

[3] National Health and Medical Research Council, 'Australian Health Ethics Committee' www.nhmrc.gov.au/australian-health-ethics-committee accessed 31 July 2018.

(NHMRC),[4] undertook a review of the scientific, ethical and regulatory landscape relating to human reproductive cloning, releasing a final report in 1998.[5] Thereafter, the lower house of the federal Parliament of Australia charged its Standing Committee on Legal and Constitutional Affairs to undertake a further inquiry, focusing on both human cloning and stem cell research, the final report of which was released in 2001.[6]

The result of these inquiries was the enactment of two key pieces of legislation: the *Prohibition of Human Cloning Act 2002*[7] and the *Research Involving Human Embryos Act 2002* (RIHE Act).[8] The title of the *Prohibition of Human Cloning Act 2002* was subsequently amended to the *Prohibition of Human Cloning for Reproduction Act 2002* (PHCR Act),[9] through the *Prohibition of Human Cloning for Reproduction and the Regulation of Human Embryo Research Amendment Act 2006*, discussed further below. The path to enactment of these statutes was not an easy one for the federal legislature. It followed one of the longest-ever parliamentary debates, with members of the political parties being given a free vote in recognition of their deeply personal views on the moral status of the embryo and the potential for these new technological developments to relieve human suffering.[10] Each of the states and territories of Australia, apart from the Northern Territory, has enacted its own legislation to ensure national consistency in respect of both acts.[11]

---

4   Australian Government, 'National Health and Medical Research Council' www .nhmrc.gov.au accessed 31 July 2018.

5   NHMRC Australian Health Ethics Committee, *Scientific, Ethical and Regulatory Considerations Relevant to Cloning of Human Beings: A Report to the Commonwealth Minister for Health and Aged Care* (Chair Professor Don Chalmers, 1998).

6   House of Representatives Standing Committee on Legal and Constitutional Affairs, *Human Cloning: Scientific, Ethical and Regulatory Aspects of Human Cloning and Stem Cell Research* (Chair Mr Kevin Andrews MP, 2001).

7   Prohibition of Human Cloning Act 2002, No. 144: An Act to prohibit human cloning and other unacceptable practices associated with reproductive technology, and for related purposes.

8   Research Involving Human Embryos Act 2002, No. 145: An Act to regulate certain activities involving the use of human embryos, and for related purposes (RIHE Act). See NHMRC, *NHMRC Embryo Research Licensing Committee: Information Kit* (NHMRC 2018).

9   Prohibition of Human Cloning for Reproduction Act 2002, No. 144: An Act to prohibit human cloning for reproduction and other unacceptable practices associated with reproductive technology, and for related purposes (PHCR Act).

10 For a review of the key elements of the debate, see D Chalmers and D Nicol, 'Embryonic Stem Cell Research: Can the Law Balance Ethical, Scientific and Economic Values?' (2003) 18 *Law and Human Genome Review* 43 and 19 *Law and Human Genome Review* 91.

11 *Human Cloning and Embryo Research Act 2004* (ACT); *Research Involving Human Embryos (New South Wales) Act 2003* (NSW); *Human Cloning for Reproduction and other Prohibited Practices Act 2003* (NSW); *Research Involving Human Embryos and Prohibition of Human*

The federal Parliament made a comprehensive review of the Acts in 2005 (the so-called *Lockhart Review*),[12] as mandated by the Acts themselves.[13] This led to the adoption of amendments in 2006 through the *Prohibition of Human Cloning for Reproduction and the Regulation of Human Embryo Research Amendment Act 2006*.[14] Aside from the change to the title of the PHCR Act, the 2006 Act also amended various substantive provisions in both Acts, reducing some of the regulatory barriers to research involving human embryos. For example, the opportunity for somatic cell nuclear transfer research was made available in Australia as a result of these amendments.

The 2006 amendments introduced a further mandatory review requirement as soon as possible after the third anniversary of the day on which it received Royal Assent.[15] Although that review, undertaken in 2010–2011, recommended some further amendments, they were never implemented. Overall, it was concluded that 'the basic structure of the legislation introduced in 2002, as amended in 2006, should remain'.[16] Since then, there have been no further mandatory requirements for review included in the legislation. Given the rate of technological change, however, it seems timely to reconsider the extent to which the provisions of both Acts remain relevant and appropriate, from technological, social, ethical and health and safety perspectives.

In parallel, concerns about the use of new genetic technologies, particularly in the context of agriculture, prompted equally impassioned debates about the benefits and risks associated with the release of genetically modified organisms (GMOs) into the environment. The first step in regulating this emergent technology was the establishment of a Recombinant DNA Monitoring Committee in 1981, which was replaced by a Genetic Manipulation

---

Cloning Act 2003 (Qld); *Research Involving Human Embryos Act 2003* (SA); *Prohibition of Human Cloning Act 2003* (SA); *Human Embryonic Research Regulation Act 2003* (Tas); *Human Cloning and Other Prohibited Practices Act 2003* (Tas); *Research Involving Human Embryos Act 2008* (Vic); *Prohibition of Human Cloning for Reproduction Act 2008* (Vic); *Human Reproductive Technology Act 1991* (WA) (as amended in 2004). Legislation has not yet been enacted in the Northern Territory.

12   Legislation Review Committee, *Legislation Review: Prohibition of Human Cloning Act 2002 and Research Involving Human Embryos Act 2002* (the Lockhart Review, after the chair, 2005).

13   Both of the original Acts specify that there should be a review as soon as possible after the second anniversary of receipt of Royal Assent. RIHE Act 2002, s 47 and PHCR Act, s 25.

14   Prohibition of Human Cloning for Reproduction and the Regulation of Human Embryo Research Amended Act 2006, No. 172: An Act to amend the *Prohibition of Human Cloning Act 2002* and the *Research Involving Human Embryos Act 2002* based on the Lockhart Review recommendations, and for related purposes.

15   RIHE Act, s 47A (as amended) and PHCR Act, s 25A (as amended).

16   Legislation Review Committee, *Legislation Review: Prohibition of Human Cloning Act 2002 and Research Involving Human Embryos Act 2002* (the Heerey Review, after the chair, 2011), 15.

Advisory Committee in 1987. However, it was not until 2000 that federal legislation was enacted, in the form of the *Gene Technology Act 2000* (GT Act).[17] As with the PHCR Act and RIHE Act, the states and territories enacted their own legislation to provide a nationally consistent regime for regulating genetic technologies.[18] The ambit of this legislation is not limited to plants and animals in agriculture, but extends into the human genome manipulation space.

Together, the PHCR Act, the RIHE Act and the GT Act form a triad of key legal instruments regulating human germline genome modification and genome editing research involving human embryos in Australia. Collectively, these statutes sensibly create a highly precautionary environment around germline genome manipulation applications, through a mixture of regulation and prohibition. At the same time, they also pose significant barriers to genome editing research involving human embryos, more so than in many other nations.

This chapter will discuss the regulation of human germline genome modification and genome editing research involving human embryos in Australia, with particular focus on these three statutes, whilst recognizing that other legislative and softer regulatory instruments also play a role in governing and limiting research in this field.

II REGULATORY ENVIRONMENT

1 *Regulatory Framework*

Australia comprises six states and two territories, as well as the federal Commonwealth of Australia, each with its own legislature and judiciary. This has resulted in a complex matrix of regulatory obligations across the country, which can potentially be both overlapping and inconsistent. Section 51 of the Australian Constitution provides an exclusive list of legislative powers of the federal (Commonwealth) Parliament.[19] There are no specific health or

---

[17]   *Gene Technology Act 2000*, No. 169: *An Act to Regulate Activities Involving Gene Technology, and for Related Purposes* (GT Act). See generally K Ludlow, 'Gene Technology Regulation and the Environment Protection and Biodiversity Conservation Act 1999 (Cth)' (2004) 30 *Monash University Law Review* 165; D Tribe, 'Gene Technology Regulation in Australia' (2012) 3 *GM Crops & Food* 21.

[18]   *Gene Technology Act 2003* (ACT); *Gene Technology (New South Wales) Act 2003* (NSW); *Gene Technology (Northern Territory) Act 2004* (NT); *Gene Technology (Queensland) Act 2016* (Qld); *Gene Technology Act 2001* (SA); *Gene Technology (Tasmania) Act 2012* (Tas); *Gene Technology Act 2001* (Vic); *Gene Technology Act 2006* (WA).

[19]   Commonwealth of Australia Constitution Act 1900.

environmental powers. However, powers relating to matters such as external affairs, interstate trade and corporations have allowed the federal Parliament to expand its legislative reach into many areas that affect the day-to-day life of Australian citizens. For example, the *Privacy Act 1988* imposes privacy obligations on federal government agencies and corporations, but leaves the states and territories to enact legislation creating privacy obligations for their own government agencies, which include most universities and public hospitals.[20] One of the difficulties this arrangement creates is that state government agencies are bound by different privacy principles from their federal counterparts and private sector organizations.

To counter further occurrences of this type of fragmentation, the various state, territory and Commonwealth governments have come together from time to time through the forum of the Council of Australian Governments[21] and agreed to enact legislation to ensure national consistency. This has been particularly evident in areas where transformational technological developments and heightened ethical and social concerns intersect. The triad of the PHCR Act, the RIHE Act and the GT Act, together with their state and territory counterparts, are examples of this approach.[22] In respect of each Act, the Commonwealth legislation is broad in its application, applying to matters involving: corporations; the course of trade or commerce between states or between Australia and other countries; the Commonwealth of Australia and Commonwealth authorities; purposes relating to the collection, compilation, analysis and dissemination of statistics; and matters otherwise within the legislative power of the Commonwealth.[23] The counterpart state and territory legislation ensures there are no regulatory gaps. Each federal Act has, at its core, concerns about human welfare. Section 3 of the GT Act reflects a concern about the health and safety of people and protection of the environment. When it comes to Section 3 of the PHCR Act and Section 3 of the RIHE

---

[20]   Privacy Act 1988, No. 119: An act to make provision to protect the privacy of individuals, and for related purposes.

[21]   Council of Australian Governments, 'Council of Australian Governments' www.coag.gov.au, accessed 31 July 2018.

[22]   See Council of Australian Governments, 'Research Involving Human Embryos and Prohibition of Human Cloning, Agreement of 31 March 2004' www.coag.gov.au/about-coag/agreements/re search-involving-human-embryos-and-prohibition-human-cloning accessed 31 July 2018. The original agreement relating to gene technology of 11 September 2001 was not made publicly available. However, in 2008 the parties publicly affirmed their commitment to the nationally consistent scheme and agreed to amend the agreement. The complete amended agreement was published on 3 July 2008. Council of Australian Governments, 'Gene Technology Agreement' www.coag.gov.au/about-coag/agreements/gene-technology-agreement accessed 31 July 2018.

[23]   PHCR Act, s 4; RIHE Act, s 4; GT Act s 13. GT Act, s 13.1(c) adds things done, or omitted to be done, by a person that may cause the spread of diseases or pests.

Act, the focus is more on ethical concerns associated with scientific developments in relation to human reproduction and the utilization of human embryos.

The prospect of clinical application of germline genome manipulation techniques is one that, as we shall see, cannot be realized without breaching the prohibitions under the current PHCR Act. However, should genetic modification of the genome of human germline advance to clinical application, the *Therapeutic Goods Act* 1989 would become relevant since it regulates the use of goods for therapeutic purposes.[24] This Act sets up a regime for registration of therapeutic goods in Australia through the Therapeutic Goods Administration, requiring sufficient evidence of quality, safety, efficacy and clinical utility prior to registration.[25] Other legislation and common law principles may also have relevance from time to time, including privacy, anti-discrimination, negligence and contract laws, but it is the RIHE Act, the PHCR Act and the GT Act that have by far the greatest influence on the practice of germline genome modification and embryonic genome editing research.

Other aspects of the regulatory environment relating to innovative reproductive technologies in Australia include ethical guidelines for research and practice and self-regulatory codes. As with the legal regime, ethical guidelines for research involving humans are consistent nationally. The NHMRC *National Statement on Ethical Conduct in Human Research* (*National Statement*) covers the overarching values and principles of ethical conduct of research and establishes the Human Research Ethics Committee system. The *National Statement* is jointly issued by the principal national funding bodies: the NHMRC and the Australian Research Council,[26] as well as Universities Australia, the peak national body representing Australian universities.[27] The latest edition of the *National Statement* was published in July 2018. Whilst it does not have the force of law, the *National Statement* is effective because universities conducting research will not be able to receive any federal funding if they do not follow it. Although the threat of withdrawal of funding does not work against the private sector, there is a widespread

---

[24] Available at www.legislation.gov.au/Details/C2018C00082 accessed 31 July 2018. Therapeutic Goods Act 1989, No. 21: An act relating to therapeutic goods.

[25] www.tga.gov.au accessed 31 July 2018. Department of Health, Therapeutic Goods Administration, 'Australian Register of Therapeutic Goods' http://tga-search.clients.funnelback.com/s/search.html ?query=&collection=tga-artg accessed 31 July 2018.

[26] Australian Government, 'Australian Research Council' www.arc.gov.au accessed 31 July 2018.

[27] Universities Australia, 'Vision & Mission' (21 January 2014) www.universitiesaustralia.edu.au /About-Us/Vision–Mission#.WxiMQC1L1AY accessed 31 July 2018.

tendency towards compliance for a number of reasons including reputation; ongoing collaboration with universities; and mandates in legislation, including the RIHE Act.

The *National Statement* sets out the key ethical principles and values that are required to be considered in the design and conduct of research involving humans. The NHMRC's Australian Health Ethics Committee also published *Guidelines on the Use of Assisted Reproductive Technology in Clinical Practice and Research* in 1982,[28] the first to do so globally.[29] These guidelines have been updated on a number of occasions, including, most recently, in 2017, through the *Ethical Guidelines on the Use of Assisted Reproductive Technology in Clinical Practice and Research*.[30] The key self-regulatory code of practice is the Fertility Society of Australia Reproductive Technology Accreditation Committee's *Code of Practice for Assisted Reproductive Technology Units*,[31] as revised in October 2017. Once again, the Code is consistent nationally. Compliance with the Code is mandatory for any units involved in the treatment of patients using ART.

The primary ethical guidelines for research involving animals are found in the *Australian Code for the Care and Use of Animals for Scientific Purposes* (8th Edition, 2013)[32] (*Animal Code*). Like the *National Statement*, the *Animal Code* is endorsed by the NHMRC, the Australian Research Council and Universities Australia. The *Animal Code* is also endorsed by the Commonwealth Scientific Industrial Research Organization (the peak Australian government research agency). The *Animal Code* establishes the Animal Ethics Committee system. As with the *National Statement*, the *Animal Code* is effective because receipt of federal funding for research is dependent on compliance with these guidelines. In the case of the *Animal Code*, however, there is further reason to comply. State and territory animal welfare legislation provides exemptions from animal cruelty offences for institutions that are compliant with the *Animal Code*.[33]

---

[28]   National Health and Medical Research Council, *Supplementary Note 4 (In Vitro Fertilisation and Embryo Transfer) to the Statement on Human Experimentation* (NHMRC, 1966).

[29]   J Leeton, 'The Early History of IVF in Australia and Its Contribution to the World (1970–1990)' (2004) 44 *Australian and New Zealand Journal of Obstetrics and Gynaecology* 495.

[30]   NHMRC, *Ethical Guidelines on the Use of Assisted Reproductive Technology in Clinical Practice and Research* (NHMRC 2017).

[31]   The Fertility Society of Australia, 'The Reproductive Technology Accreditation Committee' www.fertilitysociety.com.au/rtac/ accessed 31 July 2018.

[32]   NHMRC, *Australian Code for the Care and Use of Animals for Scientific Purposes* (8th ed., NHMRC 2013).

[33]   To provide one example, Part 4 of the *Animal Welfare Act 1993* (Tas) creates a licensing regime for animal research. Grant of licenses is conditional on compliance with a Code of Practice issued by the relevant minister. The currently approved Code of Practice is the national Animal Code. See Department of Primary Industries, Parks, Water and

Prior to the 2013 edition of the *Animal Code*, the NHMRC Animal Welfare Committee's *Guidelines for the Generation, Breeding, Care and Use of Genetically Modified and Cloned Animals for Scientific Purposes* applied.[34] However, these guidelines were revoked in 2017 on the basis that the 2013 edition of the *Animal Code* incorporated many of the recommendations therein.[35]

2 Definitions

The key definition for the purpose of this chapter is the definition of a human embryo. In Section 7 of the RIHE Act, as enacted, and Section 8 of the *Prohibition of Human Cloning Act 2002* (as the PHCR Act was entitled on enactment), the moment when two pronuclei become visible, marking the moment when fertilization can first be verified, marked the start of the human embryonic period:

> Human embryo means a live embryo that has a human genome or an altered human genome and that has been developing for less than 8 weeks since the appearance of 2 pro-nuclei or the initiation of its development by other means.[36]

This definition was significantly amended in 2006 pursuant to the recommendations of the *Lockhart Review* to make it clear that human embryo status is not achieved until the first mitotic division. Thus, currently, under Section 8 of the PHCR Act and Section 7 of the RIHE Act, a human embryo is defined as

> a discrete entity that has arisen from either:
> (a) the first mitotic division when fertilisation of a human oocyte by a human sperm is complete; or
> (b) any other process that initiates organised development of a biological entity with a human nuclear genome or altered human nuclear genome that has the potential to develop up to, or beyond, the stage at which the primitive streak appears;
> and has not yet reached 8 weeks of development since the first mitotic division.[37]

Environment, 'Animal Research' https://dpipwe.tas.gov.au/biosecurity-tasmania/animal-biosecurity/animal-welfare/animal-research accessed 31 July 2018.

[34] NHMRC Animal Welfare Committee, *Guidelines for the Generation, Breeding, Care and Use of Genetically Modified and Cloned Animals for Scientific Purposes* (NHMRC 2007).

[35] NHMRC; 'Mapping: GM Guidelines and the Code' (2017), www.nhmrc.gov.au/_files_nhmrc/file/health_ethics/animal/mapping-gm-guidelines-code.pdf accessed 31 July 2018.

[36] RIHE Act, s 7; Prohibition of Human Cloning Act 2002, s 8.

[37] Ibid.

The stated purpose of these amendments, as articulated in the *Lockhart Review*, was to allow, under license, research and training in ART clinics on various aspects of egg culture, manipulation and maintenance, while retaining a broad definition of the human embryo.[38] It will be seen later in this chapter that a sharp distinction is drawn in the legislation between embryos created by fertilization of a human oocyte by a human sperm in part (a) of the definition, and embryos created by other means in part (b). A number of the activities included in the prohibitions contained in the PHCR Act apply only to part (a) embryos, with the equivalent activities on part (b) embryos being prohibited only if they are done without a license. The stated purpose for distinguishing between these two types of embryos was so that 'the creation of human embryo clones by somatic cell nuclear transfer [c]ould be permitted, under licence, for research, training and clinical applications', in recognition of 'the wide range of diseases and conditions that stem cell research aims to help, and the burden of disease involved'.[39]

The RIHE Act limits the embryos that are available for research to: embryos falling within part (b) of the definition of embryo in Section 7 of the RIHE Act, but only to the extent that the use is a licensed use (Section 10A); and embryos falling within the part (a) definition of embryo, but only to the extent that they are created by ART (Section 11) and are excess ART embryos (Section 9). The question of how an embryo becomes an excess ART embryo is explored later in this chapter. With regard to use, excess ART embryos can only be used for licensed or exempt purposes (Section 10). Section 10.2 specifies the exempt purposes that are allowed, including: storage removal and transport; observation; allowing the embryo to succumb; uses carried out by an accredited ART centre either for diagnostic purposes on embryos not suitable for implantation, or to achieve a pregnancy in a woman other than the woman for whom the embryo was created; or other prescribed uses.

It is also relevant to note that Section 10 of the GT Act defines 'gene technology' as the modification of genes or other genetic material, but specifically excludes modifications of genes or other genetic material resulting from sexual reproduction, homologous recombination and any other prescribed technique. 'Genetically modified organism' is defined in Section 10 of the GT Act as:

(a) an organism that has been modified by gene technology; or

---

[38] Legislation Review Committee, *Legislation Review: Prohibition of Human Cloning Act 2002 and Research Involving Human Embryos Act 2002* (2005), xv.

[39] Ibid. xvii.

(b) an organism that has inherited particular traits from an organism (the initial organism), being traits that occurred in the initial organism because of gene technology; or

(c) anything declared by the regulations to be a genetically modified organism, or that belongs to a class of things declared by the regulations to be genetically modified organisms;

but does not include:

(d) a human being, if the human being is covered by paragraph (a) only because the human being has undergone somatic cell gene therapy; or

(e) an organism declared by the regulations not to be a genetically modified organism, or that belongs to a class of organisms declared by the regulations not to be genetically modified organisms.[40]

Section 10 of the GT Act also defines GMOs as genetically modified organisms.

3 Oversight Bodies

The Embryo Research Licensing Committee (ERLC) of the NHMRC administers the RIHE Act.[41] The principal tasks of the ERLC are to license the use of excess ART embryos (although not specified in the legislation, this would include use for research) and creation and use of embryos coming within part (b) of the definition of embryos in Section 7 of the RIHE Act, as well as to regulate research and training involving fertilization of human eggs and sperm, and the creation and use of animal–human hybrid embryos.[42] As with many other countries, the fourteen-day rule applies in Australia, in that the ERLC is not allowed to approve any use that would result in the development of an embryo beyond fourteen days (aside from any period where development is suspended).[43] Before issuing a license, the ERLC must first be satisfied that all the required consents are in place and that the applicant has obtained approval for the activity or project by a properly constituted HREC in accordance with the *National Statement*.[44] Then, taking into account several factors, the ERLC must decide whether the license could be issued.[45]

---

[40] GT Act, s 10.

[41] NHMRC, 'Embryo Research Licensing Committee 2015–2018' www.nhmrc.gov.au/embryo-research-licensing-committee-2015–2018 accessed 31 July 2018.

[42] RIHE Act, s 20.1.

[43] Ibid. s 20.2.

[44] Ibid. s 20.3.

[45] Ibid. s 20.4.

The Gene Technology Regulator (GTR) of the Commonwealth Department of Health administers the GT Act.[46] The GTR has a range of policy, operational, communication and advisory functions. In addition, the GTR, much like the ERLC, is responsible for issuing licenses, in this case for dealings with GMOs that either do or do not involve the intentional release of a GMO into the environment. The GT Act also creates an alternative pathway for low-risk-contained dealings with GMOs, referred to as Notifiable Low Risk Dealings, which can only be carried out in certified facilities. Such dealings include the creation and use of GMOs for research purposes. Organizations are required to establish Institutional Biosafety Committees to provide on-site scrutiny of these dealings, and they are also required to provide annual reports to the GTR.[47] Such dealings would include the creation of genetically modified animals for research and in vitro research involving genetic manipulation of human and other cells.

III SUBSTANTIVE PROVISIONS

Together, the RIHE Act and PHCR Act create a complex regulatory and prohibitory landscape for research involving genome modification of human embryos. The GT Act might add a further layer of obligations, if organisms produced through genome editing techniques such as CRISPR-Cas9 are considered GMOs. The GTR recently finalized a technical review of the Gene Technology Regulations 2001, which underpin the GT Act, which is awaiting parliamentary approval. The review is highly technical in nature, and it is not necessary for the purpose of this chapter to delve into the detailed technological analysis presented therein. The likely outcome is that some aspects of genome editing may be excluded from the licensing obligations under the GT Act.

Although the amendments to the Gene Technology Regulations are not yet in force, for the time being it seems that all forms of genome modification are probably included in the existing regulatory regime and that in the future only a small component of genome editing technology might be excluded. Although the GT Act does not impose licensing obligations for the creation of humans through somatic cell therapy,[48] germline therapy would be included, and even where the editing is undertaken on somatic cells, if the

---

[46]   Australian Government Department of Health, 'Office of the Gene Technology Regulator' www.ogtr.gov.au accessed 31 July 2018.

[47]   See generally GT Act s 73–75. Office of the Gene Technology Regulator, 'Institutional Biosafety Committee and Notifiable Low Risk Dealings' www.ogtr.gov.au/internet/ogtr/publishing.nsf/Content/ibc-1 accessed 31 July 2018.

[48]   GT Act, s 10.

modifications are introduced using a living vector, they would also be included, because they would fall within the definition of a GMO.[49] This also means that researchers using human cells for genome editing research would need to be cognizant of the Notifiable Low Risk Dealing requirements under the GT Act.

In analysing the limitations on germline genome modification and embryo research using genome editing imposed by the PHCR and RIHE Acts, it is expedient to start with the outright prohibitions. First, Section 15 of the PHCR Act criminally sanctions the intentional making of heritable alterations to the human genome.

> Offence—heritable alterations to genome
> (1) A person commits an offence if:
> (a) the person alters the genome of a human cell in such a way that the alteration is heritable by descendants of the human whose cell was altered; and
> (b) in altering the genome, the person intended the alteration to be heritable by descendants of the human whose cell was altered.[50]

Thus, Section 15 of the PHCR Act clearly criminalizes clinical applications of germline modification technologies that are intended to be inherited. This limitation would appear to apply irrespective of whether the manipulation is performed on the embryos or on gametes, irrespective of whether the embryo was created by fertilization of a human egg by a human sperm or some other technique, and irrespective of whether new genetic material is introduced. Penalties for breach are harsh, significantly harsher than those imposed in other nations: imprisonment for up to fifteen years.[51] The clear intent of this provision is to prohibit germline genetic manipulations intended to be passed on to future generations.[52] However, as will be explained later, this prohibition could also apply in the research context, where the intention for the genetic manipulation to be passed on to future generations is absent, but the intention to modify the genome in a way that *could* be inherited is present.

Section 12 of the PHCR Act criminalizes the creation of an embryo by fertilization outside the body of a woman unless this procedure is undertaken in an attempt to achieve a pregnancy in a particular woman.[53] The penalty is,

---

[49]   Ibid.

[50]   PHCR Act, s 15.

[51]   Ibid.

[52]   The Parliament of the Commonwealth of Australia, *Explanatory Memorandum: Prohibition of Human Cloning for Reproduction and Research Involving Human Embryos Amendment Act 2006*, cl 15.

[53]   PHCR Act, s 12.

again, fifteen years of imprisonment. This bars the creation of embryos by fertilization for research purposes. Because of that, there are limited avenues for embryo research in Australia. Other prohibitions are included in the PHCR Act. Section 13 criminalizes the intentional creation or development of an embryo by fertilization outside the body of a woman in circumstances where that embryo contains genetic material provided by more than two persons. This provision appears to preclude any form of genome manipulation, whether for clinical application or research, if it involves in vitro fertilization and insertion of new genetic material from another person. Once again, the penalty is considerable: imprisonment for up to fifteen years. As an example, this provision would preclude research relating to mitochondrial replacement therapy using the 'maternal spindle transfer method'.[54] The reason is that this process requires fertilizing a human egg, creating an embryo with genetic material from more than two persons.[55] In contrast, the ERLC can grant a license for research involving 'pronuclear transfer'[56] 'because the egg has already been fertilised prior to its transfer to the donor egg'.[57] Pursuant to Section 13 of the PHCR Act, however, neither technique could be used for reproductive purposes. Section 20 of the PHCR Act adds to this, creating further offences for importing, exporting or placing a prohibited embryo in the body of a woman, again with penalties of up to fifteen years' imprisonment. In each case, the mental element is knowledge or recklessness.

Arguably, Section 15 of the PHCR Act does not apply to research using germline modification in human embryos or gametes (whether clinical or basic) because there is no intention to make changes that will be passed on to future generations. However, an alternative argument could be made that Section 15 does actually apply in circumstances beyond those in which there is a clear intention for genetic modifications to be passed on to future generations. This depends on the definition of 'heritable'. The explanatory memorandum to the *Prohibition of Human Cloning for Reproduction and Research*

---

[54] As noted in a report released by the Australian Senate Community Affairs References Committee, *Science of Mitochondrial Donation and Related Matters* (2018), 31: 'Maternal spindle transfer is a technique in which the spindle shaped group of chromosomes containing the mother's nuclear DNA, known as the "maternal spindle", is extracted from one of the mother's eggs (oocytes) and transferred to an unfertilised donor egg from which the maternal spindle has been removed and that contains healthy mtDNA' (footnotes omitted).

[55] Ibid. 32.

[56] Ibid. 33: 'For pronuclear transfer, a second zygote must be created from a donor egg and the father's sperm. The two pronuclei from the first zygote are removed and transferred to the donor zygote with healthy mtDNA. The donor zygote, which needs to be at the same stage of development, has had its pronuclei removed to facilitate the transfer' (footnotes omitted).

[57] Ibid. 34.

Involving Human Embryos Act defines heritable as 'able to be passed on to subsequent generations of humans'.[58] This definition could suggest that the relevant intent is to create the genetic manipulation (which is able to be passed to future generations), not the intent to actually pass the genetic manipulation on to future generations. Support for this interpretation comes from the *Lockhart Review* itself. In recommending that the creation of embryos for research using genetic material from more than two people should be permitted, the *Lockhart Review* stated:

> creation of human embryos using the genetic material from more than two people, including *heritable alterations* to the genome or using precursor cells from a human embryo or fetus, should also all be permitted, under licence, for research to increase knowledge or treat diseases.[59]

It appears implicit in this statement that alterations can be heritable even if there is no intention for them to be inherited. On this reading, alterations to the genome of embryos created other than by fertilization would not offend Section 15, but alterations to the genome of embryos created by fertilization would, even if the alterations were made for the purpose of research. The extent to which this prohibition might apply in the research context awaits definitive statutory interpretation.

Be that as it may, it should be noted that there are some limited circumstances where research involving genetic manipulation of embryos might be allowed in Australia. For example, the introduction of genetic material from a third person may be allowed under license from the ERLC, provided that the embryo has not been created by fertilization of a human oocyte by a human sperm (i.e. the embryo falls into the part (b) definition of embryo in Section 7 of the RIHE Act).[60] Furthermore, embryos that are created by fertilization of a human oocyte by a human sperm for the purpose of ART (part (a) embryos) can be used for research under license from the ERLC if they are either unsuitable for implantation or surplus to the needs of the people who created them. This would seem to suggest that studies akin to those made to date in China (on embryos that were non-viable)[61] and the UK (embryos that were

---

[58] The Parliament of the Commonwealth of Australia, *Explanatory Memorandum: Prohibition of Human Cloning for Reproduction and Research Involving Human Embryos Amendment Act 2006*, cl 15.

[59] Legislation Review Committee, *Legislation Review: Prohibition of Human Cloning Act 2002 and Research Involving Human Embryos Act 2002* (2005), xvii (emphasis added).

[60] RIHE Act, s 10A and s 20.

[61] P Liang, et al., 'CRISPR/Cas9-Mediated Gene Editing in Human Tripronuclear Zygotes' (2015) 6 *Protein & Cell* 363; X Kang, et al., 'Introducing Precise Genetic Modifications into

surplus to the requirements of the donors)[62] could be undertaken in Australia. However, the Australian legislation is more restrictive than in either of those countries. Section 7 of the RIHE Act includes, in the definition of 'embryos unsuitable for implantation', embryos that are diagnosed by preimplantation genetic diagnosis to be unsuitable and embryos that are otherwise prescribed by regulation to be unsuitable.

As to the first, clearly, a determination on unsuitability based on the outcome of preimplantation genetic diagnosis can only be made after the embryo has undergone a sufficient number of cell divisions. If it is done too early, it might compromise the viability of what might otherwise be an embryo suitable for implantation.[63] As to the second, the chief executive officer of the NHMRC has been authorized by regulation to issue a set of objective criteria for determining when an embryo fulfils the requirements for being unsuitable for implantation.[64] This was done in 2007.[65] In addition to the diagnosis of a serious genetic condition through preimplantation diagnosis, a qualified embryologist may determine that an embryo is unsuitable for implantation if she finds evidence of a high percentage of fragmentation or degeneration or the presence of vacuoles (generally equal to or greater than 50% of the total number of cells).[66] At the one-cell stage, the criterion is the absence of two pronuclei, in other words a total failure of fertilization. Admittedly, there are other circumstances where genome modification research on early embryos that are unsuitable for implantation could be licensed. However, it is unclear how useful this research might be, given how damaged these cells need to be in order to be declared unsuitable for

---

Human 3PN Embryos by CRISPR/Cas-mediated Genome Editing' (2016) 33 *Journal of Assisted Reproduction and Genetics* 581.

[62] NME Fogarty et al., 'Genome Editing Reveals a Role for OCT4 in Human Embryogenesis' (2017) 550 *Nature* 67.

[63] Ethical guidelines on Preimplantation Genetic Diagnosis are provided in NHMRC, *Ethical Guidelines on the Use of Assisted Reproductive Technology in Clinical Practice and Research* (NHMRC 2017), 73.

[64] RIHE Act, s 7, definition of 'unsuitable for implantation'.

[65] NHMRC, *Objective Criteria for Determining Embryos That Are Unsuitable for Implantation* (Chief Executive Officer, NHMRC 2007).

[66] A vacuole is an organelle (i.e. a specialized subunit within a cell with a specific function) whose function is to hold various solutions or materials. This includes solutions that have been created and are being stored or excreted, and those that have been phagocytized, or engulfed, by the cell. An excessive number of vacuoles in an embryo are widely used as a marker for anomalies that impact negatively on successful embryo implantation. S Harbottle, C Hughes, R Cutting, S Roberts, D Brison, and on Behalf of the Association of Clinical Embryologists and the (ACE) British Fertility Society (BFS), 'Elective Single Embryo Transfer: An Update to UK Best Practice Guidelines' (2015) 18 *Human Fertility* 165.

implantation. The Chinese research on non-viable embryos suggests that the research benefits are likely to be minimal.[67]

In addition to embryos that are unsuitable for implantation, Section 9 of the RIHE Act states that an embryo can be declared to be excess if it was created by ART and it is not needed by the woman for whom it was created and her spouse (if any) at the time the embryo was created.[68] Section 20.1 of the RIHE Act authorizes the ERLC to issue a license for the use of these excess ART embryos. Pursuant to Section 9.2 of the RIHE Act, an embryo can only be declared to be an excess ART embryo if there is express written authority to that effect from the woman and her spouse. Further, Section 24.1 requires that consent must be secured not only from the woman and her spouse, but from all people falling within the definition of 'responsible person' in Section 8, which also includes the persons providing eggs or sperm and their spouses, and any other persons providing reproductive or genetic material.

It is difficult to see how a determination could be made that otherwise healthy embryos are excess to the needs of the people involved until they have already made reasonable attempts to achieve a pregnancy, or unless there are other reasons why achievement of a pregnancy is no longer within their contemplation (e.g. death, divorce and natural conception). As such, although fresh embryos that are declared to be unsuitable for implantation are available for research, frozen embryos are the only healthy embryos that could be declared excess ART embryos and, thus, potentially available for research. While it is possible to freeze zygotes and embryos at early cleavage stages, current best practice is for embryos to be frozen at much later cleavage stages, or at the blastocyst stage, making them unsuitable for the types of genome modification research currently being undertaken.

Following the *Lockhart Review*, it is now possible to create embryos other than by fertilization of a human oocyte by a human sperm, for instance by somatic cell nuclear transfer. These are the so-called part (b) embryos referred to earlier in this chapter.[69] Amongst other things, Section 20.1 of the RIHE Act allows the ERLC to license:

[67] P Liang, et al., 'CRISPR/Cas9-Mediated Gene Editing in Human Tripronuclear Zygotes' (2015) 6 *Protein & Cell* 363; X Kang, et al., 'Introducing Precise Genetic Modifications into Human 3PN Embryos by CRISPR/Cas-mediated Genome Editing' (2016) 33 *Journal of Assisted Reproduction and Genetics* 581.

[68] NHMRC, *Ethical Guidelines on the Use of Assisted Reproductive Technology in Clinical Practice and Research* (NHMRC 2017), 91–95 (ethical principles in ART research), 97–98 (research involving gametes), and 99–110 (research involving embryos).

[69] As incorporated into the definition of embryo in PHCR Act, s 8 and RIHE Act, s 7.

... (b) creation of human embryos other than by fertilisation of a human egg by a human sperm, and use of such embryos;

(c) creation of human embryos other than by fertilisation of a human egg by a human sperm that contain genetic material provided by more than 2 persons, and use of such embryos;

(d) creation of human embryos using precursor cells from a human embryo or a human fetus, and use of such embryos; ...

Section 20.1A also makes it clear that under no circumstance can the embryo, in any of the forms listed in Section 20.1, develop beyond fourteen days. Section 20.1(c) thus would appear to allow, for example, the use of nuclear transfer techniques to transfer the nucleus of a human cell that has undergone genome editing into an enucleated oocyte. However, the technical challenges involved in genome editing and nuclear transfer suggest that the chances of conducting meaningful research where both of these procedures have been undertaken is remote at best.

Section 20.1 also allows the creation and use of hybrid embryos by fertilization of an animal egg by a human sperm. However, the hybrid embryo so created is allowed to develop only up to the first mitotic division, for the purpose of testing sperm quality, and these procedures must only be undertaken in an accredited ART laboratory. Finally, as mentioned earlier, Section 20.1 allows a license to be issued for research and training involving fertilization of a human egg and human sperm for ART purposes, but, again, not beyond the first mitotic division. Section 21 provides that the ERLC must take into account a range of factors in making its determination to issue a license. These include that the number of excess ART embryos, other embryos or human eggs to be used is the minimum necessary to achieve the project goals, and that there is a likelihood of significant advance in knowledge or improvement in technologies for treatment as a result of the research, which could not reasonably be achieved by other means. It has been argued that these requirements align with principles of proportionality and necessary purpose.[70]

The NHMRC's *Ethical Guidelines on the Use of Assisted Reproductive Technology in Clinical Practice and Research* (2017) provide further guidance on research involving gametes and embryos. The guidelines set out three essential ethical criteria for licensable research activity under the RIHE:

- there is sufficient evidence that the likely benefits of the proposed research cannot be achieved without using human embryos

[70] D Chalmers, 'Stem Cell Technology: from Regulation to Clinical Applications' in B Capps and A Campbell (eds.), *Contested Cells: Global Perspectives on the Stem Cell Debate* (Imperial College Press 2010) 63, 79.

- there is proof of concept, such as success in animal studies
- the research is justifiable by its potential benefit in improving technologies for treatment of, or knowledge about, human diseases. This benefit must be sufficient in the light of the very serious moral consideration due to human embryos.[71]

The Guidelines also include specific provisions relating to research involving excess ART embryos,[72] research on embryos created by means other than by fertilization of a human egg by a human sperm[73] and research involving creation of human embryos using precursor cells from a human embryo or a human fetus.[74]

In combination, the PHCR Act and the RIHE Act (and the associated guidelines) create a highly restrictive regime in Australia. At present, there is no avenue for legitimately using embryos that have undergone germline genome modification to achieve a pregnancy in Australia. There are also very limited avenues for legitimately creating and using embryos for the purpose of clinical and basic research. However, it should be pointed out that this legislative regime relates solely to the creation and use of embryos. It does not touch research uses of embryonic stem cell lines that were in existence prior to the entry into force of the legislation. Nor does it apply to other cell lines. It should also be noted that, even for embryonic stem cell lines derived under license after the legislation entered into force, the ERLC has no remit over subsequent uses. This is because the legislation only regulates the use of living embryos, pursuant to Section 7.3 of the RIHE Act and Section 8.6 of the PHCR Act, and the process of deriving the embryonic stem cell line destroys the embryo.

The broader Australian human research ethics system created through the *National Statement* and administered by local Human Research Ethics Committees needs to be considered in this context. The ERLC relies on Human Research Ethics Committees to evaluate, approve and monitor all research approved under ERLC licenses.[75] The *National Statement* also has broader application, extending beyond ERLC-approved research to any research involving humans. The *National Statement* is divided into five sections, the first two of which provide general guidance on the core values

---

[71] NHMRC, *Ethical Guidelines on the Use of Assisted Reproductive Technology in Clinical Practice and Research* (NHMRC 2017), 100.

[72] Ibid. 102–105.

[73] Ibid. 105–109.

[74] Ibid. 109–110.

[75] RIHE Act, s 21.3(c).

and principles of respect, research merit and integrity, justice, and beneficence and the overarching themes of risk and benefit and consent. Section 3
provides guidance on the ethical considerations specific to particular research
methods or themes, including qualitative methods and databanks, and, more
relevantly, interventions including clinical and non-clinical trials, biospecimens and human genetics. Section 4 details ethical concerns specific to
participants, including women who are pregnant and the human fetus.
Finally, Section 5 provides guidance on processes of research governance
and ethical review. Like the research ethics governance regimes in many
other jurisdictions, there is always scope for improvement of the Australian
regime. Over time, it has undergone a number of reviews and restructures,[76]
and is now subject to rolling review.[77] A fully revised version of Section 3 was
published in July 2018. Human Research Ethics Committee approval is
required for all basic and clinical research undertaken at or in collaboration
with Australian universities and research institutes involving humans or
human tissue.

Likewise, research involving animals, as defined in the *Animal Code*,[78]
must be approved by local Animal Ethics Committees. The overarching
principle, upon which all decisions and actions relating to the care and use
of animals are underpinned, is respect for animals.[79] The Code requires that
the 3Rs (replacement, reduction and refinement) are applied at all stages of
animal care and research.[80] The role of the GT Act in regulating germline
genome modification research involving human embryos, human cell lines
and animals should be mentioned, although it would be impossible to do
justice to the full scope of this regulatory regime in this chapter. Briefly, the
Act regulates certain 'dealings' with GMOs and exempts others. Section 10 of
the GT Act defines 'dealings' broadly, ranging from experimentation, making,
breeding and propagating through to growing, importing, transporting and
disposing. The Act distinguishes between 'exempt dealings' and 'notifiable low

---

[76] D Chalmers, 'Research Ethics in Australia', in National Bioethics Advisory Commission,
Ethical and Policy Issues in Research Involving Humans Volume 2 (NBAC 2001).

[77] NHMRC, *National Statement on Ethical Conduct in Human Research* (first published 2007,
NHMRC 2018).

[78] NHMRC, *Australian Code for the Care and Use of Animals for Scientific Purposes* (8th edn,
NHMRC 2013), 3 (defining 'animal' as 'any live non-human vertebrate (that is, fish, amphi-
bians, reptiles, birds and mammals encompassing domestic animals, purpose-bred animals,
livestock, wildlife) and cephalopods'). The *Animal Code* also applies to teaching associated
with an educational outcome in science, field trials, product testing, diagnosis, the production
of biological products and environmental studies.

[79] Ibid. 9.

[80] Ibid. 11–12.

risk dealings',[81] neither of which require a license, and other dealings, which do require a license. Where a license is needed, dealing without a license is a criminal offence, and like the PHCR Act, the GT Act imposes significant penalties, including, in this case, up to five years imprisonment and fines of up to AUD 420,000.

'Exempt dealings' require no particular oversight, provided that there is no intention to release GMOs into the environment. The GTR describes exempt dealings as 'a category of dealings with GMOs that have been assessed over time as posing a very low risk (i.e. contained research involving very well understood organisms and processes for creating and studying GMOs)'.[82] Preclinical genome modification, whether involving humans or other animals, is unlikely to fall into the category of an exempt dealing, but it is likely to be a 'notifiable low risk dealing' because it is capable of physical containment.[83] Reliance on the protection afforded by the 'notifiable low risk dealing' classification requires that there is a suitable level of physical containment in facilities certified for this purpose. A dealing can only be considered to fall into this category after it has been assessed by an Institutional Biosafety Committee at an accredited organization.[84] Section 74 of the GT Act requires that the GTR is notified of all 'notifiable low risk dealings'. All institutions involved in animal research are also required to be compliant with ethical codes of conduct and guidelines.[85]

IV CURRENT PERSPECTIVES AND FUTURE POSSIBILITIES

Australian researchers were pioneers in the early development of ART and its translation into the clinic. They were able to work within existing legislative, self-regulatory and ethical frameworks to ensure that Australians could benefit from these developments in a safe environment. The most significant complexity, in these early days, was the lack of national consistency, leaving some of the

---

[81]   Gene Technology Regulations 2001, schedule 3.

[82]   Office of the Gene Technology Regulator, 'What Are Exempt Dealings?' (2015). www .ogtr.gov.au/internet/ogtr/publishing.nsf/Content/exemptdealclass-2 accessed 31 July 2018.

[83]   Office of the Gene Technology Regulator, 'Types of Dealings with GMOs classified as Notifiable Low Risk Dealings (NLRDs)' (8 March 2018) www.ogtr.gov.au/internet/ogtr/pub lishing.nsf/content/nlrdsSept2011-excerpt-htm accessed 31 July 2018.

[84]   For information on accreditation, see Australian Government, *Explanatory Information on the Guidelines for Accreditation of Organisations* (2013).

[85]   NHMRC, *Australian Code for the Care and Use of Animals for Scientific Purposes* (8th edn, NHMRC 2013)); NHMRC Animal Welfare Committee, *Guidelines for the Generation, Breeding, Care and Use of Genetically Modified and Cloned Animals for Scientific Purposes* (NHMRC 2007).

TABLE 20.1 *Australia: summary of the regulatory pathway for a medical product created using genome editing*

Step	Primary regulatory authorities	Primary legal framework	Main issues to be considered
Basic research (i.e. laboratory research) on human embryonic stem cells or embryos	• HUMAN RESEARCH ETHICS COMMITTEES • INSTITUTIONAL BIOSAFETY COMMITTEES • EMBRYO RESEARCH LICENSING COMMITTEE • GENE TECHNOLOGY REGULATOR	• *Prohibition of Human Cloning for Reproduction Act 2002* • *Research Involving Human Embryos Act 2002* • *Gene Technology Act 2000* • *National Statement on Ethical Conduct in Human Research* (2018) • *Ethical Guidelines on the Use of Assisted Reproductive Technology in Clinical Practice and Research* (2017)	• LABORATORY WORKER SAFETY • VALUES AND PRINCIPLES OF ETHICAL CONDUCT OF RESEARCH • TISSUE DONOR SAFETY, PRIVACY AND RIGHTS (HUMAN CELLS AND TISSUE) • ADEQUACY OF CONSENT PROCESS • ETHICAL CONCERNS ASSOCIATED WITH SCIENTIFIC DEVELOPMENTS IN RELATION TO HUMAN REPRODUCTION AND THE UTILIZATION OF HUMAN EMBRYOS • HEALTH AND SAFETY OF PEOPLE AND PROTECTION OF THE ENVIRONMENT
Preclinical studies (if done with animals)	• ANIMAL ETHICS COMMITTEES • INSTITUTIONAL BIOSAFETY COMMITTEES • GENE TECHNOLOGY REGULATOR	• *Gene Technology Act 2000* • *Australian Code for the Care and Use of Animals for Scientific Purposes* (8TH EDN, 2013)	• RESPECT FOR ANIMALS • REPLACEMENT, REDUCTION AND REFINEMENT • HEALTH AND SAFETY OF PEOPLE AND PROTECTION OF THE ENVIRONMENT
Clinical trials (investigational new drug [IND] application)	• THERAPEUTIC GOODS ADMINISTRATION • HUMAN RESEARCH ETHICS COMMITTEES • INSTITUTIONAL BIOSAFETY COMMITTEES	• *Prohibition of Human Cloning for Reproduction Act 2002* • *Research Involving Human Embryos Act 2002* • *Gene Technology Act 2000* • *Therapeutic Goods Act 1989*	• BALANCE OF ANTICIPATED RISKS AND BENEFITS TO HUMAN SUBJECTS • APPROPRIATE PROTOCOL DESIGN AND INFORMED CONSENT • ETHICAL CONCERNS ASSOCIATED WITH SCIENTIFIC DEVELOPMENTS IN

(continued)

TABLE 20.1 (continued)

Step	Primary regulatory authorities	Primary legal framework	Main issues to be considered
	• EMBRYO RESEARCH LICENSING COMMITTEE • GENE TECHNOLOGY REGULATOR	• National Statement on Ethical Conduct in Human Research (2018) • Ethical Guidelines on the Use of Assisted Reproductive Technology in Clinical Practice and Research (2017)	• RELATION TO HUMAN REPRODUCTION AND THE UTILIZATION OF HUMAN EMBRYOS • HEALTH AND SAFETY OF PEOPLE AND PROTECTION OF THE ENVIRONMENT
New medical product application (Biologic Licensing Application)	• Therapeutic Goods Administration	• Therapeutic Goods Act 1989	• Evaluation of safety and efficacy cata
Licensed medical product (post-market measures)	• Therapeutic Goods Administration	• Therapeutic Goods Act 1989	• Long-term patient safety

states of Australia with stringent ART legislation, and other states and territories with no direct legislative obligations and restrictions relating to their ART practices. This situation has now changed, with nationally consistent legislative obligations, particularly those imposed by the PHCR Act, the RIHE Act and the GT Act. Ethical and self-regulatory guidelines are also consistent nationally.

Regulatory responses to developments in cloning and stem cell technology, although initially somewhat heavy-handed in the research context, were relaxed following the *Lockhart Review* only a short time later, in recognition of the increasing willingness within the Australian community to accept that such research should be undertaken because of the promise it offers in alleviating human suffering. Despite this relaxation, legislation continues to impose significant restrictions on the capacity of Australian researchers to undertake basic genome modification research involving human embryos.

Australia is a party to the major international human rights conventions, including the International Covenant on Economic, Social and Cultural Rights and the International Covenant on Civil and Political Rights. In reporting to the United Nations on its obligations under these international instruments, Australia has affirmed an ongoing commitment to human rights: 'Australia has a long tradition of supporting human rights around the world, and was closely involved in the development of the international human rights system.'[86] Included in this is a commitment to science and health, and recognition of the right to found a family. However, in all of the debates around human cloning and embryonic stem cell research the special status of the embryo has taken centre stage.

As is the situation in many other countries, Australia is once again at a crossroad, and we must decide, collectively, how far we are willing to go in allowing genome modification research involving human embryos to progress. Currently in Australia, a 'command and control' legislative framework[87] causes significant restrictions on similar biological research, more so than in other nations. Without strong public support, it is unlikely that significant changes will be made by the legislature to Australian legislation dealing with research involving human embryos. During the *Lockhart Review*, the Australian Parliament concluded:

---

[86] Australia Government, *Common Core Document forming part of the reports of States Parties – Australia – incorporating the Fifth Report under the International Covenant on Civil and Political Rights and the Fourth Report under the International Covenant on Economic, Social and Cultural Rights* (Commonwealth of Australia, 2006).

[87] S Allan, 'Regulatory Design Strategies and Enforcement Approaches for Research involving Human Embryos and Cloning in Australia and the United Kingdom – Time for a Change' (2010) 32 *Sydney Law Review* 617.

[a]lthough a range of views was expressed about the precise moral status of preimplantation embryos in particular, there was an overall acceptance that human embryos created by the fertilisation of a human egg by a human sperm are entities of some social and ethical significance because of their association with the start of human life. Therefore, the Committee has recommended that the prohibition on the creation of an embryo by the fertilisation of a human egg by human sperm for any purpose apart from ART treatment of a woman should continue.[88]

Although it is not likely that the Australian public has changed its views on the special significance of the human embryo since the *Lockhart Review* was completed in 2005, the promise offered by new genome editing techniques may open the door for acceptance of a broader range of research opportunities, and for further consideration of the right to science, the right to health and the right to family life. It is timely for Australians to have that debate. In a recent review of the science of mitochondrial donation, the Australian Senate Community Affairs References Committee emphasized that a technological innovation of this nature should not be introduced without public consultation. While accepting that from the evidence presented so far, this may not be seen as a controversial medical treatment, 'public consultation is necessary before any full rollout of mitochondrial donation and . . . parliamentary inquiries alone will not be sufficient public consultation'.[89]

The author has recently completed a nationwide survey of attitudes towards germline gene editing with a group of colleagues.[90] The results of the survey of over 1,000 Australians suggests that there is a level of comfort with editing human and animal embryos for research purposes and to enhance human health. However, the survey also revealed that there are ongoing concerns about the consequences of genome editing. The study drew a distinction between moral concerns with the status of the human embryo and concerns about modifying the genome of future generations, and found that moral concern about the embryo was stronger than hereditary concern. This is the first survey in the published literature to ask respondents to make such a distinction. As such, it is not clear whether the special status that has so far been accorded to the human embryo by Australian legislators is out of step

[88]   Legislation Review Committee, *Legislation Review: Prohibition of Human Cloning Act 2002 and Research Involving Human Embryos Act 2002* (2005), xv.

[89]   Australian Senate Community Affairs References Committee, *Science of Mitochondrial Donation and Related Matters* (2018), 95.

[90]   C Critchley, D Nicol, G Bruce, J Walshe, T Treleaven and B Tuch, *Predicting Public Attitudes Towards Gene Editing of Germlines: The Impact of Moral and Hereditary Concern in Human and Animal Applications* (2019) 9 *Frontiers in Genetics* article 704, doi: 10.3389/fgene.2018.00704.

with legislators in other jurisdictions. It is clear that many members of the Australian community (at least those participating in this survey) continue to recognize the special status of the human embryo. The question that needs to be answered in this jurisdiction is whether this concern for the special status of the human embryo continues to outweigh other considerations, to the extent that Australia should continue to take a restrictive approach to germline genome modification and genome editing research involving human embryos.

The Regulation of Human Germline Genome Modification in Israel

Vardit Ravitsky and Gali Ben-Or

I INTRODUCTION

Israel defines itself as a 'Jewish democratic state', an expression that conveys the Jewish and liberal roots of its political, cultural and legal identity. Its political system is democratic, with a secular majority in its Parliament (*Knesset*) and strong influence of orthodox parties; and in many instances, both Jewish values and Zionist narratives play an important role in shaping public policy.[1] This includes issues related to genetics and reproduction, where cultural considerations always play a prominent role.[2]

In the Jewish tradition, attempting to cure disease and save lives is a paramount value.[3] Jewish culture encourages open-minded attitudes towards research efforts that have therapeutic goals. Jewish orthodox law, *Halacha*, usually takes a favourable and permissive view of scientific activity that has the prospects of finding cures. To illustrate, during the discussions leading to the enactment of the Israeli law pertaining to germline gene editing, which will be discussed further below, the Science Committee of the Knesset convened a meeting at the offices of the Chief Rabbis of Israel, entitled 'Genetic Cloning – Scientific Research – Moral Limitations'. In the discussions, the Chief Rabbis emphasized the importance of healing while ensuring extreme caution to avoid any harm or danger.[4]

---

[1] B Prainsack, '"Negotiating Life": The Regulation of Human Cloning and Embryonic Stem Cell Research in Israel" (2006) 36:2 *Social Studies of Science* 173.

[2] G Ben-Or and V Ravitzky, 'Cultural Values in Action: The Israeli Approach to Human Cloning' in Daphna Birenbaum-Carmeli and Yoram S. Carmeli (eds.), *Kin, Gene, Community: Reproductive Technologies among Jewish Israelis* (Berghahn Books 2010).

[3] Ibid.

[4] The Special Committee on Matters of Science and Technology Research and Development Minutes (31 March 1997 Meeting) (Unpublished, Knesset Archive). Available at the Knesset Archive and in the near future on the Israeli National Legislation Database.

Zionist narratives play a role, too, in shaping policy towards genetic and reproductive technologies. Israel was built by pioneers, who turned a small country, poor in natural resources, into an economic power-house. Science and technology have always been considered a key to economic success and future survival of the nation. Israel is known as 'start-up nation' and the Israeli Ministry of Education promotes science and technology education as a 'basis for national and social strength'.[5] This pro-science attitude echoes a long-standing Zionist narrative that links scientific and technological innovation with the notion of trans-forming the Jews into a modern nation.[6] When approaching novel technologies, legislators often share the sentiment that 'they should not interfere with scientific research at the level of the Petri dish in the lab'.[7]

Thus, unlike some states surveyed in this volume, Israel does not ban basic research involving modification of the genome of human germline cells and embryos and has a relatively simple regulatory framework. However, as many other states, it draws the line at attempting to create a 'genetically modified person'.

II THE REGULATORY FRAMEWORK

1 Constitution

As the United Kingdom, Israel does not have one single monolithic constitution. Its fundamental law is written 'chapter by chapter'. The 'establishment of a Jewish state in Eretz-Israel, to be known as the State of Israel' was declared on 14 May 1948, one day before the British Mandate in Palestine expired.[8] The Constituent Assembly of Israel was elected through the nation's first general election on 25 January 1949. While it did not draft a constitution for the newly established state, it adopted a law that turned it into the legislature of the State of Israel: the 'First Knesset'.[9]

---

5   V Ravitsky and G Ben-Or, 'The Israeli Discourse Regarding the Status of the Embryo: Embryonic Stem Cell Research and Cloning' *Bioethics in Blue and White: Toward an Israeli Approach to Bioethics and Health Law* (Bialik Institute 2014) (In Hebrew).
6   DJ Penslar, *Zionism and Technocracy: The Engineering of Jewish Settlement in Palestine, 1870–1918* (Indiana University Press 1991).
7   The Special Committee on Matters of Science and Technology Research and Development Minutes (22 December 1998 Meeting) (Unpublished, Knesset Archive).
8   *Declaration of Establishment of the State of Israel* (14 May 1948).
9   Transition Law 5709, SH 1949, 1.

The First Knesset could not reach consensus on the content of a constitution, nor on its form, or on the need for a constitution. The debate was settled by a compromise resolution, transferring the powers of the Constituent Assembly to subsequent Knessets, and empowering them to write a constitution 'by chapters', instead of one formal written document.[10] The text of this resolution, known as the 'Harari Resolution', reads as follows:

> The First Knesset instructs the Constitution, Law and Justice Committee to prepare a draft State Constitution. The Constitution will be built chapter by chapter, in such a way that each will constitute a separate Basic Law. The chapters shall be presented to the Knesset when the committee completes its work, and all the chapters together shall comprise the Constitution of the State.[11]

In 1992, the Knesset enacted the 'Basic Law: Human Dignity and Liberty' and the 'Basic Law: Freedom of Occupation' (the latter was subsequently replaced by a new version in 1994).[12] The rights contained in them are 'basic', meaning that they are fundamental. They limit future legislation, similarly to the rights enshrined in a constitution. Paragraph 8 and 4 of these basic laws state that there shall be no violation of rights except by a law befitting the values of the State of Israel, enacted for a proper purpose, and to an extent no greater than it is required or on the basis of a law by force of an explicit authorization therein. The Basic Laws do not protect freedom of speech explicitly, but there is a long-lasting tradition in Israel of freedom of speech long before the basic law enactment. There are some in Israel who claim that freedom of speech is inseparable from human dignity, thus will also be protected as a constitutional right.[13] When a question about scientific freedom arises, these two basic laws can be used to ensure its protection.

2 Ratified Treaties

In Israel, treaties are ratified by the government through a resolution, following comments from the Knesset.[14] Since Israel was ruled by the United

10 G Sapir, D Barak-Erez, and A Barak (eds.), *Israeli Constitutional Law in the Making* (Hart Studies in Comparative Public Law, UK edn, 2013) 1.

11 Knesset Plenum Assembly, No 151 of 1950.

12 Basic Law: Human Dignity and Liberty; Basic Law: Freedom of Occupation.

13 A Barak, 'The Tradition of Freedom of Expression in Israel and its Problems – Lecture' 27 Mishpatim 223, 230 (1996).

14 Rules of Procedures of the Government, www.pmo.gov.il/Secretary/Documents/takanon34 .pdf accessed 25 September 2015.

Kingdom between 1919 and 1948, its modern legal system follows the Common Law legal tradition. As in the United Kingdom, ratified treaties are not 'law of the land', but are rather an important interpretive source for the national legal system, from the Basic Laws down.

Israel ratified the Covenant on Economic, Social and Cultural Rights on 3 October 1991, with no reservations.[15] The Covenant contains several provisions that could be relevant to a discussion of human germline genome modification, including Article 15.1.b, and 15.2 through 4 (collectively known as the 'Right to Science' and the 'Rights of Science') and Article 12 (the 'Right to Health'). The right to health is also anchored in other relevant human rights treaties that have been ratified by Israel, such as the Convention of the Rights of the Child[16] and the Convention for the Elimination of all Forms of Discrimination against Women.[17] Since 1994, Israel has had a National Health Insurance Law, guaranteeing a Basic Basket of healthcare services to all its citizens.

3 The 1999 Prohibition of Genetic Intervention (Human Cloning and Genetic Manipulation of Reproductive Cells) Law

Human germline genome modification is regulated in Israel under the 1999 Prohibition of Genetic Intervention (Human Cloning and Genetic Manipulation of Reproductive Cells) Law (hereinafter referred to simply as 'the Law').[18] In the wake of the birth of the first cloned mammal in 1996, Dolly the sheep, three different bills were introduced before the Knesset. Lengthy, in-depth discussions were held, both in the Knesset plenary and in the Science and Technology Committee that prepared one of the bills for a first reading, and, afterwards, for the second and third readings. To date, the Law has been amended three times: in 2004, 2009 and 2016. The current version is set to expire on 23 May 2020.[19]

The purpose of the Law (as stated in its current version) is

---

[15]   International Covenant on Economic, Social and Cultural Rights (adopted 16 December 1966, entered into force 3 January 1976) 993 UNTS 3 (ICESCR).

[16]   Convention on the Rights of the Child (adopted 20 November 1989, entered into force 2 September 1990) 1577 UNTS 3 (CRC), art. 3.

[17]   Convention on the Elimination of All Forms of Discrimination against Women (adopted 18 December 1979, entered into force 3 September 1981) 1249 UNTS 3 (CEDAW), arts 11.f, 12, and 14.b.

[18]   Prohibition of Genetic Intervention (Human Cloning and Genetic Manipulation of Reproductive Cells) Law 1999, SH 1999, 47 (The Law); SH 2004, 340; SH 2009, 233; SH 2016, 882.

[19]   Ibid. s 8.

to prevent reproductive cloning in humans by establishing that certain kinds of genetic interventions shall not be performed on human beings in view of the moral, legal, social and scientific aspects of the prohibited forms of intervention and their implications on human dignity, and in order to asses public policy regarding those kinds of intervention in view of those aspects, considering freedom of scientific research for the advancement of medicine.[20]

Compared to the legislation regulating human germline genome modification in most developed countries, it is a concise and simple statute. In short, it prohibits two activities: human reproductive cloning[21] and the use of 'reproductive cells that have undergone a permanent intentional genetic modification (germline gene therapy) in order to cause the creation of a person' (unless a permit is granted according to Section 5).[22] The criminal sanction for those who violate the two prohibitions is up to four years imprisonment or a fine.[23]

First, regarding the prohibition of human reproductive cloning, the Law defines it as the 'the creation of a human embryo genetically identical to another, person or embryo, alive or dead', and prohibits 'insertion of a cloned embryo into the uterus or a body of a woman or womb or another body'.[24] The original draft of the law included a prohibition of reproductive cloning in animals as well, but in preliminary discussions researchers argued that animal experimentation, including on reproductive use, could be beneficial in finding cures for human diseases, including through the use of cloned animals for the creation of tissues that could have therapeutic use.[25] Therefore, the prohibition on cloning animals was removed from the bill that was submitted to the first reading.

Second, the law prohibits the use of 'reproductive cells that have undergone a permanent intentional genetic modification (germline gene therapy) in order to cause the creation of a person'.[26] Admittedly, the expression 'germline gene therapy' is somehow dated. Nowadays, scientists and legislators prefer talking about 'gene editing' or 'gene modification'. However, it still bans reproductive use of genetically edited germ cells because it bans the use of

---

[20] Ibid. s 1.

[21] Ibid. s 3.1.

[22] Ibid. s 3.2.

[23] Ibid., s 6. The fine is 'equal to six times the fine set in paragraph 61(a)(4) of the Penal Law 5737–1977: this amounts to 1,356,000 Israeli shekels which equals $ 373,862' (as calculated on 1 October 2018).

[24] Ibid. s 2.b.

[25] The Special Committee on Matters of Science and Technology Research and Development Minutes (30 June 1998 Meeting) (Unpublished, Knesset Archive).

[26] The Law, s 3.2.

'reproductive cells that have undergone a permanent intentional genetic modification'.[27]

It is interesting to note that this prohibition was not included in the first draft. It was added to the bill during the discussions in the Science and Technology Committee. Scientists who participated in the discussions argued that while cloning was still far from being feasible on human beings, the use of genetically modified germ cells for reproductive purposes, which raises a host of ethical dilemmas, could become realistic sooner.[28] Looking back almost two decades later, and in the light of recent technological developments, this was foresightful. Now that gene editing technologies have become a reality, Israel already has in place a regulatory framework.

The law defines 'reproductive cell' as human spermatozoon or ovum.[29] However, it does not define the term 'embryo'. Remarkably, and unlike many other states, the ban of the use of 'reproductive cells that have undergone a permanent intentional genetic modification' does not imply the prohibition of the creation of human embryos for research. As long as they are not used for reproductive purposes after they have been genetically manipulated, human embryos can be created ad hoc, used for research and destroyed. This is because according to Jewish tradition the embryo acquires the status of a 'formed embryo' only forty days after fertilization. Prior to that, it is considered by Jewish *Halacha* as 'mere water'.[30] That being said, Israel follows the prevailing international standard and allows research with embryos only up until the fourteenth day after fertilization (i.e. the so-called fourteen-day rule).

Although the reproductive use of genetically modified germ cells is illegal, the legislators left the door ajar to allow for hard-to-predict future developments. First, fearing a total ban might hinder medical progress, the Law includes a section that allows the minister of health to permit, through regulations, and as an exception to the overall prohibition, the performance of specific kinds of genetic interventions involving the reproductive use of germ cells that have undergone a genetic modification.[31] The rationale was to allow specific uses of germline interventions if and only when they become a safe and effective way of preventing genetic diseases in the future. Leaving

---

[27]   Ibid. s 3.2.
[28]   The Special Committee on Matters of Science and Technology research and Development Minutes (30 June 1998 Meeting) (Unpublished, Knesset Archive).
[29]   Ibid. s 2.
[30]   Babylonian Talmud Yevamot 69b.
[31]   The Law, s 5. ('... if [the Minister of Health] is of the opinion that human dignity will not be prejudiced, upon the recommendation of the Advisory committee and upon such conditions that he may prescribe, permit through regulations the performance of specific kinds of genetic intervention ...' that cause '... the creation of a person ...' with such cells).

a built-in option was deemed a simpler and more efficient solution than subsequently trying to modify the Law. It is a manifestation of the view that curing diseases is of paramount importance and should not be delayed, if progress towards it can be achieved. This unique mechanism also reflects the Israeli legislators' view of science as constantly progressing and puts an emphasis on potential therapeutic benefits. To date, the minister of health has not resorted to this mechanism and regulations have not yet been drafted, but that does not rule out future resort to it, should there be a clear justification.

Second, the law creates a moratorium on cloning and on human germline genome modification, not a permanent prohibition. When the bill was first introduced to the *Knesset* plenum, in 1999, it was explained that:

> Since the public debate on this issue is just beginning, it is hereby suggested to establish a period of five years, during which the public debate can be expanded and the ethical, legal and social implications of these develop- ments can be clarified. (. . .) To prevent the law from becoming an obstacle to scientific progress, we suggest that an advisory committee will submit to the Minister of Health a detailed annual report on these matters, thus ensuring a periodic examination of the law and its usefulness.[32]

This concept of a moratorium was inspired by the report of the US National Bioethics Advisory Commission, which recommended a three- to five-year moratorium on federal funding for cloning.[33] Legislators perceived this law as unique, since it potentially interfered with scientific freedom, and thus wanted to proceed with caution and scrutiny. Also, a moratorium, as opposed to a permanent ban, was seen as a more appropriate option to prevent the law from becoming irrelevant and outdated. Thus, the original wording of the first clause of the Law, which stated its purpose, read:

> The purpose of this law is to determine a prescribed period of five years during which no kind of genetic intervention shall be performed on human beings in order to examine the moral, legal, social and scientific aspect of such kinds of intervention and the implication of such intervention on human dignity.[34]

---

[32] Knesset Plenum Assembly, No 245 of 1998, 457–460.

[33] The National Bioethics Advisory Commission, *Cloning Human Beings – Report and Recommendations of the National Bioethics Advisory Commission*, (National Bioethics Advisory Commission, Maryland, June 1997).

[34] Prohibition of Genetic Intervention (Human Cloning and Genetic Manipulation of Reproductive Cells) Law 1999.

In January 2004, prior to the expiration of the Law, the Science Committee of the *Knesset* convened to discuss its extension. Some supported the renewal of the temporary moratorium. Others argued that the prohibition should become permanent, considering other countries passed permanent laws prohibiting cloning.[35] Following intense discussions, and aware that the debate is not settled, in Israel and in the rest of the world, the *Knesset* committed to revisiting the issue periodically and the moratorium was extended first for five years (to 2009), then for another seven years (to 2016), and then for another four years (to 23 May 2020).[36]

The 2004 amendment addressed nuclear transfer.[37] The original definition of 'human cloning' (i.e. 'the creation of a complete human being, chromosomally and genetically absolutely identical to another person or fetus, living or dead')[38] was broadened to add two elements: 'the creation of a human embryo by means of transferring the nucleus from a somatic cell into an ovum or a fertilized ovum from which the nucleus has been extracted, in order to create a person who is genetically and chromosomally identical to another person or fetus, living or dead', and 'the insertion of a cloned embryo into the uterus of a woman or another womb or body', which is mainly about the future possibility of an artificial womb, not an animal womb.

The modification was prompted by the realization that the original law did not prohibit the implantation of embryos in uterus. It was argued that, presumably, an embryo, cloned or genetically modified, could be implanted and then subsequently aborted without violating the prohibitions, since the process would not result in the creation of a 'complete human being'. Also, under the Law, as originally enacted, if the embryo was not 'absolutely identical', for instance because of differences in mitochondrial DNA, it would have been presumably legal to implant it, too. However, this was clearly not the intention of the legislator. Thus, to prevent these eventualities, the revised law explicitly prohibited 'the insertion of a cloned embryo into the uterus of a woman or another womb or body'.[39]

Interestingly, during the discussions leading to the latest extension of the moratorium, in 2016, it was proposed 'to correct the definition of cloning to include any possible and future method of human cloning . . . It was suggested to delete from the formulation of the definition any elements that refer to

---

[35] Science and Technology Committee (6 January 2004 Meeting).
[36] SH 1934, 31 March 2004, 340; SH 2212, 22 October 2009, 233; SH 2553, 25 May 2016, 882.
[37] Prohibition of Genetic Intervention (Human Cloning and Genetic Manipulation of Reproductive Cells) (Amendment) Law 2004.
[38] The Law, s 2.
[39] Ibid. ss 2 and 3.

specific technologies and methods of human cloning and give a broader definition that refers to the outcome of the proves of cloning'.[⁴⁰] Also, while the Law originally prohibited 'gene therapy', novel gene editing technologies show that numerous methods and mechanisms can now, and probably in the future, achieve the same outcome of genetic modification of the germline. Thus, in the discussion that took place in 2016, it was clarified that, while the ban on reproductive use stands, the Law does permit exceptions, with specific approval by the minister of health, to allow research on reproductive uses if they do not violate human dignity and may have therapeutic benefit. In sum, the wording of the current version of the Law returned to the intent of the original text: setting ethical boundaries and avoiding the prohibition of specific techniques.[⁴¹]

The task of enforcing the Law is entrusted to the minister of health.[⁴²] It also creates an Advisory Committee, whose tasks include the following: follow developments in medicine, science, biotechnology, bioethics and law in the field of genetic experimentations on human beings in Israel and abroad; submit to the minister and to the Science and Technology Committee of the Knesset an annual report, on the summary of these developments; and advise the minister on the matter of genetic experimentations on human beings and provide him or her with its recommendations concerning the prohibitions set out in the Law.[⁴³] The 'Advisory Committee' is the 'National Committee for Human Medical Research' (also known as the 'Supreme Helsinki Committee') with the addition of a representative of the minister of science. This committee will be discussed below.[⁴⁴]

The minister of health exercises overall supervisory and control powers and has the authority to issue regulations on specific issues. He or she also supervises the operations of the Advisory Committee and how the committee executes its authority according to the Law.[⁴⁵] Also, as we said earlier, the minister of health can draft regulations that can allow certain genetic interventions, if he or she finds they do not violate human dignity, and only if the Advisory Committee recommends to him or her to do so, and upon terms that he or she will determine.[⁴⁶]

[⁴⁰] Science and Technology Committee, 'Protocol of Meeting of 22 May 2016' (2016) http://fs .knesset.gov.il//20/Committees/20_ptv_343611.doc accessed 1 September 2018.

[⁴¹] The Special Committee on Matters of Science and Technology Research and Development Minutes (30 June 1998 Meeting) (Unpublished, Knesset Archive).

[⁴²] The Law, s 9.

[⁴³] Ibid. s 4.

[⁴⁴] Ibid. s 2.a.

[⁴⁵] The Prohibition of Genetic Intervention (Human Cloning and Genetic Manipulation of Reproductive Cells) (Advisory Committee Powers) Regulations [2006] KT 2006, 318.

[⁴⁶] The Law, ss 4.a.2, 3, 4.b, 5.a, and 9.

4 *The 2010 Ovum Donation Law*

When it comes to sourcing of germline cells, the key law in Israel is the 2010 Ovum Donation Law.[47] It was enacted by the *Knesset* in the aftermath of a scandal caused by a violation of standard procedures for ovum donation and a violation of the right of women to give informed consent. Under Section 3, ova can be removed from ovaries only for reproductive purposes. However, under Section 16, a woman who volunteers to donate ova can consent to several uses, including implantation in another woman (after fertilization), freezing for future use by herself, or research. The law also prescribes a limit on the quantity of ova intended for the use of the donor and for research, and it cannot exceed 20% of the total of ova extracted or two ova, whichever is lower.[48]

These conditions for donating ova for research are prescribed in Chapter 4 of the Ovum Donation Law. Section 27(b) specifies that a donor can consent to the use of her ova for a specific study or for specific kinds of studies that have been approved prior to her consent. Section 29 states that ova can be donated to research only with written approval of the physician in charge, and that such approval can only be given if: the study for which they are donated has been approved; the donor signed a consent form in front of the treating physician and all uses are based on her consent; and the number of ova donated to research does not exceed 20% of her extracted ova or two ova, whichever is lower.

Written consent must also indicate whether the woman agrees to have her ova used for research that would require exporting them out of Israel. Export can only occur in cases where the research project has been approved in the country where it will be conducted, based on the laws and the regulations of that country, and if the Director General of the Ministry of Health has given specific permission for export for a specific study, after consulting with the National Committee for Human Medical Research. Section 16(d) indicates that consent to freezing ova for future use or research can be given for a limited or an unlimited period of time. If the consent is for an unlimited period of time, the director general can allow the destruction of the frozen ovum after ten years.

Section 44 states explicitly that a woman may withdraw her consent any time prior to the research use she approved, and that such withdrawal of consent will not entail any civil or criminal liability. Once consent has been

---

[47] The Ovum Donation Law 2010, 5770, 2010, SH 2010, 520 (Ovum Donation Law).
[48] Ibid., s 16.c.

withdrawn for a specific use, the woman is entitled to consent to a different use with the same ova.[49] If a woman withdraws her consent to research purposes and does not consent to another use, or if she withdraws her consent for research purposes and instructs to destroy the ovum, it can be destroyed ninety days after consent has been withdrawn.[50]

Finally, Section 28 requires any hospital that runs a fertility department to report annually on the number of women who consented to donate ova for research purposes, the number of ova they consented to donate, and the number of ova that were actually used for research.

5 The 1980 Public Health (Medical Experiments on Humans) Regulations

The 1980 Public Health (Medical Experiments on Humans) Regulations address medical research involving human participants in Israel.[51] They were first published in 1980, by the Director General of the Ministry of Health, acting under the general authority given to him by the Public Health Ordinance (1940), a statute originally dating back to the era of the British Mandate in Palestine.[52] They have since been amended several times.[53]

The regulations incorporate the Declaration of Helsinki of 1964, as amended in Tokyo 1975, a set of ethical principles regarding human medical experimentation developed for the medical community by the World Medical Association that are regarded as the cornerstone document on human research ethics.[54] In Israel, there are two levels of 'Helsinki Committees': a local one in each hospital and research structure, similar to the Institutional Review Boards (IRBs) found in many countries, and a national one called the 'Supreme Helsinki Committee' (i.e. the National Committee for Human Medical Research mentioned above).[55] Under the regulations, no medical

---

[49] Ibid., s 44.b.
[50] Ibid., s 44.d.
[51] The Public Health (Medical Experiments on Humans) Regulations, KT 1980, 292 (Medical Experiments on Humans Regulations).
[52] Public Health Ordinance 1940, P.G. Supp I, p 191, art. 33.
[53] The Public Health (Medical Experiments on Humans) Regulations, KT 1982, 1272; KT 1984, 1570; KT 1984, 2646; KT 1984, 367; KT 1999, 1005.
[54] World Medical Association, 'Declaration of Helsinki: Ethical Principles for Medical Research Involving Human Subjects (as amended through 2013)' www.wma.net/policies-post/wma-declaration-of-helsinki-ethical-principles-for-medical-research-involving-human-subjects/ accessed 24 September 2018.
[55] See in this chapter, Sections III.1 and 2.

experiment on human beings will be performed unless authorized by a 'Helsinki Committee'.[56] Moreover, all clinical trials and medical research regarding genetics or regarding assisted reproductive technologies ('un-natural fertilization') in humans must be approved by a hospital IRB and subsequently by the National Committee for Human Medical Research. The Director General of the Ministry of Health delegates his or her responsibility for approving such trials to the National Committee and has a representative on the committee.

Interestingly, since 1984 there have been twelve legislative attempts to replace the Medical Experiments on Humans Regulations, with the goal of establishing a more coherent and advanced regulatory framework for multiple aspects of medical research involving human participants, including genetic research.[57] It is evident that this sensitive and important issue deserves a new and more advanced and detailed legislation. The subject is definitely very complex and includes many issues, interests and moral questions. Thus, a 2007 comprehensive bill presented by the Ministry of Health, for example, was discussed in no less than nine committee meetings (alongside other bills), but eventually was not finalized.[58] The National Committee for Human Medical Research has repeatedly urged an overall revision of the regulations of experimentation with human participants. For various reasons, to date none of these initiatives has succeeded.

6 The 1994 Animal Welfare (Animal Experimentation) Law

The 1994 Animal Welfare (Animal Experimentation) Law regulates research using animals in Israel.[59] More specific rules are contained in the Animal Welfare (Animal Experimentation) Rules, 2001.[60] None of them relate to genome editing. However, the law determines a broad set of conditions to research using animals, to align with prevailing ethical principles. These conditions would apply to any future experimentation using gene editing in animals. For example, the law requires that animal experimentation be conducted in the smallest possible number of animals required; while

[56] Medical Experiments on Human Regulations, art. 3.
[57] The list of all twelve bill proposals is available at: Israel National Legislation Database, http://main.knesset.gov.il/Activity/Legislation/Laws/Pages/LawSuggestionsSearch.aspx?t=lawsuggestionssearch&st=allsuggestions&wn=%D7%A0%D7%99%D7%A1%D7%95%D7%99%D7%99%D7%9D%20%D7%A8%D7%A4%D7%95%D7%90%D7%99%D7%99%D7%9D&wp=False&ki=-1&sb=LatestSessionDate&so=D accessed 12 July 2018.
[58] Medical Experiments Bill, SF 2001, 786.
[59] The Cruelty to Animals (Experiments on Animals) Law, No 5754 of 1994, SF 1994, 298.
[60] Rules for Cruelty to Animals (Experiments on Animals), No 5761 of 2001, KT 2001, 752.

minimizing the pain and the suffering caused to the animals; and only when no other reasonable alternatives exist.

III OVERSIGHT BODIES FOR RESEARCH ON HUMAN GERMLINE GENOME MODIFICATION

1 *Local Helsinki Committees*

The 1980 Public Health (Medical Experiments on Humans) Regulations establish two main bodies which are in charge of oversight of the regulation of human experimentation, including experiments on germline genome modification: local 'Helsinki Committees' and a national 'Supreme Helsinki Committee' (i.e. the National Committee for Human Medical Research).

Helsinki committees are chaired by a medical doctor and are composed of at least seven members: a representative of the public, who must be a clergy or a legal expert; three medical doctors who are department chairs at the hospital or, at a minimum, associate professors at a recognized medical school; one of the MDs must specialize in internal medicine and be a representative of the hospital's management.[61] Another two members can come from any background deemed necessary. The hospital appoints committee members and the Director General of the Ministry of Health approves its overall composition.[62]

2 *The National Committee for Human Medical Research*

A National Committee for Human Medical Research, also known as the 'Supreme Helsinki Committee', serves the entire state of Israel as prescribed by the 1980 Public Health (Medical Experiments on Humans) Regulations.[63] It is appointed by the Director General of the Ministry of Health. It includes at least ten members, of which at least two must be representatives of the public (one clergy and one lawyer); six professors in recognized academic institutes, of whom at least three are medical doctors at a rank of full professor; the Director General of the Ministry of Health or their representative, as long as this person is a medical doctor; and the chair of the Israeli Medical Association.[64] In reality, this committee tends to have a much wider representation. As of 2017, it was composed of twenty-four members (nine

---

[61] Second Supplement of the Medical Experimentations in Human Beings Regulations, art. 1.
[62] Ibid.
[63] The National Committee for Human Medical Research appointed under the Public Health (Medical Experiments on Humans) Regulations, No 5741–1980.
[64] Third Supplement of the Medical Experimentation in Human Beings Regulations, art. 1.

clinicians, five scientists, three ethicists, three public representatives and four lawyers).[65]

The National Committee for Human Medical Research plays a key role in interpretation, implementation and compliance monitoring of the Law.[66] The Law designates it, with the addition of a representative of the minister of science, as the 'Advisory Committee' tasked to 'follow developments in medicine, science, biotechnology, bioethics and law in the field of genetic experimentations on human beings in Israel and abroad'; 'submit to the Minister and to the Science and Technology Committee of the Knesset an annual report' summarizing these developments; 'advise the Minister on the matter of genetic experimentations on human beings'; and provide him or her with its recommendations concerning the prohibitions of human reproductive cloning and reproductive use of germline cells that have been genetically modified.[67] The National Committee for Human Medical Research also functions as the advisory committee, according to the Genetic Information Law (2000), with similar compositions and roles.

3 The Ministry of Health

The minister of health is in charge of the execution of the Law.[68] The minister exercises overall supervisory and control powers and has the authority to issue regulations on specific issues and also regarding the operation of the Advisory Committee and how it executes its authority according to the Law.[69] Also, as we said earlier, the minister of health can draft regulations that can allow certain genetic interventions, if he or she finds they do not violate human dignity, and only if the Advisory Committee recommends to him to do so, and upon terms that he will determine.[70]

IV CURRENT PERSPECTIVES AND FUTURE POSSIBILITIES

Israel's regulatory frameworks regarding modification of human germline genome is short, simple and clear. It addresses the main ethical and societal

---

[65] In 2017, the National Committee for Human Medical Research consisted of twenty-four members.

[66] The Law, s 4.

[67] Ibid. ss 2.a and 4.

[68] Ibid. s 9.

[69] The Prohibition of Genetic Intervention (Human Cloning and Genetic Manipulation of Reproductive Cells) (Advisory Committee Powers) Regulations [2006] KT 2006, 318.

[70] The Law, ss 4.a.2, 3, 4.b, 5.a, and 9.

concerns without unnecessarily hindering research. This approach will allow Israeli science to progress in coming years in this promising area and hopefully will benefit and promote medical and reproductive research conducted in Israel. Rapid developments in the field of genome modification technologies and changing attitudes of other countries will hopefully lead Israel to go ahead and advance its legislation further in these areas in the years to come.

Israel will hopefully maintain the liberal and favourable attitude that it developed since its establishment regarding science and innovation. The new legislation that is planned to regulate human experimentation will hopefully be drafted wisely, so that both human dignity and freedom of scientific research are protected appropriately, enabling research in Israel to further promote innovation in medicine and reproduction.

PART V

CONCLUSIONS

22

Toward a Human Rights Framework for the Regulation of Human Germline Genome Modification

Andrea Boggio, Cesare P. R. Romano, and Jessica Almqvist

This book presents eighteen national regulatory regimes for human germline genome modification, as well as the international legal framework within which they exist. Had this been our only aim, it would have been a worthwhile and thorough update of existing scholarly works, but hardly a novel endeavor. However, what no one has so far done is look at the existing national and international regulations through the lens of international human rights standards and in particular through the lens of two sets of internationally recognized human rights: the "right to science" and the "rights of science."

We believe international human rights standards ought to be central to the development of germline engineering law and policies for various reasons, the most cogent of which is that these rights are legally binding on states, at a minimum because they are written in treaties that have been widely ratified, or because they have become part of customary international law. No matter how technical or specific legislation regulating germline engineering is, governments cannot depart from their international human rights obligations in developing regulatory frameworks. It is not just a matter of legality. It is also a matter of legitimacy. International human rights standards are the legal articulation of widely agreed upon values. They are expression of an internationally negotiated consensus. National regulatory frameworks cannot be consistent only with some human rights obligations while neglecting others. They need to be consistent with all of them.

We are aware our claim will surprise many. The "right to science" and the "rights of science" have been rarely invoked in the context of the discussion of the regulation of human genome modification. Moreover, the collective understanding of the normative content of the right to science – that is, what exactly are states' specific obligations generated by these rights – is limited, if compared to other human rights. Nonetheless, it is time to take a hard look at current national regulatory standards to ask whether they meet international human rights standards, in particular the "right to science" and the "rights of science."[1] This is the goal of this chapter. Here, we analyze current national regulatory standards of the selected eighteen countries in light of the five foundational principles that a reading of international bioethics law combined with international human rights standards suggests. They are: (i) freedom of research; (ii) benefit sharing; (iii) solidarity; (iv) respect for dignity; and (v) the obligation to respect and to protect the rights and individual freedoms of others.

We identified these principles by looking at key international bioethics instruments and in particular at the three UNESCO declarations – on the human genome and human rights (1997), human genetic data (2003), and bioethics and human rights (2005)) – while also taking into account the key provisions of the International Bill of Rights that concern science, including Article 15 of the International Covenant on Economic, Social and Cultural Rights (ICESCR) and Article 27 of the Universal Declaration of Human Rights (UDHR).[2] Whether current national regulatory standards respect these five principles, and thus meet international human rights standards, is the key question we raise in this chapter.

This chapter is divided in three sections: (I) evidence, where we summarize what emerges from a legal and comparative analysis of the national chapters included in this volume; (II) analysis, in which we discuss the extent to which the current national regulatory standards are consistent with the five foundational principles we identified; and (III) recommendations, where we offer our vision of an international governance framework that promotes science and technological development while being mindful and respectful of international human rights standards as well as the different sensitivities with which citizens from different parts of the world approach this complex problem.

As the readers will notice, our analysis focuses mainly on the first two principles: freedom of research and benefit sharing. This is because the evidence gathered in the first section points to problems precisely with these two principles, which reflect the primary goals of the right to science and the rights of science. We present a short discussion of the other three – solidarity, respect for dignity, and the obligation to respect and to protect the rights and individual freedoms of others – in the recommendations section. Of course,

---

[1]   *See, also*, A Boggio, BM Knoppers, J Almqvist and CPR Romano, "The Human Right to Science and the Regulation of Human Germline Engineering," *The CRISPR Journal*, vol. 2 (2019): 134–142; A Boggio, "Would a Gene-editing Ban Fit Human-rights Law?," *Nature*, vol. 569 (2019): 630.

[2]   See, in this book, Chapter 2, Section 2.1 and Section 2.3.

we hasten to say that the views expressed in this chapter do not necessarily reflect the views of the authors who contributed the national chapters to this book.

I EVIDENCE

1 *Basic Research*

In this study, we defined "basic research" as in vitro or ex vivo studies of germline tissue of humans, animals, or of the two in combination, done to understand the biological mechanisms of germline genome modification. Basic research on germline genome modification can be done using either gametes (sperm and oocytes) or embryos.

a Basic Research Using Gametes

Among the countries we studied, the regulation of research using gametes is relatively underdeveloped. Very few have rules that apply specifically to the use of sperm and oocyte in basic research. In this regard, the Swiss Federal Constitution is an exception. It prohibits any "interference with the genetic material of human reproductive cells," including gametes.[3] In Singapore, regulations provide that research with oocytes must be treated in the same way as research with embryos.[4] Protocols of research on oocytes are subject to the full ethical review and the preapproval of an institutional review board.

None of the countries surveyed in this book prohibits the in vitro modification of gametes for research purposes. This includes "gametogenesis," the in vitro derivation of gametes from iPSCs using gene editing techniques. Japan and the United Kingdom are among the few countries in the world that have enacted specific regulation for gametogenesis.[5] In both, the regulations permit gametogenesis and basic research involving germ cells derived from stem cells but prohibit the fertilization of iPSCs-derived gametes.[6] Several countries (e.g. Australia, Germany, Spain, and Singapore) prohibit clinical applications with gametes used in research.[7]

---

[3] See in this book, Chapter 15, p. 587.
[4] See in this book, Chapter 19, pp. 523–524.
[5] See in this book, Chapter 16, p. 446, and Chapter 7, p. 231.
[6] Ibid.
[7] See in this book, Chapter 20, p. 544; Chapter 8, p. 253; Chapter 13, p. 372; Chapter 19, pp. 531–532.

b Basic Research Using Embryos

The situation is more complex when basic research is done on embryos. Technically, CRISPR-based interventions are more efficient if a CRISPR/Cas9 tool is injected at the time of fertilization. This way, the likelihood of "mosaicism" in the resulting edited embryos[8] or off-target mutations is lower than when CRISPR/Cas9 tools are used at later stages of development. The second-best option is to intervene on one-cell embryos (zygotes). Although off-target mutations may still occur, mosaicism is relatively under control. After cell divisions or "cleavages," controlling how CRISPR-based interventions affect the embryos is more arduous.

Currently, basic research with CRISPR-based interventions at fertilization stage and one-cell stages is possible only in a handful of countries that permit the creation of "research embryos," that is to say, embryos that are intended to be used only for research but not reproductive purposes. Of the eighteen countries surveyed in this book, only seven permit the fertilization of an egg for research purposes (i.e. Belgium, China, Israel, Singapore, Sweden, the United Kingdom, and several jurisdictions in the United States). Of the other countries, ten prohibit scientists from creating embryos for research (i.e. Canada, France, Germany, Italy, Japan Mexico, the Netherlands, South Korea, Spain, and Switzerland). One, Australia, has restrictions so extensive that they amount to a de facto prohibition.[9]

It is important to note that, even where producing research embryos is permitted, research with gametes and embryos is still tightly regulated. All countries surveyed have adopted in some way (law, regulation, or guideline) the so-called fourteen-day rule, which prohibits experimenting on embryos fourteen days after fertilization. In addition, scientists must obtain approvals from a regulatory agency or an independent body. These approvals are granted only upon showing that the statutory requirements are met.

In Belgium, research embryos can be produced only as a last resort, that is, when the research goal cannot be achieved by other means, including resorting to supernumerary embryos.[10] Additionally, basic research must pursue a therapeutic objective, be based on the most recent scientific knowledge, meet the requirements of a correct methodology of scientific research, and be

---

[8] The term "mosaicism" describes a situation in which different cells in the same individual have different numbers or arrangements of chromosomes.

[9] Australia's regulatory framework is complex. Dianne Nicol concludes that there are "very limited avenues for legitimately creating and using embryos for the purpose of clinical and basic research." See, in this book, Chapter 20, p. 560. In Spain, embryos created via somatic cell nuclear transfer are not considered "embryos" and thus can be created for research purposes.

[10] See in this book, Chapter 9, p. 274.

carried out in an approved research facility and under the supervision of a person who possesses certain credentials.[11] The statutory regulator is the Federal Commission, which preapproves and oversees basic research with research and supernumerary embryos (and with gametes used to derive embryos).[12]

In Singapore, research embryos can be produced only if scientists demonstrate "strong scientific merit" and "potential medical benefit" of the research, the lack of acceptable alternatives to achieve the research goals, and obtain approval from a regulatory agency.[13] Similar standards must be satisfied in Sweden for basic research that uses gametes or embryos that can be traced back to a living or deceased donor: respect human dignity, human rights and fundamental freedoms, promotion of new knowledge, scientific value of the research, alternative ways to achieve the intended outcome; data protection issues, and researcher's credentials.[14]

In the United Kingdom, basic research on human germline genome modification can be carried out only after the regulatory authority has issued a license, contingent upon meeting various statutory requirements.[15] These include: informed consent of tissue donors, showing that the use of human embryos is necessary and not merely desirable, an independent research ethics committee's approval of the research, inspection of the research facilities, and a positive review of the research proposal by peers.[16]

In the United States, the situation is more complex because of its federal system. The creation of research embryos is regulated both federally and at state level, with important differences. While some states allow the creation of research embryos, others prohibit it.[17] Federal law does not prohibit the creation of research embryos per se, but federal funds cannot be used to support research where scientists edit the genomes of human embryos.[18] Preapproval of research by a review body that assesses the risks and benefits of the research is typically needed. Yet, independent scientists and fertility clinics that refuse federal funds are not bound by these federal requirements. They are only subject to the rules of the specific state/s in which they operate.[19]

---

[11] Ibid.
[12] Ibid., p. 270.
[13] See in this book, Chapter 19, p. 521.
[14] See in this book, Chapter 10, p. 300.
[15] See in this book, Chapter 7, pp. 226–229.
[16] Ibid.
[17] See in this book. Chapter 4, pp. 118–119.
[18] Ibid., pp. 120–122.
[19] Ibid., p. 113.

Of the eleven remaining countries discussed in this book, Australia, Canada, France, Japan, South Korea, Mexico, the Netherlands, and Spain permit research on supernumerary IVF embryos, that is, embryos that were produced during an assisted reproduction procedure and are no longer wanted, or cannot be implanted because not viable. Unsurprisingly, the seven countries that permit the creation of research embryos allow also research with supernumerary IVF embryos (i.e. Belgium, China, Israel, Singapore, Sweden, the United Kingdom, and several jurisdictions in the United States).[20] Research with supernumerary embryos is subject to limitations similar to those discussed above for research embryos: informed consent of tissues donors, need for research protocols to be preapproved, ethical oversight, need to show scientific rationale for the use of embryos, meeting research standards, and compliance with the fourteen-day rule.

Some countries have set up additional requirements that limit research with supernumerary IVF embryos further.[21] For instance, in Australia, embryos can be used in research only when they are "unsuitable" for assisted reproduction.[22] This means that an embryo must have undergone a sufficient number of cell divisions to determine that it cannot be used for reproduction. As Dianne Nicol notes in her chapter, those embryos are not particularly useful for gene editing research "given how damaged these cells need to be declared unsuitable for implantation."[23] In South Korea, supernumerary embryos can be used only in research that targets certain rare or incurable diseases enumerated by law.[24] In Mexico, supernumerary embryos can be used in research that benefits a particular embryo (e.g. to eliminate or improve disease of the embryo) but without altering the embryo's genotype. In France, the requirements for using IVF supernumerary embryos are so stringent that Blasimme, Caminiti, and Vayena report that "till 31 December 2015, out of the 20,000 embryos offered by couples to research, less than 10 percent have been made available for research."[25]

Of the countries surveyed, those with the most restrictive laws are Germany, Italy, and Switzerland. They prohibit research with embryos. However, research bans are not absolute. In Switzerland, embryos can be used in vitro to derive hESCs, but not to do experiments. As Blasimme, Caminiti, and

---

[20]   See, in this book, Chapter 4, p. 272–277; Chapter 21, p. 573; Chapter 19, pp. 522–524; Chapter 10, p. 307; Chapter 7, pp. 219–221(although not discussed explicitly); and Chapter 4, pp. 122–125.

[21]   In this case, the regulation of research with gametes is not relevant because the gametes were procured according to the rules regulating assisted reproduction.

[22]   See in this book, Chapter 20, pp. 556–557.

[23]   Ibid., p. 557.

[24]   See, in this book, Chapter 18, p. 590.

[25]   See, in this book, Chapter 14, p. 396.

Vayena report, embryos cannot be edited, and not even tested, for nonmedical reasons.[26] "Given such rules," the authors conclude, "it is not possible to perform genome editing of embryos in Switzerland for basic research purposes."[27] In these three countries, as well as in Mexico, embryos can be manipulated, but only as long as the purpose is a therapeutic benefit for the specific manipulated embryo. What these countries seem to permit are in vivo and in vitro manipulations of embryos[28] to correct genetic variations that would determine the birth of a child carrying a genetic disease.[29] However, at this point, scientists are still far from being able to engage in these sorts of germline interventions with confidence, and it is unclear how they can hone their skills if they cannot practice. Besides, it is unclear whether these statutes truly permit research aimed at editing the variations present in that embryo and, if so, what standard scientists would have to satisfy before their research is approved. As the authors of the chapters point out, the conclusion that this research is possible is merely speculative, since there are no reports of governmental authorities having permitted it, nor of scientists having engaged in this kind of research without sanction.[30]

That being said, the regulatory framework of most countries neither prohibits nor permits germline genome modifications expressly, creating uncertainties for researchers that we will discuss later in this chapter. In some cases, while silent as to whether researchers can modify gametes and embryos, the regulatory frameworks prohibit using modified gametes and embryos to achieve reproduction. If one follows the general legal principle by which "everything which is not forbidden is allowed," the conclusion can be drawn that since the regulators excluded *some* goals of germline engineering, particularly clinical research and applications, they did not exclude *other* goals of germline engineering, particularly acquiring knowledge and basic research. This is the conclusion that was reached by the authors of chapters on the six countries that allow the creation of research embryos (Belgium, Israel, Singapore, Sweden, the United Kingdom, and several jurisdictions in the United States).

---

[26] See, in this book, Chapter 15, p. 429.

[27] Ibid.

[28] With the exception of Switzerland, where only in vivo manipulations are permitted. See, Ibid., p. 430.

[29] See, in this book, Chapter 8, p. 264; Chapter 12, p. 355; and Chapter 15, pp. 403–404.

[30] It is not clear how these exceptions must be interpreted. We invite readers to read the respective sections of the national chapters for more nuanced arguments about the meaning of these clauses.

Overall, the picture that emerges from our comparative analysis of the regulation of basic research with embryos and gametes is that this is an area filled with prohibitions and restrictions. In the second part of this chapter, we will discuss whether these regulations are excessively restrictive given that states must ensure the freedom indispensable for scientific research and the right of everyone to enjoy the benefits of scientific progress. However, before we move to that analysis, we need to address two more issues: the regulation of clinical research and applications.

2 *Clinical Applications*

Because the regulation of clinical applications is comparatively easier to navigate than that of clinical research, let us address first the end of the translational pipeline. In this book, we defined "clinical application" of human germline genome modification as the use of these techniques on patients in a clinical setting. Most countries surveyed in this book prohibit unequivocally the provision of germline engineering therapies in a clinical setting. There are statutory prohibitions in Australia, Canada, France, Germany, Israel, Japan, the Netherlands, South Korea, Spain, Sweden, Switzerland, and the United Kingdom. Other countries have achieved the same result through regulatory mechanisms. For instance, in Singapore, clinical research and applications are not allowed as a result of a moratorium issued by the Bioethics Advisory Committee in 2005.[31] In China, a technical specification of standards for assisted reproduction, issued in 2003 by the Ministry of Health, prohibits "gene manipulation of human gametes, zygotes or embryos for reproductive purposes."[32] In the United States, while no federal law expressly prohibits clinics from providing germline editing services, the federal legislature has prohibited the federal agency from accepting applications to begin clinical research.[33] This also means that no gene editing applications can be offered to patients in a clinical setting, since the regulators' premarket approval is a prerequisite to offering clinical applications.[34]

A few regulatory frameworks leave the door open, intentionally or accidentally, to the possibility that, in some cases, germline engineering might be used in a clinical context. This is the case of Belgium, and also, counterintuitively,

---

[31] See in this book, Chapter 19, p. 524.
[32] See in this book, Chapter 17, p. 495.
[33] See in this book, Chapter 4, p. 120.
[34] Ibid., p. 113. Yet, as we have seen, independent scientists and fertility clinics that refuse federal funds are not bound by these federal requirements.

of Italy and Mexico.[35] This conclusion is reached if one keeps in mind the rationale of those national regulatory frameworks, which is to prevent embryos from being "harmed" during research. Arguably, interventions that improve the well-being of the embryo are lawful. This interpretation of the Belgian, Italian, and Mexican statutes has not been tested in courts, so the extent to which clinical applications are actually permitted is unclear, but the possibility is intriguing.

Similarly, France permits the study of germline engineering techniques whose primary aim is therapeutic (e.g. preventing a genetic disease) rather than altering the descendants of the treated embryo. "If that is correct," Blasimme, Caminiti, and Vayena conclude, "the use of genome editing technologies on human embryos that will likely result in germline modifications may not be a priori forbidden."[36]

Singapore and Israel appear to leave room for some procedures as long as their safety and effectiveness is demonstrated. Specifically, Singapore seems open to certain types of genetic germline modification technologies to prevent the transmission of mitochondrial diseases, including ooplasmic transfer, pronuclear transfer, and maternal spindle transfer.[37] De Miguel Beriain and Casabona argue that Spanish law does not ban clinical applications nor basic and clinical research using germline modification technologies as long as no new genetic material is introduced intentionally into the genome of the embryos.[38]

3 Clinical Research

The regulation of clinical research is less clear-cut, and, although in the translational pipeline it comes before clinical application, we present it after the discussion on clinical applications because, in some countries, the ban on clinical research is the corollary of a ban on clinical applications. Clinical research involves experimenting on a living person, testing therapies on patients. Clinical research on human germline genome modification would involve modifying germline tissue of the research subject in vivo, or transferring to a research subject gametes or embryos that were modified ex vivo (i.e. by transferring a modified embryo in the uterus), to test the safety and efficacy of germline genome engineering. All countries surveyed in this volume

---

[35] See in this book, Chapter 9, p. 274; Chapter 12, pp. 355–356; Chapter 5, p. 140.
[36] See in this book, Chapter 15, p. 397.
[37] See in this book, Chapter 19, p. 524.
[38] See in this book, Chapter 13, pp. 376–378.

prohibit clinical research. In some countries, the ban is blank or absolute. In other countries, exceptions to the prohibition are contemplated expressly, or are revealed by statutory interpretation.

The countries with blank or absolute prohibitions are: Canada, Japan, Singapore, South Korea, Switzerland, the United States,[39] and the European Union. In these countries, all clinical research involving human genome germline modifications is prohibited. In others, the ban is not absolute. For instance, in Israel, the minister of health could authorize, through regulations, clinical research on and clinical use of genetically modified germline cells, as long as it does not violate human dignity and may have therapeutic benefit.[40] According to Ravitsky and Ben-Or, the exception was designed as "a simpler and more efficient solution than subsequently trying to modify the Law" in the event germline-based therapeutic options become available.[41] Mexico permits clinical applications that have a positive therapeutic effect for the embryo.[42] It follows logically that clinical research testing the safety and effectiveness of a procedure that is lawful should also be lawful. While this is a reasonable interpretation, the fact that the relevant statutes authorize clinical applications but not research cannot be ignored. What the legislator may have envisioned is that clinical applications that have been tested and approved in a different jurisdiction may then be offered to patients in their country. Or, it might be simply an accidental omission caused by hasty legal drafting.

Limits to the bans can also be identified by means of statutory interpretation. For instance, in Australia only clinical applications of germline modification technologies that cause modification that are "intended to be inherited" are prohibited.[43] Can germline modifications be tested on humans if there is no intent to pass on the modifications to the offspring of the research subject? This could be the case when the research subject has agreed to terminate the pregnancy after data for the clinical trial are collected. However, in her chapter, Dianne Nicol proposes a more restrictive reading of the statute: the prohibition of clinical applications "could also apply in the research context," Nicol argues, "where the intention for the genetic

---

[39] In the United States, clinical research cannot be carried out not as result of a legislative ban, but rather because the legislative branch has barred the FDA from receiving any application for clinical research using germline genome modification. See in this book, Chapter 4, p. 120.

[40] See in this book, Chapter 21, pp. 572–574.

[41] Ibid., p. 574.

[42] See in this book, Chapter 5, p. 140.

[43] See in this book, Chapter 20, p. 554.

manipulation to be passed on to future generations is absent, but the intention to modify the genome in a way that *could* be inherited is present."[44] In South Korea, since the law prohibits clinical research with a therapeutic goal, one could surprisingly argue that clinical research without a therapeutic goal (e.g. enhancement or aesthetic reasons) is allowed.[45] In Japan, although clinical research using germline genome editing is largely prohibited, editing that does not involve "the administration of a gene or cells" is not prohibited. Acutely, Ishii points out that this could be done if editing is performed using a messenger RNA (mRNA) rather than by inserting a plasmid harboring a gene of template DNA.[46]

In China, clinical trials involving human genome germline modifications seem to fall in a legislative vacuum, and therefore there is some uncertainty as to what is prohibited. Research with human subjects is subject to regulations that have incorporated the main international standards for biomedical research. The *Guiding Principles for Human Gene Therapy Research and Quality Control of Preparation* allow only genetic therapy using somatic, but not germline, cells.[47] However, it is unclear whether the *Guiding Principles* allow gene therapy on human embryos and whether germline genome modifications can be clinically tested on humans. In the wake of Dr. He Jiankui's controversial revelations, the Chinese regulatory and funding agencies and various professional bodies issued statements condemning Dr. He's actions. In a joint statement, the Chinese Society for Stem Cell Research and the Committee of Genome Editing, Genetics Society of China concluded that "we believe the research led by He is strongly against ... the Chinese regulations."[48] An investigating task force set up by the Health Commission of China in Guangdong Province released a preliminary report on January 21, 2019, stated that He had violated government bans.[49]

The situation in Europe is even more complex due to the stratification of regulatory instruments. First, of the nine European states surveyed in this book, three (France, Switzerland, and Spain) are bound to prohibit

[44] Ibid.
[45] See in this book, Chapter 18, p. 513.
[46] See in this book, Chapter 16, p. 459.
[47] See in this book, Chapter 17, p. 491.
[48] Committee of Genome Editing, Genetics Society of China and Chinese Society for Stem Cell Research, Statement "Condemning the Reproductive Application of Gene Editing on Human Germline," November 27, 2018. www.cscb.org.cn/news/20181127/2988.html accessed December 11, 2018.
[49] XINHUA, Guangdong Releases Preliminary Investigation Result of Gene-Edited Babies, January 21, 2019, www.xinhuanet.com/english/2019–01/21/c_137762633.htm accessed January 25, 2019.

interventions to modify the human genome for the purpose of introducing modifications in the genome of any descendants by virtue of their ratification of the Oviedo Convention.[50] Then, eight out of nine are members of the European Union. Currently, EU Regulation 536/2014 includes a blank prohibition under which EU member states cannot approve clinical trials involving the modification of the human genome germline.[51] However, before 2014, clinical trials were prohibited by a directive. As a general rule, directives are not self-executing, and member states must adopt their own laws to reach the policy goals set by a directive. As a result, some EU countries had no national legislation or had adopted national laws prohibiting clinical research. The Netherlands is the only European country surveyed in this book without a statute prohibiting clinical research on germline editing.[52] The key Dutch statute prohibits "deliberately modifying the genetic material of the nucleus of human germ cells with which a pregnancy will be established."[53] As van Beers, de Kluiver and Maas note, "These words suggest that human genetic modification is prohibited only for reproductive purposes, and only where nuclear DNA is concerned."[54]

All other countries have statutes, some of which contain language that may be interpreted as granting certain exceptions to the pre-2014 EU-mandated ban on clinical research on germline modifications. The German and Swedish statutes expressly prohibit germline interventions that are therapeutic.[55] The authors of those chapters point that that certain human germline genome editing interventions without a therapeutic purpose might fall outside the scope of the statute.[56]

On the opposite side of the spectrum, Belgium, France, and Italy permit only germline interventions that are therapeutic. Belgium and Italy have statutory language similar to Mexico, that is, they permit clinical applications that have a positive therapeutic effect for the embryo. As we have seen, one could reasonably argue that clinical research testing clinical applications that are beneficial to the embryo is permitted.[57] The French Civil Code includes an exception to the ban on clinical research allowing for research activities

[50]   See in this book, Chapter 13, p. 362; Chapter 14, p. 382; and Chapter 15, p. 412.
[51]   See in this book, Chapter 6, p. 194.
[52]   See in this book, Chapter 11, p. 325.
[53]   Ibid.
[54]   Ibid.
[55]   See in this book, Chapter 8, pp. 253–256; and Chapter 10, pp. 304–305.
[56]   Ibid., p. 264 and pp. 304–305.
[57]   See in this chapter, Section I.2, pp. 592–593.

aimed at preventing or treating genetic diseases and not at modifying the genetic traits of a person.[58]

In the United Kingdom, which soon might be no longer part of the European Union, the key statute does not set up a mechanism to evaluate and possibly authorize the clinical research on new technologies or treatments. To address this legislative void, Lawford Davies draws a parallel with the 2017 approval of the clinical use of mitochondrial donation using a pronuclear transfer. Thirteen years after the submission of the proposal for clinical research, the regulators authorized the research, but under the strict oversight of the agency and with the obligation that researchers apply for permission for each patient and monitor patients' health scrupulously in follow-up sessions. "Should clinical application of human genome germline modification become technically feasible," Lawford Davies concludes, "it is highly likely that a similar process of review and consultation will unfold."[59]

II ANALYSIS

This is what we learned from the analysis of the selected national regulatory frameworks. However, each of those states, as any other state, has international legal obligations that frame and constrain their national legal frameworks, including a set of obligations deriving from two specific branches of international law: international human rights law and international bioethics law. The international context in which we carry out our analysis should be clear to the readers by now.[60] However, it is worth reiterating here the key rights that inform our analysis: the "right to science," also known as the right of everyone to benefit from scientific progress (benefit sharing), and the "rights of science," of which the right to engage in scientific research (scientific freedom) is an essential component.

Both international human rights law and international bioethics law agree that freedom of research must be respected. Respecting freedom of research requires states to refrain from interfering directly or indirectly with it,[61] and avoiding taking measures that hinder or prevent the enjoyment of this

---

[58] See in this book, Chapter 14, pp. 386, 396–399.

[59] See in this book, Chapter 7, p. 597.

[60] See in this book, in general, Chapter 2.

[61] See UN Committee on Economic, Social and Cultural Rights (UNCESCR), General Comment No. 14 (2000) on the Right to the Highest Attainable Standard of Health (Article 12 of the International Covenant on Economic, Social and Cultural Rights), UN doc. E/C.12/2000/4, August 11, 2000, paras. 33; General Comment No. 17 (2005) on the Right of Everyone to Benefit from the Protection of the Moral and Material Interests Resulting from Any Scientific, Literary or Artistic Production of Which He or She is the Author (Article 15.1.c of the

right.[62] Simply put: scientists must be allowed to engage in scientific inquiries freely. However, the "right to science" and the "rights of science" are not absolute rights. They can be limited. Restrictions on the enjoyment of these rights are allowed only if they are consistent with international human rights standards. Specifically, they require that three conditions are met: (1) any restriction must be prescribed by law (condition of legality); (2) any restriction must pursue a legitimate aim (condition of legitimacy); and (3) any restriction must be limited to what is necessary to fulfill that aim, and be the result of a careful balancing of interests (condition of proportionality).[63] Governments bear the burden of showing that the restrictions they impose do not violate international human rights standards.[64] Until this burden is met, states must avoid imposing restrictive measures that interfere with the rights to scientific freedom and benefit sharing.

In the following subsections, we will discuss and critically examine the most important limits imposed on the effective enjoyment of the human rights to science in the area of human germline engineering. We will pay special attention to the question whether these limits are consistent with international human standards and states' obligations related to these rights.

1 Restrictions Must Be Prescribed by Law (Condition of Legality)

According to Article 4 of the ICESCR, limitations to scientific freedom must be "determined by law" (condition of legality). In broad terms, this requirement entails that the "limitation should have a basis specifically in domestic law consistent with the Covenant; the law must be adequately accessible; the relevant domestic law must be formulated with sufficient precision," and the "law must not be arbitrary, unreasonable, discriminatory or incompatible with the principle of interdependence of all human rights."[65] According to the requirement of clear and precise laws, captured by the principle of legal certainty, limitations are determined by law only when they are sufficiently

---

International Covenant on Economic, Social and Cultural Rights), UN doc., E/C.12/GC/17, January 12, 2006, paras. 28

[62] Analogously, see UNCESCR, General Comment No. 13 (Twenty-first session, 1999) on the Right to Education (Article 13 of the International Covenant on Economic, Social and Cultural Rights), UN doc. E/C.12/1999/10, December 8, 1999, para. 47.

[63] O de Schutter, *International Human Rights Law*, Cambridge University Press (2010), p.288.

[64] Analogously, UN Commission on Human Rights, Siracusa Principles on the Limitation and Derogation Provisions in the International Covenant on Civil and Political Rights, UN doc. E/CN.4/1987/17, para. 12.

[65] M Ssenyonjo, *Economic, Social and Cultural Rights in International Law*, Hart Publ. (2016), 2nd edition, p. 152.

clear to allow a reasonable person to regulate her conduct based on that law.[66] The empirical evidence presented in the national chapters shows that, in many countries, the laws regulating research on human germline genome modification are excessively vague. This raises the question whether laws in place actually provide a sufficient degree of legal certainty as required by human rights standards. Before analyzing this question, let us lay out the requirements of this principle within the international human rights framework.

The principle of legal certainty is a "general principle of law common to civilized nations,"[67] that is to say a legal principle that can be found in the legal system of several, if not all, "civilized nations."[68] Indeed, it is a well-established legal concept, found both in the Civil Law and in the Common Law legal traditions. In Europe, the concept of legal certainty has been recognized as one of the general principles of European Union law by the European Court of Justice since the 1960s. It is found in all European continental legal systems, those that follow the Romano-Germanic (Civil) legal tradition.[69] In the Common Law tradition, legal certainty is often explained in terms of citizens' ability to organize their affairs in such a way that does not break the law. In the United States, the principle of legal certainty is understood as "fair warning" and the "void for vagueness."[70] In both legal traditions, legal certainty is regarded as grounding value for the legality of legislative and administrative measures taken by public authorities.[71] The principle is also given importance in the context of the UN work on the promotion of the rule of law at the national and international levels. Here the rule of law requires legal certainty, and both are an essential condition for the full realization of human rights.[72]

---

[66] Siracusa Principles (n 64), para. 17, according to which "legal rules limiting the exercise of human rights shall be clear and accessible to everyone".

[67] Charter of the United Nations and Statute of the International Court of Justice, June 26, 1945, Art. 38.1.c. 59 Stat. 1031 [the Charter], 1055 [ICJ Statute], T.S. No. 993 [I.C.J. Statute at 25], 3 Bevans 1153 [I.C.J. Statute at 1179].

[68] For a comparative law discussion of the principle, see M Fenwick; M Siems; W Stefan (eds.), *The Shifting Meaning of Legal Certainty in Comparative and Transnational Law*, Hart Publ. (2017). The authors show how widespread recognition of the principle is, while pointing out that pinning down what legal certainty means and when it is violated remains difficult.

[69] For instance, one can find it in the German legal system as "Rechtssicherheit," in France as "sécurité juridique," in Spain as "seguridad juridica," in Italy as "certezza del diritto," in the Benelux countries as "rechtszekerheid," in Sweden as "rättssäkerhet," in Poland as "pewność prawa," in Finland as "oikeusvarmuus." D Chalmers, *European Union Law: Text and Materials*, Cambridge University Press (2006), p. 454.

[70] E Claes, W Devroe, B Keirsblick, *Facing the Limits of the Law*, Springer (2009), p. 93.

[71] Ibid., pp. 92–93.

[72] For example, Declaration of the High-Level Meeting of the General Assembly on the Rule of Law at the National and International Levels, UN doc. A/RES/67/1, November 30, 2012, para.

Invariably, human rights bodies resort to the principle of legal certainty to determine the legitimacy of restrictions on human rights. While there is no instrument that speaks directly to the limitations of scientific freedom, the UN Human Rights Committee has applied the principle of legal certainty to a cognate freedom: the freedom of expression. According to the Human Rights Committee's General Comment No. 34 on Article 19 of the International Covenant on Civil and Political Rights (Freedoms of Opinion and Expression), "a norm, to be characterized as a 'law,' must be formulated with sufficient precision to enable an individual to regulate his or her conduct accordingly."[73] Restrictions to freedom of expression "shall only be such as are provided by law and are necessary: (a) For respect of the rights or reputations of others; (b) For the protection of national security or of public order (*ordre public*), or of public health or morals."[74] The parallel with Article 4 of the Covenant on Economic, Social and Cultural Rights is striking. In both cases, restrictions on freedom, whether of expression or of research, must be provided by "law."

Unclear laws are particularly problematic when they provide for criminal sanctions. At the international level, the European Court of Human Rights has asserted repeatedly the paramount importance of legal certainty in connection with criminal laws.[75] Because several countries surveyed in this book have chosen to regulate some, if not all, aspect of activities modifying the genome of human germline cells through criminal law, legal certainty is paramount. Australia, Canada, China, France, Germany, Israel, Italy, Mexico, the Netherlands, South Korea, Spain, Sweden, Switzerland, and the United Kingdom criminalize certain activities connected with using

---

7. For a definition of the rule of law mentioning specifically legal certainty, see www.un.org /ruleoflaw/what-is-the-rule-of-law/ accessed January 23, 2019.

[73] UN Human Rights Committee, General Comment no. 34, Article 19, Freedoms of Opinion and Expression, September 12, 2011, CCPR/C/GC/34, para. 25.

[74] International Covenant on Civil and Political Rights (adopted December 16, 1966, entered into force March 23, 1976) 999 UNTS 171 (ICCPR), Art. 19.3.

[75] C Grabenwarter, *European Convention on Human Rights: A Commentary*, Hart Publ. (2014), p. 68, pp, 178–181, p. 191, and p. 263. Several decisions of the European Court of Human Rights address the issue. E.g. see *Kolevi v. Bulgaria*, no. 1108/02, Judgment, November 5, 2009, page 29, § 174; *Baranowski v. Poland*, no. 28358/95, Judgment, March 28, 2000, page 12, § 52; *Cantoni v. France*, no. 17862/91, Judgment, November 11, 1996, page 14, § 29; *C.R. v. the United Kingdom*, no. 20190/92, Judgment, November 22, 1995, page 12, § 33; *S.W. v. the United Kingdom*, no. 20166/92, Judgment, November 22, 1995, page 13, § 35; *Kokkinakis v. Greece*, no. 14307/88, Judgment, May 25, 1993, pages 17–18, §; Streletz, *Kessler and Krenz v. Germany*, nos. 34044/96, 35532/97 and 44801/98, Judgment, March 22, 2001, page 26, § 50; *Malone v. the United Kingdom*, no. 8691/79, Judgment, August 2, 1984, pages 27–28, § 67; Valenzuela Contreras v. Spain, no. 58/1997/842/1048, Judgment, July 30, 1998, page 15, § 46.

human embryos and/or modifying the human genome. Yet, as we also have seen, some of these criminal prohibitions lack in clarity and precision. Granted, one could argue that what is not prohibited is permitted, and, thus, unless research or clinical activity is expressly prohibited, it is lawful. Loopholes abound. Nonetheless, scientists are unlikely to take advantage of them and move ahead with innovative research when the risk is to be criminally prosecuted.

In many jurisdictions, the limitations to scientific freedom are contained in laws and regulations that are unnecessarily vague. For example, many regulatory frameworks do not address research on the human germline expressly, and therefore do not allow scientists to be sufficiently confident that their research can be done lawfully. Several fail to give scientists reasonable notice of exactly what is permitted and prohibited. Authors of the chapters on Canada, Italy, Mexico, the Netherlands, Singapore, Spain, and Sweden identify key aspects of the regulation of basic research as "unclear." The authors of the chapters on China, France, Mexico, Spain, and Sweden talk about "uncertainties." They give us examples of instances where definitions and substantive provisions have not been updated, despite the advent of CRISPR, which has transformed our understanding of what constitutes "gene therapy,"[76] or of instances where new advancements are not expressly regulated, as in the case of in vitro gametogenesis.

Indecipherable laws and regulations have a chilling effect on scientific freedom. Faced with muddy regulatory frameworks, scientists likely refrain from doing something that is not expressly prohibited. Some authors explicitly acknowledge the chilling effect vague regulatory frameworks have on research. Song and Isasi conclude that, in China, "obscurity in the breadth and scope of normative instruments, paired with blurred jurisdictional boundaries between governmental actors, have created what it seems to be an unstable regulatory environment where accountability is uncertain, with chilling effects on research."[77] De Miguel Beriain and Casabona note that, in Spain, "the prevailing view amongst scholars is that any intervention seeking to modify the human genome that is not for preventive, diagnostic, or therapeutic purposes is prohibited ... has chilling effect on Spanish researchers, who, currently, are not engaging in research in this direction."[78] Timo Faltus point out that, in Germany, the ban on human germline genome

---

[76] J Doudna and E Charpentier, "The New Frontier of Genome Engineering With CRISPR-Cas9" *Science*, vol. 346, no. 6213 (2014): 1258096.

[77] See in this book, Chapter 17, p. 497.

[78] See in this book, Chapter 13, p. 372.

modification "has chilling effects on the funding of borderline research (i.e. research on asexually produced embryos, tripronuclear embryos), too."[79]

Consider Article 15.2 of the International Covenant on Economic, Social and Cultural Rights, which recites: "The steps to be taken by the States Parties to the present Covenant to achieve the full realization of this right shall include those necessary for the conservation, the development, and the diffusion of science and culture." We believe that whenever laws lack the necessary precision, thus inhibiting scientific freedom, governments have failed to take the steps "necessary for . . . the development of science." These steps "must be deliberate, concrete and targeted" toward the full realization of this right.[80] Incomplete and unclear statutes fail to comply with the principle of legal certainty and therefore cannot be considered to be truly "determined by law." States must take steps to ensure that scientists are in a position to tell with reasonable precision whether their research is lawful. As we will discuss in the last section, the best way to meet international legal standards is for governments to enact legislation that regulates research on human genome germline modifications expressly and clearly.

2 Restriction Must Pursue a Legitimate Aim (Condition of Legitimacy)

Clarity is not sufficient. Restrictions must also be justified by the pursuit of a legitimate aim (condition of legitimacy). In this regard, Article 4 of the ICESCR specifies that the rights the Covenant recognizes may be subject "only in so far as this may be compatible with the nature of these rights and solely for the purpose of promoting the general welfare in a democratic society."[81] What this proviso means and consequently requires is somewhat unsettled. The notion of "general welfare" has been understood as "furthering the well-being of the people as a whole."[82] The expression "in

---

[79] See in this book, Chapter 18, pp. 250–251.

[80] See the following UNCESCR General Comments: No. 3: The Nature of States Parties' Obligations (Art. 2, para. 1, of the Covenant), UN doc. E/1991/23, December 14, 1990, para. 2; General Comment 13 (n 62), para. 43; General Comment No. 14 (n 61), para. 30; General Comment No. 17 (n 61), para. 25.

[81] International Covenant on Economic, Social and Cultural Rights (adopted December 16, 1966, entered into force January 3, 1976) 993 UNTS 3 (ICESCR), art. 4. At the time of writing, the UN Committee on Economic, Social and Cultural Rights has not yet adopted a General Comment concerning the interpretation of art. 4.

[82] Limburg Principles on the Implementation of the International Covenant on Economic, Social and Cultural Rights adopted in Maastricht on June 2–6, 1986, UN Commission on Human Rights, Note verbale dated December 5, 1986 from the Permanent Mission of the Netherlands to the United Nations Office at Geneva addressed to the Centre for Human Rights ("Limburg Principles"), January 8, 1987, E/CN.4/1987/17 para. 52.

a democratic society" should be construed as imposing a further restriction to the application of limitations by requiring the state to demonstrate that the limitations do not impair the democratic functioning of the society.[83] According to the Committee on Economic, Social and Cultural Rights, at the very least it demands that a state ensures that limitations on economic, social and cultural rights are "necessary and proportionate and do not interfere with the core minimal content of the rights."[84] The requirement that any limitation must be "necessary in a democratic society" implies the existence of a "pressing social need" or a "high degree of justification" for the limitation in question.[85]

Even if public safety, order, health, or morals are not mentioned explicitly as grounds that justify limitations of the Covenant rights, they are generally understood as providing valid grounds for limiting not only civil and political rights, but also economic, social, and cultural rights.[86] However, although concerns with morals, safety, health, or order may be aspects of the "general welfare of a democratic society," to limit a right legitimately on these nonexplicit grounds, it must be clear that the protection of these concerns is necessary for the promotion of welfare *in a democratic society*.

Thus, for example, a legislator that invokes safety or health concerns to justify restrictions has the burden of explaining how the balancing between the individual right to health and the right to health and safety of the many, has been achieved in conformity with the proportionality test. In meeting this test, it must be recalled that "public health" may be invoked as a ground "to allow a state to take measures dealing with a serious threat to the health of the population or individual members of the population. These measures must be specifically aimed at preventing disease or injury or providing care for the sick and injured."[87] Public safety "cannot be used for imposing vague or arbitrary limitations and may only be invoked when there exist adequate safeguards and effective remedies against abuse."[88] Likewise, a state may invoke public morality as a ground for restricting rights.

---

[83] Ibid., paras. 53–54.

[84] UNCESCR, Concluding Observations, Vietnam, C/C.12/VNM/CO/2–4 (December 15, 2014), para. 8. Also see A Müller, "Limitations to and Derogations from Economic, Social and Cultural Rights," *Human Rights Law Review*, vol.9, no. 4 (2009): 577–601, referring to the Principles on the Limitations and Derogation from Provisions in the ICCPR, E/CN.4/ 1985/4 (1985).

[85] M Ssenyonjo (n 65), p. 152.

[86] O De Schutter (n 63), p. 291.

[87] Siracusa Principles (n 64), para. 25.

[88] Ibid., paras. 33–34.

However, even if enjoying a certain margin of appreciation, it "must demonstrate that the limitation in question is essential to the maintenance of respect for fundamental values of the community,"[89] and that these values have been identified and discussed through a democratic process that takes into account the voices and interests of particularly vulnerable groups and minorities.

The rules protecting the rights of research subjects, above all their right to free and informed consent, are a typical example of a legitimate restriction, consistent with the need to protect the human rights of others, a legitimate goal. Another limitation accepted in democratic societies to promote the general welfare is that scientific research must be done responsibly. Scientists have an individual and collective duty to act responsibly. Just because you can do something, it does not mean that you will do it and damn the consequences, as Dr. He Jiankui did.[90] Scientists must adhere to the rules of good research conduct, and the scientific community has the duty of, but also the right to, self-regulation, that is, to regulate the scientific enterprise to ensure the integrity of the research process and the minimization of misconduct.

Restrictions must not be arbitrary, lest the condition of legitimacy would be violated. When limitations are arbitrary or unwarranted, the freedom indispensable for scientific research is not respected. For instance, while Italy bans the creation of embryos for research, Italian scientists are reported to import them from abroad to carry out their research.[91] It is hard to explain how the different protection afforded to "national embryos" and "foreign embryos" can be reconciled with the stated purpose of protecting the dignity of the embryo. Even if freedom of research may be restricted for reasons of public morality, as has been said, a state that invokes it "must demonstrate that the limitation in question is essential to the maintenance of respect for fundamental values of the community."[92] In the Italian case, it is unclear what that fundamental value would actually be, given the disparity of treatment between embryos created in the national territory and those coming from abroad.

---

[89]   Ibid., para. 27.

[90]   See in this book, Preface, p. xxix.

[91]   See, generally, B Forest, "Three Courageous Italian Scientists – An Example for Louisiana," World Congress for Freedom of Scientific Research, April 19, 2010, www.freedomofresearch.org/three-courageous-italian-scientists-an-example-for-louisiana-by-barbara-forrest/, accessed March 7, 2017.

[92]   Ibid., para. 27.

3 Restrictions Must Be Limited to What is Necessary to Fulfill Legitimate Aims, and be the Result of a Careful Balancing of Interests (Condition of Proportionality)

Restrictions must not only be the result of reasonably clear laws adopted democratically for legitimate goals. They must also be proportional, limited to what is necessary to fulfill those legitimate goals, and be the result of a careful balancing of interests (condition of proportionality). Total bans and the so-called ne plus ultra prohibitions violate the condition of proportionality.

a The Prohibition of the Creation of Embryos

Of the eighteen countries surveyed in this book, only seven permit the fertilization of an egg for research purposes (i.e. Belgium, China, Israel, Singapore, Sweden, the United Kingdom, and several jurisdictions in the United States). Of the other countries, ten prohibit scientists to create embryos for research (i.e. Canada, France, Germany, Italy, Japan Mexico, the Netherlands, South Korea, Spain, and Switzerland). One, Australia, has restrictions so extensive they amount to a de facto prohibition.[93]

We believe that protecting scientific freedom entails permitting the creation of research embryos. As deWert and colleagues noted, "Only in countries where the creation of embryos for the exclusive purpose of research is allowed could [gene editing] be applied at earlier stages and with fresh oocytes and embryos."[94] Research on supernumerary IVF embryos is only a second best, because of the limited supply of embryos and the fact that, likely, these embryos are either not viable or affected by various disorders. Further, freedom of research encompasses the ability to modify the genome of gametes as well as grow stem cells clonally and expand them into many millions of cells, allowing detailed screening for off-target events before an embryo is made.[95]

We are not advocating unlimited freedom to create any embryos for research. The six jurisdictions that permit the creation of research embryos show that it is possible to strike a balance between the needs of science and ethical concerns. There, the creation of research embryos is limited by various rules, including the requirement to obtain consent from tissue donors,

---

[93] See, comments on Australia in footnote 9.

[94] G de Wert et al., "Responsible Innovation in Human Germline Gene Editing. Background Document to the Recommendations of ESHG and ESHRE," *European Journal of Human Genetics*, vol. 26, no. 4 (2018): pp. 450–470, 453.

[95] Mosaicism is not a problem when the gametes carry the genome edits because these will be present in all cells of the resulting embryo.

approval oversight, and the "14-day" rule. We believe that these limitations are compatible with human rights standards as their rationale is to protect other human rights (the rights of the research subjects) and are enacted democratically. Ethical approvals and oversight ensure that the research is carried out responsibly, respecting the sensitivities of the societies where it is done. This regulatory approach, in our opinion, is the one that best balances freedom and the limits of this research because the limitations to scientific research are both appropriate and narrowly tailored.

b Ne Plus Ultra Prohibitions

Several countries allow the translational pipeline to advance only up to a certain point. They might allow basic research but prohibit clinical research. We believe these ne plus ultra, or blank, prohibitions are difficult to reconcile with everyone's right to "benefit from scientific and technological progress" and the principle of benefit sharing, even when lawful limitations to these rights are taken into account.[96] Noticeably, Article 15.2 of the Covenant requires governments to "take steps ... to achieve the full realization of this right shall include those necessary for ... the diffusion of science." If everyone is to truly enjoy the benefits of scientific progress, biomedical knowledge must be allowed to be translated into clinical applications, unless there are legitimate grounds for limiting the right. Claims to benefit sharing are particularly strong when knowledge might lead to developing new medical treatments that make it possible to cure or even prevent diseases that otherwise would be incurable.

In situations such as these, it is doubtful that generic bans meet the legitimacy and proportionality tests. When discussing limitations on the right to health, the UN Committee on Economic, Social, and Cultural Rights has noted that when several types of limitations are available, the least restrictive alternative must be adopted.[97] In this context, it also noted that the limitations should be of limited duration and be subject to review.[98] Even if some aspects of the bans may be justified, it seems important to consider whether they could at least be narrowed down.

As we have seen earlier in this chapter, Canada, Japan, Singapore, South Korea, Switzerland, the United States,[99] and the European Union have

---

[96]   ICESCR, Articles 2.2 and 15.1.b.
[97]   See, analogously, UNCESCR, General Comment No. 14 (n 61), para. 29.
[98]   Ibid.
[99]   In the United States, clinical research is not possible not because of a statutory ban, but rather because the legislative branch has barred the FDA from receiving any application for clinical research using germline genome modification. See in this book Chapter 4, p. 120.

adopted blank prohibitions of clinical research involving human genome germline modifications. In our judgment, blank prohibitions to translate basic research into clinical research, which, if safety and efficacy are proven, can lead to offering clinical applications to patients, conflict with the right to science contained in the Covenant.[100] The prohibition to test new cures, or methods to prevent deadly or severely impairing diseases that are otherwise incurable, can hardly be said to "promote the general welfare in a democratic society." A more balanced approach that respects the proportionality test is needed. Israel is a good example. There, the law prohibits clinical research but leaves the door open to cases in which testing germline engineering may be warranted. The power to authorize clinical trials under exceptional circumstances is given to the minister of health, who can adopt a regulation greenlighting experimenting germline engineering on humans.[101] This approach is similar to the one recommended by influential ethical statements, such as those of the National Academies of Sciences, Engineering, and Medicine and the Nuffield Council on Bioethics.[102] These statements reflect a shift in opinion from a blank prohibition to the permissibility of translational pathways to germline editing.

We do not advocate giving researchers carte blanche. They would have to adhere to widely accepted standards for clinical research and follow robust preclinical evidence supporting the clinical promise of modification of the human germline. The National Academies of Sciences, Engineering, and Medicine recommends, among others, that clinical trials using heritable genome editing be permitted only in "the absence of reasonable alternatives ... to prevent a serious disease or condition ... on genes that have been convincingly demonstrated to cause or to strongly predispose to that disease or condition ... and [with] reliable oversight mechanisms."[103] Admittedly, clinical experimentation for reproductive purposes seems to be premature at this point in history. However, while it would be untimely to do germline genome editing with the intent of bringing that possible future child to birth, blank prohibitions of clinical research fall short of international human rights standards. They inhibit a conversation about what clinical research should and could look like, and

---

[100] Similar considerations are put forth by R Yotova, The Regulation of Genome Editing and Human Reproduction under International Law, EU Law and Comparative Law, Nuffield Council on Bioethics (2017): p. 29.

[101] See in this book Chapter 21, pp. 572–574.

[102] Nuffield Council on Bioethics, *Genome Editing and Human Reproduction*, London: Nuffield Council on Bioethics (2018); National Academies of Sciences, Engineering, and Medicine, Human *Genome Editing: Science, Ethics, and Governance*, The National Academies Press (2017).

[103] Ibid., pp. 189–190.

whether it can be carried out promoting the "general welfare in a democratic society." As Bryan Cwik persuasively argued prior to Dr. He Jiankui's stunt, "It's important to consider seriously what would be required for the conduct of ethically sound clinical trials of [gene editing]. Human germline gene editing raises a new set of ethical issues that are extremely difficult to resolve by current ethical guidelines and regulations."[104]

4 Obsolete Regulatory Frameworks Violate the Conditions of Legality, Legitimacy, and Proportionality

We believe obsolete regulatory frameworks fail to meet the conditions of legality, legitimacy, and proportionality. Even if they may have met them in the past, restrictive measures on the right to science and the rights of science, such as any other Covenant right, must be reviewed on a regular basis in the light of changing circumstances, lest they would not be any longer "adopted by law," "necessary," or "proportionate."

Human rights courts and other bodies "constantly stress that they interpret human rights in accordance with changing structures, values and priorities of societies."[105] Article 4 requires states to adopt and upkeep laws that are appropriate, in the sense of being abreast with new scientific developments. States have an ongoing obligation to revise laws as science and technology advance and to ensure that, when progress is substantial and clear enough, a broad public dialogue takes place to ensure existing regulations reflect current societal values.[106] If they do not, or do not any longer, then these regulations cannot be considered anymore as promoting the welfare of the democratic society under present-day conditions: they have become obsolete and must be reformed.

Indeed, as all the other rights recognized in the Covenant, the obligations created by the right to science are not necessarily fulfilled once and for all by "one-time" measures. Under Article 2.1 of the Covenant, states must take steps

---

[104] B Cwik, "Designing Ethical Trials of Germline Gene Editing," *New England Journal of Medicine*, vol. 377, no. 20 (2017): pp. 1911–1913.

[105] A Müller, "Limitations to and Derogations from Economic, Social and Cultural Rights," *Human Rights Law Review*, vol.9 no. 4 (2009): 557–601, at 560–561.

[106] See, A Chapman, "Towards an Understanding of the Right to Enjoy the Benefits of Scientific Progress and its Applications" *Journal of Human Rights*, vol. 8, no. 1 (2009): 1–36, at 17–18 (the author states her agreement with the findings the National Bioethics Advisory Commission, pointing out that "limits on freedom of inquiry must be carefully set, must be justified and should be reevaluated on an ongoing basis"); National Bioethics Advisory Commission, Cloning Human Beings: Report and Recommendations of the National Bioethics Advisory Commission, NBAC (1997).

to discharge their obligations with a view to "achieving progressively the full realization of the rights recognized in the present Covenant by all appropriate means, including particularly the adoption of legislative measures." As repeatedly noted by the Committee on Economic, Social, and Cultural Rights, the "progressive realization" of the rights recognized in the Covenant means that states parties have a "specific and continuing obligation to move as expeditiously and effectively as possible" toward the full realization of these rights, logically including the right to science and the rights of science.[107]

In the case of the regulatory frameworks of heritable genome editing, only Japan has adopted a regulatory framework in recent times (in 2014). Only a handful of other countries (e.g. France, the Netherlands, South Korea, Sweden, and the United Kingdom) have undertaken formal policy discussions on germline engineering in the past five years. The others surveyed in this book have laws that were drafted, debated, and enacted in the 1980s, 1990s, and 2000s, well before the advent of CRISPR.[108] To wit, Mexico adopted its key statute regulating basic research on germline engineering in 1982 and 2002;[109] Germany in 1991;[110] China adopted various instruments between 1993 and 2003;[111] Switzerland in 1998 and 2003;[112] Australia and the Netherlands in 2002;[113] Canada,[114] France,[115] Italy,[116] and South Korea in 2004;[117] Spain in 2007.[118]

These prohibitions or restrictions have not yet been reexamined in light of the recent advancements in gene editing technology. The advent of CRISPR has fundamentally changed the cost-benefit analysis.[119] While in the 1990s and

[107] See UNCESCR, General Comment No. 3 (n 80), para. 9; General Comment 13 (n 62), para. 44; General Comment No. 14 (n 61), para. 31; General Comment No. 17 (n 61), para. 26.

[108] CRISPR-Cas9 first appeared in the scientific literature in 2012. See, M Jinek and others, "A Programmable Dual-RNA–Guided DNA Endonuclease in Adaptive Bacterial Immunity," *Science*, vol. 337 (2012): 816–821.

[109] The 1982 law was amended in 2011. See in this book Chapter 5, pp. 137, 143.

[110] See in this book Chapter 8, p. 244.

[111] The regulations of research with human subjects were adopted in 2016. See in this book Chapter 8, pp. 476–477.

[112] See in this book Chapter 15, pp. 413–415.

[113] See in this book Chapter 20, p. 544, and Chapter 11, pp. 317–319.

[114] See in this book Chapter 3, p. 86.

[115] In 2013, the French Parliament changed the default rule from a ban (on using human embryos for research excepting supernumerary embryos) to permissibility (of research with supernumerary embryos upon prior approval and under tight oversight). See in this book Chapter 14, p. 386.

[116] See in this book Chapter 12, pp. 344–346.

[117] See in this book Chapter 18, p. 501.

[118] See in this book Chapter 13, p. 363.

[119] G Daley, R Lovell-Badge and J Steffann, "After the Storm – A Responsible Path for Genome Editing," *New England Journal of Medicine*, vol. 380 (2019): 897–899.

2000s the costs of human germline genome modification were clear while benefits were speculative, now the benefits are coming into focus. The regulatory frameworks adopted before the advent of CRISPR must be reexamined to be up-to-date with new scientific developments if the goal of achieving progressively the full realization of the human right to science, both scientific freedom and benefit sharing, is to be reached. Where no public legislative debate took place in recent times, in spite of new important scientific developments, can it be argued that the purpose of those restrictions is still the protection of the "general welfare in a democratic society"?

Even the sacrosanct fourteen-day rule, the current prevailing universal standard, should be open to rediscussion should our understanding of what happens around that threshold change, or our values change.[120] The fourteen-day rule was adopted about thirty years ago as an acceptable compromise between those who believe human life begins at fertilization and those who believe the early stages of development do not yet constitute a human life. Since then, it is widely considered to be an acceptable balance between the moral imperatives of religious beliefs and the need to advance science.[121] The fourteen-day rule is a "legal and regulatory line in the sand that has for decades limited in vitro human-embryo research to the period before the 'primitive streak' appears."[122] That being said, the fourteen-day rule "was never intended to be a bright line denoting the onset of moral status in human embryos."[123] Instead, it has been a "theoretical [line respected] until now because scientists have been technologically incapable of moving past the 14-day threshold."[124] However, recent developments have raised the question of further extending the possibility of researching on embryos beyond fourteen days. Until 2016,

---

[120]   FS Collins and S Gottlieb, "The Next Phase of Human Gene-Therapy Oversight" *New England Journal of Medicine*, vol. 379 (2018): 1393, 1395; G Cavaliere, "A 14-day Limit for Bioethics: the Debate Over Human Embryo Research," *BMC Medical Ethics*, vol. 18 (2017): 38; I Hyun, A Wilkerson & J Johnston, "Embryology Policy: Revisit the 14-day Rule," *Nature*, vol. 533 (2016): 169–171.

[121]   J Harris, "It's Time to Extend the 14-Day Limit for Embryo Research," The Guardian, May 6, 2016.

[122]   Revisit the 14-day Rule, (n 120), p. 170. The appearance of the "primitive streak" (i.e. a transient structure whose formation marks the start of the process in which the inner cell mass in converted into the three germ layers: ectoderm, mesoderm, and endoderm) signals that the individuality of an embryo is assured.

[123]   Ibid.

[124]   M Kaplan, "Call to Re-Examine '14 Day Rule' Limiting In-Vitro Human Embryo Research," Case Western Reserve University of Medicine, Press Release May 4, 2016, https://casemed .case.edu/newscenter/news-release/newsrelease.cfm?news_id=302 accessed March 15, 2017.

culturing human embryos in vitro had never gone beyond nine days.[125] Since then, human embryos were sustained in vitro for twelve to thirteen days.[126]

Obsolescence of regulatory frameworks is certainly not a new problem in science and technology law and policy, or a problem only of science and technology law and policy. As often happens with disruptive scientific and technological breakthroughs, lawmakers are struggling to adjust the regulatory frameworks with these developments. Elen Stokes refers to this problem as one of "inherited rules." "New technologies," she points out, "do not always elicit new regulatory responses. More often than not, policymakers deal with new technologies by deferring to existing regulatory regimes."[127] However, the fact that this problem occurs "more often than not" does not make it acceptable. Indeed, it directs attention to the fact that the Sisyphean task of meeting human rights obligations is a never-ending enterprise that requires legislative bodies to be well informed about new developments with a view to revise and update laws accordingly.

CRISPR is a significant scientific and technological advancement that has accelerated the timeline of clinical applications based on germline engineering becoming available to patients. It is a game changer, one that puts in question how governments have regulated human genome germline modifications in the past, and that calls "for a broad public dialogue about these technologies and their applications."[128] Obsolete legislation may not reflect how the public values the benefits and risks of heritable genome editing. Governments must engage legislatures, ministerial bodies, national science councils, and other venues for public engagement to ensure that regulatory frameworks adopted years before the advent of CRISPR are adjusted to how to best promote the welfare in a democratic society considering the opportunities offered by new technology and scientific progress here and now, not a decade ago.[129]

III RECOMMENDATIONS

To conclude, as a recommendation, we would like to sketch what we believe a regulatory framework for human genome germline modifications that is

---

[125] Ibid.
[126] Revisit the 14-day rule (n 120), p. 169.
[127] E Stokes, "Nanotechnology and the Products of Inherited Regulation," *Journal of Law and Society*, vol. 39 (2012): 93–112.
[128] *Human Genome Editing* (n. 102), p. 163.
[129] Ibid.

informed by international human rights law and, more specifically, the right to science and the rights of science should look like.

To begin, we believe the primary responsibility for regulating heritable gene editing falls on (legitimately elected) governments rather than international organizations or civil society bodies. International law creates obligations that national governments must discharge to ensure progressively the full realization of human rights in the area of scientific and technological progress, not least the human right to science. International organizations or civil society bodies can play an important governance role in supporting the implementation of these obligations. However, they cannot substitute the role national governments are expected to play.

As with the rest of the rights recognized in the Covenant, the right to science and the rights of science imply different sets and levels of obligations. The key obligations governments have in this regard are to "respect, protect, and fulfill" everyone's rights to contribute to scientific progress (scientific freedom) and to enjoy such progress (benefit sharing).[130] As discussed in the previous section, the "obligation to respect" requires that governments do not interfere in the enjoyment of the right to science unless they have a legitimate reason for doing so, one that is based on science and actual risks, as opposed to political opportunity and speculation. However, just as important to our analysis is the "obligation to fulfill." This obligation translates into the creation of a legal framework and a regulatory environment that is conducive to the effective enjoyment of the right to science, both scientific freedom and benefit sharing.[131] The same obligation requires states to "adopt appropriate legislative, administrative, budgetary, judicial, promotional and other measures" toward the full realization of the rights to science.[132] At a minimum, governments are expected to adopt legislative measures that allow a person to exercise or enjoy scientific freedom and benefit sharing effectively. As we discussed in the previous section, this requires, inter alia, that the legal framework is sufficiently certain and up-to-date.[133] In taking legislative measures,

---

[130]   See the following general comments by the UNCESCR, General Comment 13 (n 62), para. 46; General Comment No. 14 (n 61), para. 33; General Comment No. 17 (n 61), para. 28. According to these, the obligation to protect requires states also to prevent third parties from interfering in the enjoyment of the right to science. This issue will not be discussed in this chapter.

[131]   Analogously, UNCESCR, General Comment No. 13 (n 62), para. 46.

[132]   Analogously, UNCESCR, General Comment No. 14 (n 61), para. 33; and General Comment No. 17 (n 61), para. 28. Also see UNCESCR, General Comment No. 3 (n 80), para. 9.

[133]   Article 2.1 of the ICESCR limits state duties to "taking steps, individually and through international assistance and co-operation, especially economic and technical, to the maximum of its available resources, with a view to achieving progressively the full realization of

governments are expected to engage legislatures, ministerial bodies, national science councils, and other venues for public engagement to ensure that regulatory frameworks reflect current values in their respective societies.

From a policy perspective, legislative measures in the biomedical field must guarantee, as a default rule, the freedom to engage in basic and clinical research and to make safe and effective treatment, therapies, and other applications available to patients in a clinical setting.[134] In regulating heritable gene editing, legislative measures must guarantee, again as the default rule, scientists' freedom to use CRISPR, and any other gene editing tools that might be invented in the future, to create and modify human gametes and embryos and identify reasonable opportunities for translational pathways of therapies to cure heritable genetic disorders.

We recognize that this is a controversial area of science and that not all societies are willing, at least for now, to move forward with heritable gene editing, even if the goal is strictly therapeutic. Human rights law accommodates this diversity of viewpoints by establishing that the human right to science, which incorporates scientific freedom and benefit sharing, is not absolute. As it has already been mentioned several times before, according to Article 4 of the Covenant rights can be restricted by law for the purpose of promoting the general welfare in a democratic society. Governments may, and in certain cases must, restrict scientific freedom and benefit. They can certainly ban applications of gene editing techniques to enhance humans or for cosmetic reasons, if they democratically and lawfully decide to do so.

We believe blank prohibitions, such as those banning all research on human embryos and all clinical research, to be in violation of international law. Limitations based on safety health considerations are easier to defend since preclinical research has so far failed to show that germline engineering is sufficiently safe to be experimented on humans, due to the risk of off-target mutations and mosaicism. However, states must discharge the burden of proving an actual risk to health and safety and explain how and why the health of the many trumps the right of those who are sick to be cured. In addition, the only lawful prohibitions are those determined by law, and law must be sufficiently clear. Given the transformative nature of CRISPR, prohibitions that date back a decade or two cannot be considered to have been truly

the rights recognized in the present Covenant by all appropriate means, including particularly the adoption of legislative measures."

[134] The rules governing access to clinical applications are better analyzed within the framework of the right to health, which exceeds the scope of our analysis. On the international human rights to health, see, in general, J Tobin, *The Right to Health in International Law*, Oxford University Press (2012).

democratically accepted. These issues need to be debated again and, only if a broad agreement is reached in favor of prohibiting this kind of research applied to humans, as happened in the case of the fourteen-day rule, limitations will be acceptable.

We want to stress that we do not argue against the need for restrictions on freedom of research and benefit sharing, when necessary to ensure respect for the fundamental values of the community. Such restrictions might allow national governments to accommodate considerations of ethical or religious diversity, and give some margin for societies to choose the appropriate speed at which they wish to participate in innovation. The international human rights framework allows countries to choose, in consideration of their available resources, as long as they use their resources to the maximum, to be at the forefront of innovation by developing a regulatory framework favorable to advances in the area of gene editing. This is another reason to entrust national, rather than international, lawmakers to find the right balance between the right to science and the rights of science and their limitations. The goal is to promote science and technological development while being mindful and respectful of international human rights standards as well as the different sensitivities with which citizens from different parts of the world approach this complex problem. That being said, national policies must fulfill article 2 of the Covenant, which requires the *progressive* realization of the rights.

National policies must also reflect the recognition, contained in Article 15.3, of the "benefits to be derived from the encouragement and development of international contacts and co-operation in the scientific and cultural fields." International cooperation is particularly important in germline genome engineering. The scientific and technical complexities of this field demand scientific efforts that transcend national boundaries and often involve scientists from multiple countries. These efforts may take the form of collaborations among researchers across borders, pooling and sharing resources and expertise, and validation of results with scientists traveling to other countries to attend meetings, to visit labs, to lecture, or to access resources and expertise. National policies must enable international cooperation, especially since only a few countries have chosen to be at the forefront of innovation in this field. When clinical applications become reality, international cooperation will foster exchanges that ensure the sharing of benefits to patients of countries that have chosen a more conservative approach. International bodies have a role to play as facilitators of regulatory harmonization, as custodians of knowledge of best practices and current regulations, and as promoters of a global conversation on how innovation can be balanced against other considerations.

National policies must also ensure that other human rights are protected. The rights of research participants are particularly important in this area. In accordance with Article 7 of the International Covenant on Civil and Political Rights, any research must be carried out in accordance with international standards of research involving human subjects, including the right to free and informed consent. In addition, research preapproval and oversight, which are commonplace in biomedical research, are necessary to ensure a responsible exercise of scientific freedom. To the extent possible, legal frameworks must be narrowly tailored and be limited to setting out basic guarantees, leaving space to the scientific community for self-regulation. Here, we see an important role for scientific societies, expert bodies, and ministerial bodies to design "soft law" instruments, such as guidelines and ethical standards, in accordance with international human rights standards. These instruments might then be embraced by funding agencies or professional bodies, thus empowering either government agencies or the scientific community to monitor best practices. The ideal framework must also incorporate considerations of solidarity. In the context of heritable gene editing, solidarity translates into rules guaranteeing, at minimum, fairness in access, nationally and transnationally, to the clinical applications of germline engineering, outlawing any form of genetic discrimination, and ensuring intergenerational equity.[135] These are important goals, which cannot be fully articulated within the limits of these conclusions. It suffices to say that, if proper legal mechanisms are in place to ensure that these policy goals are achieved, some of the objections to heritable gene editing would be defeated and the public would likely be more willing to support its legalization. Considerations of solidarity can inform policies with regard to cost of treatments (if treatments are expensive and thus unaffordable to some), to reproductive tourism (if treatments are only available in certain countries and thus only available to those who can afford to travel to foreign countries), and genetic enhancement (if everybody can "enhance" their offspring, at least fairness is no longer an issue).

Finally, as to the elephant in the room, the governance framework must be respectful of human dignity. That is required by all existing international regulatory standards. However, what respecting human dignity entails is a question that we do not intend to settle here. The concept has always been and remains undefined, and probably humanity will never agree on a clear

[135] ICESCR, Article 2.2; UNCESCR, General Comment No. 20: Non-Discrimination in Economic, Social and Cultural Rights (art. 2, para. 2) U.N. Doc. E/C.12/GC/20 (2009). More in-depth analysis of the intersection between equity, right to health, and right to science would be necessary, but it exceeds the scope of this book.

definition of it. That being said, we believe that it is essential to distinguish between different concepts of dignity. The concept of human dignity in the sense of autonomy, rank, or status of human beings should not be confused with other uses of dignity. When Christian theology stresses the absolute worth and sacredness of human life from the time of conception, it stretches the concept beyond what is accepted in international human rights. International human rights law's understanding of dignity is limited to the protection of the autonomy, rank, and status of human beings and to the furtherance of the highest attainable level of health. Although international law upholds the rights to science and to health and, in this sense, is inclined to promote scientific progress and applications that strengthen the protection of these rights, it does not settle the question of the status of embryos as such, or whether it is contrary to dignity to interfere with gametes or embryos. From an international human rights perspective, once a person is born, genetically modified or not, she has the same rights and freedoms as all others. If her genes have been modified, she is no less of a person with dignity than someone whose genes have not been modified.

The question whether the current regulatory framework for human germline genome engineering should be eased is giving rise to ethical and moral disagreements in pluralistic societies. On one extreme, there are believers and religious authorities that demand full respect for the absolute worth and sacredness of human life from time of conception. Indeed, these demands often translate into claims not just about the need to ban human genome germline modification, but also stem cell research and other experiments involving the use of human embryos, and beyond, to abortion and end-of-life issues. At the other extreme, there are those who defend the need for scientific progress to cure serious diseases and human suffering, even if it requires modification of the germline of embryos. For them the duty to make use of novel scientific tools to assist people to attain the highest attainable standard of health made possible by new scientific developments overshadows any other consideration.

In a democratic society, which is a society characterized by pluralism, tolerance, and broad-mindedness, no particular view is acceptable a priori. No one has the right to impose their view on the rest through laws and regulations or fait accompli. Moral and ethical disagreement and the state of evolving technology demand some form of debate, possibly open, inclusive and transparent. International human rights law, including the right to science and the rights of science, should frame and inform these debates. The rights in focus in this volume point to some fundamental interests of humans that may not have been considered fully, such as the universal right to enjoy the benefits of scientific research and all its applications, as well as the right to scientific freedom within the limits established by international human rights law. The same rights also point to the need for democratic debates concerning how to meet the international obligations that flow from these rights. Let the debate begin.

Index

618

Printed in the United States
by Baker & Taylor Publisher Services

Printed in the United States
by Baker & Taylor Publisher Services